AF793457

Mantu K. Ghosh

HPLC Methods on Drug Analysis

Springer-Verlag
Berlin Heidelberg New York
London Paris Tokyo
Hong Kong Barcelona Budapest

Dr. Mantu Kumar Ghosh
4420 Coesfeld, Germany

Library of Congress Cataloging in Publication Data
Ghosh, Mantu K. (Mantu Kumar), 1942–
HPLC methods on drug analysis / Mantu K. Ghosh.
Includes index.

ISBN-13: 978-3-540-53824-0 e-ISBN-13: 978-3-642-76506-3
DOI: 10.1007/978-3-642-76506-3

1. High performance liquid chromatography — Laboratory manuals.
2. Drugs — Analysis — Laboratory manuals. I. Title.
[DNLM: 1. Chromatography, High Pressure Liquid — methods — laboratory manuals.
2. Drugs — analysis laboratory manuals. QV 25 G427h]
RS189.5.H54G47 1992
615'. 1901--dc20
DNLM/DLC

This work is subject to copyright. All rights are reserved, whether the whole or part of the material is concerned, specifically the rights of translation, reprinting, re-use of illustrations, recitation, broadcasting, reproduction on microfilms or in other ways, and storage in data banks. Duplication of this publication or parts thereof is only permitted under the provisions of the German Copyright Law of September 9, 1965, in its current version and a copyright fee must always be paid. Violations fall under the prosecution act of the German Copyright Law. Neither the author nor the publisher is responsible for the reproducibility of the described methods.

© Springer-Verlag Berlin Heidelberg 1992

The use of registered names, trademarks, etc. in this publication does not imply, even in the absence of a specific statement, that such names are exempt from the relevant protective laws and regulations and thereof free for general use.

Typesetting: Macmillan India Ltd. Bangalore 25.
52/3020-543210-Printed on acid-free paper.

Dedicated to the memory of my parents

Contents

Preface

The dramatic development of chromatographic techniques, specially high performance or high pressure liquid chromatography (HPLC) has made possible the easy analysis of organic compounds, including drugs and drug components, for last two decades.

This rapid increase and improvement of analytical methodology with HPLC has enabled researchers and scientists to cope with other scientific and instrumental developments in their fields of work.

Thousands of impressive and original scientific publications, text books and monographs describe the techniques for drug analysis with high performance liquid chromatography. However, no concise presentation of the general properties of the drugs and their HPLC methodology exists together in the market. This work contains the general properties necessary for the analysis of 232 drugs as well as the HPLC methods for many other drugs and drug components. It is hoped that it will fill a gap and provide a precise survey of the HPLC methods for drug analysis. It is intended as an immediate guide in the laboratory and will be of help to the scientists, researchers and technicians in the field of analysis.

This book describes approx. 1300 HPLC methods, in a precise form, for the analysis of some hundreds of drugs and drug components from raw materials, pharmaceuticals and biological samples under 232 headings. Emphasis is laid on the important points of the general properties of the described drugs, extraction, apparatus, column, eluent, temperature, injection size, detection, retention times and references whereever possible. The tabular presentation is very easy to understand with minimum knowledge of chemistry as well as of HPLC for the quantitative analysis of drugs.The preservatives used in drugs are considered, since they are also a part of the pharmaceutical formulations in most cases, and some autocoids used for diagnostic purposes or treatment are also described here.

I must express my love for and gratitude to my wife, Manikuntala Ghosh, B.Sc. for her direct help in the preparation of the manuscript. I highly appreciate and thank Mrs. K. Rausse, Muenster, for her patience in typing the manuscript. I also take this opportunity to thank many of my friends for their inspiration for this work to become a reality.

Coesfeld, Germany, M.K. Ghosh

Abbreviations

A	ampere
Afs	ampere full scale
α	alpha
appr.	approximately
aq	aqueous
AU	absorbance units
AUFS	absorbance units full scale
β	beta
BP	British Pharmacopoeia
°C	degree Celsius, centigrade
CAS	chemical abstract service
Ci	curie
cm	centimeter
CNS	central nervous system
D	dextro (rotatory)
δ	delta
EC	electrochemical
Ed	editor or edited by
Edn	edition
EDTA	ethylenediaminetetraacetic acid
em	emission
et al.	et alli, 'and others' for more than three co-authors
eV	electron volt
ex	excitation
°F	degree Fahrenheit
FID	flame ionization detector
fg	femtogram ($= 10^{-15}$ g)
fmol	femtomol
FS	full scale
fsd	full scale deflection
g	gram
γ	gamma
GC	glassy carbon (electrode)
h	hour
HP	Hewlett-Packard
HPLC	high-performance liquid chromatography; also, high-pressure liquid chromatography
i (*or* μ)	ionic strength
ID	inner diameter
in	inch
insol	insoluble
IU	international unit
k'	(column) capacity factor or capacity ratio
kg	kilogram
kPa	kilopascal
l	liter
lb, lbs	pound, pounds
LC	liquid chromatography
LDC	Laboratory Data Control
L	levo (rotatory)
m	milli ($= 10^{-3}$); also, meter
m	*meta-*
M	molar (mol/l) or model (apparatus)
max	maximum
mg	milligram
min	minute
ml	milliliter
mm	millimeter
mM	millimole
mol	mole
MPa	mega pascal
mV	millivolt
μ	micro ($= 10^{-6}$); also, ionic strength
μg	microgram ($= 10^{-6}$ g)
μl	microliter ($= 10^{-6}$ l)
μm	micrometer (10^{-6} m $= 10^{-4}$ cm)
n	nano ($= 10^{-9}$); normal in organic names, e.g., *n*-butylamine
N	normal (solution); nitrogen, e.g., N-Acetylcysteine
NF	National Formulary
ng	nanogram ($= 10^{-9}$ g)
nm	nanometer ($= 10^{-9}$ m)
nM	nanomol

No	number
NP	normal phase
o-	*ortho*-
O	attachment to oxygen
p-	*para*-
Pa	pascal
pg	picogram (= 10^{-12} g)
pH	hydrogen ion exponent
PLRP	polymeric reversed-phase
pmol	picomol
ppb	parts per billion
ppm	parts per million
PRP	polymeric reversed-phase
psi	pounds per square inch
R	racemic (isomer)
RCM	radial compression module
RCSS	radial compression separation system
RE	reference electrode
ref.	reference
RI	refractive index
RP	reversed-phase
RT	retention times
S	stereo (isomer)
s	second
sol	solution or soluble
SP	Spectra-Physics
λ	lambda (wavelength, nm)
tert.-	tertiary, e.g., *tert*.-butanol
USP	United States Pharmacopeia
UV	ultraviolet
V	volt
VIS	visible
vs	versus, against
v/v	volume per volume
w/v	weight per volume
w/w	weight per weight
$>$	greater than; faster than
$<$	less than; slower than
°	degree(s) Celsius, centigrade

List of the Described Drugs

Explanatory Notes

Headings
The names of the described drugs or related substances have been arranged in alphabetical order using their generic (INN = International Nonproprietary Names), trivial, or simple chemical names as monograph headings. The American style of spelling has been adopted. Each monograph is divided into two parts. Part I contains the physical data of the drugs and Part II the HPLC methods for analysis, in a systematic order.

CAS-No
Chemical Abstract Service (CAS) registry numbers have been given as additional information.

Chemical Names
The nomenclature defines the IUPAC names (International Union of Pure and Applied Chemistry), or the alternative or simple chemical names. In the case of some drugs of microbial origin, the name of the micro-organism is mentioned.

Empirical Formula and Molecular Mass
Empirical formulas of the drugs have been taken from different pharmacopeias and Martindale: The Extra Pharmacopoeia where their molecular masses are calculated using the revised table of Atomic Weights released by IUPAC in 1977.

Proprietary Names
The three most common proprietary names used in different countries are given in alphabetical order, where available.

Clinical Use
The major therapeutic indication adopted by most of the pharmacology books is included under this heading.

Solubility
Solubilities at room temperature in some common solvents used for analysis are given in the monographs and they have been collected from published data. Where no figure is given, the following definitions are used:

Very soluble	1 in less than 1
freely soluble	1 in 1 to 1 in 10
soluble	1 in 10 to 1 in 30
sparingly soluble	1 in 30 to 1 in 100
slightly soluble	1 in 100 to 1 in 1000
very slightly soluble	1 in 1000 to 1 in 10000
practically insoluble	1 in more than 10000

The words water, ether, ethanol, aq HCl and aq NaOH refer to purified water, diethyl ether, ethanol 95%, 0.1 N HCl and 0.1 N NaOH, respectively.

Light Absorption
The wavelength of maximum absorption (λ_{max}) of a compound is given where available. The absorption maximum is calculated from E 1%, 1 cm (extinction: 1% solution, 1 cm light path) in a solvent.

HPLC
The methods are numbered.

Sample
The forms of the samples described in the articles are taken without any specifications. The word raw material is used when no special form is mentioned.

Extraction
Under this heading the method has been given in a concise form mentioning the type of extraction system (liquid-liquid or solid-phase), extractant as well as the pH, where available.

Apparatus
Only the LC system or pumps, detectors and injectors and their manufacturers are mentioned under this heading. The sensitivity of the apparatus as well as the reproducibility of the method are not given.

Column and Pre-Column
The type of columns, trade names, carbon atoms (RP), particle size and the length and inner diameter are quoted. This does not imply any guarantee of the reproducibility of any analytical parameters. Ion-pair chromatography has not been specified separately but as a reversed-phase system.

Mobile Phase
The mobile phase has been taken as described by the authors. In the case of buffer, acid, alkali or salt solutions, where the solvent is not mentioned, it is assumed to be water.

Flow-Rate
This defines the amount of total flow through the column in ml or μl per minute or second, unless otherwise stated. In the case of preparative chromatography it is expressed as cm or mm per second or minute. The back pressure of the column is expressed in psi, kPa, MPa, when this parameter is available.

Temperature
The Centigrade (Celsius) thermometric scale is used in expressing the temperature of the analytical column, unless otherwise stated.

Internal Standard
In most of the cases, the internal standard used has been mentioned.

Derivatization
Derivatization methods are not described due to the concise presentational form. Only the pre- or post-column system and the name of the derivatizing reagent is mentioned.

Injection Size
Injection size is given in μl, μg, ng or pg injected into the column. μl following a range of concentration indicates the linearity and only one concentration refers to a demonstrable chromatogram.

Detection
Under this heading the detector types and wavelengths in case of ultraviolet and fluorescence detectors are mentioned following the absorption range, where available. The electrode, reference electrode and voltage used are given in case of an electro-chemical detector. For other detectors no scale data are provided.

Retention Times
It applies to the total retention time, which is needed by a component to migrate from column inlet to the column outlet (sample injection to detector) and it is given in

a) *Retention times (RT) in minute*
b) *Capacity factor or capacity ratio (k′)*

$$k'_1 = \frac{t_1 - t_0}{t_0}, \qquad k'_2 = \frac{t_2 - t_0}{t_0}$$

Where t_0 is the retention time of the solvent, and t_1 and t_2 are the retention times of compound 1 and 2 respectively. k' is a measure of solute retention and quoted whereever necessary.

c) *Relative retention (selectivity, α)*

It is the ratio between two capacity factors, known as selectivity or separation factor (α).

$$\alpha = \frac{k'_2}{k'_1}$$

α depends on the parameters of the column as well as on the temperature. α measures the ability of a column to separate compound 1 and 2 due to different affinities and therefore their retentions.

References

1. Anon. (1983): The Merck Index, 10th edn. Merck & Co. Inc., Rahway
2. Anon. (1980): British Pharmacopoeia. London Her Majesty's Stationary Office, London
3. Anon. (1985): The United States Pharmacopeia XXI & The National Formulary XVI. United States Pharmacopeial Convention Inc., Rockville
4. Auterhoff H and Kovar K-A (1985): Identifizierung von Arzneistoffen, 5th edn. Wissenschaftliche Verlagsgesellschaft mbH Stuttgart
5. Gilman AG et al (1985): Goodman and Gilman's The Pharmacological Basis of Therapeutics, 6th edn. Macmillan Publishing Company, New York
6. Martindale W (1982): Martindale, The Extra Pharmacopoeia, 28th edn. The Pharmaceutical Press London.
7. Meyer V (1984): Praxis der Hochleistungsfluessigchromatographie, 3rd edn. Diesterweg Salle Sauerlaender, Frankfurt and Salzburg
8. Moffat AC et al (1986): Clarke's Isolation and Identification of Drugs, 2nd edn. The Pharmaceutical Press London
9. Tsuji K and Morozowich W (1978): GLC and HPLC Determination of Therapeutic Agents, Part 1. Marcel Dekker Inc., New York and Basel

1 Acetylcysteine

CAS-No	616-91-1
Chemical name	*N*-Acetyl-L-cysteine
Empirical formula	$C_5H_9NO_3S$
Molecular mass	163.2
Proprietary names	Airbron; Fluimucil; Mucomyst
Clinical use	Mucolytic/Paracetamol Antidote
Solubility	Water 1 in 8; ethanol 1 in 2; chloroform insoluble; ether insoluble

---

HPLC 1

Sample	Dosage forms (aqueous solutions)
Apparatus	LDC Constametric LC-III pump and UV detector
Column	Reversed-phase; Spherisorb ODS 10 µm, C_{18}, 200 × 5 mm ID
Mobile phase	0.5% w/v Aqueous sodium perchlorate, pH 2.0 with ortho-phosphoric acid
Flow-rate	2 ml min^{-1}
Temperature	60 °C
Internal standard	L-Tyrosine
Injection size	20 µl, 0.5 mg ml^{-1}
Detection	UV, λ = 215 nm; 0.1–0.2 AUFS
Recording	Hewlett-Packard 3356 computer
Retention times, min	L-Cystine 1.3; L-Cysteine 1.4; N-Acetylcysteine 2.7; N,S-Diacetylcysteine 7.6; L-Tyrosine 7.8; N,N'-Diacetylcystine 8.3
Ref.	Farquhar J, Finlay G et al (1985) J. Pharm. Biomed. Anal. 3:279

HPLC 2

Sample	Dosage forms (solutions)
Apparatus	Varian 5020 LC and UV 50 detector
Column	Reversed-phase; Lichrosorb 7 µm, C_8, 250 × 4.6 mm ID
Mobile phase	A. Acetonitrile/0.05 M phosphate buffer, pH 3.0 (45:55, v/v) B. Acetonitrile/0.05 M phosphate buffer, pH 3.0 (57:43)

Flow-rate	1 ml min^{-1}		
Temperature	Ambient		
Internal standard	Toluic acid		
Derivatization	Pre-column; with ethacrynic acid		
Injection size	10 µl, 5.4–10.8 µg ml^{-1}		
Detection	UV, λ = 266 nm; 0.05 AUFS		
Retention times, min	Mobile phase	A	B
	Cysteine	2.8	
	Acetylcysteine (ethacrynic acid adducts)	3.4	3.1
	Toluic acid		4.4
	Ethacrynic acid (estimated from drawing)	10.5	5.6
Ref.	Cavrini V, Gatti R et al (1987) Chromatographia 23:680		

2 Acetylsalicylic Acid

CAS-No	50-78-2			
Chemical name	2-Acetoxybenzoic acid			
Empirical formula	$C_9H_8O_4$			
Molecular mass	180.2			
Proprietary names	Aspirin; Caprin; Solprin			
Clinical use	Analgesic			
Solubility	Water	alcohol	chloroform	ether
	1 in 300	1 in 5 to 7	1 in 17	1 in 20
Light absorption	λ_{max} = 230 nm in aqueous acid			

HPLC 1

Sample	Human urine
Extraction	Diluting with water
Apparatus	LDC Constametric III LC pump, Milton Roy Spectromonitor III and Rheodyne 7125 injector valve
Column	Reversed-phase; Spherisorb ODS-5, 250 × 4.6 mm ID
Mobile phase	Water/methanol/acetic acid (71:25:4), pH 2.5
Flow-rate	1.2 ml min^{-1}
Temperature	Ambient
Internal standard	*o*-Anisic acid
Injection size	100 µl, 2.5–10 µg ml^{-1}
Detection	UV, λ = 245 nm

Retention times, min	Gentisic acid 5.8 Acetylsalicylic acid 11.1 Salicyluric acid 7.5 Salicylic acid 20.8 Internal standard 9.8
Ref.	Ogunbona FA (1986) J. Chromatogr. Biomed. Appl. 377:471

HPLC 2

Sample	Tablets
Apparatus	Waters M 6000A pump, Pye-Unicam LC-3 UV detector and Valco injector
Column	Reversed-phase; Zorbax C_8, 7 μm, 250 × 4.6 mm ID
Mobile phase	Methanol/water/1 M phosphoric acid (59:36:5)
Flow-rate	1 ml min^{-1}
Temperature	20 °C
Internal standard	Secobarbital
Injection size	20 μl, 0–25 μg ml^{-1}
Detection	UV, λ = 240 nm; 0.08 AUFS
Recording	Pye-Unicam PD 88 computing integrator and Kipp & Zonen BD 40 recorder
Retention times, min	(estimated from drawing) Acetylsalicylic acid 5.2 Salicylic acid 7.0 Acetylsalicyl salicylic acid 8.7 Internal standard 10.2 Acetylsalicylic anhydride 13.0 Salicylsalicylic acid 14.0
Ref.	Verstraeten A, Roets E and Hoogmartens J (1987) J. Chromatogr. 388:201

HPLC 3

Sample	Solid dosage forms
Apparatus	Waters LC system with ultraviolet detector
Column	Reversed-phase; Spherisorb 5 μ, C_8, 30 × 4.6 mm ID
Mobile phase	Methanol/water/85% phosphoric acid (70:30:3, v/v)
Flow-rate	3.0 ml min^{-1}
Temperature	Ambient
Injection size	10 μl, 0.25 mg ml^{-1}
Detection	UV, λ = 205 nm
Retention times, min	(estimated from drawing) Acetylsalicylic acid 0.67 Salicylic acid 1.05
Ref.	Heidemann DR, Schulenberg ES and Smith WH (1987) J. Assoc. Off. Anal. Chem. 70:964

HPLC 4

Sample	Urine
Extraction	Ethyl acetate under acidic or alkaline condition
Apparatus	Hewlett-Packard 1090 HPLC with photodiode array UV detector
Column	Reversed-phase; Hypersil ODS C_{18}, 5 µm, 200 × 4.6 mm ID or Reversed-phase; Lichrosorb C_{18} 5 µm, 200 × 4.6 mm ID
Mobile phase	A. 0.05 M Phosphate buffer (containing 6.9 g NaH_2PO_4,H_2O and 1.59 g propylamine hydrochloride per liter), pH 3.0 B. Acetonitrile Gradient: B 0% to 15% in 2 min; B 15% to 80% in 18 min
Flow-rate	1 ml min^{-1}
Temperature	Ambient (22 °C)
Injection size	5 µl
Detection	Diode array UV, λ = 275 nm
Retention times, min	Acetaminophen 3.45 Acetylsalicylic acid 7.16
Ref.	Cooper SF, Massé R and Dugal R (1989) J. Chromatogr. Biomed. Appl. 489:65

HPLC 5

Sample	Benzoic acid analogue
Apparatus	HPLC low-pressure ternary gradient system
Column	Reversed-phase; Lichrosorb C_8, 250 × 4 mm ID
Mobile phase	A. Methanol B. Aqueous tetrabutylammonium hydroxide (1 g l^{-1}) Gradient: A 10% for 3.5 min; A 10% to 25% in 2.5 min; A 25% to 50% in 9 min
Flow-rate	1 ml min^{-1}; 1800 psi
Injection size	20 µl
Detection	UV, λ = 254 nm; 0.64 AUFS
Retention times, min	(estimated from drawing) Aminobenzoic acid 3.29 *p*-Hydroxybenzoic acid 4.14 Acetylsalicylic acid 5.82 Salicylic acid 6.38 Benzoic acid 8.88
Ref.	Merck Application Manual on HPLC, TLC and HPTLC (1986). E. Merck, Darmstadt, p 56

3 Acyclovir

CAS-No	59277-89-3
Chemical name	9-(2-Hydroxyethoxymethyl)guanine
Empirical formula	$C_8H_{11}N_5O_3$
Molecular mass	225.2
Proprietary names	Zovirax; Zovirax I.V.
Clinical use	Antiviral

HPLC 1

Sample	Blood plasma from dog
Extraction	Centrifugation
Apparatus	Spectra-Physics 8100 LC, 8110 autosampler and Schoeffel FS 970 fluorescence detector
Column	Spheron Micro 300, 12.5 µm, 150 × 3.2 mm ID
Pre-column	Spheron Micro 300, 16 µm, 20 × 4.6 mm ID
Mobile phase	0.1 M Phosphoric acid and 0.1 M sodium sulfate, pH 1.8
Flow-rate	1.0 ml min^{-1}
Temperature	45 °C
Injection size	10 µl
Detection	Fluorescence, λ = 285/370 nm
Recording	Spectra-Physics 4000 data system
Retention time, min	Acyclovir 1.5
Ref.	Salamoun J, Sprta V et al (1987) J. Chromatogr. Biomed. Appl. 420:197

HPLC 2

Sample	Human serum
Extraction	Deproteinization with 5% perchloric acid
Apparatus	Waters ALC/GPC M 204 LC, 712 WISP autosampler and 450 variable-wavelength detector
Column	Reversed-phase; Nucleosil C_{18}, 5 µm, 200 × 4 mm ID
Mobile phase	Methanol/water (5:95, v/v)
Flow-rate	1.0 ml min^{-1}
Temperature	Ambient
Injection size	25 µl, 1.25–20.0 µg ml^{-1}
Detection	UV, λ = 250 nm
Retention time, min	Acyclovir 8.00
Ref.	Cronqvist J and Nilsson-Ehle I (1988) J. Liq. Chromatogr. 11:2593

HPLC 3

Sample	Human blood plasma and urine
Extraction	Deproteinization with 5% ammonium sulfate and 0.15 M barium hydroxide
Apparatus	Spectra-Physics 8000 HPLC system and 8310 UV detector
Column	Reversed-phase; ODS 5 μm, 250 × 4.6 mm ID
Mobile phase	0.005 M Sodium acetate and 0.0025 M heptanesulfonic acid, sodium salt, pH 6.5
Flow-rate	1.2 ml min^{-1} (plasma); 1.5 ml min^{-1} (urine)
Temperature	50 °C (plasma); 35 °C (urine)
Injection size	60 μl (plasma) 15 μl (urine)
Detection	UV, λ = 254 nm; 0.005 AUFS
Retention times, min	Acyclovir 6.07 (plasma) 22.67 (urine)
Ref.	Land G and Bye A (1981) J. Chromatogr. Biomed. Appl. 224:51

HPLC 4

Sample	Serum
Extraction	Deproteinization with mobile phase and trifluoroacetic acid
Apparatus	LDC Constametric III LC pump, Waters M 710B autosampler and 440 UV detector
Column	PRP-1, 10 μm, 150 × 4.2 mm ID
Mobile phase	Methanol/0.1 M hydrochloric acid/0.02 M sodium heptanesulfonate/0.25 M sodium chloride (10:10:20:60)
Flow-rate	1.5 ml min^{-1}
Temperature	Ambient
Injection size	100 μl, 0.5–12.0 μg ml^{-1}
Detection	UV, λ = 254 nm
Recording	Spectra-Physics 4100 computing integrator
Retention time, min	Acyclovir 3.2
Ref.	Smith RL and Walker DD (1985) J. Chromatogr. Biomed. Appl. 343:203

4 Allopurinol

CAS No	315-30-0
Chemical name	1*H*-Pyrazolo[3,4-*d*]pyrimidin-4-ol
Empirical formula	$C_5H_4N_4O$
Molecular mass	136.1

Proprietary names	Alloprin; Lopurin; Urosin
Clinical use	Uricosuric/Xanthine oxidase inhibitor
Solubility	Water and alcohol very slightly soluble; ether insoluble; aq NaOH soluble
Light absorption	λ_{max} = 250 nm in aqueous acid (by dissolving in alkali)

HPLC 1

Sample	Blood plasma
Apparatus	LKB M 2150 LC pump and Waters M 440 UV detector
Column	Reversed-phase; Hypersil-ODS C_{18}, 5 µm, 100 × 4.6 mm ID
Mobile phase	20 mM Na_2HPO_4, $2H_2O$, pH 2.0 with ortho-H_3PO_4
Flow-rate	2 ml min^{-1}
Internal standard	Floxuridine
Injection size	20 µl
Detection	UV, λ = 254 nm
Retention times, min	(estimated from drawing) Allopurinol 3.9; Oxipurinol 3.5; Internal standard 6.8
Ref.	Hung CT, Zoest AR and Perrier DG (1986) J. Liq. Chromatogr. 9:2471

HPLC 2

Sample	Human blood plasma
Apparatus	Waters M 6000A LC pump and 440 UV detector
Column	Reversed-phase; µ-Bondapak C_{18}, 10 µm, 300 × 3.9 mm ID
Pre-column	Bondapak C_{18}, 37–50 µm, 23 × 3.9 mm ID
Mobile phase	0.05 M Phosphate buffer, pH 6.0
Flow-rate	2.5 ml min^{-1}
Internal standard	Acetaminophen
Injection size	20 µl
Detection	UV, λ = 254 nm; 0.01–0.2 AUFS
Retention times, min	Uric acid 2.2; Allopurinol 7.2; Oxipurinol 6.7; Internal standard 18.0
Ref.	Nissen P (1982) J. Chromatogr. Biomed. Appl. 228:382

HPLC 3

Sample	Biological fluids
Extraction	Centrifugation (serum) and diluting (urine) with water
Apparatus	Gynkotek M 600/200 LC pump, Uvikon M 720 LC (Kontron) UV detector and Kontron ASI 45 injector

Column	Reversed-phase; SAS-Hypersil 5 μm, 300 × 4.1 mm ID
Mobile phase	0.1 M Citric acid, monohydrate/0.2 M Na_2HPO_4/water (190:810:2000, v/v)
Flow-rate	2 ml min^{-1}; 3915 psi
Temperature	Ambient
Injection size	20 μl
Detection	UV, λ = 252 nm; 0.02 AUFS
Recording	Spectra Physics 4100 computing integrator and Kipp & Zonen BD-9 recorder
Retention times, min	Oxipurinol 7.8 Allopurinol 10.0
Ref.	Breithaupt H and Goebel G (1981) J. Chromatogr. Biomed. Appl. 226:237

HPLC 4

Sample	Human blood plasma and urine
Extraction	Deproteinization with 20% trichloroacetic acid (serum) and diluting the urine with water
Column	Reversed-phase; Brownlee C_{18}, 5 μm, 220 × 4.6 mm ID
Pre-column	Reversed-phase; Brownlee C_{18}, 5 μm
Mobile phase	0.02 M Sodium acetate, pH 7.2 for plasma and pH adjusted to 5.0 with acetic acid for urine samples
Flow-rate	1.3 ml min^{-1}
Injection size	25–50 μl
Detection	UV, λ = 254 nm
Recording	Hewlett-Packard 3390A integrator
Retention times, min	(estimated from drawing) Oxipurinol 10.00 Allopurinol 13.33
Ref.	Failler JM, Farinotti R and Dauphin A (1985) Ther. Drug Monit. 7:324

HPLC 5

Sample	Blood plasma
Extraction	Deproteinization with 4.4 M perchloric acid
Column	Reversed-phase; μ-Bondapak C_{18}, 250 × 2.9 mm ID
Mobile phase	50 mM Potassium dihydrogen phosphate, pH 4.6
Flow-rate	1.5 ml min^{-1}
Temperature	25 °C
Injection size	20 μl
Detection	UV, λ = 254 and 280 nm; 0.01 AUFS
Retention time, min	Allopurinol 12.12
Ref.	Wung WE and Howell SB (1980) Clin. Chem. 26:1704

5 Amiodarone

CAS-No	1951-25-3; 19774-82-4 (hydrochloride)		
Chemical name	2-Butyl-3-benzofuranyl 4-[2-(diethylamino)ethoxy]-3,5-diiodophenyl ketone		
Empirical formula	Amiodarone $C_{25}H_{29}I_2NO_3$	Hydrochloride $C_{25}H_{29}I_2NO_3$, HCl	
Molecular mass	645.3	681.8	
Proprietary names	Atlansil; Cordarone; Cordarone X		
Clinical use	Antiarrhythmic/Antianginal		
Solubility,	Water	ethanol	chloroform
Amiodarone	soluble	soluble	freely soluble
Hydrochloride	very slightly soluble	soluble	freely soluble
Light absorption	λ_{max} = 241 nm in aqueous acid		

HPLC 1

Sample: Blood plasma

Column: A. Normal phase; Spherisorb S5W, silica 250 × 8 mm ID
B. Normal phase; Spherisorb S5, ODS 1, 250 × 8 mm ID

Mobile phase: For A: Methanolic perchloric acid (0.04%, v/v)
For B: Methanolic perchloric acid (0.02%, v/v)

Flow-rate: 4 ml min^{-1}

Injection size: 100 µl, 10 µg ml^{-1}

Detection: UV, λ = 240 nm

Retention times, min: (estimated from drawing)

Column	A.	B.
Didesethylamiodarone	6.6	5.1
Desethylamiodarone	7.5	5.8
Amiodarone	10.8	11.8
Desiodoamiodarone	12.0	11.0
Didesiodoamiodarone	13.2	10.0

Ref. Flanagan RJ, Bhamra RK et al (1988) J. Liq. Chromatogr. 11:1015

HPLC 2

Sample: Blood plasma

Extraction: Methanol with 1% triethylamine on a cyano-bonded silica (CN)-minicolumn

Apparatus: Beckman M 110A LC pump, Bioanalytical Systems LC 6 UV detector and Altex loop injector or Hewlett-Packard 1081B autosampler

Column	Reversed-phase; Spherisorb-octyl 5 µm, 150 × 3.2 mm ID
Mobile phase	Acetonitrile/10 mmol phosphate buffer, pH 3.5 with H_3PO_4 (62:38, v/v) containing 0.5 ml l^{-1} triethylamine
Flow-rate	0.9 ml min^{-1}; 7000 kPa
Temperature	Ambient
Internal standard	L8040 (from Labaz Brussels)
Injection size	20 µl containing 250 ng ml^{-1}
Detection	UV, λ = 254 and 240 nm
Recording	Hewlett-Packard 3390A recording integrator

Retention times, min

Verapamil	0.38	Nifedipine	1.46
Norverapamil	0.48	Warfarin	1.56
Captopril	0.85	Quinidine	1.68
Labetalol	1.05	Diltiazem	2.10
Tocainide	1.16	Desethylamiodarone	6.88
Furosemide	1.21	Bepridil	6.84
Disopyramide	1.23	Amiodarone	9.71
Procainamide	1.26	L8040 (IS)	11.87
Propranolol	1.42		

Ref. Pollak PT, Carruthers SG and Freeman DJ (1986) Clin. Chem. 32:890

HPLC 3

Sample	Blood plasma
Extraction	Methyl tert.-butyl ether at pH 6.0 with buffer
Column	Reversed-phase; C_{18} DB (Supelcosil)
Mobile phase	Acetonitrile/10 mM potassium dihydrogen phosphate containing 20 mM tetramethyl-ammonium chloride/85 % H_3PO_4 (90:9:1, v/v)
Flow-rate	0.8 ml min^{-1}
Temperature	50 °C
Internal standard	L8040 (from Labaz Brussels)
Injection size	200 µl, 1 µg ml^{-1}
Detection	UV, λ = 254 nm
Retention times, min	(estimated from drawing) L8040 (IS) 5.7; Desethylamiodarone 5.5; Amiodarone 6.2
Ref.	Shipe JR (1984) Clin. Chem. 30:1259

HPLC 4

Sample	Serum
Extraction	1-Chlorobutane
Apparatus	LDC Constametric III HPLC system, Rheodyne M 7125 injector and Spectromonitor III UV detector (LDC)

Column	Nitrile phase; 5 μ, 250 × 4.5 mm ID
Mobile phase	Acetonitrile/0.02 M phosphate buffer, pH 3.0 (60:40)
Flow-rate	2.0 ml min^{-1}
Internal standard	Fenethazine
Injection size	20 μl, 0.1–5.0 μg ml^{-1}
Detection	UV, λ = 238 nm; 0.01 AUFS
Recording	Chromatography control module (LDC)
Retention times, min	Fenethazine (IS) 3.9 Desethylamiodarone 7.5 Amiodarone 9.3
Ref.	Hutchings A, Spragg BP and Routledge PA (1986) J. Chromatogr. Biomed. Appl. 382:389

HPLC 5

Sample	Serum and tears
Extraction	Hexane
Apparatus	Altex M 110A LC pump, 210 manual injector and 115-10 UV detector
Column	Reversed-phase; Ultrasphere Octyl 5 μm, 150 × 4.6 mm ID
Mobile phase	10% 0.01 M Ammonium acetate, pH 4.0 and 90% of methanol/acetonitrile (50:50, v/v)
Flow-rate	2.0 ml min^{-1}; 1500 psi
Temperature	Ambient
Internal standard	Trifluoperazine
Injection size	25 μl, 50 ng-10 μg ml^{-1}
Detection	UV, λ = 240 nm
Retention times, min	Trifluoperazine 5.4 N-Desethylamiodarone hydrochloride 6.4 Amiodarone 7.2
Ref.	Muir KT, Kook KA et al (1986) J. Chromatogr. Biomed. Appl. 374:394

---

6 Amitriptyline

CAS-No	50-48-6; 549-18-8 (hydrochloride)	
Chemical name	3-(10,11-Dihydro-5*H*-dibenzo[*a*,*d*]cyclohepten-5-ylidene)-*NN*-dimethylpropylamine	
Empirical formula	Amitriptyline	Hydrochloride
	$C_{20}H_{23}N$	$C_{20}H_{23}N$, HCl
Molecular mass	277.4	313.9
Proprietary names	Amitid; Domical; Triavil	

Clinical use	Antidepressant				
Solubility,	Water	alcohol	acetone	chloroform	ether
Hydrochloride	1 in 1	1 in 1.5	1 in 56	1 in 1.2	insoluble
Light adsorption	λ_{max} = 239 nm in aqueous acid				

HPLC 1

Sample	Body fluids and organ extracts
Extraction	Dichloromethane
Apparatus	Hewlett-Packard 1084B LC with ultraviolet detector
Column	Reversed-phase; Lichrosorb C_8, 5 µm, 250 × 4.0 mm ID
Mobile phase	I. Acetonitrile/tetraethylammonium phosphate buffer (containing 3.5 ml 85% H_3PO_4, 900 ml water and 20% tetra-ethylammonium hydroxide solution for pH 2.5 and make 1 l with water) (40:60) II. Acetonitrile/tetraethylammonium phosphate buffer solution as before (30:70), for amoxapine, loxapine and their metabolites III. Acetonitrile/methanol/tetraethylammonium phosphate buffer solution as before (11:11:78), for nomifensine and its metabolites
Flow-rate	1 ml min^{-1}
Injection size	50 µl containing 10–1000 ng
Detection	UV, λ = 200 nm

Retention times, min

Mobile phase	I	II	III
10-Hydroxy-desipramine	2.2		
10-Hydroxy-amitriptyline	2.5		
Nomifensine	2.6		
2-Hydroxy-desipramine	2.7		
2-Hydroxy-imipramine	2.9		
2-Hydroxy-iminodibenzyl	3.2		
Amoxapine	3.4		
8-Hydroxymianserin	3.5		
Hydroxymaprotilin	3.8		
N-Desmethyldoxepin	3.9		
N-Desmethylmianserin	4.2		
Doxepin	4.3		
7-Hydroxyamoxapine		4.4	
Mianserin	4.8		
N-Bidesmethyl-amitriptyline	4.9		
4-Hydroxynomifensine			4.9
N-Desmethyl-maprotilin	5.0		
Noxiptiline	5.0		
Protriptyline	5.0		

Desipramine	5.1		
8-Hydroxyamoxapine		5.4	
N-Bidesmethyl-trimipramine	5.7		
7-Hydroxyloxapine		5.7	
4-Hydroxy-3-methoxynomifensine			5.7
Imipramine	5.7		
Nortriptyline	5.7		
N-Bidesmethyl-dimetacrine	5.8		
Maprotilin	5.8		
N-Desmethyl-trimipramine	6.2		
7-Methylmianserin	6.2		
Amitriptyline	6.4		
8-Hydroxyloxapine		6.6	
N-Bidesmethyl-clomipramine	6.8		
Trimipramine	7.0		
Amitriptyline oxide	7.1		
3-Hydroxy-4-methoxynomifensine			7.7
N Desmethyl-clomipramine	7.8		
Dimetacrine	7.9		
Melitracen	8.5		
Clomipramine	8.7		
Dibenzepine	10.5		
Dimethylacridan	11.8		
Iminodibenzyl	12.0		
Amoxapine		15.0	
Loxapine		18.3	
Nomifensine			22.3
Lofepramine	27.0		

Ref. Wedy L, Bohn G. and Blaschke G (1988) Fresenius Z. Anal. Chem. 332:266

HPLC 2

Sample	Blood plasma
Extraction	n-Heptane with 1.5% isoamyl alcohol under alkaline condition
Apparatus	Waters M 6000A LC pump, U6K injector or WISP 710A autosampler and 440 ultraviolet detector
Column	Reversed-phase; Supelcosil 5 μm, LC-1, 250 × 4.6 mm ID
Mobile phase	0.1 M Acetate buffer, pH 4.2/acetonitrile (70:30) with 0.005 M sodium heptanesulfonate and 0.01 M triethylamine
Flow-rate	1.8–2.0 ml min^{-1}
Temperature	Ambient
Internal standard	Loxapine
Injection size	50 ng

Detection	UV, λ = 254 nm; 0.0005 AU	
Retention times, min	(estimated from drawing)	
	trans-10-OH-Nortriptyline	5.0
	trans-10-OH-Amitriptiline	5.5
	cis-10-OH-Nortriptiline	6.3
	cis-10-OH-Amitriptyline	7.3
	Loxapine	13.0
	Nortriptyline	16.6
	Amitriptyline	19.0
Ref.	Suckow RF and Cooper TB (1982) J. Chromatogr. Biomed. Appl. 230:391	

HPLC 3

Sample	Blood plasma
Extraction	Hexane/isoamyl alcohol (98:2, v/v)
Column	Normal phase: Varian Si-10, 300 × 4 mm ID
Mobile phase	Methanol/ammonium hydroxide/ammonium nitrate (95:3:2, v/v)
Flow-rate	1.7 ml min^{-1}
Temperature	25 °C
Injection size	25 µl
Detection	UV, λ = 254 nm; 0.02 AUFS
Ref.	Fazio A, Spina E and Pisani F (1987) J. Liq. Chromatogr. 10:223 (review)

HPLC 4

Sample	Blood plasma	
Extraction	Hexane containing 5% isoamyl alcohol	
Apparatus	Waters M 6000A LC pump, U6K injector and 441 UV detector	
Column	Normal phase; Zorbax Sil, 250 × 4.6 mm ID	
Mobile phase	Ammonium hydroxide/methanol (2:998, v/v)	
Flow-rate	1.5 ml min^{-1}	
Internal standards	N-Desmethylclomipramine HCl (I) and desipramine HCl (II)	
Injection size	50 µl, 30–300 ng, ml^{-1}	
Detection	UV, λ = 214 and 254 nm; 0.005–0.01 AUFS	
Recording	Waters M 730 data module	
Retention times, min	Amitriptyline	4.0
	M_1-Metabolite (amitriptyline)	4.1
	2-Hydroimipramine	4.3
	Doxepin	4.7
	Imipramine	5.0
	M_2-Metabolite (nortriptyline)	8.2

	Desmethyldoxepin 9.0 Nortriptyline 9.0 Internal standard (I) 10.4 2-Hydroxydesipramine 12.0 Desipramine 13.3
Ref.	Sutfin TA, D'Ambrosio R and Jusko WJ (1984) Clin. Chem. 30:471

HPLC 5

Sample	Tablet formulations
Apparatus	Varian M 2010 LC pump with 2050 spectrophotometric detector
Column	Reversed-phase; Supelcosil LC-8-DB, 5 μm, 250 × 4.6 mm ID
Mobile phase	0.02 M Ammonium acetate in acetonitrile/methanol/water (45:15:40), pH 5.0 with acetic acid
Flow-rate	1.5 ml min^{-1}
Temperature	Ambient
Internal standard	Naphthalene
Injection size	10 μl containing 0.9–2.6 μg
Detection	UV, λ = 254 nm; 0.2 AUFS
Recording	Varian M 4290 integrator
Retention times, min	Impurity-I 3.32 Internal standard 7.66 Amitriptyline 5.37 Impurity-II 8.93 Perphenazine 6.50
Ref.	Jalal IM, Sa'Sa SI and Khalil HS (1988) J. Liq. Chromatogr. 11:1531

7 Amobarbital

CAS-No	57-43-2; 64-43-7 (sodium salt)				
Chemical name	5-Ethyl-5-isopentylbarbituric acid				
Empirical formula	Amobarbital	Sodium Salt			
	$C_{11}H_{18}N_2O_3$	$C_{11}H_{17}N_2NaO_3$			
Molecular mass	226.3	248.3			
Proprietary names	Amytal; Eunoctal; Stadadorm				
Clinical use	Hypnotic/Sedative				
Solubility,	Water	alcohol	chloroform	ether	aq NaOH
Amobarbital	1 in 1500	1 in 5	1 in 20	1 in 6	soluble
Sodium salt	1 in < 1	1 in 2	insoluble	insoluble	
Light absorption	λ_{max} = 240 nm in 0.05 M borax buffer, pH 9.2				

HPLC 1

Sample	Body fluids
Extraction	Serum: Deproteinization with 40% perchloric acid Urine: With diethyl ether under acidic condition
Apparatus	Waters M 510 LC pump, 490 UV/VIS detector and WISP 712 autosampler
Column	Reversed-phase; μ-Bondapak 10 μm, C_{18}, 300 × 3.9 mm ID
Mobile phase	Acetonitrile/water (35:65, v/v)
Flow-rate	1 ml min^{-1}
Temperature	Ambient
Injection size	20 μl, 2–20 μg ml^{-1}
Detection	UV, λ = 198 nm; 0.1–0.01 AUFS
Recording	Waters data-control M 840
Retention times, min	Amobarbital 11.20 Aprobarbital 13.60
Ref.	Rop PP, Spinazzola J et al (1988) J. Chromatogr. Biomed. Appl. 427:172

HPLC 2

Sample	Urine
Extraction	On-line sample clean-up on 10 μm styrene-divinyl benzene copolymer PRP-1 (pre-column) using buffer, water and methanol
Apparatus	Kontron LC M 410 pumps, MCS 670 tracer valve-switching unit and Kratos (Ramsey) M 757 ultraviolet detector
Column	Reversed-phase; C_{18}, 5 μm, 100 × 3.0 mm ID
Pre-column	PRP-1, 2 × 4.6 mm ID and 5 × 3.0 mm ID (Aminex A-28)
Mobile phase	0.1 M Acetate buffer, pH 5.0/methanol (50:50) after washing with different buffer solutions
Flow-rate	2 ml min^{-1}
Injection size	25 μl, 1 μg ml^{-1}
Detection	UV, λ = 220 nm; 0.01 AUFS
Retention times, min	(estimated from drawing) Butobarbital 4.8 Amobarbital 7.9 Hexobarbital 5.9 Secobarbital 9.1
Ref.	De Jong J, Nielen MWF et al (1986) J. Chromatogr. Biomed. Appl. 381:431

HPLC 3

Sample	Rat tissue
Apparatus	Varian M 5000 HPLC with UV-100 absorbance detector and Waters M 710B WISP autosampler
Column	Reversed-phase; Alltech C_{18}, 10 μm, 250 × 4.6 mm ID

Mobile phase	Acetonitrile/4 mM potassium phosphate buffer, (50:50, v/v), pH 4.0
Flow-rate	1.2 ml min^{-1}
Internal standard	Thiopental
Injection size	0.1 and 0.2 μg
Detection	UV, λ = 290 nm
Retention times	(relative) (retention time of thipental = 4.09 min)

Phenobarbital	0.26	Secobarbital	0.62
Heptabarbital	0.49	Thiopental	1.00
Amobarbital	0.50	Methohexital	1.16
Pentobarbital	0.50	Thiamylol	1.26
Hexobarbital	0.59		

Ref. Ebling WF, Mills-Williams L et al (1989) J. Chromatogr. Biomed. Appl. 490:339

HPLC 4

Sample	Urine
Extraction	Ethyl acetate
Apparatus	Bio-Rad Labs M 1330 HPLC and 1306 UV detector
Column	Reversed-phase; Econosphere C_{18}, 5 μm, 250 × 4.6 mm ID
Pre-column	Reversed-phase; C_{18}, 30–40 μm, 20 × 2 mm ID
Mobile phase	20% (v/v) Acetonitrile in 0.025 M sodium phosphate buffer, pH 6.5
Flow-rate	1.4 ml min^{-1}
Temperature	25 and 35 °C
Internal standard	Butalbital
Injection size	20 μl containing 6.8 nmol ml^{-1}
Detection	UV, λ = 198 nm
Recording	Hewlett-Packard 3393A integrator
Retention times, min	

Temperature	25 °C	35 °C
Amobarbital A	14.2	12.9
Amobarbital B	15.9	14.2
Butalbital	22.8	19.4
Amobarbital	42.7	35.6

Ref. Soine PJ and Soine WH (1987) J. Chromatogr. Biomed. Appl. 422:309

HPLC 5

Sample	Test samples
Apparatus	Waters ALC/GPC M 201 LC with M 6000 pumps, U6K injector, 600 programmer and Varian UV detector
Column	A. Du Pont strong anion exchange (SAX), acid B. Du Pont strong anion exchange (SAX), base C. Du Pont strong cation exchange (SCX), acid

	D. Du Pont ETH Permaphase
	E. Waters Corasil II
Mobile phase	For A: 0.01 M Citric acid solution
	For B: 0.01 M Sodium borate + 0.01 M sodium nitrate
	For C: 0.01 M Citric acid solution
	For D: Water
	For E: Chloroform
Injection size	1.0 µl, 0.1–1 mg ml^{-1}
Detection	UV, $\lambda = 245$ nm
Retention times	(retention positions)

Column	A	B	C	D	E
Barbital	1	1	1	1	8
Diallylbarbital	2	2	3	2	
Aprobarbital	3	3	4	3	5
Butabarbital	4	5	11	5	
Butethal	5	4	5	4	6
Butalbital	6	5	6	7	1
Hexobarbital	7	6	8	8	
Phenobarbital	8	7	2	6	7
Amobarbital	9	8	7	9	4
Pentobarbital	10	9	9	10	2
Secobarbital	11	10	10	11	3

Ref. Henry D, Block JH et al (1976) J. Med. Chem. 19:619

8 Amoxicillin

CAS-No	26787-78-0; 34642-77-8 (sodium salt); 61336-70-7 (trihydrate)		
Chemical name	*D*(–)-α-Amino-*p*-hydroxybenzylpenicillin		
Empirical formula	Amoxicillin	Sodium Salt	Trihydrate
	$C_{16}H_{19}N_3O_5S$	$C_{16}H_{18}N_3NaO_5S$	$C_{16}H_{19}N_3O_5S, 3H_2O$
Molecular mass	365.4	387.4	419.4
Proprietary names	Amoxil; Clamoxyl; Trimox		
Clinical use	Antibiotic		

Solubility,	Water	alcohol	methanol	ether
Sodium salt	1 in < 1			
Trihydrate	1 in 400	1 in 1000	1 in 200	insoluble

Light absorption $\lambda_{max} = 247$ nm in aqueous alkali

HPLC 1

Sample	Raw materials (embonate)
Apparatus	Hewlett-Packard M 1090 LC with 85B computer, Pye-Unicam 4020 UV detector and 5 µl injection loop
Column	Reversed-phase; HP 79915 MO C_8, 10 µm, 200 × 4.6 mm ID
Mobile phase	1/15 M Sörensen phosphate buffer, pH 7.0/methanol (62:38)
Flow-rate	1.5 ml min^{-1}
Temperature	40 °C
Internal standard	Penicillin V potassium
Injection size	5 µl, 42 µg ml^{-1} (embonate)
Detection	UV, λ = 240 nm
Recording	Hewlett-Packard 3380A computing integrator
Retention times, min	Amoxicillin 1.75 Embonic acid 3.72–4.35 Ampicillin 2.89 Penicillin V 5.85
Ref.	Saesmaa T (1988) J. Chromatogr. 455:415

HPLC 2

Sample	Blood plasma
Extraction	Methanol, using Baker-10 SPE TM ion-pair extraction column
Apparatus	Waters M 45 LC pump, 441 ultraviolet detector and 710B sample processor
Column	Reversed-phase; Nucleosil C-18, 5 µm, 150 × 4.6 mm ID
Mobile phase	Methanol/0.005 M tetrabutylammonium hydrogen sulfate containing 0.05 M NaH_2PO_4, H_2O (pH 6.0 with NaOH) (31:69, v/v)
Flow-rate	1.0 ml min^{-1}
Temperature	Ambient
Internal standard	Benzoic acid, 0.025 mM
Injection size	10 µl containing 0.1 µg ml^{-1}
Detection	UV, λ = 229 nm
Recording	Spectra-physics 4100 computing integrator
Retention times, min	Amoxicillin 2.75 Internal standard 5.88
Ref.	Jonkman JHG, Schoenmaker R and Hempenius J (1985) J. Pharm. Biomed. Anal. 3:359

HPLC 3

Sample	Raw materials
Apparatus	Varian M 5000 LC, UV-50 variable-wavelength detector and Valco automatic loop injector
Column	Reversed-phase; bonded ODS-5, MCH-5, 150 × 4.0 mm ID

Mobile phase	I. 0.01 M H_3PO_4/acetonitrile (20:80) II. 0.01 M H_3PO_4/acetonitrile (20:80) containing 0.002 M KCl in aqueous phase
Flow-rate	1 ml min^{-1}
Injection size	200 μl containing 2 μg
Detection	UV, λ = 220 nm
Retention times, min	Mobile phase I. II. Amoxicillin 9.18 4.12 Ampicillin 9.98 4.32
Ref.	Moats WA and Leskinen L (1987) J. Chromatogr. 386:79

HPLC 4

Sample	Blood plasma
Apparatus	Waters M 6000A LC pump with 720 system controller, WISP autosampler and 480 ultraviolet detector
Column	Reversed-phase; RCM, Z-module C_{18}, 10 μm, 100 × 8.0 mm ID
Pre-column	Guard-Pak, C_{18}
Mobile phase	Aqueous acetic acid, pH 3.0/acetonitrile (750:250, v/v)
Flow-rate	4.0 ml min^{-1}; 1000 psi
Injection size	20 μl
Detection	UV, λ = 280 nm
Recording	Waters M 730 data module
Retention times, min	Acetaminophen 1.17 Chloramphenicol 3.40 Fluorocortisone 1.89 Amoxicillin 4.50 Aspirin 2.54 Hydrocortisone 5.18
Ref.	El-Yazigi A, Yusuf A and Al-Humaidan A (1987) Clin. Chem. 33:1814

HPLC 5

Sample	Pharmaceutical dosage forms (capsules) and bulk drugs
Apparatus	Spectra-Physics 8000 LC with 8300 UV detector
Column	Reversed-phase Brownlee C_8, 10 μm, 250 × 4.6 mm ID
Mobile phase	Methanol/0.05 M phosphate buffer, pH 5.0 (6:94, v/v)
Flow-rate	1.0 ml min^{-1}
Temperature	30 °C
Internal standard	Phenoxyacetic acid, 0.8 mg ml^{-1}
Injection size	10 μl, 0.5–2.5 mg ml^{-1}
Detection	UV, λ = 254 nm
Recording	Spectra-Physics data system
Retention times	(relative retention) Amoxicilloic acid 0.76

	6-Aminopenicillanic acid	0.83
	Amoxicillin	1.00
	Methylparaben	1.93
	Internal standard	2.22
	Ethylparaben	3.97
	Propylparaben	7.65
Ref.	Lebelle MJ, Wilson WL and Lauriault G (1980) J. Chromatogr. 202:144	

9 Amphetamine

CAS-No	300-62-9; 60-13-9 (sulfate)		
Chemical name	(±)-α-Methylphenethylamine		
Empirical formula	Amphetamine	Sulfate	
	$C_9H_{13}N$	$(C_9H_{13}N)_2$, H_2SO_4	
Molecular mass	135.2	368.5	
Proprietary names	Anfetamina; Centramina; Simpatina		
Clinical use	CNS-Stimulant		
Solubility,	Water	ethanol	chloroform and ether
Amphetamine	1 in 50	very soluble	very soluble
Sulfate	1 in 9	1 in 515	insoluble
Light absorption	λ_{max} = 257 nm in aqueous acid		

HPLC 1

Sample	Raw materials
Apparatus	Spectra-Physics 3500B LC pump and 770 ultraviolet detector
Column	Reversed-phase; Bondapak 10 µm, C_{18}, 300 × 3.9 mm ID
Mobile phase	Methanol/acetic acid/triethylamine/water (15:1.5:0.5:93)
Flow-rate	1.5 ml min^{-1}
Temperature	Ambient
Injection size	10 µl, 1.0 mg ml^{-1}
Detection	UV, λ = 254 nm; 0.16 AUFS
Recording	Hewlett-Packard 3380A integrator
Retention times, min	(estimated from drawing)

Hydroxyamphetamine	5.3	Methamphetamine	13.0
Phenylpropanolamine	6.6	Phentermine	18.8
Ephedrine	8.0	Mephentermine	21.0
Amphetamine	11.0		

Ref. Roos RW and Lau-Cam CA (1986) J. Chromatogr. 370:403

HPLC 2

Sample	Raw materials (enantiomers)
Apparatus	Waters M 6000A LC pump, U6K injector and 440 UV absorbance detector
Column	Reversed-phase; Zorbax ODS 5 μm, C_{18}, 150 × 4.6 mm ID
Pre-column	Whatman ODS, Co: Pell, 30 to 38 μm, 70 × 2.1 mm ID
Mobile phase	Methanol/water (60:40)
Flow-rate	1.5 ml min^{-1}
Temperature	Ambient
Derivatization	Pre-column; with 4-nitrophenylsulfonyl-(S)-propyl chloride
Injection size	5 μl
Detection	UV, $\lambda = 254$ nm; 0.2 AUFS
Retention times, min	(estimated from drawing) S-Amphetamine 12.2 R-Amphetamine 14.0
Ref.	Barksdale JM and Clark CR (1985) J. Chromatogr. Sci. 23:176

HPLC 3

Sample	Raw materials (enantiomers)
Apparatus	Waters M 6000A LC pump, U6K injector and 440 UV absorption detector
Column	Normal phase; Supelcosil LC-Si, 5 μm, 150 × 4.6 mm ID
Pre-column	Wahtman HC Pellosil 30–38 μm, 70 × 2.1 mm ID
Mobile phase	Chloroform/n-heptane (80:20)
Flow-rate	1.4 ml min^{-1}
Temperature	Ambient
Derivatization	Pre-column; with 4-nitrophenylsulfonyl-(S)-propyl chloride
Injection size	5 μl
Detection	UV, $\lambda = 254$ nm; 0.2 AUFS
Retention times, min	(estimated from drawing) R-Amphetamine 3.8 S-Amphetamine 4.5
Ref.	Barksdale JM and Clark CR (1985) J. Chromatogr. Sci. 23:176

HPLC 4

Sample	Raw materials
Apparatus	Waters M 6000A LC pump, U6K injector and 440 UV detector
Column	Reversed-phase; μ-Bondapak C_{18}, 300 × 3.9 mm ID

Pre-column	Whatman Co: Pell ODS, 70 × 2.1 mm ID		
Mobile phase	Water/acetonitrile/acetic acid (59:40:1)		
Flow rate	1.5 ml min^{-1}		
Temperature	Ambient		
Detection	UV, λ = 254 and 280 nm		
Retention times, min	(estimated from drawing)		
	UV, λ =	254 nm	280 nm
	Phenylpropanolamine	7.5	8.0
	Ephedrine	8.5	9.0
	Pseudoephedrine	10.5	11.0
	Phenmetrazine	16.0	17.0
	Amphetamine	18.5	18.5
	Methamphetamine	24.5	24.5
	Phentermine	30.5	31.0
Ref.	Noggle Jr FT and Clark CR (1984) J. Assoc. Off. Anal. Chem. 67:687		

HPLC 5

Sample	Test solution containing dextropropoxyphene, amphetamine sulfate, pholcodine and protriptyline			
Column	Normal phase; Spherisorb S5W, 250 × 4.5–4.9 mm ID			
Mobile phase	2700 ml Methanol and 300 ml aqueous ammonium nitrate buffer (containing 27 g NH_4NO_3 and 90 ml concentrated ammonia, specific gravity 0.880 and 900 ml water), pH 9.3–9.5			
Flow-rate	2.0 ml min^{-1}			
Temperature	Ambient (19–33 °C)			
Internal standard	Protriptyline			
Injection size	5 µl			
Detection	UV, λ = 254 nm			
Retention times, min	Dextropropoxyphene 1.7		Pholcodine	3.5
	Amphetamine 2.7		Protriptyline	4.7
Ref.	Gill R, Osselton MD et al (1987) J. Chromatogr. 386:65			

---

10 Amphotericin B

CAS-No	1397-89-3
Chemically	Mixture of antifungal polyenes produced by Streptomyces nodosus
Empirical formula	$C_{47}H_{73}NO_{17}$
Molecular mass	924.1
Proprietary names	Ampho-Moronal; Fungilin; Fungizone
Clinical use	Antifungal

Solubility	Water	alcohol	dimethyl sulfoxide	methanol
	insoluble	insoluble	1 in 20	1 in 625

HPLC 1

Sample	Human serum
Extraction	Methanol
Apparatus	Waters LC M 6000A pump, 440 absorbance detector and 710B sample processor
Column	Reversed-phase; μ-Bondapak C_{18}, 10 μm; 300 × 3.9 mm ID
Pre-column	Vydac-201 RP, 30–40 μm, 50 × 3.9 mm ID
Mobile phase	Methanol/acetonitrile/0.0025 M EDTA (500:350:200, v/v)
Flow-rate	1.6 ml min^{-1}
Temperature	Ambient
Injection size	150 μl, 1.0 μg ml^{-1}
Detection	UV, λ = 405 nm; 0.005–0.2 AUFS
Recording	Waters data module M 730
Retention time, min	Amphotericin B 4.16
Ref.	Brassinne C, Laduron C et al (1987) J. Chromatogr. Biomed. Appl. 419:401

HPLC 2

Sample	Urine
Extraction	Absolute methanol on Sep-Pak C_{18} column
Apparatus	Trirotar LC and Uvidec 100-VI UV detector
Column	Reversed-phase; μ-Bondapak C_{18}, 10 μm, 300 × 4 mm ID
Mobile phase	10 mM Acetate buffer, pH 7.0/acetonitrile/methanol/tetrahydrofuran (50:25:20:5, v/v)
Flow-rate	1.5 ml min^{-1}
Temperature	25–27 °C
Injection size	80 μl, 4 ng ml^{-1}
Detection	UV, λ = 405 nm; 0.0025 AUFS
Retention time, min	Amphotericin B 14.0 (estimated from drawing)
Ref.	Kobayashi K, Sakoguchi T et al (1987) J. Chromatogr. Biomed. Appl. 417:439

HPLC 3

Sample	Raw materials
Apparatus	Beckman M 344 LC and Hitachi M 100–40 UV/VIS detector
Column	Reversed-phase; Ultrasphere-ODS 5 μm, 150 × 4.6 mm ID

Pre-column	Reversed-phase; Ultrasil-ODS 5 μm, 45 × 3.2 mm ID		
Mobile phase	Methanol/0.05 M citrate buffer, pH 5.3 (75:25)		
Flow-rate	1 ml min^{-1}		
Injection size	5 μl		
Detection	UV, λ = 305 and 405 nm		
Retention times, min	(amphotericin B)	305 nm	405 nm
	Tetraenes component	1.7	
	Pentaenes component	3.4	
	Tetraene or		
	Pentaene component	8.6	
	Heptaene component		4.2
	Heptaene component		8.1
Ref.	Kelly PM (1988) J. Chromatogr. 437:221		

HPLC 4

Apparatus	Waters ALC/GPC M 200 LC
Column	Reversed-phase; μ-Bondapak C_{18}, 300 × 4 mm ID
Mobile phase	Methanol/0.005 M sodium EDTA (8:2, v/v)
Flow-rate	2.5 ml min^{-1}
Detection	UV, λ = 364, 384 and 407 nm; 0.01 AUFS
Retention time, min	Amphotericin B 4.16
Ref.	Nilsson-Ehle I, Yoshikawa TT et al (1977) J. Inf. Dis. 135:414

HPLC 5

Column	Reversed-phase; Vydac
Mobile phase	Water/methanol/tetrahydrofuran (420:90:45)
Flow-rate	0.9 ml min^{-1}; 1000 psi
Injection size	1 μl
Detection	UV, λ = 350 nm; 0.004 A
Retention time, min	Amphotericin B 10.2 (estimated from drawing)
Ref.	Tsuji K (1978) in: Tsuji K (ed) GLC and HPLC Determination of Therapeutic Agents, Part 2. Marcel Dekker Inc., New York and Basel, p 739

HPLC 6

Sample	Aliquot containing amphotericin
Extraction	Methanol
Column	μ-Bondapak C_{18}
Mobile phase	0.005 M Disodium EDTA and methanol (20:80, v/v)
Flow-rate	2.5 ml min^{-1}
Temperature	Ambient
Injection size	300 μl
Detection	UV, λ = 405 nm
Ref.	Ristuccia PA (1987) J. Liq. Chromatogr. 10:241

11 Ampicillin

CAS-No	69-53-4; 69-52-3 (sodium salt); 7177-48-2 (trihydrate)			
Chemical name	(6*R*)-6-(α-D-Phenylglycylamino)penicillanic acid			
Empirical formula	Ampicillin	Sodium Salt	Trihydrate	
	$C_{16}H_{19}N_3O_4S$	$C_{16}H_{18}N_3NaO_4S$	$C_{16}H_{19}N_3O_4S, 3H_2O$	
Molecular mass	349.4	371.4	403.4	
Proprietary names	Binotal; Penbritin; Polycillin			
Clinical use	Antibiotic			
Solubility,	Water	ethanol	acetone	chloroform
Ampicillin	1 in 170	insoluble	insoluble	insoluble
Sodium salt	1 in 2		1 in 50	slightly soluble
Trihydrate	1 in 150	insoluble	insoluble	insoluble
Light absorption	λ_{max} = 257 nm in aqueous acid			

HPLC 1

Sample	Human urine
Apparatus	Beckmann M 114 HPLC pump with 165 ultraviolet detector
Column	Reversed-phase; Synchropack 6.5 μm, 250 × 4.1 mm ID
Mobile phase	0.01 M Sodium acetate, pH 4.0/methanol (9:1)
Flow-rate	1 ml min^{-1}
Temperature	Ambient
Internal standard	Cefoxitin sodium, 1 mg ml^{-1}
Injection size	20 μl
Detection	UV, λ = 225 nm; 0.2 AUFS
Recording	Spectra-Physics 4270 integrator-plotter
Retention times, min	Ampicillin 8.0 Internal standard 10.0
Ref.	Ibrahim E-SA, Abdel-Hamid ME et al (1988) Anal. Lett. 21:423

HPLC 2

Sample	Raw materials
Apparatus	Varian M 5000 LC, UV-50 detector and Valco automatic loop injector
Column	A. Hamilton PRP-1; 10 μm styrene-divinyl benzene copolymer, 250 × 4.1 mm ID B. Reversed-phase; Micropak SPC-18, 4 μm, 150 × 4.6 mm ID
Pre-column	Special, for B
Mobile phase	I. For A: 0.01 M ortho-H_3PO_4/acetonitrile Gradient: 85:15 to 40:60 in 20 min II. For B: as I.

	III. For B: 0.01 M Tetramethylammonium chloride in 0.01 M ortho-H_3PO_4/acetonitrile. Gradient: 85:15 to 40:60 in 20 min
Flow-rate	1 ml min^{-1}
Injection size	200 µl containing 2 µg
Detection	UV, λ = 220 nm; 0.2 AUFS
Retention times, min	(estimated from drawing)

Mobile phase/column	I/A	II/B	III/B
Amoxicillin	3.3	5.4	3.2
Ampicillin	4.4	7.2	4.2
Cephapirin	5.6	10.7	7.9

Ref. Moats WA (1986) J. Chromatogr. 366:69

HPLC 3

Sample	Raw materials
Apparatus	Varian M 5000 LC, UV-50 variable-wavelength detector and Valco automatic loop injector
Column	Normal phase; Micropak Si-5, 5 µm, silica, 250 × 4.0 mm ID
Mobile phase	I. Acetonitrile/0.01 M H_3PO_4, pH 1.6 (80:20) II. Acetonitrile/0.002 M $NH_4H_2PO_4$, pH 4.6 (80:20) III. Acetonitrile/0.01 M H_3PO_4 + 0.002 M KCl, pH 1.6 (80:20) IV. Acetonitrile/0.002 M KH_2PO_4, pH 4.6 (80:20)
Flow-rate	1 ml min^{-1}
Injection size	200 µl containing 2 µg
Detection	UV, λ = 220 nm
Capacity factor, *k'*	

Mobile phase	I.	II.	III.	IV.
Ampicillin	4.08	5.39	3.02	6.02
Procaine	4.52	15.26	3.39	8.05

Ref. Moats WA and Leskinen L (1987) J. Chromatogr. 386:79

HPLC 4

Sample	Blood plasma and broncho-alveolar lavage fluid
Extraction	Deproteinization with trichloroacetic acid 10%
Apparatus	Spectra-Physics 8700 HPLC and Hewlett Packard 1046A fluorescence detector
Column	Reversed phase; Spherisorb-5 ODS, 250 × 4 mm ID
Mobile phase	Methanol/water (60:40, v/v) containing Pic RB 7 (0.005 M 1-heptanesulfonic acid; pH 3.7 with acetic acid)
Flow-rate	1 ml min^{-1}
Temperature	50 °C
Derivatization	Pre-column; with mercury(II) chloride
Internal standard	Amoxicillin
Injection size	20 µl, 1–100 µg ml^{-1}

Detection	Fluorescence, $\lambda = 345/425$ nm
Retention times, min	Ampicillin 3.45
	Amoxicillin 4.67
Ref.	Rosseel MT, Bogaert MG and Valcke YJ (1989) Chromatographia 27:243

HPLC 5

Sample	Raw materials (anhydrous and trihydrate forms and sodium salts)
Apparatus	Spectra-Physics 3500 HPLC system with 8200 UV detector or UV detector from Waters or Tracor
Column	Reversed-phase; Partisil PXS-525 ODS, 250 × 4.6 mm ID
Pre-column	Reversed-phase; ODS 40–60 μm, 100 × 4.6 mm ID
Mobile phase	15 ml of 1.0 M KH_2PO_4, 1.0 ml of 1.0 M acetic acid and 80 ml acetonitrile in 1 liter water
Flow-rate	1.5 ml min^{-1}
Temperature	Ambient
Internal standard	Caffeine, 0.2 mg ml^{-1}
Injection size	15 μl, 0.7–36 μg
Detection	UV, $\lambda = 254$ nm
Recording	Electronic integrator from Hewlett-Packard or Spectra-Physics
Retention times, min	(estimated from drawing)
	Amipicillin 6.0
	Internal standard 13.0
Ref.	Margosis M (1982) J. Chromatogr. 236:469

---

12 Ascorbic Acid

CAS-No	50-81-7			
Chemical name	2,3-Didehydro-L-*threo*-hexono-1,4-lactone			
Empirical formula	$C_6H_8H_6$			
Molecular mass	176.1			
Proprietary names	Cebion; Megascorb; Redoxon			
Clinical use	Vitamin			
Solubility	Water	alcohol	methanol	ether
	1 in 3 to 3.5	1 in 25	1 in 10	insoluble
Light absorption	$\lambda_{max} = 243$ nm in aqueous acid			

HPLC 1

Sample	Raw materials
Apparatus	Spectra-Physics 8000 HPLC with ultraviolet detector
Column	Waters carbohydrate analysis type, 10 μm, 300 × 3.9 mm ID

Mobile phase	20% Aqueous acetonitrile containing 0.003 M (final) potassium phosphate buffer (pH 6.0)
Flow-rate	2.0 ml min^{-1}
Temperature	25 °C
Injection size	10 μl, 250 μg ml^{-1}
Detection	UV, λ = 265 nm; 0.4 AUFS
Retention times, min	(estimated from drawing) D-Isoascorbic acid 3.6 L-Ascorbic acid 4.5
Ref.	Geigert J, Hirano DS and Neidleman SL (1981) J. Chromatogr. 206:396

HPLC 2

Sample	Human milk, blood plasma and leukocytes
Extraction	Deproteinization with 12.5% trichloroacetic acid
Apparatus	LKB Bromma M 2150 HPLC pump, 2151 variable-wavelength monitor or Perkin-Elmer Series 4 pump and LC-90 UV/VIS detector
Column	Reversed-phase; Nucleosil 7 C_{18}
Mobile phase	2 mM Tetrabutylammonium hydroxide in water, pH 2.92
Flow-rate	1.5 ml min^{-1}
Injection size	20 μl
Detection	UV, λ = 254 nm; 0.002 AUFS
Retention times, min	L-Ascorbic acid 2.9 and 4.0
Ref.	Nyyssoenen K, Pikkarainen S et al (1988) J. Liq. Chromatogr. 11:1717

HPLC 3

Sample	Blood plasma
Extraction	Centrifugation, filtration and dilution with isotonic saline
Apparatus	Waters M 510 LC pump, 721 controller, 450 UV detector and WISP autosampler
Column	Anion-exchange; Ultrasil-AX 10 μm, 250 × 4.6 mm ID
Mobile phase	A. Water B. 0.3 M Potassium dihydrogen phosphate, pH 4.4 Gradient: B 0% to 19% (1.9 ml–2.2 ml min^{-1}) in 7 min; B 19% to 65% (2.2 ml–3 ml min^{-1}) in 1 min; hold 3 min (3.0 ml–4.0 ml min^{-1}); B 65% to 0% in 4 min (4.0 ml–3.0 ml min^{-1}); hold 4 min (3.0 ml–1.9 ml min^{-1})
Detection	UV, λ = 254 nm
Retention times, min	Adenine 1.44 Ascorbic acid 6.20 Hypoxanthine 2.20 Ascorbate-2-PO_4 11.06
Ref.	Moore GL and Fishman RM (1987) J. Chromatogr Biomed. Appl. 419:95

HPLC 4

Sample	Human urine
Extraction	Filtration through Dowex 50W-X8 (H^+) column
Apparatus	Jasco Trirotar-III LC pump, autosampler and FP-550F spectrofluorimeter
Column	Hydrophilic gel; Asahipak GS-320, 50 × 7.6 mm ID
Mobile phase	4.5 g Tartaric acid, 1.5 g disodium EDTA, 1 g β-thiodiglycol in 2 l water, pH 3.0–3.03 with 4 M NaOH solution
Flow-rate	1 ml min^{-1}
Temperature	30 °C
Derivatization	Post-column; with benzamidine hydrochloride
Injection size	50–250 µl containing 20 µM
Detection	Fluorescence, λ = 325/400 nm
Retention times, min	(estimated from drawing) Glucose 37.0 Diketogulonic acid 40.3 Diketogluconic acid 40.3 Dehydroisoascorbic acid 42.4 Dehydroascorbic acid 43.8 Ascorbic acid 48.0 Isoascorbic acid 49.4
Ref.	Seki T, Yamaguchi Y et al (1987) J. Chromatogr. 385:287

HPLC 5

Sample	Body fluids
Apparatus	Shimadzu LC-4A pump and electrochemical detector
Column	Asahipak GS-320H, 250 × 7.6 mm ID
Mobile phase	0.1 M Na_2HPO_4 containing 0.3 M NaCl, pH 7.0
Flow-rate	1.0 ml min^{-1}
Temperature	25 °C ± 1 °C
Injection size	10 µl, 10 µg ml^{-1}
Detection	EC, GC electrode, + 800 mV vs Ag/AgCl RE
Retention times, min	Ascorbic acid 7.6 Uric acid 14.3
Ref.	Yoshiura M and Iriyama K (1986) J. Liq. Chromatogr. 9:177

HPLC 6

Sample	Goat milk
Apparatus	Waters M 510 LC pump, Rheodyne M 7125 injector and Hewlett-Packard 1040A photodiode array detector
Column	Reversed-phase; 5 µm, C_{18}, 8 mm ID with Waters radial RCM-100

Pre-column	μ-Bondapak C_{18}
Mobile phase	0.0005 M Tetrabutylammonium phosphate (containing Waters one low-UV PIC A reagent vial in 1000 ml water)/methanol (99:1)
Flow-rate	2.0 ml min^{-1}; 2000 psi
Injection size	20 μl
Detection	UV, λ = 254 nm
Recording	Hewlett-Packard 9000 Series 300 technical computer
Retention time, min	Ascorbic acid 2.648
Ref.	Lavigne C, Zee JA et al (1987) J. Chromatogr. 410:201

HPLC 7

Sample	Biological fluids
Extraction	Deproteinization with metaphosphoric acid
Apparatus	Beckman M 100A LC pump, manual injector and Perkin-Elmer M 650-10M spectrophotofluorimeter
Column	Reversed-phase; ODS 3 μm, 75 × 4.6 mm ID
Pre-column	PRP-1; 10 μm, 70 × 2.0 mm ID
Mobile phase	0.15 mM Hexadecyltrimethylammonium bromide, 0.5 mM sodium acetate and 0.15 mM disodium EDTA, pH 4.0 with CH_3COOH
Flow-rate	0.5 ml min^{-1}
Derivatization	Post-column; with 4,5-dimethyl-*o*-phenylenediamine dihydrochloride
Internal standard	Isoascorbic acid
Injection size	20 μl, 1–40 μg ml^{-1}
Detection	Fluorescence, λ = 365/440 nm
Recording	Hewlett-Packard 3390 integrator
Retention times, min	Dehydroascorbic acid 5.6
	Ascorbic acid 15.5
	Isoascorbic acid (IS) 19.9
Ref.	Lopez-Anaya A and Mayersohn M (1987) Clin. Chem. 33:1874

---

13 Atropine

CAS-No	51-55-8; 5908-99-6 (sulfate, monohydrate)	
Chemical name	(1*R*,3*r*,5*S*)-Tropan-3-yl(±) tropate	
Empirical formula	Atropine	Sulfate
	$C_{17}H_{23}NO_3$	$(C_{17}H_{23}NO_3)_2,H_2SO_4,H_2O$
Molecular mass	289.4	694.8
Proprietary names	Atropinol; Liotropina; Mydricaine	
Clinical use	Anticholinergic	

Solubility,	Water	ethanol	choroform	ether
Atropine	1 in 400	1 in 3	1 in 1	1 in 60
Sulfate	1 in < 1	1 in 4	insoluble	insoluble
Light absorption	λ_{max} = 252 nm and 258 nm in aqueous acid			

HPLC 1

Sample	Raw materials
Apparatus	Spectra-Physics 3500B LC pump and 770 ultraviolet detector
Column	Reversed-phase; Bondapak 10 m, C_{18} 300 × 3.9 mm ID
Mobile phase	Methanol/acetic acid/triethylamine/water (15:1.5:0.5:83)
Flow-rate	1.5 ml min^{-1}
Temperature	Ambient
Injection size	10 µl, 1.0 mg ml^{-1}
Detection	UV, λ = 230 nm; 0.08 AUFS
Recording	Hewlett-Packard 3380A integrator
Retention times, min	(estimated from drawing)

Homatropine	6.0	Tropic acid	10.8
Scopolamine	8.0	Atropine methyl	13.6
Methscopolamine	9.7	Atropine	17.2

Ref. Roos RW and Lau-Cam CA (1986) J. Chromatogr. 370:403

HPLC 2

Sample	Atropine injection
Apparatus	Waters M 6000 LC pump, Rheodyne M 7120 injector and LDC M 1203 ultraviolet detector
Column	Cyano-phase; Nucleosil 5 CN 5 µm, 50 × 4.65 mm ID and amino-phase; Nucleosil 5 NH_2, 5 µm, 50 × 4.65 mm ID connected in series
Mobile phase	0.05 M Sodium acetate buffer, pH 5.0/methanol (3:1, v/v)
Flow-rate	1.2 ml min^{-1}
Injection size	5 µl
Detection	UV, λ = 254 nm; 0.0032 AU
Recording	Spectra-Physics Autolab Minigrator
Retention times, min	(estimated from drawing) Apoatropine 4.24
	Atropine 2.27 Tropic acid 7.27
Ref.	Lund U and Hansen SH (1978) J. Chromatogr. 161:371

HPLC 3

Sample	Alkaloids
Apparatus	HPLC low-pressure ternary gradient system

Column	Reversed-phase; Lichrosorb C_8, 250×4 mm ID
Mobile phase	Acetonitrile/aqueous sodium dodecylsulfate (30:70, v/v)
Flow-rate	2.5 ml min^{-1}; 2500 psi
Injection size	20 μl containing 2 μg
Detection	UV, λ = 220 nm; 0.08 AUFS
Retention times, min	(estimated from drawing) Papaverine 4.05
	Sparteinine 0.92 Atropine 5.50
	Hordenine 2.67 Pilocarpine 14.97
	Noscapine 3.19 Scopolamine 17.42
Ref.	Merck Application Manual on HPLC, TLC and HPTLC (1986). E. Merck, Darmstadt, p 60

HPLC 4

Sample	Raw materials
Apparatus	Waters M 6000A LC pump, 440 UV detector and Valco sample loop
Column	Normal phase; Lichrosorb Si 100, 10 μm, 200 ×4.0 mm ID
Mobile phase	Sodium phosphate buffer, pH 2.2, ionic strength 0.1, + 1.9% (v/v) 1-pentanol
Flow-rate	1.6 mm sec^{-1}; 7.7 MPa
Temperature	25 °C
Injection size	10 μl
Detection	UV, λ = 254 nm; 0.001 A
Retention times, min	(estimated from drawing)
	Scopolamine hydrobromide 3.95
	Atropine sulfate 5.00
	Benactyzine hydrochloride 5.80
	Adiphenine hydrochloride 6.70
	Methylatropine nitrate 8.00
	Oxypyrronium bromide 9.90
	Oxyphenonium bromide 12.50
Ref.	Crommen J (1979) J. Chromatogr. 186:705

14 Bacampicillin

CAS-No	50972-17-3; 37661-08-8 (hydrochloride)	
Chemically	Semisynthetic penicillin	
Empirical formula	Bacampicillin	Hydrochloride
	$C_{21}H_{27}N_3O_7S$	$C_{21}H_{27}N_3O_7S$, HCl
Molecular mass	465.5	502.0
Proprietary names	Ambaxin; Penglobe; Spectrobid	
Clinical use	Antibiotic	

Solubility,	Water	ethanol	chloroform	ether
Hydrochloride	1 in 15	1 in 7	1 in 10	insoluble
Light absorption	λ_{max} = 310 nm in aqueous alkali			

HPLC 1

Sample: Raw materia!s (ampicillin esters)
Apparatus: LDC/Miltron Roy minipump with pulse dampener, Valco injector and Altex M 153 UV detector
Column: Reversed-phase; Du Pont Zorbax C_8, 7 μm, 250 × 4.6 mm ID
Flow-rate: 1.0 ml min^{-1}
Temperature: 25 °C
Injection size: 10 μl
Detection: UV, λ = 254 nm; 0.16 AUFS
Recording: Kipp & Zonen BD 40 recorder and Pye-Unicam integrator DP88

Retention times, min
Mobile phase: Organic modifier/water/0.2 M phosphate buffer, pH 7.0 [*x*:(95-*x*):5]

Organic modifier	*x*	Retention times Talampicillin		Bacampicillin		Pivampicillin
Methanol	65	8.5	9.2	12.7	13.4	20.5
Methanol	60	14.7	16.3	23.6	25.4	41.4
tert.-Butanol	23	15.6	18.2	23.8	25.8	55.2
Acetonitrile	40	17.3	18.5	20.8	20.8	32.6
Tetrahydrofuran	35	18.0	19.6	20.0	20.0	33.4
2-Propanol	28	18.2	21.0	27.6	30.6	62.0
Ethanol	40	18.7	21.4	29.9	32.2	61.7

Ref.: Van Schepdael A, Roets E et al (1986) J. Chromatogr. 370:149

HPLC 2

Sample: Raw materials (ampicillin esters)
Apparatus: LDC/Milton Roy minipump with pulse dampener, Valco injector and Altex M 153 UV detector
Column:
A. Reversed-phase; μ-Bondapak C_{18}, 10 μm, 250 × 4.6 mm ID
B. Reversed-phase; Hypersil ODS, 5 μm, 250 × 4.6 mm ID
C. Reversed-phase; Spherisorb S5, C_8, 5 μm, 250 × 4.6 mm ID
D. Reversed-phase; Zorbax C_8, 7 μm, 250 × 4.6 mm ID

Mobile phase:
I. Ethanol/water/0.2 M phosphate buffer, pH 7.0 (40:55:5)

	II. Ethanol/water/0.2 M phosphate buffer, pH 7.0 (42.5:52.5:5)
Flow-rate	1.0 ml min^{-1}
Temperature	25 °C
Injection size	10 µl
Detection	UV, λ = 254 nm; 0.16 AUFS
Recording	Kipp & Zonen BD 40 recorder and Pye-Unicam integrator DP88
Retention times, min	

	Talampicillin		Bacampicillin		Pivampicillin	
Mobile phase	I.	II.	I.	II.	I.	II.
A. µ-Bondapak	12.5	9.9	16.7	12.9	31.7	22.9
	13.9	10.9	18.2	13.9		
B. Hypersil	11.9	10.1	19.3	15.7	41.2	30.9
	13.7	11.5	21.0	16.9		
C. Spherisorb	14.7	12.8	19.6	17.1	38.9	31.1
	15.9	13.7	21.7	18.8		
D. Zorbax	18.7	12.4	29.9	23.3	61.7	43.7
	21.4	13.9	32.2	25.2		

Ref.	Van Schepdael A, Roets E et al (1986) J. Chromatogr. 370:149

HPLC 3

Sample	Rabbit serum after administration of suppositories
Extraction	Diluting 5–10 folds with water
Apparatus	Shimadzu LC-5A pump and SPD-2A detector
Column	TSK-Gel; ODS-80TM, 150 × 4.0 mm ID
Mobile phase	I. 0.01 M Phosphate buffer, pH 6.0/methanol (70:30) for ampicillin
	II. 0.01 M Phosphate buffer, pH 6.0/methanol (40:60) for bacampicillin
Flow-rate	1 ml min^{-1}
Detection	UV, λ = 225 nm; 0.02 AUFS
Retention times, min	(estimated from drawing)

Mobile phase	I.	II.
Ampicillin	11.0	
Bacampicillin		8.0

Ref.	Kawashima S, Hirose Y et al (1988) Chem. Pharm Bull. 36:1463

15 Bacitracin

CAS-No	1405-87-4; 1405-89-6 (bacitracin zinc)
Chemically	Polypeptide produced by an organism of the licheniformis group of Bacillus subtilis

Empirical formula	Bacitracin A		Bacitracin Zinc	
	$C_{66}H_{103}N_{17}O_{16}S$			
Molecular mass	1411			
Proprietary names	Bacitin; Orobicin; Polyfax			
Clinical use	Antibiotic			
Solubility,	Water	ethanol	chloroform	ether
Bacitracin	freely soluble	soluble	insoluble	insoluble
Bacitracin zinc	1 in 196	1 in 500	insoluble	1 in 153
Light absorption	λ_{max} = 252 nm in aqueous acid (zinc bacitracin)			

HPLC 1

Sample	Raw materials
Apparatus	LDC modular LC, high-pressure Milton Roy Minipump (max 2500 psi) with a pulse dampener, ultragradient mixer and FID and UV detector
Column	Reversed-phase; μ-Bondapak C_{18}, 300 × 4.6 mm ID
Mobile phase	A. 50 ml Absolute methanol, 850 ml water and 100 ml 100 mM phosphate buffer, pH 4.5 B. 500 ml Absolute methanol, 200 ml water, 200 ml acetonitrile and 100 ml 100 mM phosphate buffer, pH 4.5 Covex gradient: 5% Methanol to 50% methanol and 20% acetonitrile both with 10 mM phosphate buffer, pH 4.5 in 1 hour
Flow-rate	1 ml min^{-1}; 1000 psi
Temperature	Ambient
Injection size	2 μl, 5 mg ml^{-1}
Detection	Pye LC M2 moving wire flame ionization detector, range 32, FS 1.0; UV, λ = 254 nm
Retention times, min	(estimated from drawing)

Bacitracin X	12.5	Bacitracin A	36.5
Bacitracin D	23.0	Bacitracin C	39.0
Bacitracin E	30–31.5	Bacitracin G	43.0
Bacitracin B_1	34.0	Bacitracin F_2	46.5
Bacitracin B_2	35.0	Bacitracin F_1	47.5

Ref.	Tsuji K and Robertson JH (1975) J. Chromatogr. 112:663

HPLC 2

Sample	Raw materials
Apparatus	Shimadzu LC-5A HPLC system with SPD-2AM UV detector
Column	Reversed-phase; Capell Pak C_{18}, 5 μm, 150 × 4.6 mm ID

Mobile phase	Methanol/0.04 M aqueous disodium hydrogen-phosphate (6:4)
Flow-rate	1.3 ml min^{-1}
Injection size	Bacitracin A 50–1000 ng Bacitracin F 65–1000 ng
Detection	UV, λ = 234 nm
Retention times, min	Bacitracins (25) 1–25.0 Bacitracin A 11.0 Bacitracin F 24.0
Ref.	Oka H, Ikai Y et al (1989) J. Chromatogr. 462:315

HPLC 3

Sample	Raw materials
Apparatus	LDC liquid chromatograph, Milton Roy mini-pumps with LDC 709 pulse dampener and 1285 UV monitor
Column	Reversed-phase; Bondapak C_{18}, Corasil, 1000 × 2.1 mm ID
Mobile phase	A. Absolute methanol/water/0.1 M potassium phosphate buffer, pH 4.5 (50:750:200, v/v) B. Absolute methanol/acetonitrile/water/0.1 M phosphate buffer, pH 4.5 (400:200:200:200, v/v) Gradient: B 0% to B 100% in 35 min
Flow-rate	0.9 ml min^{-1}; 1000–1500 psi (column) and 2 ml min^{-1} (detector)
Temperature	Ambient
Injection size	4 µl
Detection	UV, λ = 254 nm
Retention times, min	Bacitracins (22) 1–37.0 Bacitracin F_1 4.3 Bacitracin F_2 6.6 Bacitracin B 31.3 Bacitracin A 33.2 Bacitracin C 36.7
Ref.	Tsuji K, Robertson JH and Bach JA (1974) J. Chromatogr. 99:597

HPLC 4

Sample	Raw materials of bacitracin A
Apparatus	Waters M 6000A LC pump, 660 solvent programmer and 440 UV detector
Column	Reversed-phase; Vydac C_{18}, 300 × 3.6 mm ID
Mobile phase	(1) 0.01 M Pi (monosodium phosphate), (2) 0.05 M Pi and (3) 0.01 M HSA (1-heptane-sulfonic acid sodium salt), and each with 80% MeOH (methanol), 60% AcCN (acetonitrile) and 40% THF (tetrahydrofuran), and pH each 2.0, 3.4 and 6.6

Flow-rate	1.0 ml min^{-1}
Temperature	Ambient
Injection size	1–10 µl
Detection	UV, λ = 254 nm; 0.2 A
Recording	Computing integrator Supergrator-1 (Columbia Scientific)
Retention times, min	Mobile phase

Buffer	Modifier	pH 2.0	3.4	6.6
0.05 M Pi	AcCN	3.26	3.66	3.02
0.01 M Pi	THF	3.48	4.46	4.17
	AcCN	3.64	3.88	2.76
	MeOH	4.45	5.11	3.42
0.01 M HSA	AcCN	3.77	5.42	3.74
	MeOH	6.68	7.56	6.11
	THF	7.37	9.61	5.52

Ref. Aszalos A and Aquilar A (1984) J. Chromatogr. 290:83

16 Beclomethasone

CAS-No	4419-39-90; 5534-09-8 (dipropionate)	
Chemical name	9α-Chloro-16β-methylprednisolone	
Empirical formula	Beclomethasone	Dipropionate
	$C_{22}H_{29}ClO_5$	$C_{28}H_{37}ClO_7$
Molecular mass	408.9	521.0
Proprietary names	Beclovent; Propaderm; Viarox	
Clinical use	Corticosteroid	

Solubility, Dipropionate	Water	alcohol	chloroform	acetone
	insoluble	1 in 60	1 in 8	freely soluble

Light absorption: λ_{max} = 239 nm in dehydrated alcohol

HPLC 1

Sample	Beclomethasone dipropionate drug substances
Column	Nitrile phase; Spherisorb CN 5 µm, 200 × 5 mm ID
Mobile phase	Acetonitrile/water (40:60, v/v)
Flow-rate	2 ml min^{-1}
Temperature	Ambient
Injection size	5 µl
Detection	UV, λ = 239 nm; 0.5 AUFS
Recording	Data processor
Retention times, min	Beclomethasone 2.0; Beclometehasone-21-monopropionate 2.8; Beclomethasone-

	17-monopropionate	2.5	Beclomethasone-dipropionate 3.5

Ref.	Mulholland M and Rudd DR (1985) Anal. Proc. 22:140

HPLC 2

Sample	Beclomethasone Dipropionate cream
Column	Reversed-phase; Spherisorb ODS 5 µm, 100 × 5 mm ID
Mobile phase	Methanol and water (70:30)
Flow-rate	2.0 ml min^{-1}
Temperature	60 °C
Internal standard	Testosterone propionate
Detection	UV, λ = 238 nm
Retention times, min	Beclomethasone 17-propionate appr. 1.5 Beclomethasone dipropionate appr. 2.0
Ref.	Anonym (1980): British Pharmacopoeia, Addendum 1983. London Her Majesty's Stationary Office, p 268

17 Benzalkonium Chloride

CAS-No	8001-54-5
Chemically	Mixture of alkylbenzyldimethylammonium chlorides of the general formula $[C_6H_5 \cdot CH_2 \cdot N(CH_3)_2 \cdot R]Cl$, in which R is a mixture of the alkyls from C_8H_{17} to $C_{18}H_{37}$
Proprietary names	Morpan BC; Roccal; Sabol
Clinical use	Topical Anti-infective/Preservative
Solubility	Very soluble in water, ethanol and acetone and insoluble in ether
Light absorption	λ_{max} = 262 nm in water

HPLC 1

Sample	Blood plasma
Extraction	Methanol/ethyl acetate (1:1) containing 0.01% ammonium chloride on Baker C_{18} extraction column
Apparatus	Waters M 6000A LC pump, U6K injector and 481 UV detector
Column	Cyano-phase; Spherisorb
Mobile phase	Acetonitrile/0.161 M propionate buffer, pH 5.4 (containing 75 ml 10% Na_2CO_3 in 1.5 l water and mixed with 12 ml propionic acid and diluted to 2 l with water) (90:10)
Flow-rate	2 ml min^{-1}

Temperature	Ambient
Injection size	50 μl, 5–25 ng ml^{-1}
Detection	UV, $\lambda = 214$ and 254 nm
Retention times, min	Benzalkonium 5.9–6.3 (estimated from drawing)
Ref.	Bleau G and Desaulniers M (1989) J. Chromatogr. Biomed. Appl. 487:221

HPLC 2

Sample	Test substance
Column	Reversed-phase; Chromspher C_{18}, 5 μm, 200 × 3 mm ID
Mobile phase	40% Methanol and 60% pentane sulfonic acid (PSA) solution (0.5 g PSA and 0.5 ml acetic acid in 500 ml water)
Flow-rate	0.4 ml min^{-1}; 2000 psi
Injection size	15 μl, 3 mg ml^{-1}
Detection	UV, $\lambda = 257$ nm
Ref.	Chrompack Application Notes (1987) Chrompack International BV, Middelburg

HPLC 3

Sample	Standard solution and cream base
Column	β-Cyclobond I (ict), 250 × 4.6 mm ID
Mobile phase	Methanol/water (60:40, v/v)
Flow-rate	1.0 ml min^{-1}
Detection	UV, $\lambda = 220$ nm; 0.1 AUFS
Retention time, min	Benzalkonium chloride 1.0
Ref.	ict Application notes on Cyclobond HPLC Columns. itc, Frankfurt, Basel and Vienna

18 Benzyl Alcohol

CAS-No	100-51-6	
Chemical name	Phenylmethanol	
Empirical formula	C_6H_5, CH_2OH	
Molecular mass	108.1	
Clinical use	Preservative	
Solubility	Water	alcohol, chloroform, ether and fixed oils
	1 in 25	miscible

HPLC 1

Sample	Aromatic solutes
Column	Reversed-phase; Nucleosil C_{18}, 10 μm, 300 × 4.6 mm ID

Mobile phase	A. Water/methanol (50:50) B. Water/tetrahydrofuran (68:32) C. Methanol/tetrahydrofuran/water (35:10:55) D. Methanol/tetrahydrofuran/water (10:25:65)
Flow-rate	1.5 ml min^{-1}
Detection	UV, $\lambda = 254$ nm
Retention times, min	(estimated from drawing)

Mobile phase	A.	B.	C.	D.
Benzyl alcohol	5.48	4.29	4.76	4.76
Phenol	5.48	7.86	5.95	7.62
3-Phenylpropanol	11.90	7.38	10.24	9.05
2,4-Dimethylphenol	15.95	20.24	17.14	20.95
Benzene	17.38	17.14	17.14	18.10
Diethylphthalate	20.48	12.86	17.14	15.24

Ref. Schoenmakers PJ, Drouen ACJH et al (1982) Chromatographia 15:688

HPLC 2

Sample	Test mixture
Column	Reversed-phase; ODS-Hypersil 5 μm, 100 × 4.6 mm ID
Mobile phase	Water/acetonitrile (60:40)
Flow-rate	3 ml min^{-1}
Temperature	40 °C
Injection size	5 μl
Detection	UV
Retention times, min	(estimated from drawing)

Benzyl alcohol	0.60	Diethyl phthalate	1.25
Benzaldehyde	0.75	4-Chloro-nitrobenzene	2.00
Dimethyl phthalate	0.95	Nitrobenzene	2.70

Ref. Hewlett-Packard Analytical Supplies Catalog and Chromatography Reference Guide (1988) Hewlett-Packard, USA, p 130

HPLC 3

Sample	Pharmaceutical cream
Extraction	Water
Apparatus	Altex M 380 LC pump, Waters M 710B autosampler and Pye-Unicam 4020 UV detector
Column	Reversed-phase; Ultrasphere-octyl 5 μm, 150 × 4.6 mm ID
Mobile phase	Methanol/water (60:40), pH 3.5 with acetic acid
Flow-rate	1.0 ml min^{-1}
Injection size	20 μl, 125–1000 μg ml^{-1}
Detection	UV, $\lambda = 260$ nm
Recording	Waters M 730 integrator data module

Retention times, min	Benzyl alcohol 4.5 Benzaldehyde 6.5
Ref.	Gormand D, Brazier JL et al (1986) J. Chromatogr. 355:345

HPLC 4

Sample	Benzyl alcohol in colognes
Apparatus	HPLC low-pressure ternary gradient system
Column	Reversed-phase; Superspher 60, C_8, 125 × 4 mm ID
Pre-column	Reversed-phase; C_8, 25 × 4 mm ID
Mobile phase	Methanol/water (65:35, v/v)
Flow-rate	1.0 ml min^{-1}; 3750 psi
Injection size	5 µl, appr. 2 mg ml^{-1}
Detection	UV, λ = 254 nm; 0.04–0.01 AUFS
Ref.	Merck Application Manual on HPLC, TLC and HPTLC (1986). E. Merck, Darmstadt, p 61

19 Benzylpenicillin

CAS-No	61-33-6; 6130-64-9 (procaine salt, monohydrate); 69-57-8 (sodium salt)			
Chemical name	(6*R*)-6-(2-Phenylacetamido)penicillanic acid			
Empirical formula	Benzylpenicillin $C_{16}H_{18}N_2O_4S$	Procaine Salt $C_{16}H_{18}N_2O_4S$, $C_{13}H_{20}N_2O_2$, H_2O	Sodium Salt $C_{16}H_{17}N_2NaO_4S$	
Molecular mass	334.4	588.7	356.4	
Proprietary names	Gonopen; Novopen; Wycillin			
Clinical use	Antibiotic			
Solubility,	Water	ethanol	chloroform	ether
Benzylpenicillin	insoluble	soluble		soluble
Procaine salt	1 in 200	1 in 30	1 in 60	
Sodium salt	very soluble	soluble	insoluble	insoluble
Light absorption	λ_{max} = 257 nm in water (sodium salt)			

HPLC 1

Sample	Penicillins
Apparatus	Haskel Engineering M 26980-4 LC pump, Altex M 153 ultraviolet detector and Rheodyne 70–10 injector
Column	Reversed-phase; Lichrosorb C_{18}, 10 µm, 300 × 4.6 mm ID
Mobile phase	Methanol/0.01 M sodium dihydrogen phosphate (35:65)
Flow-rate	1 ml min^{-1}; 1320 psi
Temperature	Ambient

Injection size	8 µl containing 1.4 µg
Detection	UV, λ = 225 nm; 0.04 AUFS
Retention times, min	(estimated from drawing) Ampicillin 11.2 Penicillin V 22.0 Penicillin G 12.5 Oxacillin 25.5 Methicillin 14.0
Ref.	White ER and Zarembo JE (1981) J. Antibiotics 34:836

HPLC 2

Sample	Blood plasma and urine
Extraction	Serum: With diethyl ether or with dichloromethane Urine: Diluting with Sörensen buffer
Apparatus	Kratos Spectroflow M 400 LC pump, 773 Spectroflow UV detector and Rheodyne M 7125 injector
Column	Reversed-phase; MOS-Hypersil 5 µm, C_8, 100 × 3.0 mm ID
Mobile phase	0.005 M Sodium acetate buffer, pH 5.0/ acetonitrile/methanol (62:12:26) with 0.003 M tetrabutylammonium bromide
Flow-rate	1 ml min^{-1}
Temperature	22 °C
Injection size	20 µl, 50 µg ml^{-1}
Detection	UV, λ = 231 nm
Retention times, min	Benzylpenicillin 4.00 Probenecid 11.00
Ref.	Van Gulpen C, Brokerhof AW et al (1986) J. Chromatogr. Biomed. Appl. 381:365

HPLC 3

Sample	Serum
Extraction	Dichloromethane
Apparatus	Kontron M 620 HPLC pump, Rheodyne M 7125 injector and Uvikon 720-LC variable-wavelength detector (Kontron)
Column	Reversed-phase; Spherisorb ODS, 5 µm, 250 × 4.6 mm ID
Pre-column	Whatman Co:Pell ODS, 50 × 4.6 mm ID
Mobile phase	Acetonitrile/10 mM ammonium acetate buffer, pH 6.4 (14:86, v/v)
Flow-rate	2 ml min^{-1}
Temperature	Ambient
Internal standard	Phenoxymethylpenicillinic acid
Injection size	20 µl, 0.5–100 µg ml^{-1}
Detection	UV, λ = 208 nm
Recording	Hewlett-Packard 3390A integrator

Retention times, min	(estimated from drawing) Benzylpenicillin 4.0 Internal standard 6.1–7.0
Ref.	Soto-Otero R, Mendez-Alvarez E et al (1987) Biomed. Chromatogr. 2:177

HPLC 4

Sample	Mixture of penicillins
Column	Reversed-phase; Micropak CH, 250 × 2 mm ID
Mobile phase	20% Methanol in water
Flow-rate	0.5 ml min^{-1}
Detection	UV, λ = 254 nm; 1.0 AUFS
Retention times, min	(estimated from drawing) Penicillin G 3.7 6-Amino-penicillanic acid 1.0 Penicillin V 5.5
Ref.	Ottake T and Yaguchi M: Varian Liquid Chromatography at work 23. Varian, California

HPLC 5

Sample	Raw materials
Column	Reversed-phase; Porasil B C_{18}, 1000 × 2.1 mm ID
Mobile phase	Methanol/0.05 M ammonium carbonate (30:70)
Flow-rate	0.5 ml min^{-1}; 500 psi
Detection	UV, λ = 254 nm; 0.04 AUFS
Retention times, min	(estimated from drawing) Penicillin G 13.5 Penicillin V 24.0
Ref.	White ER, Carroll MA et al (1975) J. Antibiotics 28:205

HPLC 6

Sample	Swine tissues
Extraction	Deproteinization with acetonitrile under alkaline condition and then with methylene chloride under acidic condition
Apparatus	Varian M 5000 LC and UV-50 variable-wavelength detector
Column	Reversed-phase; Micropak MCH-10, 10 μm, 300 × 4.6 mm ID
Mobile phase	A. 0.01 or 0.02 M H_3PO_4 B. Acetonitrile Gradient: B 20% to 60% in 20 min
Flow-rate	1 ml min^{-1}
Injection size	200 μl containing 0–6 μg

Detection	UV, λ = 220 nm
Ref.	Moats WA, Harris EW and Steele NC (1986) J. Agric. Food Chem. 34:452

20 Betamethasone

CAS-No	378-44-9; 2152-44-5 (valerate)			
Chemical name	9α-Fluoro-16β-methylprednisolone			
Empirical formula	Betamethasone	Valerate		
	$C_{22}H_{29}FO_5$	$C_{27}H_{37}FO_6$		
Molecular mass	292.5	476.6		
Proprietary names	Betnesol; Colestone; Valisone			
Clinical use	Corticosteroid			
Solubility,	Water	alcohol	chloroform	acetone
Betamethasone	insoluble	1 in 75	1 in 1100	sparingly soluble
Valerate	insoluble	1 in 12 to 16	1 in 2	freely soluble
Light absorption	λ_{max} = 240 nm in ethanol			

HPLC 1

Sample	Corticosteroid mixtures
Apparatus	Kronton M 410 LC pump, Pye-Unicam LC UV spectro-photometer detector and Rheodyne M 7125 injection valve
Column	Normal phase; Lichrosorb Si 60, Nucleosil 50-5, Zorbax SIL, Partisil 5, Spherisorb S5 W, Lichrosorb Si 100, Nucleosil 100-5 and Hypersil: 120 × 4.6 mm ID Reversed-phase; Lichrosorb RP-18, Nucleosil 5 C_{18}, Hypersil ODS, Zorbax ODS, Partisil ODS-3 and Spherisorb S5 ODS: 120 × 4.6 mm ID
Mobile phase	For Normal phase: Methanol/tetrahydrofuran/water/0.2 M phosphate buffer, pH 7.0 (25:13:57:5) containing 2.5 mM cetyltrimethylammonium bromide. For Reversed-phase: Acetonitrile/water (37:63)
Flow-rate	1 ml min^{-1}
Injection size	10–20 µl, 0.05%–0.1% solution
Detection	UV, λ = 254 nm; 0.01 AU
Recording	Hewlett-Packard 3353A laboratory data system and Kipp & Zonen BD-8 recorder

Retention times, min	Cortisone acetate, hydrocortisone, hydrocortisone acetate, prednisolone, prednisolone acetate, corticosterone, triamcinolone acetonide, methylprednisolone, betamethasone, fluocinolone acetonide, dexamethasone and desoxycortone acetate, are column specific
Ref.	Helboe P (1986) J. Chromatogr. 366:191

HPLC 2

Sample	Serum and urine
Extraction	Dichloromethane
Apparatus	Varian M 8500 HPLC and Waters M 440 and Spectra-Physics 770 UV detectors
Column	Reversed-phase; C_8, 10 µm, 250 × 4.6 mm ID
Mobile phase	Methanol/water (60:40)
Flow-rate	1.25 ml min^{-1}; 700 psi
Internal standard	Hydrocortisone 21-acetate
Injection size	100 µl containing 10–300 ng
Detection	UV, λ = 240 and 254 nm
Retention times, min	Hydrocortisone 5.3; Internal standard 7.2; Betamethasone 6.5; Betamethasone-21-acetate 9.0
Ref.	Petersen MC, Nation RL and Ashley JJ (1980) J Chromatogr. Biomed. Appl. 183:131

HPLC 3

Sample	Dosage forms (ointments)
Extraction	Dimethyl sulfoxide
Apparatus	Altex M 100A LC pump, Rheodyne M 7120 injector and Pye LC-3 UV detector
Column	Normal phase; Spherisorb 5 µm, 250 × 4.6 mm ID
Mobile phase	Ethyl acetate/chloroform/methanol (71:28:1) saturated with water
Flow-rate	1 ml min^{-1}; 870 psi
Injection size	20 µl, 20–120 ng ml^{-1}
Detection	UV, λ = 260 nm; 0.32 AUFS
Retention times, min	Betamethasone-21-valeate 3.1
	Hydrocortisone-21-butyrate 3.2
	Betamethasone-17-valeate 3.7
	Hydrocortisone-17-butyrate 4.1
	Betamethasone 4.8
	Hydrocortisone 5.9
Ref.	Li Wan PO A, Irwin WJ and Yip YW (1979) J. Chromatogr. 176:399

HPLC 4

Sample	Raw materials
Column	Reversed-phase; ODS 5 to 10 µm, 300 × 4 mm ID
Mobile phase	Acetonitrile/water (appr. 37:63)
Temperature	Ambient
Internal standard	Propylparaben, 0.125 mg ml^{-1}
Injection size	5 and 25 µl, 0.1 mg ml^{-1}
Detection	UV, λ = 254 nm or 240 nm; 0.6 FS
Retention times, min	Betamethasone 3.3 Propylparaben 5.4
Ref.	Anonym (1985): The United States Pharmacopeia XXI and The National Formulary XVI. United States Pharmacopeial Convention Inc., Rockville, p 107

21 Bleomycin Sulfate

CAS-No	11056-06-7 (bleomycin); 9041-93-4 (sulfate)		
Chemically	Sulfates of a mixture of glycopeptides including bleomycin A_2 and B_2 produced by Streptomyces verticillus or by other means		
Proprietary names	Blenoxane; Bleomicina; Verbublen		
Clinical use	Antineoplastic		
Solubility,	Water and methanol	ethanol	ether
Bleomycin	very soluble	slightly soluble	insoluble
Sulfate	very soluble		
Light absorption	λ_{max} = 244–248 nm and 289–294 nm in water		

HPLC 1

Sample	Copper(II)-chelated bleomycin; bulk drugs Blenoxane and bleomycin hydrolase
Apparatus	Waters M 6000A and 45 HPLC pumps, 660 solvent programmer, 440 UV detector and U6K injector
Column	Reversed-phase; µ-Bondapak C_{18}, 10 µm, 300 × 3.9 mm ID
Pre-column	Reversed-phase; Corasil C_{18}, 30–38 µm, 23 × 3.9 mm ID
Mobile phase	A. Water containing 5.0 mM sodium pentanesulfonic acid, and 0.5% (v/v) glacial acetic acid, pH 4.3 with concentrated ammonium hydroxide

B. Methanol containing 5.0 mM sodium pentanesulfonic acid and 0.5% (v/v) CH_3COOH
Linear gradient: B 28% to 48% in 45 min

Flow-rate: 1.5 ml min^{-1}; 2200 psi
Injection size: 10–20 μl containing 50 pmol to 10 nmol
Detection: UV, λ = 280 nm
Recording: Waters M 730 data module
Retention times, min: (estimated from drawing) (Bleomycin = BLM)

Samples	Blenoxane mixture	Blenoxane mixture	Cu(II) -BLMs	BLM-hydrolase
Cu(II)-BLM A_2	12.0	11.5		
CU(II)-BLM B_2	19.2	17.5		
Cu(II)-BLM DM-A_2	36.8	33.5		
In(III)-BLM A_2		15.0		
In(III)-BLM B_2		20.5		
In(III)-BLM DM-A_2		36.8		
Free Cu(II)			2.4	
Bleomycinic acid			5.0	
BLM B_1			9.5	
BLM A_2			12.0	11.5
BLM A_5			13.8	
BLM B_2			18.7	17.0
BLM B_4			22.7	
BLM B_6			26.0	
BLM CHP			28.7	
BLM PEPP			31.8	
BLM DM-A_2			34.0	
BLM-desamido-A_2				10.0
BLM-desamido-B_2				15.8
BLM-desamido-DM-A_2				31.7
BLM-DM-A_2				33.5

Ref.: Klett RP, Chovan JP and Danse IHR (1984) J. Chromatogr. Biomed. Appl. 310:361

HPLC 2

Sample: Different cobalt-bleomycin complexes (Co-blm)
Extraction:
A. Co-blm A_2, demethylated by heating at 120 °C overnight
B. Blm dA_2 chelated with cobalt
C. Preparation after 24 hours standing at room temperature

Column: Reversed-phase; Nucleosil C_{18}, 10 μm
Mobile phase: 1% Ammonium acetate/methanol (60:40, v/v)
Flow-rate: 2 ml min^{-1}
Injection size: 1 mg
Detection: UV, λ = 254 nm

Retention times, min	(estimated from drawing)			
	Orange Co-blm A_2	3.0	Orange Co-blm dA_2	9.6
	Co-blm A_1 form II	4.8	Co-blm dA_2 form II	12–12.8
	Co-blm A_1 form I	5.8	Co-blm dA_2 form I	16–17.0

Ref. Vos CM, Schipper D et al (1982) J. Antibiotics 35:837

HPLC 3

Sample	Bleomycin formulations
Apparatus	Waters M 6000 LC pump with 660 solvent programmer and 440 UV detector
Column	Reversed-phase; μ-Bondapak C_{18}, 10 μm, 250 × 4.6 mm ID
Mobile phase	A. 0.005 M 1-Pentanesulfonic acid in 0.5% acetic acid deaerated aqueous solution, pH 4.3 with concentrated NH_4OH B. Methanol Linear gradient: Methanol 10% to 40% in 60 min and hold upto 75 min
Flow-rate	1.2–1.8 ml min^{-1}; 1600 psi
Injection size	3–9 μl
Detection	UV, $\lambda = 254$ nm
Recording	Microprocessor (Columbia Instruments Supergrator I)
Retention times, min	(estimated from drawing) (Bleomycin = Bleo)

Bleomycinic acid	36.0	Bleo B_5	52.0
Bleo A_2	37.5	Bleo B_4	58.5
Bleo B_2	46.0	Bleo DM-A_2	68.5

Ref. Aszalos A, Crawford J et al (1981) J. Pharm. Sci. 70:878

HPLC 4

Sample	Raw materials of bleomycin S
Apparatus	Waters M 6000A LC pump, 660 solvent programmer and 440 UV detector
Column	Reversed-phase; Vydac C_{18}, 300 × 3.6 mm ID
Mobile phase	(1) 0.01 M Pi (monosodium phosphate), (2) 0.05 M Pi and (3) 0.01 M HSA (1-heptane-sulfonic acid sodium salt), and each with 80% MeOH (methanol), 60% AcCN (acetonitrile) and 40% THF and pH each 2.0, 3.4 and 6.6 (tetrahydrofuran)
Flow-rate	1.0 ml min^{-1}
Temperature	Ambient
Injection size	1–10 μl
Detection	UV, $\lambda = 254$ nm; 0.2 A

Recording	Computing integrator
Retention times, min	(bleomycin S)

Mobile phase		pH		
Buffer	Modifier	2.0	3.4	6.6
0.01 M Pi	THF	2.84	3.18	3.10
	AcCN	2.97	3.24	3.21
	MeOH	3.46	3.53	4.10
0.05 M Pi	AcCN	3.25	3.46	3.02
0.01 M HSA	AcCN	3.06	3.80	3.48
	THF	4.63	5.47	3.77
	MeOH	5.26	5.48	5.11

Ref.	Aszalos A and Aquilar A (1984) J. Chromatogr. 290:83

22 Bupivacaine Hydrochloride

CAS-No	2180-92-9 (bupivacaine); 14252-80-3 (hydrochloride, monohydrate)	
Chemical name	(±)-(1-Butyl-2-piperidyl)formo-2′,6′-xylidide	
Empirical formula	Bupivacaine	Hydrochloride
	$C_{18}H_{28}N_2O$	$C_{18}H_{28}N_2O$, HCl, H_2O
Molecular mass	288.4	342.9
Proprietary names	Carbostesin; Duracaine; Marcaine	
Clinical use	Local Anesthetic	

	Water	alcohol	acetone, chlorform and ether
Solubility, Hydrochloride	1 in 25	1 in 8	slightly soluble

Light absorption	λ_{max} = 263 nm in aqueous acid

HPLC 1

Sample	Serum
Extraction	n-Hexane under alkaline condition
Apparatus	Varian LC-5560 pump and Varian UV-200 variable-wavelength detector
Column	Enantiopak (LKB), α_1-acid-glycoprotein, 100 × 4 mm ID
Mobile phase	9%, 2-Propanol in 0.008 M sodium phosphate buffer with 0.1 M sodium chloride
Flow-rate	0.3 ml min^{-1}
Temperature	30 °C
Internal standard	Diazepam
Injection size	20 µl, 0.5–2.5 µg ml^{-1}

Detection	UV, λ = 215 nm; 0.005 AUFS
Recording	Varian Vista M 402 data system
Retention times, min	Internal standard 15.7 R(+)Bupivacaine 21.0 S(−)Enantiomer 28.3
Ref.	Lee EJD and Ang SB (1987) J. Chromatogr. Biomed. Appl. 420:203

HPLC 2

Sample	Dosage forms (injections)
Column	Reversed-phase; ODS 5–10 μm, 300 × 4 mm ID
Mobile phase	Acetonitrile/phosphate buffer, pH 6.8 (65:35), pH 7.7 with 1 M H_3PO_4
Flow-rate	2.0 ml min^{-1}
Internal standard	Dibutyl phthalate
Injection size	Appr. 20 μl
Detection	UV, λ = 263 nm
Retention times	(relative) Bupivacaine hydrochloride 1.0 Internal standard 1.2
Ref.	Anonym (1985): The United States Pharmacopeia XXI and The National Formulary XVI, Supplement 1, 1984. United States Pharmacopeial Convention Inc., Rockville, p 1706

HPLC 3

Sample	Blood serum
Extraction	Acetonitrile on a Sep-Pak ODS cartridge
Apparatus	Du Pont M 8800 HPLC pump and Bioanalytical Systems electrochemical detector (LC-4B amperometer)
Column	Normal phase; silica 10 μm, 300 mm
Pre-column	30 mm Silica cartridge
Mobile phase	Acetonitrile/40 mM ammonium phosphate, pH 7.8 (22–25:75–78)
Flow-rate	2.0 ml min^{-1}
Injection size	50 μl
Detection	EC, GC electrode, + 1.2 V vs Ag/AgCl RE
Retention times, min	(estimated from drawing) Glycine xylidide 3.5 Lidocaine 6.0 Monoethylglycine xylidide 5.2 Bupivacaine 8.0
Ref.	Halbert MK and Baldwin RP (1984) J. Chromatogr. Biomed. Appl. 306:269

23 Caffeine

CAS-No	58-08-2; 5743-12-4 (hydrate)			
Chemical name	1,3,7-Trimethylxanthine			
Empirical formula	Caffeine	Hydrate		
	$C_8H_{10}N_4O_2$	$C_8H_{10}N_4O_2$, H_2O		
Molecular mass	194.2	212.2		
Proprietary names	Caffeine and Sodium Benzoate; No Doz; Pro-Plus			
Clinical use	Xanthine-Stimulant			
Solubility,	Water	ethanol	chloroform	aq HCl
Caffeine	1 in 60	1 in 130	1 in 7	soluble
Hydrate	1 in 60	1 in 110	soluble	
Light absorption	λ_{max} = 273 nm in aqueous acid			

HPLC 1

Sample	Urine
Extraction	Ethyl acetate/chloroform/isopropanol (45:45:10) under alkaline condition
Apparatus	Gradient HPLC System, ultraviolet detector and Rheodyne loop injector
Column	Reversed-phase; 5-μm Hypersil ODS, 250 × 4.5 mm ID
Mobile phase	A. 1% Tetrahydrofuran in 10 mM acetate buffer, pH 4.8 B. 15% Acetonitrile and 1% tetrahydrofuran in 10 mM acetate buffer, pH 4.8 Gradient: 100% A for 5 min; B 5, 10 and 2% per min; each step for 5 min
Flow-rate	1.5 ml min^{-1}
Internal standard	Proxyphylline, 10 mg l^{-1}
Injection size	50 μl, 10 mg l^{-1}
Detection	UV, λ = 280 nm
Retention times, min	

Uric acid	3.6	1,7-Dimethyluric-acid	12.0
7-Methyluric acid	5.3		
7-Methylxanthine	6.0	1,7-Dimethylxanthine	12.7
1-Methyluric acid	6.8	Theophylline	13.5
3-Methylxanthine	7.2	1,3,7-Trimethyl-uric acid	14.6
1-Methylxanthine	8.6		
1,3-Dimethyluric acid	9.5	Caffeine	16.7
Theobromine	9.7	Proxyphylline	18.0

Ref. Scott NR, Chakraborty J and Marks V (1986) J. Chromatogr. Biomed. Appl. 375:321

HPLC 2

Sample	Test solution containing caffeine, imipramine hydrochloride, morphine hydrochloride, methyl-amphetamine hydrochloride and protriptyline
Apparatus	HPLC equipments with ultraviolet detector
Column	Normal phase; Spherisorb S5W, 250 × 4.5–4.9 mm ID
Mobile phase	2700 ml Methanol and 300 ml aqueous ammonium nitrate buffer (containing 27 g NH_4NO_3 and 90 ml concentrated ammonia, specific gravity 0.880 in 900 ml water), pH 9.3–9.5
Flow-rate	2.0 ml min^{-1}
Temperature	Ambient (19–33 °C)
Internal standard	Protriptyline
Injection size	5 µl
Detection	UV, λ = 254 nm
Recording	Electronic integrator
Retention times, min	Caffeine 1.7; Imipramine 2.5; Morphine 3.1; Methylamphetamine 4.0; Protriptyline 4.6
Ref.	Gill R, Osselton MD et al (1987) J. Chromatogr. 386:65

HPLC 3

Sample	Horse serum
Apparatus	Waters M 6000 LC pump and 440 UV detector
Column	Reversed-phase; octyl Hi-Chrom 5 µm, 250 × 4 mm ID
Mobile phase	3% Tetrahydrofuran in 10 mM disodium hydrogen-phosphate, pH 6.5 with ortho-phosphoric acid
Flow-rate	2.0 ml min^{-1}
Temperature	Ambient
Injection size	60–120 µl
Detection	UV, λ = 280 nm; 0.1 AUFS
Retention times, min	Theobromine 2.5; Paraxanthine 4.1; Theophylline 5.0; Caffeine 7.8; Enprofylline 10.4
Ref.	Grgurinovich N (1986) J. Chromatogr Biomed. Appl. 380:431

HPLC 4

Sample	Tablets
Apparatus	Perkin-Elmer LC Series 2 and LC-55B ultraviolet detector
Column	Reversed-phase; 10 µm ODS, C_{18}, 250 × 4.6 mm ID

Mobile phase	Gradient: First methanol/water (30:70), and then 30% methanol to 95% methanol in 13 min
Flow-rate	1.8 ml min^{-1}
Temperature	Ambient
Internal standard	Phenacetin
Injection size	10 µl
Detection	UV, λ = 270 nm; 0.2–0.4 AUFS
Recording	Perkin-Elmer strip-chart-recorder and integrator
Retention times, min	(estimated from drawing)

4-Aminophenol	3.2	Phenacetin	10.2
Paracetamol	4.2	4-Chloroacetanilide	10.8
Caffeine	6.0	Propyphenazone	12.8

Ref. Mamolo MG, Vio L and Maurich V (1985) J. Pharm. Biomed. Anal. 3:157

---

24 Carbamazepine

CAS-No	298-46-4
Chemical name	5*H*-Dibenz[*b*,*f*]azepine-5-carboxamide
Empirical formula	$C_{15}H_{12}N_2O$
Molecular mass	236.3
Proprietary names	Lexin; Tegretol; Telesmin
Clinical use	Anticonvulsant

Solubility

Water and ether	alcohol	chloroform	acetone
Insoluble	1 in 10	1 in 10	soluble

Light absorption: λ_{max} = 237 nm 285 nm in methanol

HPLC 1

Sample	Tablets
Apparatus	Varian M 5000 LC pump with UV detector
Column	5 µm-Diol, 250 × 4.5 mm ID (Merck)
Mobile phase	Acetonitrile/methanol/0.05% acetic acid (5:5:90, v/v)
Flow-rate	1 ml min^{-1}
Internal standard	Antipyrine, 0.015 mg ml^{-1}
Injection size	5–10 µl, 0.001 mg ml^{-1}
Detection	UV, λ = 230 nm
Recording	Varian CDS 401 data station
Retention times, min	10,11-Dihydrocarbamazepine 2.62
	Carbamazepine 3.14
	10-Bromocarbamazepine 4.76
Ref.	Cyr TD, Matsui F et al (1987) J. Assoc. Off. Anal. Chem. 70:836

HPLC 2

Sample	Serum
Extraction	Advanced automated sample processor (AASP), on-line
Apparatus	Gradient HPLC system and Spectra-Physics 4-spectrophotometer
Column	Reversed-phase; Shandon-ODS, Hypersil 5 μm, 250 × 4.6 mm ID
Mobile phase	A. Phosphate buffer, pH 4.0/acetonitrile (9:1) B. Phosphate buffer, pH 4.0/acetonitrile (4:6) Gradient: B 15% to 50% in 7 min; hold 2 min; B 50% to 15% in 0.1 min; run 3.9 min
Flow-rate	1.5 ml min^{-1}
Temperature	70 °C
Injection size	100 μl diluted sample
Detection	UV, λ = 205 nm
Recording	Shimadzu C-R3A integrator

Retention times, min

2-Ethyl-2-phenylmalone diamide	3.65
Ethosuximide	4.24
Primidone	4.96
10,11-Dihydro-10,11-dihydroxycarbamazepine	6.03
Phenobarbital	7.50
N-Desmethylmethsuximide	8.02
Carbamazepine-10,11-epoxide	8.47
Phenytoin	10.61
Carbamazepine	11.00

Ref. Juergens U (1986) J. Chromatogr. 371:307

HPLC 3

Sample	Serum
Extraction	Ultrafiltration
Apparatus	LDC Constametric III HPLC pump, Rheodyne M 7120 injector and Kratos M 773 UV detector
Column	Reversed-phase; 5 μm, C_{18}, 150 × 4 mm ID
Mobile phase	240 ml Methanol, 160 ml acetonitrile and 10 ml 0.1 M potassium phosphate, pH 3.7 and water to make 1 liter
Flow-rate	2.0 ml min^{-1}
Temperature	Ambient
Injection size	20 μl, 1.8–8.4 μg ml^{-1}
Detection	UV, λ = 210 nm

Retention times, min (estimated from drawing)

Primidone	1.7	Phenytoin	7.2
Phenobarbital	3.0	Carbamazepine	8.0

Ref. Bock JL and Ben-Ezra J (1985) Clin. Chem. 31:1884

HPLC 4

Sample	Serum
Extraction	Methanol on Bond-Elut column with Vac-Elut apparatus
Apparatus	Perkin-Elmer Series 3 LC pump with LC-85 ultraviolet detector and Rheodyne M 7125 injector
Column	Reversed-phase; C_{18}, 5 μm, 125 × 4.6 mm ID
Mobile phase	Methanol/acetonitrile/phosphate buffer (20 mM KH_2PO_4, pH 3.7 with H_3PO_4) (35:13.5:51.5, v/v)
Flow-rate	3.0 ml min^{-1}
Temperature	50 °C
Internal standard	Cyclopal
Injection size	5 μl
Detection	UV, $\lambda = 210$ or 195 nm
Recording	Perkin-Elmer Sigma 10 data system

Retention times, min		
	Caffeine	0.72
	Cefamandole	0.73
	Phenylethylmelanamide	0.78
	Ethosuximide	0.82
	Lidocaine	0.87
	Primidone	0.93
	5-Hydroxyphenyl-5-phenylhydantoin	0.98
	Sulfamethazole	0.98
	Chloramphenicol	1.02
	Nirvanol	1.17
	Phenobarbital	1.25
	Phensuximide	1.28
	Carbamazepine epoxide	1.35
	Butabarbital	1.57
	Butalbital	1.80
	Mephenytoin	1.88
	Methsuximide	1.88
	2-Phenyl-2-methylsuccimide	1.88
	Phenytoin	2.03
	Pentobarbital	2.37
	Carbamazepine	2.40
	Glutethimide	2.43

Ref. Kabra PM, Nelson MA and Marton LJ (1983) Clin. Chem. 29:473

HPLC 5

Sample	Human blood plasma and urine
Extraction	Methyl tert.-butyl ether under alkaline condition
Column	Reversed-phase; C_{18}, 5 μm, 150 × 4.6 mm ID
Mobile phase	Acetonitrile/water (28:72)
Flow-rate	2.0 ml min^{-1}

Temperature	37 °C
Internal standard	2-Methylcarbamazepine
Injection size	20 µl
Detection	UV, λ = 212 nm
Retention times, min	(estimated from drawing)
	trans-10,11-Dihydroxycarbamazepine 1.33
	2,2-Hydroxycarbamazepine 1.78
	Carbamazepine-10,11-epoxide 2.44
	Carbamazepine 4.89
	Internal standard 9.33
Ref.	Chelberg RD, Gunawan S and Treiman DM (1988) Ther. Drug Monit. 10:188

---

25 Carbenicillin

CAS-No	4697-36-3; 4800-94-6 (sodium salt)			
Chemical name	α-Carboxybenzylpenicillin			
Empirical formula	Carbenicillin	Sodium Salt		
	$C_{17}H_{18}N_2O_6S$	$C_{17}H_{16}N_2Na_2O_6S$		
Molecular mass	378.4	422.4		
Proprietary names	Microcillin; Pyopen; Rexcilina			
Clinical use	Antibiotic			
Solubility,	Water	alcohol	chloroform	ether
Sodium salt	1 in 1.2	1 in 25	insoluble	insoluble
Light absorption	λ_{max} = 259 nm in aqueous acid			

HPLC 1

Sample	Dosage forms (injectable)
Apparatus	Waters M 510 LC pump, 481 variable wavelength detector and U6K injector
Column	Reversed-phase; µ Bondapak C_{18}, 10 µm, 300 × 3.9 mm ID
Mobile phase	I. Acetonitrile/0.01 M KH_2PO_4 (12:88, v/v) II. Acetonitrile/methanol/0.01 M KH_2PO_4/0.02 M octanesulfonic acid, sodium salt in 0.2% glacial acetic acid (12:7:41:40, v/v)
Flow-rate	1 min^{-1}
Internal standard	Propylthiouracil
Injection size	20 µl, 40–130 µg ml^{-1}
Detection	UV, λ = 225 nm; 0.1 AUFS
Recording	Waters M 730 integrating recorder

Retention times, min	Mobile phase I. II. Ticarcillin 4.58 Carbenicillin 5.08 Propylthiouracil 9.66 Carbenicillin (diastereoisomers) 5.48, 6.33 Ticarcillin (diastereoisomers) 4.80, 5.45 Degradation products after storing 24 hours at room temperature Ticarcillin 4.62 Carbenicillin 5.12 Degradation product I 6.37 Degradation product II 7.04 Internal standard 9.70
Ref.	Mopper B (1988) J. Assoc. Off. Anal. Chem. 71:390

HPLC 2

Sample	Serum and urine
Extraction	Ultrafiltration
Apparatus	Jasco Bip-I pump, VL-614 loop injector and Uvidec 100-V spectrophotometer
Column	Reversed-phase; Develosil ODS-5, 5 μm, 150 × 4.6 nm ID
Mobile phase	A. 0.0001 M NaH_2PO_4 (containing 0.005 M tetrabutylammonium bromide and sodium thiosulfate)/acetonitrile (1.8:1, v/v) B. As in A but 0.01 M NaH_2PO_4
Flow-rate	1.0 ml min^{-1}
Temperature	40 °C and ambient
Derivatization	Pre-column; with 2 M 1,2,4-triazole and 10^{-3} M $HgCl_2$, pH 9.0 at 60 °C
Injection size	30–60 μl
Detection	UV, λ = 327, 328 and 331 nm
Recording	Shimadzu C-R3A integrator
Retention times, min	(estimated from drawing) Carbenicillin 4.5 (328 nm, 40 °C) (A) Ticarcillin 4.5 (327 nm, 40 °C) (A) Sulbenicillin 6.0 (331 nm, ambient) (B) (estimated from drawing)
Ref.	Haginaka J and Wakai J (1985) Analyst. 110:1185

HPLC 3

Sample	Raw materials
Apparatus	Varian M 5000 LC, UV-50 detector and Valco automatic loop injector
Column	Organic Polymer; Hamilton PRP-1, 10 μm, styrene-divinyl benzene copolymer, 250 × 4.1 mm ID

Mobile phase	Gradient: 0.01 M ortho-Phosphoric acid/acetonitrile (85:15 to 40:60 in 20 min)
Flow-rate	1 ml min^{-1}
Injection size	200 μl containing 2 μg
Detection	UV, λ = 220 nm; 0.2 AUFS
Retention times, min	(estimated from drawing)

Carbenicillin	12.0	Penicillin V	16.2
Methicillin	13.0	Cloxacillin	20.0
Penicillin G	14.3	Nafcillin	20.5
		Dicloxacillin	22.2

Ref. Moats WA (1986) J. Chromatogr. 366:69

HPLC 4

Apparatus	Waters ALC/GPC M 204 HPLC with 440 UV detector
Column	Reversed-phase; μ-Bondapak C_{18}, 30.4 × 4.0 mm ID
Mobile phase	Methanol/0.01 M tetrabutylammonium bromide (4:7)
Flow-rate	3.0 ml min^{-1}; 3500 psi
Injection size	10 μl
Detection	UV, λ = 254 nm
Retention times, min	(estimated from drawing)
	Carbenicillin diastereomers 5.0, 6.0
Ref.	Yamaoka K, Narita S et al (1979) J. Chromatogr. 168:187

HPLC 5

Sample	Bulk products
Apparatus	LDC Model II G Constant-flow-pump and spectromonitor III
Column	Reversed-phase; Spherisorb ODS 5 μm, 150 × 4.6 mm ID
Mobile phase	37% or 35% Methanol with 0.05 M potassium dihydrogen phosphate (v/v) containing 0.1% (w/v) tetrabutylammonium bromide, pH with 10% H_3PO_4
Flow-rate	1.2 and 1.3 ml min^{-1}
Detection	UV, λ = 220 and 254 nm; 0.1–0.5 AUFS
Recording	Shimadzu E-1A electronic integrator
Retention times, min	(estimated from drawing)

Mobile phase with methanol	37%				35%
Flow-rate, min^{-1}	1.2 ml				1.3 ml
UV, λ =	220 nm				254 nm
pH, mobile phase	3.00	3.35	3.7	4.35	3.35
Carbenicillin	8.0	10.0	8.5	6–8	13.0
Penicillin G					27.0
Degradation products					5–12.0

Ref. Twomey PA (1981) J. Pharm. Sci. 70:824

26 Cefamandole

CAS-No	34444-01-4; 42540-40-9 (nafate)	
Chemically	Cephalosporin antibiotic	
Empirical formula	Cefamandole	Nafate
	$C_{18}H_{18}N_6O_5S_2$	$C_{19}H_{17}N_6NaO_6S_2$
Molecular mass	462.5	512.5
Proprietary names	Kefadol; Mandokef; Mandol	
Clinical use	Antibiotic	
Solubility,	Water, methanol	organic solvents
Nafate	freely soluble	insoluble
Light absorption	λ_{max} = 272 nm in methanol	

HPLC 1

Sample	Serum
Extraction	Deproteinization with acetonitrile
Apparatus	Waters M 6000A HPLC pump, 440 ultraviolet detector and U6K sample injector
Column	Reversed-phase; RCM C_{18}, 100 × 8 mm ID
Pre-column	Waters RCSS Guard-Pak
Mobile phase	Methanol/acetic acid, 7.5 ml l^{-1} (30:70, v/v), pH 5.5 with triethylamine
Flow-rate	3.0 ml min^{-1}
Internal standard	4′-Nitroacetanilide
Injection size	25 µl, 100 µg ml^{-1}
Detection	UV, λ = 254 nm

Retention times, min

Vancomycin	1.1	Cefazolin	3.6
Moxalactam	1.5	Cefoxitin	4.2
Procainamide	1.6	Sulfamethoxazole	4.3
Acetaminophen	1.9	Tobramycin	5.5
Cimetidine	2.0	Cephapirin	6.7
N-Acetylprocainamide	2.3	Cefamandole	8.2
Cephalexin	2.6	Chloramphenicol	9.5
Theophylline	2.8	4′-Nitroacet-anilide	12.4
Cefaclor	3.0	Cephalothin	13.0
Cefotaxime	3.0	Miconazole	14.1

Ref. Danzer LA (1983) Clin. Chem. 29:856

HPLC 2

Sample	Serum
Extraction	Chloroform/butanol (3:1)
Apparatus	Varian M 5000 HPLC System with UV-50 variable-wavelength detector

Column	Phenyl; Bondapak, 300 × 4 mm ID
Mobile phase	A. 10 mM Ammonium acetate, pH 4.2 B. 10 mM Ammonium acetate, pH 4.2 in methanol/water (50:50) and adjusted to pH 7.0, 5.0, 4.6, 4.2 and 4.0 Linear gradient: Hold 85% A and 15% B for 3.5 min; B 15% to 100% in 5.5 min; hold 7.0 min
Flow-rate	2.0 ml min^{-1}
Temperature	Ambient
Internal standard	Cefazolin
Injection size	100 µl containing 2.5 µg
Detection	UV, λ = 250 nm
Retention times, min	

pH, mobile phase	7.0	5.0	4.6	4.2	4.0
Ticarcillin	1.4	2.0	2.6	3.5	8.7
Moxolactam	2.1	2.8	3.4	4.4[a]	5.6
Carbenicillin	2.4	4.9	5.2	5.5	6.0
Salicylate	–	–	3.8	3.8	4.0
Cefazolin	6.6	9.2	9.2	9.2	9.9
Cephalothin	8.4	10.3	10.3	10.3	11.1
Theophylline	8.7	8.8	8.8	8.7	8.8
Cefotaxime	8.8	8.8	8.8	8.7	9.0
Chloramphenicol	10.9	10.9	10.8	10.7	9.8
Cefamandole	11.1	11.1	11.1	11.1	11.2
Nafcillin	11.5	12.8	13.3	15.1	19.5

Ref.	Annesley T, Wilkerson K et al (1984) Clin. Chem. 30:908

HPLC 3

Sample	Serum, urine and dialysis fluid
Extraction	Deproteinization with acetonitrile and then with dichloromethane
Apparatus	Beckman M 100A LC pump, 504 autosampler and 160 UV detector
Column	Reversed-phase; Ultrasphere ODS 5 µm, 250 × 4.6 mm ID
Pre-column	Whatman, Co:Pell 10 µm ODS
Mobile phase	Methanol/water (31:69, v/v) containing 500 µl phosphoric acid, 20 mmol sodium sulfate and 200 µl triethylamine per liter, pH 6.0 with NaOH
Flow-rate	1.1 ml min^{-1}; 3500 psi
Injection size	20 µl
Detection	UV, λ = 254 nm
Recording	Hewlett-Packard 3390 A integrator
Retention time, min	Cefamandole 12.0
Ref.	Bliss M and Mayersohn M (1986) Clin. Chem. 32:197

HPLC 4

Sample	Biological materials
Extraction	First, chloroform/1-pentanol (3:1) under acidic condition, and then with phosphate buffer, pH 7.0
Apparatus	Waters M 6000A HPLC, WISP 710A autosampler and 450 UV detector
Column	Reversed-phase; μ-Bondapak C_{18}, 10 μm, 300 × 3.9 mm ID
Mobile phase	0.01 M Acetate buffer, pH 4.8/methanol (85:15)
Flow-rate	1.5 ml min^{-1}; 2175 psi
Temperature	Ambient
Injection size	10–50 μl, 0.15–5 μg ml^{-1}
Detection	UV, λ = 270 nm; 0.02 AUFS

Retention times, min

	UV	RT
Deacetylcephalothin	240 nm	2.70
Cefuroxime	254 nm	3.10
Cefoxitin	245 nm	3.30
Cefotaxime	234 nm	3.45
Cefazolin	275 nm	4.20
Cefamandole	270 nm	6.15
Cephalothin	240 nm	8.00
Cefoperazone	240 nm	8.35

Ref. Brisson AM and Fourtillan JB (1981) J. Chromatogr. Biomed. Appl. 223:393

27 Cefazolin Sodium

CAS-No	25953-19-9 (cefazolin); 27164-46-1 (sodium salt)		
Chemically	Cephalosphorin antibiotic		
Empirical formula	Cefazolin	Sodium Salt	
	$C_{14}H_{14}N_8O_4S_3$	$C_{14}H_{13}N_8NaO_4S_3$	
Molecular mass	454.5	476.5	
Proprietary names	Ancef; Gramaxin; Kefzol		
Clinical use	Antibiotic		
Solubility,	Water	methanol	chloroform and ether
Sodium salt	freely sol	slightly sol	insoluble

HPLC 1

Sample	Ophthalmic ointment
Apparatus	Waters M 6000A LC pump, 440 UV detector and 710B WISP autosampler

Column	Reversed-phase; μ-Bondapak C_{18}, 10 μm, 300 × 3.9 mm ID
Mobile phase	0.05 M Ammonium acetate buffer, pH 6.8/ methanol (70:30)
Flow-rate	1 ml min^{-1}
Temperature	Ambient
Internal standard	Salicylic acid
Injection size	20 μl, 0.18 mg ml^{-1}
Detection	UV, λ = 254 nm; 0.05 AUFS
Recording	HP 3357 data collection system (Hewlett-Packard)
Retention times, min	Internal standard 4.8 Cefazolin sodium 5.9
Ref.	Martin CF, Takahashi LT et al (1987) J. Chromatogr. 402:376

HPLC 2

Sample	Plasma and urine
Extraction	Chloroform after protein precipitation with acetonitrile
Apparatus	Waters M 45 LC pump, 710B automatic injector and 480 ultraviolet detector
Column	Reserved-phase; Radial Pak 10 μm, Bondapak C_{18}, 100 × 8 mm ID
Mobile phase	Water/methanol/triethylamine (750:250:4, v/v) pH 3.0 with ortho-phosphoric acid
Flow-rate	4.0 ml min^{-1}
Temperature	Ambient
Injection size	5–20 μl, 8 μg ml^{-1}
Detection	UV, λ = 270 nm
Retention times, min	Ceftriaxone 4.2 Cefazolin 6.8
Ref.	Demotes-Mainard FM, Vincon GA et al (1988) J. Pharm. Biomed. Anal. 6:407

HPLC 3

Sample	Raw materials
Column	Reversed-phase; ODS 10 μm, 300 × 4 mm ID
Mobile phase	Buffer, pH 3.6 (containing 0.90 g Na_2HPO_4 and 1.298 g citric acid, monohydrate per liter water)/acetonitrile (9:1)
Flow-rate	2 ml min^{-1}
Internal standard	Salicylic acid
Injection size	Appr. 10 μl
Detection	UV, λ = 254 nm
Retention times	(relative) Internal standard 0.70 Cefazolin 1.00

Ref.	Anonym (1985): The United States Pharmacopeia XXI and The National Formulary XVI, Supplement 1, 1984. United States Pharmacopeial Convention Inc., Rockville, p 1708

28 Cefixime

CAS-No	79350-37-1	
Chemically	Cephalosporin antibiotic	
Empirical formula	Cefixime	Sodium Salt
	$C_{16}H_{15}N_5O_7S_2$	$C_{16}H_{13}N_5Na_2O_7S_2$
Molecular mass	453.4	497.4
Proprietary name	Fujisawa	
Clinical use	Antibiotic	

HPLC 1

Sample	Biological fluids
Extraction	Diluting with Soerensen buffer, pH 7.4
Apparatus	LDC Constametric III-G LC pump, ultraviolet detector and Spectra-Physics 8780 XR autosampler
Column	Reversed-phase; Nucleosil 5 µm, C_{18}, 200 × 4 mm ID
Mobile phase	Methanol/57.4 mM K_2HPO_4 (15:85), pH 5.2 with ortho-phosphoric acid
Flow-rate	1 ml min^{-1}
Temperature	Ambient
Injection size	20 µl, 1–250 µg ml^{-1}
Detection	UV, $\lambda = 230$ nm
Recording	Computing integrator M 301 (LDC)
Retention time, min	Cefixime 10.0 (estimated from drawing)
Ref.	Knoeller J, Koenig W et al (1988) J. Chromatogr. Biomed. Appl. 427:257

HPLC 2

Sample	Serum, urine and buffer-solution
Apparatus	LDC Constametric III-G pump and ultraviolet detector
Column	Reserved-phase; Nucleosil 5, C_{18}, 200 × 4 mm ID
Mobile phase	Methanol/43 mM K_2HPO_4 (15:85, v/v), pH 5.2 with H_3PO_4
Flow-rate	1 ml min^{-1}
Injection size	20 µl, 5–250 µg ml^{-1}

Detection	UV, λ = 230 nm; 0.2 AUFS
Recording	Computing integrator 301 (LDC)
Retention time, min	Cefixime 12.0
Ref.	Knoeller J and Schoenfeld W et al (1987) J. Chromatogr. 389:312

HPLC 3

Sample	Human serum
Extraction	Deproteinization with acetonitrile
Apparatus	Waters HPLC M 590 pump, 481 ultraviolet detector and 710B autosampler
Column	Reversed-phase; Ultrasphere C_8, 5 μm, 150 × 4.6 mm ID
Pre-column	Waters RCSS silica Guard-Pak
Mobile phase	Methanol/12.5 mmol sodium dihydrogen phosphate (20:80), pH 2.6 with concentrated H_3PO_4
Flow-rate	2.0 ml min^{-1}
Temperature	Ambient
Injection size	50–80 μl, 0.1–10 μg ml^{-1}
Detection	UV, λ = 240 nm
Retention times, min	Cefadroxil 3.5 Cephalexin 15.0
	Cefaclor 6.1 Cephadrine 21.0
	Cefixime 10.8 (estimated from drawing)
Ref.	Mcateer JA, Hiltke MF et al (1987) Clin. Chem. 33:1788

29 Cefotaxime

CAS-No	63527-52-6; 64485-93-4 (sodium salt)	
Chemically	Caphalosporin antibiotic	
Empirical formula	Cefotaxime	Sodium Salt
	$C_{16}H_{17}N_5O_7O_2$	$C_{16}H_{16}N_5NaO_7S_2$
Molecular mass	455.5	477.4
Proprietary name	Claforan	
Clinical use	Antibiotic	
Solubility,	Water	organic solvents
Sodium salt	Freely soluble	insoluble
Light absorption	λ_{max} = 235 nm in methanol	

HPLC 1

Sample	Blood plasma of rats
Extraction	Deproteinization with methanol

Apparatus	Waters M 45 LC pump and 440 ultraviolet detector			
Column	Reversed-phase; 10 μm, C_8, 250 × 4.6 mm ID			
Mobile phase	20 mM Sodium phosphate, pH 4.5/methanol (77:23)			
Flow-rate	1 ml min^{-1}			
Temperature	25 °C			
Injection size	20 μl, 2.5–500 μg ml^{-1}			
Detection	UV, λ = 254 nm; 0.05 AUFS			
Retention times, min	Desacetylcefotaxime 5.4	Cefotaxime 14.6		
	Ceftriaxone 7.6	Sodium barbital 15.4		
	Cephradine (IS) 11.6	Caffeine 21.6		
Ref.	Hakim L, Bourne DWA and Triggs EJ (1988) J. Chromatogr. Biomed. Appl. 424:111			

HPLC 2

Sample	Biological materials
Extraction	Deproteinizing with acetonitrile
Apparatus	Waters M 6000 LC pump, 480 ultraviolet detector and U6K injector
Column	Reserved-phase; Brownlee 5 μm, C_{18}, 100 × 4.6 mm ID
Pre-column	Brownlee C_{18}, 30 × 4.6 mm ID
Mobile phase	Acetonitrile/phosphate buffer, pH 7.0/water (20:5:75, v/v) with tetrapentylammonium bromide 3.8 g l^{-1}
Flow-rate	2 ml min^{-1}
Injection size	20 μl, 1–50 μg ml^{-1}
Detection	UV, λ = 254 nm; 0.01–0.1 AUFS
Recording	Waters data module
Retention times, min	Desacetylacefotaxime 3.4
	Cefotaxime 6.0
Ref.	Hary L and Andrejak M (1987) J. Chromatogr. Biomed. Appl. 419:396

HPLC 3

Sample	Serum and bile
Extraction	Deproteinization with 0.4 mol perchloric acid
Apparatus	Waters M 6000A HPLC pump, U6K injector, WISP 710A autosampler and 440 UV detector
Column	Reversed-phase; Lichrosorb C_{18}, 7 μm, 250 × 4 mm ID
Pre-column	Reversed-phase; Corasil C_{18}, 37–50 μm, 22 × 4 mm ID
Mobile phase	20 mM NaH_2PO_4/methanol/acetonitrile (83:7:10, v/v)
Flow-rate	1.5 ml min^{-1}
Injection size	20–25 μl

Detection	UV, λ = 254 nm
Recording	Waters M 730 data module
Retention times, min	Cefotaxime 4.8
	Desacetylcefotaxime 7.5
Ref.	Kees F, Strehl E et al (1981) Arzneim.-Forsch./Drug Res. 31:362

HPLC 4

Sample	Serum and urine
Extraction	Deproteinization with acidic methanol
Apparatus	Gynkotek M 300C LC pump, Waters M 440 UV detector, amperometric detector and Rheodyne loop injector
Column	Reversed-phase; Hypersil ODS 5 μm, 200 × 4.6 mm ID
Mobile phase	0.1 M Acetate buffer, pH 4.1/methanol (83:17) with 0.01 M NaBr
Flow-rate	1.8 ml min^{-1}; 1740 psi
Temperature	20 °C ± 2 °C
Injection size	At least 3–4 ng
Detection	UV, λ = 254 nm; EC, + 0.4 V vs Ag/AgCl RE
Retention times, min	(estimated from drawing) UV/EC
	Desacetylcefotaxime 4.0
	Cefotaxime 14.0
Ref.	Fabre H and Kok WTh (1988) Anal. Chem. 60:136

HPLC 5

Sample	Blood plasma and urine
Extraction	Plasma: Deproteinization with 0.45 N H_3PO_4 or 10% trichloroacetic acid
	Urine: By dilution
Apparatus	Hewlett-Packard 1081B HPLC with UV detector
Column	Reversed-phase; Lichrosorb C_{18}, 5 μm, 120 × 4.7 mm ID
Pre-column	Whatman Co: Pell ODS, 100 × 4.7 mm ID
Mobile phase	Methanol/aqueous salt solution (containing 0.01 M tetrabutylammonium hydrogen sulfate, 0.02 M KH_2PO_4 and 0.02 M K_3PO_4) (20:80, v/v)
Flow-rate	1 ml min^{-1}
Temperature	Ambient; 40 °C (cefotiam)
Internal standard	Cephalexin
Injection size	20 μl
Detection	UV, λ = 254 nm; 0.01 AUFS
Recording	Spectra-Physics 4100 computing integrator
Retention times, min	Desacetylcefotaxime 4.0 Cephalexin 14.0
	Cefotaxime 10.0 Cefotiam 23.0

Ref.	Lecaillon JB, Rouan MC et al (1982) J. Chromatogr. Biomed. Appl. 228:257

30 Ceftriaxone

CAS-No	73384-59-5; 74578-69-1 (disodium salt)	
Chemically	Cephalosporin antibiotic	
Empirical formula	Ceftriaxone	Sodium Salt
	$C_{18}H_{18}N_8O_7S_3$	$C_{18}H_{16}N_8Na_2O_7S_3$
Molecular mass	554.6	598.5
Clinical use	Antibiotic	

HPLC 1

Sample	Blood plasma, urine and saliva
Extraction	Saliva and serum with acetonitrile, and urine by diluting with water
Apparatus	Kontron M 410 LC pump, Pye-Unicam 4020 UV/VIS detector, Kontron ASI 45 sample processor and Rheodyne injector
Column	Amino phase ; Lichrosorb-NH_2, 5 µm, 250 × 4 mm ID
Pre-column	Corasil, 37–50 µm, 100 × 4 mm ID and Corasil, 37–50 µm, 30 × 4 mm ID
Mobile phase	A. Acetonitrile/water/10% (w/v) $(NH_4)_2Co_3$ (70:26:4) B. As in A but 5% $(NH_4)_2CO_3$ (saliva)
Flow-rate	1.5 ml min^{-1}; 1450 psi
Injection size	20–100 µl
Detection	UV, $\lambda = 274$ and 245 nm
Recording	Perkin-Elmer Sigma-10 chromatographic computer and Pye-Unicam 8252 recorder
Retention times, min	Ceftriaxone 7.5 (A); 6.5 (B)
Ref.	Ascalone V and Dal Bo L (1983) J. Chromatogr. Biomed. Appl. 273:357

HPLC 2

Sample	Sodium salt
Column	Reversed-phase; ODS 5 µm, 150 × 4 mm ID
Mobile phase	3.2 g Tetraheptylammonium bromide in 400 ml acetonitrile, and add 44 ml buffer, pH 7.0 (containing 13.6 g K_2HPO_4 and 4.0 g KH_2PO_4 in 1000 ml water, pH 7.0 with H_3PO_4 or 10 N KOH) and 4 ml buffer, pH 5.0 (containing 25.8 g sodium citrate in 1500 ml H_2O, pH 5.0 with citric acid and water to 2000 ml) and make 1 liter with water

Flow-rate	2 ml min^{-1}
Injection size	Appr. 20 µl
Detection	UV, λ = 270 nm
Ref.	Anonym (1985): The United States Pharmacopeia XXI and The National Formulary XVI, Supplement 5, 1987. United States Pharmacopeial Convention Inc., Rockville, p 2362

HPLC 3

Sample	Blood plasma and milk from cattle
Extraction	Protein precipitation with acetonitrile, and then with methylene chloride
Apparatus	Kontron M 414 LC pump with column switching system, ASI-45 autosampler and Kratos M 773 LC-UV detector
Column	Reversed-phase; MOS Hypersil C_8, 5 µm, 125 × 4 mm ID
Pre-column	Reversed-phase; Lichrosorb C_{18}, 4 × 4 mm ID
Mobile phase	300 ml Acetonitrile, 1 g tetraoctylammonium bromide, 10 ml ethanol and 12 ml 0.1 M EDTA, pH 8.0 and make 1 liter with water
Flow-rate	1.5 ml min^{-1}
Temperature	50 °C
Internal standard	Ro 14-1761 (Hoffman-La Roche)
Injection size	20–40 µl
Detection	UV, λ = 274 nm
Recording	Spectra-Physics 4100 computing integrator
Ref.	Jordan J-C and Ludwig BM (1986) J. Chromatogr. 362:263

31 Cephalexin

Cas-No	15686-71-2(anhydrous); 23325-78-2 (monohydrate)		
Chemically	Cephalosporin antibiotic		
Empirical formula	$C_{16}H_{17}N_3O_4S$, H_2O		
Molecular mass	365.4		
Proprietary names	Ceporex; Keflex; Oracef		
Clinical use	Antibiotic		
Solubility	Water	alcohol and chloroform	0.2% HCl
	1 in 100	insoluble	1 in 30
Light absorption	λ_{max} = 258 nm in aqueous acid		

HPLC 1

Sample: Raw materials

Column: Reversed-phase; Hypersil MOS 5 μm, C_8, 160 × 5 mm ID

Mobile phase:
I. 0.2% Sodium pentanesulfonate (NaPS) in 1% CH_3COOH/methanol (67:33)
II. 0.2% NaPS in 1% CH_3COOH/acetonitrile (80:20)
III. 0.005 M Tetrabutylammonium phosphate (PIC-A)/methanol (70:30)
IV. 0.005 M PIC-A/acetonitrile (80:20)

Injection size: 10 μl, 0.5 mg ml^{-1}

Detection: UV, λ = 254 nm; 0.01 AUFS

Retention times, min

	RT	RT relative to cephalexin			
Mobile phase	II.	I.	II.	III.	IV.
Flow-rate, ml min^{-1}	1.0	1.5	1.0	1.5	1.0
7-Aminodesacetoxy cephalosporinic acid	2.5				
Phenylglycine	2.5				
3-Hydroxy-4-methyl-2-thiphenone	4.0				
Cephalexin	7.5	1.00	1.00	1.00	1.00
L-cephalexin	4.9	0.63	0.71	0.84	1.04
Δ^2-cephalexin	6.3	1.18	0.88	1.45	1.35
7-α-cephalexin		1.00	1.03	1.08	1.14

Ref.: Mason B and Tranter J (1981) Anal. Proc. 18:310

HPLC 2

Sample: Serum and urine

Extraction: Centrifugation and dilution

Apparatus: Waters M 6000 LC pump, 440 ultraviolet detector

Column: Reversed-phase; Pecosphere 5 μm, C_{18}, 150 × 4.6 mm ID

Pre-column: Co: Pell ODS 30–38 μm, 30 × 2 mm ID

Mobile phase: Acetonitrile/0.5 M $(NH_4)H_2PO_4$, pH 3 with H_3PO_4/water (100:90:810, v/v)

Flow-rate: 1.7 ml min^{-1}

Temperature: Ambient

Internal standard: Ceftazidime

Injection size: 10–25 μl containing at least 3 ng

Detection: UV, λ = 254 nm

Recording: Waters M 740 data module

Retention times, min:

Cephalexin	appr. 4.0
Internal standard	appr. 8.0

Ref.	Emm TA, Leslie J et al (1988) J. Chromatogr. Biomed. Appl. 427:162

HPLC 3

Sample	Cephalosphorins
Apparatus	Haskel Engineering M 26980-4 LC pump, Altex M 153 ultraviolet detector and Rheodyne M 70-10 injector
Column	Reversed phase; Lichrosorb C_{18}, 10 μm, 300 × 4.6 mm ID
Mobile phase	Methanol/0.01 M sodium dihydrogen phosphate (17:83)
Flow-rate	1.3 ml min^{-1}; 1015 psi
Temperature	Ambient
Injection size	10 μl, containing 2.2 μg
Detection	UV, λ = 254 nm; 0.04 AUFS
Retention times, min	(estimated from drawing) Cephalexin 11.7 Cephaloglycin 18.3 Cefazolin 12.5 Cephalothin 23.4 Cephradine 15.3
Ref.	White ER and Zarembo JE (1981) J. Antibiotics 34:836

HPLC 4

Sample	Serum
Apparatus	Beckman M 110A LC pump and Kratos M 773 UV detector and Perkin-Elmer M 420B autoinjector
Column	Waters Resolve, 5 μm silica, 100 × 8 mm ID
Mobile phase	0.1% Aqueous phosphoric acid/acetonitrile (90:10, v/v)
Flow-rate	1–3 ml min^{-1}
Injection size	20 μl, 5 μg ml^{-1}
Detection	UV, λ = 220 nm
Recording	Hewlett-Packard 3357 laboratory computer
Retention times, min	Cephalexin 5.9 Cephradine 7.4
Ref.	Adamovics JA (1987) J. Pharm. Biomed. Anal. 5:267

HPLC 5

Sample	Human blood plasma, urine and saliva
Extraction	Methanol
Apparatus	LDC Constametric II G LC pump with UV detector
Column	Reversed-phase; μ-Bondapak C_{18}, 10 μm, 300 × 3.9 mm ID
Mobile phase	Methanol/water (20:80) containing 0.5% acetic acid

Flow-rate	2.0 ml min^{-1} (plasma and saliva); 1.7 ml min^{-1} (urine)
Internal standard	Cephaloglycine
Injection size	50 µl, 0.2–50 µg ml^{-1}
Detection	UV, λ = 254 nm; 0.005–0.08 AUFS
Retention times, min	Cephalexin 3.0 Internal standard 4.0
Ref.	Nahata MC (1981) J. Chromatogr. Biomed. Appl. 225:532

HPLC 6

Sample	Blood plasma and urine
Extraction	Plasma: Deproteinization with 0.45 N H_3PO_4 or 10% trichloroacetic acid Urine: By dilution
Apparatus	Hewlett-Packard 1081B HPLC with UV detector
Column	A. Reversed-phase; Lichrosorb C_8, 5 µm, 150 × 4.7 mm ID B. Reversed-phase; Lichrosorb C_{18}, 5 µm, 150 × 4.7 mm ID C. Reversed-phase; Lichrosorb C_8, 5 µm, 100 × 7.5 mm ID D. Reversed-phase; Lichrosorb C_8, 5 µm, 100 × 7.5 mm ID
Pre-column	Whatman Co: Pell ODS, 100 × 4.7 mm ID
Mobile phase	For A: Methanol/aqueous salt solution [containing 0.1 THBS (tetrabutylammonium hydrogen sulfate) and 0.09 M K_3PO_4] (15:85, v/v) For B: Methanol/aqueous salt solution (containing 0.02 M THBS, 0.024 M K_3PO_4 and 0.016 M KH_2PO_4) (23:77, v/v) For C: Methanol/aqueous salt solution (containing 0.002 M H_3PO_4) (28:72, v/v) For D: Methanol/aqueous salt solution (containing 0.001 M H_3PO_4) (30:70, v/v)
Flow-rate	1 ml min^{-1}; 2 ml min^{-1} (cefroxadin)
Temperature	Ambient; 40 °C (for Cefotiam)
Injection size	20 µl
Detection	UV, λ = 254 nm; 0.002 and 0.004 AUFS
Recording	Spectra-Physics 4100 computing integrator

Retention times, min

Column	A	B	C	D
Cefsulodin	8			
Cefroxadin	14		12	14
Cefuroxime		12		
Cephalexin	16	19	18	18
Cefotiam	26	23		

Ref. Lecaillon JB, Rouan MC et al (1982) J. Chromatogr. Biomed. Appl. 228:257

HPLC 7

Sample	Cephalexin epimers
Apparatus	Spectra-Physics 8100 HPLC with 8400 UV detector
Column	Reversed-phase; Zorbax TM 5 µm, C_8, 250 × 4.6 mm ID
Mobile phase	A. Methanol/0.1 M NaBr (40:60), pH 2.5 with H_2SO_4 B. Methanol/0.1 M $NaClO_4$ (40:60), pH 2.5 with H_2SO_4 C. Methanol/0.1 M ortho-H_3PO_4 (40:60), pH 2.5 with NaOH D. Methanol/0.1 M NaH_2PO_4 (5:95), pH 2.5 with H_2SO_4 E. Methanol/0.1 M $NaClO_4$ (5:95), pH 2.5 with H_2SO_4 F. Acetonitrile/0.1 M $NaClO_4$ (20:80), pH 2.5 with H_2SO_4 G. Acetonitrile/10 mM tetrabutylammonium hydroxide (20:80), pH 7.0 with ortho-H_3PO_4
Flow-rate	1.5–2.0 ml min^{-1}
Temperature	40 °C
Detection	UV, λ = 254 nm
Recording	Spectra-Physics 4100 computing integrator
Retention times, min	(estimated from drawing)

Mobile phase	A.	B.	C.	D.	E.	F.	G.
β-Isomer	7.0	7.0	6.7	3.0	4.1	5.0	5.3
α-Isomer	7.4	8.0	7.5	3.7	5.5	5.5	6.3

Ref. Kennedy JH and Olsen BA (1987) J. Chromatogr. 389:369

32 Cephalothin

CAS-No	153-61-7; 58-71-9 (sodium salt)			
Chemical name	7-(2-Thienylacetamido)caphalosporanic acid			
Empirical formula	Cephalothin $C_{16}H_{16}N_2O_6S_2$		Sodium Salt $C_{16}H_{15}N_2NaO_6S_2$	
Molecular mass	396.4		418.4	
Proprietary names	Cepovenin; Keflin; Restin			
Clinical use	Antibiotic			
Solubility,	Water	alcohol	chloroform	ether
Sodium salt	1 in 3.5	1 in 700	insoluble	insoluble
Light absorption	λ_{max} = 237 nm in water (sodium salt)			

HPLC 1

Sample	Cephalosporins
Apparatus	Haskel Engineering M 26980-4 LC pump, Altex 153 ultraviolet detector and Rheodyne M 70–10 injector
Column	Reversed-phase; Lichrosorb 5 μm, C_8, 150 × 4.6 mm ID
Mobile phase	Methanol/0.01 M sodium hydrogen phosphate (15:85)
Flow-rate	1.0 ml min^{-1}; 1320 psi
Temperature	Ambient
Injection size	20 μl containing 3.1 μg
Detection	UV, λ = 254 nm; 0.08 AUFS
Retention times, min	(estimated from drawing)

Cephacetrile	5.0	Cephradine	19.0
Cephalexin	12.5	Cephaloglycin	21.0
Cefazolin	15.0	Cephalothin	29.0

Ref.	White ER and Zarembo JE (1981) J. Antibiotics 34:836

HPLC 2

Apparatus	Laboratory Data Control Constametric I pump and Spectromonitor I
Column	Reversed-phase; μ-Bondapak C_{18}, 250 mm
Mobile phase	Acetonitrile/0.1 N acetate buffer, pH 6.0 (25:75)
Flow-rate	2.0 ml min^{-1}
Temperature	40 °C
Detection	UV, λ = 270 nm; 0.04 AUFS
Retention time, min	Cephalothin 2.4
Ref.	Koup JR, Brodsky B et al (1978) Antimicrob. Ag. Chemother. 14:439

HPLC 3

Apparatus	Waters ALC/GPC M 200 LC with 6000 solvent delivery system and 440 UV detector
Column	Reversed-phase; μ-Bondapak C_{18}, 300 × 4 mm ID
Mobile phase	Methanol/acetic acid (40:60)
Flow-rate	2.5 ml min^{-1}
Detection	UV, λ = 254 nm; 0.01 AUFS
Recording	Hewlett-Packard 3380A computing integrator
Retention times, min	Deacetylcephalothin 4.44
	Cephalothin 9.99
Ref.	Nilsson-Ehle I, Yoshikawa TT et al (1978) Antimicrob. Ag. Chemother. 13:221

HPLC 4

Apparatus	Varian LCS 1000 LC with UV detector
Column	Reeve Angel AS Pellionex-SAX resin; 91.44 × 1.8 mm ID
Mobile phase	0.01 M Sodium nitrate
Flow-rate	0.43 ml min^{-1}; 350 psi
Temperature	50 °C
Detection	UV, λ = 254 nm; 0.04 and 0.32 A
Ref.	Rolewicz TF, Mirkin BL et al (1977) Clin. Pharmacol. Therap. 22:928

33 Cephradine

CAS-No	38821-53-3			
Chemically	Cephalosporin antibiotic			
Empirical formula	$C_{16}H_{19}N_3O_4S$			
Molecular mass	349.4			
Proprietary names	Eskacef; Sefril; Velosef			
Clinical use	Antibiotic			
Solubility	Water	methanol	acetone	alcohol
	1 in 100	1 in 70	sparingly soluble	insoluble
Light absorption	λ_{max} = 257 nm in aqueous acid			

HPLC 1

Sample	Raw materials
Column	Cation-exchange; sulfonated fluorocarbon polymer, 30–50 μm, 1000 × 2 mm ID
Pre-column	Cation-exchange; sulfonated fluorocarbon polymer, 30–50 μm, 200 × 2 mm ID
Mobile phase	20 ml 1 N Acetic acid and 48.4 g anhydrous sodium sulfate in 1900 ml water, pH 4.7 with 1 N NaOH and make 2000 ml with water
Flow-rate	0.5 ml min^{-1}
Internal standard	Sulfonamide derivative
Injection size	Appr. 5 μl
Detection	UV, λ = 254 nm
Ref.	Anonym (1985) The United States Pharmacopeia XXI and The National Formulary XVI. United States Pharmacopeial Convention Inc., Rockville, p 182

HPLC 2

Sample	Cephalosporins
Apparatus	HPLC low-pressure ternary gradient system
Column	Reversed-phase; Superspher 100, C_{18}, 125 × 4 mm ID
Mobile phase	Methanol/0.01 M potassium dihydrogen phosphate, pH 3.3 (20:80, v/v)
Flow-rate	1 ml min^{-1}; 2500 psi
Injection size	20 μl containing 9 μg
Detection	UV, λ = 278 nm; 0.08 AUFS
Retention times	Cefadroxil, cephalexin and cephradine, in ascending order (from drawing)
Ref.	Merck Application Manual on HPLC, TLC and HPTLC (1986). E. Merck, Darmstadt, p 82

HPLC 3

Sample	Cephalosporins
Column	Reversed-phase; ODS-Sil-X-II, 1000 × 2.1 mm ID
Mobile phase	5% Methanol, 95% 0.05 M ammonium carbonate
Flow-rate	0.5 ml min^{-1}; 800 psi
Detection	UV, λ = 254 nm; 0.08 AUFS
Retention times, min	(estimated from drawing)

Cephaloridine	5.0	Cephaloglycin	23.0
Cefazolin	8.5	Cephalothin	30.0
Cephradine	14.0		

Ref.	White ER, Carroll MA et al (1975) J. Antibiotics 28:205

HPLC 4

Sample	Test substances and blood plasma from rats
Apparatus	Lewa FLM/50 HPLC pump with pulse dampening system and Perkin-Elmer LC 55 UV detector
Column	Reversed-phase; Lichrosorb C_8, 10 μm, 250 × 3 mm ID
Mobile phase	0.05 M Phosphate buffer, pH 7.0/methanol (5:1, v/v)
Flow-rate	3.2 ml min^{-1}
Temperature	Ambient
Injection size	10 μl, 1 mg ml^{-1}
Detection	UV, λ = 220 nm
Recording	Hewlett-Packard 3352B computer-system
Retention times, min	(estimated from drawing)

Cephalexin	2.0
Cephradine	3.0

Ref.	Hartmann V and Roediger M (1976) Chromatographia 9:266

34 Chlorambucil

CAS-No	305-03-3
Chemical name	4-[4-Bis(2-chloroethyl)aminophenyl]butyric acid
Empirical formula	$C_{14}H_{19}Cl_2NO_2$
Molecular mass	304.2
Proprietary names	Chloraminophene; Leukeran; Linfolysin
Clinical use	Antineoplastic
Solubility	Water insoluble; alcohol 1 in 1.5; acetone 1 in 2; aq NaOH soluble
Light absorption	λ_{max} = 258 nm in methanol

HPLC 1

Sample	Blood (with chlorambucil analogues)
Extraction	Deproteinization with methanol
Apparatus	Waters M 6000A LC pump, 720 system controller, 440 ultraviolet detector and WISP 710B sample processor
Column	Reversed-phase; Novapak 4 µm, C_{18} Radial-Pak, 100 × 8 mm ID
Pre-column	Reversed-phase; µ-Bondapak C_{18}
Mobile phase	Methanol/water/tetrabutylammonium hydroxide (70:25:5, v/v), pH 7.4 with ortho-H_3PO_4
Flow-rate	3 ml min^{-1}; 9 MPa
Temperature	Ambient
Injection size	10 µl
Detection	UV, λ = 254 and 280 nm
Recording	Waters M 730 data module
Retention times, min	(estimated from drawing) Metabolites (8) 2–8.0 Chlorambucil 3.6
Ref.	Workman P and Oppitz M et al (1987) J. Chromatogr. Biomed. Appl. 422:315

HPLC 2

Sample	Blood plasma
Extraction	Over Sep-Pak C_{18} RP column with cold methanol under acidic condition after deproteinization with perchloric acid
Apparatus	Pye-Unicam 4000 LC system, 4020 UV, 4022 electrochemical and Kratos FS-970 fluorescence detectors

Column	Reversed-phase; Spherisorb ODS 5 μm, 250 × 4.6 mm ID
Mobile phase	Methanol/water (80:20, v/v)
Flow-rate	1.3 ml min^{-1}
Temperature	40 °C
Injection size	200 μl
Detection	UV, λ = 260 nm; fluorescence, λ = 260/350 nm and EC, GC electrode, 0.92 V vs AgCl RE
Recording	Pye-Unicam CDP4 computing integrator
Retention times, min	Chlorambucil 4.2
	Melphalan 9.5
Ref.	Adair CG, Thorburn Burns D and Harriott M (1986) Anal. Proc. 23:30

HPLC 3

Sample	Blood plasma
Extraction	Deproteinization with acetonitrile
Apparatus	Waters M 270 LC with gradient controller, 710B WISP autosampler and Beckman M 155 UV detector
Column	Reversed-phase; Radial-Pak C_{18}, 10 μm
Mobile phase	Acetonitrile/0.2% acetic acid buffer (65:35, v/v)
Flow-rate	1 ml min
Injection size	50–75 μl, 0.5–5 μg ml^{-1}
Detection	UV, λ = 263 nm
Recording	Waters M 730 data module
Retention time, min	Chlorambucil 8.0
Ref.	Ahmed AE, Koenig M and Farrish Jr. HH (1982) J. Chromatogr. Biomed. Appl. 233:392

HPLC 4

Sample	Human blood plasma
Extraction	Ethyl acetate
Apparatus	Waters ALC/GPC M 204 HPLC with 6000A pumps, 660 solvent programmer, U6K injector and 440 UV detector
Column	Reversed-phase; μ-Bondapak C_{18}
Mobile phase	Linear gradient: Methanol/0.175 M acetic acid (60:40, v/v) to 100% methanol in 10 min
Flow-rate	2 ml min^{-1}
Injection size	50 μl containing 5–1000 ng
Detection	UV, λ = 254 and 280 nm; 0.02 or 0.05 AUFS
Retention times, min	Phenyl acetic mustard 5.95
	Chlorambucil 7.83
Ref.	Newell DR, Hart LI and Harrap KR (1979) J. Chromatogr. Biomed. Appl. 164:114

HPLC 5

Sample	Tablet formulations
Column	Reversed-phase; ODS 5 to 10 μm, 250 or 300 × 2 mm ID
Mobile phase	500 ml Alcohol with 1 ml glacial acetic acid and make 1 liter with water
Temperature	Ambient
Internal standard	Propylparaben
Injection size	10–20 μl
Detection	UV, λ = 254 nm
Ref.	Anonym (1985): The United States Pharmacopeia XXI and The National Formulary XVI. The United States Pharmacopeial Convention Inc., Rockville, p 187

HPLC 6

Sample	Rats blood plasma
Extraction	n-Hexane
Apparatus	Waters M 6000A LC pump, Perkin-Elmer LC-85B UV detector, LS-4 Spectrofluorimeter and Rheodyne M 7125 injector
Column	A. Normal phase; cyano, 250 × 4.6 mm ID B. Reversed-phase; CP-Spher C_{18}, 100 × 3 mm ID
Mobile phase	For A: n-Hexane/2-propanol (98:2, v/v) For B: Tetrahydrofuran/water (73:27, v/v)
Flow-rate	1.8 ml min^{-1} (A); 1.2 ml min^{-1} (B)
Injection size	20 μl
Detection	UV, λ = 254 nm; fluorescence, λ = 275/335 nm
Recording	Perkin-Elmer Sigma 10 data station

Retention times, min

Column	A.	B.
Pro chlorambucil 6	4.8	7.2
Pro chlorambucil 5	5.0	5.5
Pro chlorambucil 4	5.4	4.4
Pro chlorambucil 3	5.8	3.5

Ref. Saraiva Goncalves JC, Razzouk C et al (1989) J. Chromatogr. Biomed. Appl. 494:389

---

35 Chloramphenicol

CAS-No	56-75-7; 982-57-0 (sodium succinate)
Chemical name	2,2-Dichloro-*N*-[α*R*,β*R*)-β-hydroxy-α-hydroxymethyl-4-nitrophenethyl]acetamide

Empirical formula	Chloramphenicol	Sodium Succinate		
	$C_{11}H_{12}Cl_2N_2O_5$	$C_{15}H_{15}Cl_2N_2NaO_8$		
Molecular mass	323.1	445.2		
Proprietary names	Chloromycetin; Leukomycin; Ophthochlor			
Clinical use	Antibiotic			
Solubility,	Water	alcohol	acetone	aq HCl, aq NaOH
Chloramphenicol	1 in 400	1 in 2.5	freely sol	slightly soluble
Sodium succinate	very soluble	1 in 1		
Light absorption	λ_{max} = 271 nm in dehydrated alcohol and 278 nm in water (palmitate salt)			

HPLC 1

Sample	Dosage forms and blood
Extraction	Diluting in mobile phase after deproteinization with perchloric acid (blood)
Apparatus	LDC Constametric LC-III pump with M 3000 ultraviolet and Bioanalytical System LC-4B electrochemical detectors
Column	Reversed-phase; 5 µm-ODS, 250 × 4.6 mm ID
Mobile phase	20 mM Monochloroacetic acid, 1 mM Na_2EDTA, 14.6 mM sodium acetate in water/n-propanol (92:8)
Flow-rate	1.5 ml min^{-1}
Temperature	Ambient
Injection size	200 µl
Detection	UV, λ = 275 nm; 0.02 AUFS; EC, GC electrode vs Ag/AgCl RE

Retention times, min

(estimated from drawing)	UV	EC
Nitrophenylamino-propanedione	3.2	4.8
Chloramphenicol	12.8	14.1
Nitrosochloramphenicol	16.0	16.5
Dehydrochloramphenicol	18.5	20.0

Ref. Abou-Khalil S, Abou-Khalil WH et al (1987) J. Chromatogr. Biomed. Appl. 417:111

HPLC 2

Sample	Biological materials
Extraction	Dichloroethane
Apparatus	Waters M 45 LC pump, Cecil M 212 UV/VIS detector and Rheodyne M 7125 injector
Column	Reversed-phase; Hypersil 5 µm, MOS C_8, 250 × 4.4 mm ID
Pre-column	Co:Pell ODS, 50 × 4.4 mm ID

Mobile phase	Acetonitrile/acetate buffer (containing 50 ml 1 M NaOH plus 58 ml 1 M CH_3COOH, diluted to 1 liter), pH 5.8 (2:3)
Flow-rate	1.8 ml min^{-1}
Temperature	Ambient
Internal standard	Methoxycarbamazepine
Injection size	10–20 µl, 2.5–20 µg ml^{-1}
Detection	UV, λ = 280 nm; 0.2 AUFS
Retention times, min	Chloramphenicol sodium succinate 1.0 Chloramphenicol 1.5 Internal standard 2.2
Ref.	Berry DJ (1987) J. Chromatogr. 385:337

HPLC 3

Sample	Blood plasma
Apparatus	Waters M 6000A LC pump, 720 system controller, WISP autosampler and ultraviolet detector
Column	Reversed phase; Z-Module, Radial Pak C_{18}, 10 µm, 100 × 8 mm ID
Pre-column	Guard-Pak TM, C_{18}
Mobile phase	750 ml Aqueous acetic acid, pH 3 and 250 ml acetonitrile
Flow-rate	4 ml min^{-1}; 1000 psi
Internal standard	Methylprednisolone
Injection size	10–20 µ, 1–50 µg ml^{-1}
Detection	UV, λ = 280 nm
Recording	Waters data module M 730
Retention times, min	Chloramphenicol 3.4 Chloramphenicol succinate 4.7, 6.3 Internal standard 9.8
Ref.	El-Yazigi A, Yusuf A and Al-Humaidan A (1987) Clin. Chem. 33:1814

HPLC 4

Sample	Blood serum
Extraction	Protein precipitation by centrifugation
Column	A. Reversed-phase; C_{18}, 3 µm, 100 × 1 mm ID B. Reversed-phase; C_{18}, 5 µm, 100 × 1 mm ID
Mobile phase	For A. Acetate/acetonitrile/tetrahydrofuran (85:15:1.5) For B. Acetate/acetonitrile (80:20)
Flow-rate	80 µl min^{-1} (A); 60 µl min^{-1} (B)
Temperature	50 °C (A); Ambient (B)
Internal standard	Tolybarb
Injection size	0.5 µl
Detection	UV, λ = 254 nm; 0.002 AUFS

Retention times, min	Column	A.	B.
	Chloramphenicol	4.0	6.0
	Internal standard	8.0	14.0
Ref.	Wong SH-Y (1987) Clin. Chem. 33:1021		

HPLC 5

Sample	Blood and cerebrospinal fluid
Extraction	Ethyl acetate under acidic and alkaline conditions
Apparatus	Waters M 6000A LC pump and 440 UV detector
Column	Reversed-phase: μ-Bondapak C_{18}, 300 × 3.9 mm ID
Pre-column	Bondapak C_{18}/Corasil 37–50 μm
Mobile phase	2.04 g Monobasic potassium phosphate in 675 ml water, 3.5 ml 0.9 M H_3PO_4, 200 ml methanol and 125 ml acetonitrile
Flow-rate	2.0 ml min^{-1}
Injection size	10–20 μl, appr. 30 μg ml^{-1}
Detection	UV, λ = 254 nm; 0.005 AUFS
Recording	Hewlett-Packard 3390A integrator
Retention time, min	Chloramphenicol 8.0
Ref.	Sood SP, Green VI and Bailey CL (1987) Ther. Drug Monit. 9:347

36 Chlordiazepoxide

CAS-No	58-25-3; 438-41-5 (hydrochloride)			
Chemical name	7-Chloro-2-methylamino-5-phenyl-3*H*-1,4-benzodiazepine 4-oxide			
Empirical formula	Chlordiazepoxide $C_{16}H_{14}ClN_3O$	Hydrochloride $C_{16}H_{14}ClN_3O,HCl$		
Molecular mass	299.8	336.2		
Proprietary names	Contol; Equibral; Librium			
Clinical use	Tranquillizer			
Solubility,	Water	alcohol	ether	chloroform
Chlordiazepoxide	insoluble	1 in 50	1 in 130	slightly soluble
Hydrochloride	1 in 10	1 in 40	insoluble	insoluble
Light absorption	λ_{max} = 246 nm in aqueous acid			

HPLC 1

Sample	Tablet formulations
Apparatus	LDC M 396/2396 LC pump, 1285 UV detector and Rheodyne M 7010 loop injector
Column	Reversed-phase; μ-Bondapak C_{18}, 300 × 3.9 mm ID

Mobile phase	0.01 M Sodium lauryl sulfate in tetrahydrofuran/methanol/Britton-Robinson buffer (with 10.5 ml 0.2 N NaOH to 100 ml of a solution containing 0.04 M CH_3COOH, 0.04 M H_3PO_4 and 0.04 M boric acid, pH 2.5) (120:30:150)
Flow-rate	1 ml min^{-1}
Internal standard	Sulfanilamide
Injection size	20 µl, 0.1 mg ml^{-1}
Detection	UV, $\lambda = 254$ nm
Retention times, min	Sulfanilamide 3.0
	Chlordiazepoxide 5.5
	Amitriptyline 8.0
Ref.	Burke D and Sokoloff H (1980) J. Pharm. Sci. 69:138

HPLC 2

Sample	Serum
Extraction	Diethyl ether and 1 M borate buffer
Apparatus	Waters M 510 dual-piston pump and Beckman M 163 UV detector
Column	Reversed-phase; Ultrasphere C_{18}, 5 µm, 150 × 2 mm ID
Pre-column	Waters, 2 µm -Guard-Pak
Mobile phase	Methanol/acetonitrile/0.056 M sodium acetate buffer, pH 4.0 with glacial acetic acid (50:5:45, v/v)
Flow-rate	0.3 ml min^{-1}; 2000 psi
Internal standard	Demoxepam
Injection size	20 µl
Detection	UV, $\lambda = 228$ nm; 0.002 AUFS
Recording	Perkin-Elmer LCI-100 integrator
Retention times, min	(estimated from drawing)

Demoxepam	3.2	Chlordiazepoxide	6.3
Desmethylchlor-diazepoxide	4.4	Desmethyldiazepam	8.2
		Temazepam	8.8
Oxazepam	5.2	Diazepam	13.5

Ref. Lau CE, Dolan S and Tang M (1987) J. Chromatogr. Biomed. Appl. 416:212

HPLC 3

Sample	Blood plasma and urine
Extraction	Diethyl ether
Apparatus	Spectra-Physics 3500B HPLC with 770 UV detector
Column	Reversed-phase; Chrompack C_8, 5 µm, 100 × 4.6 mm ID
Mobile phase	Water/methanol/acetonitrile (500:450:50, v/v)
Flow-rate	1.6 ml min^{-1}; 3674 psi
Internal standard	Flunitrazepam

Injection size	100 μl containing at least 30 ng ml^{-1}
Detection	UV, λ = 230 nm
Retention times	(relative, k')

Pentobarbital	5.58	Oxydiazepam	13.30
Flunitrazepam	6.50	N-Desmethyl-	
Chlordiazepoxide	10.07	diazepam	17.00
Oxazepam	10.67	Diazepam	21.80

Ref. Vree TB, Baars AM et al (1979) J. Chromatogr. Biomed. Appl. 162:605

HPLC 4

Sample	Raw materials
Column	Reversed-phase; ODS 5 to 10 μm, 300 × 3.9 mm ID
Mobile phase	Water/tetrahydrofuran/methanol (5:4:1)
Flow-rate	1 ml min^{-1}
Internal standard	Sulfanilamide
Injection size	Appr. 20 μl, 0.1 mg ml^{-1}
Detection	UV, λ = 254 nm
Retention times, min	Internal standard 3.00 Chlordiazepoxide 6.00
Ref.	Anonym (1985): The United States Pharmacopeia XXI and The National Formulary XVI. United States Pharmacopeial Convention Inc., Rockville, p 193

HPLC 5

Sample	Benzodiazepine mixture
Apparatus	HPLC low-pressure ternary gradient system
Column	Reversed-phase; Lichrosorb C_{18}, 250 × 4 mm ID
Mobile phase	Acetonitrile/water (35:65, v/v)
Flow-rate	0.7 ml min^{-1}; 3000 psi
Injection size	25 μl containing 10 μg
Detection	UV, λ = 254 nm; 0.04 AUFS
Retention times	Oxazepam, chlordiazepoxide, N-desmethyl-diazepam and diazepam, in ascending order (from drawing)
Ref.	Merck Application Manual on HPLC, TLC and HPTLC (1986). E. Merck, Darmstadt, p 72

37 Chloroquine

CAS-No	54-05-7; 50-63-5 (phosphate); 132-73-0 (sulfate, anhydrous)

Chemical name	7-Chloro-4-(4-diethylamino-1-methylbutyl-amino)quinoline		
Empirical formula	Chloroquine	Phosphate	Sulfate
	$C_{18}H_{26}ClN_3$	$C_{18}H_{26}ClN_3$, $2H_3PO_4$	$C_{18}H_{26}ClN_3, H_2SO_4$, H_2O
Molecular mass	319.9	515.9	436.0
Proprietary names	Aralen; Nivaquine; Resochin		
Clinical use	Antımalarial		
Solubility,	Water	alcohol	chloroform and ether
Chloroquine	very sligthly soluble		soluble
Phosphate	1 in 4	very sligthly soluble	insoluble
Sulfate	1 in 3	insoluble	very sparingly sol
Light absorption	λ_{max} = 343 nm in aqueous acid		

HPLC 1

Sample	Human blood plasma, saliva and urine
Extraction	Hexane under alkaline condition, and then with mobile phase
Apparatus	Varian M 5000 LC, manual loop valve injector and UV detector
Column	Micropak MCH 10, 10 µm, 300 × 4.0 mm ID
Mobile phase	Methanol/0.2 M sodium dihydrogen phosphate (50:50) with 75 mM perchloric acid, pH 3.8
Flow-rate	1 ml min^{-1}
Temperature	Ambient
Internal standard	Papaverine
Injection size	10 µl
Detection	UV, λ = 254 nm; 0.005–0.01 AUFS
Retention times, min	(estimated from drawing) Chloroquine 6.8 Desethylchloroquine 5.2 Internal standard 10.6
Ref.	Ogunbona FA, Onyeji CO et al (1986) J. Chromatogr. Biomed. Appl. 380:425

HPLC 2

Sample	Human blood, blood plasma, red blood cells and urine
Extraction	Methylene dichloride under alkaline condition and re-extracted with acid aqueous phase
Apparatus	Waters M 6000 HPLC pump, 440 absorbance detector, and WISP 710B sample processor
Column	Reversed-phase; Novapak C_{18}, 5 µm, 150 × 3.9 mm ID

Mobile phase	45 mM Potassium dihydrogen phosphate buffer, (pH 3.0 with ortho-phosphoric acid) containing 12% acetonitrile
Flow-rate	0.6 ml min^{-1}; 957–1044 psi
Temperature	Ambient
Injection size	100–200 µl
Detection	UV, λ = 254 or 340 nm; 0.005–0.01 AUFS
Retention times, min	Monodesethylchloroquine 4.5
	Monodesethylamodiaquine dihydrochloride 5.0
	Chloroquine diphosphate 6.0
	Amodiaquine dihydrochloride 7.0
Ref.	Pussard E, Verdier F and Blayo M-C (1986) J. Chromatogr. Biomed. Appl. 374:111

HPLC 3

Sample	Blood
Extraction	Methyl tert.-butyl ether/hexane (1:1) under alkaline condition
Apparatus	Waters M 6000A HPLC pump, 660 solvent programmer, U6K injector, Perkin-Elmer M 650S fluorescence detector and Varian UV detector
Column	Normal phase; Zorbax Sil 5–6 µm, 250 × 4.6 mm ID
Mobile phase	A. Hexane/methyl tert.-butyl ether/methanol/diethylamine (37.25:37.25:25.0:0.5)
	B. As in A but (46:46:7.5:0.5)
Flow-rate	1 ml min^{-1}; 986 psi
Internal standard	Quinoline-derivate
Injection size	30 µl containing at least 5 ng ml^{-1}
Detection	Fluorescence, λ = 320/380 nm; UV, λ = 340 nm (A) and 254 nm (B)
Recording	Varian CDS-111 integrator
Retention times, min	(estimated from drawing)

Mobile phase	A.	B.
λ =	340 nm	254 nm
Chloroquine	4.7	11.4
Internal standard	5.8	
Desethylchloroquine	6.9	
Amodiaquine		6.2
Pyrimethamine		7.0
Quinine		18.7

Ref.	Patchen LC, Mount DL et al (1983) J. Chromatogr. Biomed. Appl. 278:81

HPLC 4

Sample	Human blood plasma and urine
Extraction	Etyhlene dichloride under alkaline condition

Apparatus	LDC M 712-74 LC pump, Waters M 440 UV detector and Schoeffel FS 970 fluorimeter
Column	Reversed-phase; Nucleosil C_{18}, 5 or 10 μm, 200 × 4 mm ID
Mobile phase	Acetonitrile/phosphate buffer (40:60), pH 3.0 and 75 mM perchlorate
Flow-rate	0.8–1.0 ml min^{-1}
Temperature	Ambient
Detection	UV, λ = 254 and 340 nm Fluorescence, λ = 335/370 nm
Retention times, min	(estimated from drawing) Desmethylchloroquine 6.5 Chloroquine 7.9
Ref.	Bergqvist Y and Frisk-Holmberg M (1980) J. Chromatogr. 221:119

38 Chlorothiazide

CAS-No	58-94-6; 7085-44-1 (sodium salt)
Chemical name	6-Chloro-2*H*-1,2,4-benzothiadiazine-7-sulfonamide 1,1-dioxide
Empirical formula	Chlorothiazide $C_7H_6ClN_3O_4S_2$; Sodium Salt $C_7H_5ClN_3NaO_4S_2$
Molecular mass	295.7; 317.7
Proprietary names	Clotride; Diuret; Saluric
Clinical use	Diuretic

Solubility,	Water	ethanol	acetone	chloroform	aq NaOH
Chlorothiazide	insoluble	1 in 650	1 in 100	insoluble	soluble
Sodium salt	soluble				

Light absorption λ_{max} = 292 nm in aqueous alkali

HPLC 1

Sample	Blood plasma and urine
Extraction	Ethyl acetate under acidic condition
Apparatus	Altex M 110 LC pump, 100–10 UV detector and Altex 210 injection valve
Column	Reversed-phase; Micropak MCH-10, 300 × 4 mm ID
Mobile phase	For Urine: 4% Acetonitrile in 0.01 M sodium perchlorate, pH 4.6 with 0.01 M perchloric acid For Plasma: 15% Methanol in 0.01 M acetic acid
Flow-rate	2.5 ml min^{-1}
Internal standard	Hydrochlorothiazide
Injection size	20 μl

Detection	UV, λ = 228 nm (plasma) and 278 nm (urine)		
Retention times, min	Sample	plasma	urine
	Chlorothiazide	3.5	10.5
	Hydrochlorothiazide	4.6	13.5
Ref.	Barbhaiya RH, Phillips TA and Welling PG (1981) J. Pharm. Sci. 70:291		

HPLC 2

Sample	Test mixture of diuretics
Apparatus	HPLC low-pressure ternary gradient system
Column	Reversed-phase; Lichrosorb C_{18}, 125 × 4 mm ID
Mobile phase	Acetonitrile/water (30:70, v/v)
Flow-rate	1.5 ml min^{-1}; 2800 psi
Injection size	10 µl containing 5 µg
Detection	UV, λ = 220 nm; 0.04 AUFS
Retention times	Chlorothiazide, hydrochlorothiazide, cyclothiazide and bendroflumethiazide, in ascending order (from drawing)
Ref.	Merck Application Manual on HPLC, TLC and HPTLC (1986). E. Merck, Darmstadt, p 70

HPLC 3

Sample	Tablet formulations	
Column	Reversed-phase; Corasil C_{18}, 35–50 µm, 122 cm	
Mobile phase	Absolute methanol/0.5% ammonium chloride (55:45)	
Flow-rate	2 ml min^{-1}; 2000 psi	
Temperature	25 °C	
Injection size	250 ng	
Detection	UV, λ = 254 nm	
Retention times, min	Chlorothiazide	2.2
	Reserpine	4.7
	Phenanthrene	14.7
Ref.	de Silva JAF (1978) in: Tsuji K (ed) GLC and HPLC Determination of Therapeutic Agents, Part 2. Marcel Dekker Inc., New York and Basel, p 581	

HPLC 4

Sample	Raw materials
Column	Reversed-phase; ODS 5–10 µm, 250 × 4.6 mm ID
Mobile phase	0.1 M Dibasic sodium phosphate and acetonitrile (9:1), pH 3.0 with phosphoric acid
Flow-rate	1.2 ml min^{-1}
Internal standard	Sulfonamide derivative
Injection size	Appr. 20 µl
Detection	UV, λ = 254 nm

Retention times	(relative) Internal standard 0.9 Chlorothiazide 1.0
Ref.	Anonym (1985): The United States Pharmacopeia XXI and The National Formulary XVI. United States Pharmacopeial Convention Inc., Rockville, p 199

39 Chlorpheniramine

CAS-No	132-22-9; 42882-96-2(±); 113-92-8 (maleate)	
Chemical name	2-[*p*-Chloro-α-(2-dimethylaminoethyl)benzyl] pyridine	
Empirical formula	Chlorpheniramine	Maleate
	$C_{16}H_{19}ClN_2$	$C_{16}H_{19}ClN_2,C_4H_4O_4$
Molecular mass	274.8	390.9
Proprietary names	Chloramin; Haymine; Lorphen	
Clinical use	Antihistaminic	

	Water	alcohol	chloroform	ether
Solubility, Maleate	1 in 4	1 in 10	1 in 10	slightly soluble

Light absorption: λ_{max} = 265 nm in aqueous acid

HPLC 1

Sample	Tablets
Apparatus	Perkin-Elmer Series 3B LC pump, LC-75 UV detector and 420 autosampler
Column	Cyano phase; Zorbax CN 10 µm, 250 × 4.2 mm ID
Pre-column	Silica 20 µm Guard-Pak
Mobile phase	18.0 g Brij 35; 3.46 g sodium lauryl sulfate and 45 ml n-propanol in 1 liter water, pH 2.25 with H_3PO_4
Flow-rate	1.5 ml min^{-1}; 2000 psi
Temperature	65 °C
Injection size	20 µl, 0.019–0.124 mg ml^{-1}
Detection	UV, λ – 220 nm
Recording	Sigma 15 data system
Retention time, min	(estimated from drawing) Chlorphenipamine 13.7
Ref.	Biemer TA (1987) J. Chromatogr. 410:206

HPLC 2

Sample	Blood
Extraction	Carbon tetrachloride on Extrelut column (E. Merck), pH 9.5

Apparatus	Shimadzu LC-3A pump and RF-500 Fluorescence detector
Column	Reversed-phase; Zorbax ODS, 250 × 4.6 mm ID
Pre-column	ODS, 50 × 2.1 mm ID
Mobile phase	Methanol/water (84:16, v/v)
Flow-rate	1.7 ml min^{-1}
Temperature	50 °C
Derivatization	Pre-column; with benzyl chloroformate
Injection size	100 µl, 0–40.0 ng ml^{-1}
Detection	Fluorescence, λ = 350/500 nm
Retention times, min	Chlorpheniramine derivative 3.5 Chlorpheniramine 4.0 (estimated from drawing)
Ref.	Miyamoto Y (1987) J. Chromatogr. Biomed. Appl. 420:63

HPLC 3

Sample	Commercial cough preparations
Column	Reversed-phase; Phenyl/Corasil
Mobile phase	Acetonitrile/water/0.1 M $(NH_4)_2CO_3$
Flow-rate	1 ml min^{-1}
Detection	UV, λ = 254 nm
Retention times, min	Acetaminophen, ephedrine, dextromethorphan, methapyriline, pyrilamine and chlorpheniramine, in ascending order (from drawing)
Ref.	Wittmer DP and Pidacks C (1979) in: Tsuji K (ed) GLC and HPLC Determination of Therapeutic Agents, Part 3. Marcel Dekker Inc., New York and Basel, p 1211

HPLC 4

Sample	Tablets
Column	Cation-exchange; Zipax SCX
Mobile phase	0.02 M $(NH_4)_2HPO_4$ in 36% dioxane/water, apparent pH 8.2
Flow-rate	1.1 ml min^{-1}; 100 psi
Temperature	Ambient
Injection size	20 µl
Detection	UV, λ = 254 nm; 0.001 AU
Retention times, min	(estimated from drawing) Phenylpropanolamine 2.5 Ephedrine 4.5 Chlorpheniramine 9.0
Ref.	Mollica JA and Stober HC (1978) in: Tsuji K (ed) GLC and HPLC Determination of Therapeutic Agents, Part 2, Marcel Dekker Inc., New York and Basel, p 691

HPLC 5

Sample	Blood plasma, saliva and urine
Extraction	Diethyl ether or hexane under alkaline condition
Apparatus	Waters M 6000A LC pump, U6K sample injector and 440 UV detector
Column	Reversed-phase; μ-Bondapak C_{18}, 10 μm, 300 × 3.9 mm ID
Mobile phase	A. Blood plasma and saliva: Acetonitrile/0.075 M $NH_4H_2PO_4$ in 0.16% H_3PO_4 (20:80, v/v) B. Urine: Acetonitrile/0.05 M $NH_4H_2PO_4$ in 0.11% H_3PO_4 (20:80, v/v)
Flow-rate	2 ml min^{-1}
Internal standard	Brompheniramine maleate
Injection size	50 μl, 5–40 ng ml^{-1}
Detection	UV, λ = 254 nm; 0.002 AUFS
Retention times, min	

Sample	plasma, saliva	urine
Chlorpheniramine	6.60	6.60
Internal standard	7.80	8.00
Didesmethylchlorpheniramine		4.40
Desmethylchlorpheniramine		5.24

Ref. Athanikar NK, Peng GW et al (1979) J. Chromatogr. Biomed. Appl. 162:367

HPLC 6

Sample	Cold medicine
Column	Reversed-phase; C_{18}, 5 μm, 250 × 4.6 mm ID
Mobile phase	A. 0.15 M Potassium dihydrogen phosphate, pH 3.0 B. Acetonitrile/tetrahydrofuran (75:25) Gradient: B 15% to 100% in 30 min
Detection	UV
Retention times, min	(estimated from drawing)

Phenylpropanolamine	3.28
Chlorpheniramine	8.06
Aspirin	9.10

Ref. Vydac HPLC Application Notes on columns and Separation Materials (1987–88). Vydac, California, p 18

HPLC 7

Sample	Biological fluids
Apparatus	Spectra-Physics 8770 LC pump and 8300 UV detector

Column	Phenyl phase; μ-Bondapak Rad-Pak 10 μm, Z-module, 100 × 8 mm ID
Mobile phase	Water/acetonitrile/methanol (90 : 8 : 2, v/v) containing triethylamine, pH 2.55 with H_3PO_4
Flow-rate	4 ml min^{-1}
Temperature	Ambient (25 °C)
Injection size	20 μl
Detection	UV, λ = 254 nm; 0.01 AUFS
Retention times, min	Amopyroquine 4.0 Chlorpheniramine 6.0
Ref.	Coleman MD, Edwards G et al (1987) J. Chromatogr. Biomed. Appl. 414:242

40 Chlorpromazine

CAS-No	50-53-3; 69-09-0 (hydrochloride)			
Chemical name	2-Chloro-10-(3-dimethylaminopropyl)-phenothiazine			
Empirical formula	Chlorpromazine $C_{17}H_{19}ClN_2S$		Hydrochloride $C_{17}H_{19}ClN_2S,HCl$	
Molecular mass	318.9		355.3	
Proprietary names	Contomin; Largactil; Megaphen			
Clinical use	Tranquillizer			
Solubility,	Water	ethanol	ether	aq HCl
Chlorpromazine	insoluble	1 in 2	1 in 1	soluble
Hydrochloride	1 in 0.4	1 in 1.4	insoluble	
Light absorption	λ_{max} = 255 nm in aqueous acid			

HPLC 1

Sample	Human blood plasma
Extraction	Methanol under acidic condition on extraction column
Apparatus	Waters M 6000A HPLC pumps with 710B auto-sampler and 440 UV detector
Column	Reversed-phase; 5 μm, C_8, 250 × 4.6 mm ID
Pre-column	Supelcosil, 40-μm, C_8 bonded silica, 50 × 4.6 mm ID
Mobile phase	Methanol/water (55 : 45) containing 275 mM methylamine hydrochloride and 25 mM Na_2HPO_4, pH 6.75
Flow-rate	2 ml min^{-1}
Temperature	40 °C

Internal standard	Nortriptyline
Injection size	50 µl, 10 ng ml^{-1}
Detection	UV, $\lambda = 254$ nm; 0.005 AUFS
Recording	Hewlett-Packard 3390 integrator and IBM PC
Retention times, min	(estimated from drawing) Metabolites (19) 3–19 Internal standard 11.2 Chlorpromazine 20.9
Ref.	Smith CS, Morgan SL et al (1987) J. Chromatogr. Biomed. Appl. 423 : 207

HPLC 2

Sample	Pharmaceutical preparations
Apparatus	Altex LC with M 110A pump with Perkin-Elmer LS-5 Fluorescence detector
Column	Amino phase; Lichrosorb-NH_2 5 µm, 250 × 4 mm ID
Mobile phase	Acetonitrile/0.0001 M HCl, pH 4.00 (with 0.01% sodium metabisulfite, 0.01% D-isoascorbic acid and 0.06% NaCl), (83 : 17, v/v)
Flow-rate	1 ml min^{-1}
Internal standard	Quinine hydrochloride 25 µg ml^{-1}
Injection size	50 µl, 50–250 µg ml^{-1}
Detection	Fluorescence, $\lambda = 280/450$ nm and 280/385 nm (chlorpromazine sulfoxide)
Retention times, min	(estimated from drawing) Chlorpromazine 3.0 Quinine (IS) 3.8 Chlorpromazine sulfone 3.6 Chlorpromazine sulfoxide 4.5
Ref.	Chagonda LFS and Millership JS (1988) Analyst. 113:233

HPLC 3

Sample	Human blood plasma and urine
Apparatus	Waters ALC/GPC M 204 HPLC with electrochemical detector
Column	Reversed-phase; Nucleosil-5 µm, C_{18}, 150 × 4 mm ID
Mobile phase	Pyridine/tetrahydrofuran/acetonitrile/0.1 M acetate buffer, pH 3.5 (0.1 : 1 : 68.9 : 30, v/v) containing 20 mM $CaClO_4$
Flow-rate	0.7 ml min^{-1}; 1200–1300 psi
Injection size	5 µl containing 5 ng
Detection	EC, GC electrode, 0.95 V vs Ag/AgCl RE
Retention times, min	Levomepromazine 7.0 Chlorpromazine 8.0 Thioridazine 9.5

Ref.	Murakami K, Murakami K et al (1982) J. Chromatogr. Biomed. Appl. 227: 103

HPLC 4

Sample	Swine Kidneys
Extraction	Acidic acetonitrile on Sep-Pak C_{18} cartridge
Apparatus	Millipore M 6000A HPLC with WISP 710B auto-sampler, UV-VIS detector and fluorescence detector
Column	Reversed-phase; Bondapak C_{18}, 300 × 3.9 mm ID
Pre-column	Bondapak C_{18}, 37–50 μm, 10 × 2.1 mm ID
Mobile phase	2.46 g Anhydrous sodium acetate, 450 ml water and 550 ml acetonitrile, pH 6.5 with acetic acid
Flow-rate	1.2 ml min^{-1}
Injection size	50 μl, 0.2 μg ml^{-1}
Detection	UV λ = 240 nm; 0.002 AUFS Fluorescence, λ = 246/351 nm
Retention time, min	Chlorpromazine 25.5 (estimated from drawing)
Ref.	Keukens HJ and Aerts MML (1989) J. Chromatogr. 464 : 149

HPLC 5

Sample	Blood plasma
Extraction	Isopropanol/n-pentane/dichloromethane (5 : 49 : 46) under alkaline condition
Apparatus	HPLC system with Rheodyne M 7125 injector and Waters M 481 UV detector
Column	Cyano phase; Ultrasphere 5 μm, 250 × 2 mm ID
Mobile phase	4% 0.06 M Sodium acetate and 4% 0.06 M ammonium acetate in acetonitrile (92%) containing 0.05% diethylamine
Flow-rate	0.5 ml min^{-1}; 1682 psi
Temperature	Ambient
Internal standard	Mesoridazine
Injection size	200 μl, 5–50 ng ml^{-1}
Detection	UV, λ = 248 nm; 0.02–0.16 AUFS
Recording	Shimadzu CRO-3A integrator
Retention times, min	(estimated from drawing)

Chlorpromazine (CPZ)	3.29
7-Hydroxychlorpromazine	4.74
N,N-Didesmethyl-CPZ	5.11
N-Monodesmethyl-CPZ	6.75
7-Hydroxy-N-monodesmethyl-CPZ	8.21
CPZ-sulfoxide	11.50
CPZ-N-oxide	12.77
N,N-Didesmethyl-CPZ-sulfoxide	14.60
Mesoridazine (IS)	16.05
N-Monodesmethyl-CPZ-sulfoxide	20.99

Ref.	Midha KK, Hubbard JW et al (1987) Ther. Drug Monit 9 : 358

HPLC 6

Sample	Pharmaceutical dosage forms
Apparatus	Spectra-Physics 3500B LC with Du Pont M 837 UV detector and Varian fluorescence detector
Column	Amino phase; Lichrosorb NH_2 10 μm, 250 × 4.6 mm ID and μ-Bondapak NH_2, 300 × 4 mm ID
Mobile phase	Acetonitrile/benzene/water (16 : 4 : 1), water containing 0.01% sodium metabisulfite and 0.01% d-araboascorbic acid as antioxidants
Flow-rate	1.0 ml min^{-1}; 140 psi
Temperature	Ambient
Injection size	100 μl, 10–500 μg ml^{-1}
Detection	UV, λ = 345 nm Fluorescence, λ = 280/385 or 450 nm
Retention times	Chlorpromazine, chlorpromazine-sulfoxide and chlorpromazine-sulfone, in ascending order (from drawing)
Ref.	Takahashi DM (1980) J. Pharm. Sci. 69 : 184

HPLC 7

Sample	Test samples
Apparatus	Spectra-Physics 3500B LC pump and 770 ultra-violet detector
Column	Reversed-phase; μ-Bondapak C_{18}, 10 μm, 300 × 3.9 mm ID
Mobile phase	Methanol/acetic acid/triethylamine/water (70.0 : 1.5 : 0.5 : 28.0)
Flow-rate	1.5 ml min^{-1}
Temperature	Ambient
Injection size	10 μl, 0.25 mg ml^{-1}
Detection	UV, λ = 254 nm; 0.32 AUFS
Recording	Hewlett-Packard 3380A integrator
Retention times, min	(estimated from drawing)

Mesoridazine	5.0	Thioridazine	9.5
Promethazine	5.3	Prochlorperazine	11.0
Acetophenazine	5.5	Butaperazine	12.0
Chlorpromazine	7.5	Thiethylperazine	14.0

Ref.	Roos RW and Lau-Cam CA (1986) J. Chromatogr. 370 : 403

41 Chlorpropamide

CAS-No	94-20-2			
Chemical name	1-(4-Chlorobenzenesulphonyl)-3-propylurea			
Empirical formula	$C_{10}H_{13}ClN_2O_3S$			
Molecular mass	276.7			
Proprietary names	Diabinese; Insulase; Promide			
Clinical use	Antidiabetic			
Solubility	Water	alcohol	ether	chloroform
	insoluble	1 in 12	1 in 200	1 in 9
Light absorption	λ_{max} = 232 nm in methanolic acid			

HPLC 1

Sample: Tablet formulations

Column:
A. Reversed-phase; Zorbax ODS 5–6 μm, 250 × 4.6 mm ID
B. Reversed-phase; Bondapak C_{18}
C. Reversed-phase; ASI C_{18}, 10
D. Reversed-phase; Alltech C_{18}, 10

Mobile phase: 1% Acetic acid/acetonitrile (54:46; 52:48; 50:50)

Flow-rate: 1.2–1.5 ml min^{-1}

Injection size: 20 μl containing 0.2–2.0 μg

Detection: UV, λ = 240 nm

Recording: Computing integrator

Retention times, min:

Column	solvent ratios	flow-rate ml min^{-1}	RT
B. Bondapak C_{18}	52/48	1.2	4.10
B. Bondapak C_{18}	52/48	1.5	4.25
D. Alltech C_{18} 10	52/48	1.5	4.27
C. ASI C_{18} 10	50/50	1.5	4.41
A. Zorbax	50/50	1.4	4.80
B. Bondapak C_{18}	54/46	1.5	5.32

Ref.: Everett RL (1986) J. Assoc. Off. Anal. Chem. 69:519

HPLC 2

Sample: Tablet formulations

Apparatus: Du Pont M 820 LC with UV detector

Column: Hydrocarbon polymer (HCP); 1000 × 2.1 mm ID

Mobile phase:
A. 0.01 M Monobasic sodium citrate containing 15% methanol, pH 4.4
B. 0.01 M Monobasic sodium citrate containing 10% methanol

Flow-rate: 0.36 ml min^{-1}; 500 psi

Temperature: Ambient

Injection size: 1 μl containing 1.5 μg

Detection	UV, λ = 254 nm; 0.04–0.08 AU
Retention times, min	(estimated from drawing)

Mobile phase	A.	B.
Chlorpropamide	6.50	7.19
Tolazamide	7.25	8.75
Tolbutamide	8.50	11.25
Acetohexamide	11.30	15.94

Ref. Beyer WF (1979) in: Tsuji K (ed) GLC and HPLC Determination of Therapeutic Agents, Part 3. Marcel Dekker Inc., New York and Basel, p 1065

HPLC 3

Sample	Tablets
Apparatus	Du Pont M 820 HPLC with UV detector
Column	Normal phase; Si 60, 1000 × 2.1 mm ID
Mobile phase	Methanol
Flow-rate	0.5 ml min^{-1}
Internal standard	Bis(dodecyl)phthalate
Injection size	3 µl
Detection	UV, λ = 254 nm
Retention times, min	Internal standard 2.8
	Chlorpropamide 4.8
Ref.	Molins D, Wong CK et al (1975) J. Pharm. Sci. 64:123

HPLC 4

Sample	Raw materials
Column	Reversed-phase; ODS 5, 10 µm, 250 × 4.6 mm ID
Mobile phase	Acetonitrile/diluted glacial acetic acid (1 in 100) (50:50, v/v) or as per suitability
Flow-rate	1.5 ml min^{-1}
Injection size	Appr. 20 µl
Detection	UV, λ = 240 nm
Ref.	Anonym (1985): The United States Pharmacopeia XXI and The National Formulary XVI, Supplement 4, 1986. United States Pharmacopeial Convention Inc., Rockville, p 2147

---

42 Chlorprothixene

CAS-No	113-59-7; 6469-9-8 (hydrochloride)
Chemical name	3-(2-Chloro-9*H*-thioxanthen-9-ylidene)-*N*,*N*-dimethyl-1-propanamine

Empirical formula	Chlorprothixene		Hydrochloride		
	$C_{18}H_{18}ClNS$		$C_{18}H_{18}ClNS$, HCl		
Molecular mass	315.9		352.3		
Proprietary names	Taractan; Truxal; Truxaletten				
Clinical use	Tranquillizer				
Solubility,	Water	ethanol	acetone	chloroform	ether
Chlorprothixene	1 in 1700	1 in 29	1 in 18	1 in 2	1 in 14
Hydrochloride	freely sol				
Light absorption	λ_{max} = 230 nm in aqueous acid				

HPLC 1

Sample	Blood plasma
Extraction	Heptane/isoamyl alcohol (99:1) under alkaline condition
Apparatus	Waters M 6000A HPLC pump with U6K injection and 440 UV detector
Column	Cyano phase; Supelcosil, LC-PCN, 250 × 4. 6 mm ID
Mobile phase	Acetonitrile/0.02 M potassium dihydrogen phosphate, pH 4.5 (60:40)
Flow-rate	2.0 ml min^{-1}; 1400 psi
Temperature	40 °C
Internal standard	Thioridazine
Injection size	50 µl containing 5–50 ng
Detection	UV, λ = 229 nm; 0.005 AUFS
Retention times, min	Chlorprothixene sulfoxide 4.8
	N-Desmethyl chlorprothixene 6.9
	Chlorprothixene 7.6
	Internal standard 8.5
Ref.	Brooks MA, Didonato G and Blumenthal HP (1985) J. Chromatogr. Biomed. Appl. 337:351

HPLC 2

Sample	Tabletts
Apparatus	LDC Constametric IIG pump, M 1203 UV monitor III and Latek M 7120 sample injector
Column	Reversed-phase; µ-Bondapak C_{18}, 300 × 3.9 mm ID
Mobile phase	Methanol/water (700:300, v/v) containing Waters PIC B5
Flow-rate	2.0 ml min^{-1}; 2400 psi
Internal standard	Doxepin hydrochloride
Injection size	10 µl containing 2.5–20 µg
Detection	UV, λ = 254 nm; 0.016 AUFS

Retention times, min	(estimated from drawing) Internal standard 2.0 Chlorprothixene 3.5
Ref.	Thoma K and Albert K (1984) Arch. Pharm. 317:133

HPLC 3

Sample	Blood serum or plasma
Apparatus	Varian M 5000 LC with Bioanalytical Systems LC-4A amperometric detector
Column	Cyano phase; Lichrosorb CN-10, 300 × 4 mm ID
Mobile phase	45% Acetonitrile and 55% 0.02 M KH_2PO_4
Flow-rate	2.0 ml min^{-1}
Internal standard	2,4-Dichlorpromazine
Injection size	50 µl
Detection	EC, GC electrode, + 0.9 V vs Ag/AgCl RE
Retention times, min	

Acetaminophen	1.5	Imipramine	4.9
Phenothiazine 5-oxide	1.8	Methotrimeprazine	5.0
Phenothiazine-3	1.9	Trimeprazine	5.0
Phenothiazine	2.1	Chlorpromazine	5.3
Acetophenazine	4.2	Triflupromazine	5.3
Desipramine	4.5	Chlorprothixene	5.5
Perphenazine	4.7	Internal standard	5.7
Fluphenazine	4.8	Trifluoperazine	5.7
Promazine	4.8	Clonazepam	5.8
Promethazine	4.8	Prochlorperazine	5.9
		Thioridazine	6.0

Ref.	Wallace JE, Shimek Jr. EL et al (1981) Anal. Chem. 53:960

HPLC 4

Sample	Raw materials
Apparatus	HPLC with photodiode array
Column	A. Reversed-phase; Superspher 100 C_{18}, 5 µm, 125 × 4 mm ID B. Reversed-phase; Chromspher C_{18}, 5 µm, 100 × 3 mm ID
Mobile phase	A. 0.001 M Phosphate buffer (containing 0.1376 g KH_2PO_4 l^{-1} and pH 3.2 with 1% H_3PO_4) and 0.05% nonylamine B. Acetonitrile Gradient: Acetonitrile 10% for 2 min; acetonitrile 10% to 50% in 15 min; acetonitrile 50% to 70% in 10 min; hold 3 min
Flow-rate	1.5 ml min^{-1}

Temperature	19–20 °C
Injection size	10 µl, 0.4 mg ml^{-1}
Detection	UV, λ = 220 nm
Ref.	Bogusz M, Neidl-Fischer G and Aderjan R (1988) J. Anal. Toxicol. 12:325

HPLC 5

Sample	cis- and trans-Isomers of tricyclic neuroleptic drugs
Apparatus	Altex M 100A LC pump, Rheodyne M 7120 injector and Cecil M 212 UV monitor
Column	Reversed-phase; Spherisorb 5 µm, 250 × 4.6 mm ID
Mobile phase	Ethyl acetate/methanol/3% w/v ammonia (85:15:1, v/v)
Flow-rate	1.0 ml min^{-1}; 1015 psi
Injection size	20 µl
Detection	UV, λ = 260 nm; 0.5 AUFS

Retention times, min

Isomers	cis		trans
Chlorprothixene	6.04		8.22
Flupenthixol	7.09		8.05
Clopenthixol	8.02		8.89
Dothiepin	8.54		9.05
Doxepin	8.84		9.66
Amitriptyline		8.70	
Chlorpromazine		9.20	
Imipramine		12.45	
Promazine		16.00	
Thiothixene		20.00	
Nortriptyline		27.70	

Ref. Li Wan PO A and Irwin WJ (1979) J. Pharm. Pharmacol. 31:512

43 Chlortetracycline

CAS-No	57-62-5; 64-72-2 (hydrochloride)	
Chemical name	7-Chlorotetracycline	
Empirical formula	Chlortetracycline	Hydrochloride
	$C_{22}H_{23}ClN_2O_8$	$C_{22}H_{23}ClN_2O_8$,HCl
Molecular mass	478.9	515.3
Proprietary names	Aureomycin; Chlortet; Clorciclina	
Clinical use	Antibiotic	

Solubility, Chlortetracycline	Water	alcohol	ether	aq NaOH
	very slightly soluble	slightly soluble	insoluble	

Hydrochloride	1 in 75–110	1 in 250–560	insoluble	soluble
Light absorption	$\lambda_{max} = 266$ nm in aqueous acid			

HPLC 1

Sample	Premixes and veterinary products
Apparatus	Instrument specialistics M LC-825 pump and 226 UV detector
Column	Reversed-phase; Alltech Econosphere C_8, 5 µm, 150 × 4.6 mm ID
Pre-column	Brownlee C_8, 15 × 3.2 mm ID, and silicagel, 250 × 4.5 mm ID
Mobile phase	Acetonitrile/0.1 M phosphate buffer (containing 13.8 g NaH_2PO_4, 2.28 g EDTA and 8 ml N,N-dimethyloctylamine per liter, pH 8.0 with NaOH) (30 : 70)
Flow-rate	1.5 ml min^{-1}
Injection size	20 µl
Detection	UV, $\lambda = 280$ nm
Recording	Hewlett-Packard 3390A integrator
Retention times, min	(estimated from drawing) epi-Chlortetracycline 3.2 Chlortetracycline 4.8
Ref.	Chappell GS, Houglum JE and Kelley WN (1986) J. Assoc. Off. Anal. Chem. 69 : 28

HPLC 2

Sample	Blood plasma
Extraction	Protein precipitation with 1.25 M perchloric acid or with ethyl acetate
Apparatus	Spectra-Physics 8000B HPLC, Rheodyne M 7126 injector and Pye-Unicam LC-UV 3 detector
Column	A. XAD-2, 6–7 µm, 100 × 3.0 mm ID B. PRP-1, 10 µm, 100 × 3.0 mm ID
Pre-column	XAD-2 or PRP-1, 250 × 4.6 mm ID
Mobile phase	For A. Acetonitrile/dichloromethane/0.2 M acetate buffer, pH 3.6 plus 0.025 M EDTA (15:1:85, v/v) For B. Acetonitrile/dichloromethane/0.05 M acetate buffer, pH 3.6 plus 0.025 M EDTA (12.5:1:87.5, v/v)
Flow-rate	0.6 ml min^{-1}
Temperature	25 °C
Injection size	20 µl
Detection	UV, $\lambda = 268$ or 357 nm

Retention times, min	(estimated from drawing)	Column	
		A.	B.
	iso-Chlortetracycline	1.0	1.7
	epi-Chlortetracycline	1.6	1.7

	A.	B.
epi-Demethylchlortetracycline	1.6	1.6
Chlortetracycline	1.8	2.5
Demethylchlortetracycline	1.8	1.8
epi-anhydro-Demethylchlor-tetracycline	3.0	8.7
epi-anhydro-Chlortetracycline	5.0	15.3
anhydro-Demethylchlortetra-cycline	8.0	17.5
anhydro-Chlortetracycline	17.0	40.0

Ref. Reeuwijk HJEM and Tjaden UR (1986) J. Chromatogr. 353:339

HPLC 3

Sample: Meat (beef)

Extraction: McIlvaine buffer, pH 5.5/methanol (3:7) and defatted with petroleum and eluted with methanol on Sep-pak C_{18} cartridge

Column: Reversed-phase; YMC-Pak ODS, 150 × 6 mm ID

Mobile phase: 1 M Imidazole buffer containing 50 mM magnesium acetate/methanol (77:23)

Flow-rate: 0.8 ml min^{-1}

Temperature: 40 °C

Injection size: 50 µl containing 100 ng

Detection: Fluorescence, $\lambda = 380/520$ nm

Retention times, min: (estimated from drawing)

		Tetracycline	7.23
Oxytetracycline	5.4	Chlortetracycline	16.92

Ref. Yoneda Y, Furushiro N and Andoh R (1989) J. Food Hyg. Soc. Jap. 30:42

HPLC 4

Sample: Raw materials

Apparatus: Varian M 5000 LC, UV-50 detector and Valco automatic loop injector

Column:
A. PLRP-S; 5 µm, styrene-divinyl benzene copolymer, 150 × 4.6 mm ID
B. Reversed-phase; Micropak SPC-18, 4 µm, 150 × 4.6 mm ID

Pre-column: Special

Mobile phase:
I. For A: Gradient, 0.01 M ortho-H_3PO_4/acetonitrile (85:15 to 40:60 in 20 min)
II. For B: 0.01 M Tetramethylammonium chloride in 0.01 M ortho-H_3PO_4, and other conditions as in I.
III. For B: 0.01 M Sodium heptane sulfonate in 0.01 M ortho-H_3PO_4, and other conditions as in I.

Flow-rate	1 ml min^{-1}
Injection size	200 µl containing 2 µg
Detection	UV, λ = 355 nm; 0.2–0.5 AUFS
Retention times, min	(estimated from drawing)

Mobile phase	I/A	II/B	III/B
Oxytetracycline	6.9	6.7	12.0
Tetracycline	8.6	8.8	13.2
Chlortetracycline	11.3	12.0	16.4

Ref. Moats WA (1986) J. Chromatogr. 366:69

HPLC 5

Sample	Bulk drugs
Apparatus	Spectra-Physics 8700 XR LC pumps, Waters M 440 UV detector and Valco loop injector
Column	Reversed-phase; Zorbax C_8, 7 µm, 250 × 4.6 mm ID
Mobile phase	I. Dimethyl sulfoxide/1 M perchloric acid/water (55:5:40) II. Dimethyl sulfoxide/1 M perchloric acid/water (32:5:63) III. Dimethyl sulfoxide/1 M perchloric acid/ water (60:5:35)
Flow-rate	1 ml^{-1}; 1000–2500 psi
Temperature	35 °C
Injection size	20 µl
Detection	UV, λ = 280 nm
Recording	Hewlett-Packard 3393A integrator
Retention times, min	(estimated from drawing)

Mobile phase	I.	II.	III.
(EISOCTC) 4-Epiisochlor-tetracycline	5.5		
(TC) Tetracycline	6.5	9.8	
(DMCTC) 6-Demethyl-chlortetracycline	8.0	14.5	
(ISOCTC) Isochlortetracycline	8.5	18.0	
(ECTC) 4-Epichlortetracycline	9.8	15.2	
(CTC) Chlortetracycline	12.0	22.5	
(ATC) Androtetracycline	19.0		9.8
(EATC) Epiandrotetracycline			8.2
(ETC) Epitetracycline		8.0	
(EACTC) 4-Epianhydro-chlortetracycline			23.5
(ACTC) Anhydrochlor-tetracycline	52.7		26.0

Ref. Khan NH, Roets E et al (1989) J. Pharm. Biomed. Anal. 7:339

44 Chlorthalidone

CAS-No	77-36-1			
Chemical name	2-Chloro-5-(1-hydroxy-3-oxoisoindolin-1-yl) benzenesulphonamide			
Empirical formula	$C_{14}H_{11}ClN_2O_4S$			
Molecular mass	338.8			
Proprietary names	Hygroton; Novothalidone; Urolin			
Clinical use	Diuretic			
Solubility	Water	alcohol	methanol	aq NaOH
	insoluble	1 in 150	1 in 25	soluble
Light absorption	λ_{max} = 275 nm in ethanol			

HPLC 1

Sample	Tablets
Apparatus	Waters LC with M 6000A pumps, 440 ultraviolet detector and 710 autosampler
Column	Reversed-phase; Zorbax ODS, 5–6 μm, 250 × 4.6 mm ID
Mobile phase	Methanol/water (50:50)
Flow-rate	1 ml min^{-1}
Temperature	23 °C
Internal standard	Caffeine
Injection size	25 μl, 50 μg ml^{-1}
Detection	UV, λ = 254 nm; 0.2 AUFS
Retention times, min	Chlorthalidone 4.1
	Internal standard 5.8
Ref.	Walters SM and Stonys DB (1983) J. Chromatogr. Sci. 21:43

HPLC 2

Sample	Human blood plasma
Extraction	Diethyl ether/isopropanol (19:1)
Apparatus	Waters M 501 HPLC and Shimadzu SPD-6A UV detector
Column	Reversed-phase; 10 μm, C_8, 220 × 4.6 mm ID
Pre-column	C_8, 15 × 3.2 mm ID
Mobile phase	0.05 M Sodium dodecyl sulfate in phosphate buffer, pH 5.8/propanol (95:5)
Flow-rate	1.3 ml min^{-1}
Internal standard	Xipamide
Injection size	100 μl, 50–800 ng ml^{-1}
Detection	UV, λ = 235 nm

Retention times, min	Chlorthalidone	8.4
	Xipamide	13.0
Ref.	Dadgar D and Kelly MT (1988) Analyst. 113:1223	

HPLC 3

Sample	Tablet formulations	
Apparatus	Varian M 2010 LC, Rheodyne M 7125 injection loop and Varian M 2050 Spectrophotometric detector	
Column	Supelcosil LC-8-DB, 5 μm, 250 × 4.6 mm ID	
Mobile phase	1.0 mM Ammonium acetate and 2.0 mM octane-sulfonic acid sodium salt in acetonitrile/water (25:75), pH 3.5 with glacial acetic acid	
Flow-rate	1.5 ml min^{-1}	
Temperature	Ambient	
Internal standard	Methylparaben	
Injection size	10 μl containing 1.25–3.75 μg	
Detection	UV, λ = 254 nm; 0.2 AUFS	
Recording	Varian M 4290 integrator	
Retention times, min	Hydrochlorothiazide	3.17
	Related compound I	3.53
	Atenolol	5.16
	Chlorthalidone	5.93
	Internal standard	8.40
	Related compound II	10.63
Ref.	Sa'sa SI, Jalal IM, Khalil HS (1988) J. Liq. Chromatogr. 11:1673	

HPLC 4

Sample	Biological fluids	
Extraction	Deproteinization of serum with 0.33 N perchloric acid, and then with 50% acetonitrile in 0.01 M sodium acetate on a Bond-Elut RP extraction column	
Column	Reversed-phase; RCM-100 C_{18}, 10 μm	
Mobile phase	0.01 M Aqueous sodium acetate with acetonitrile (80:20, v/v)	
Flow-rate	2.0 ml min^{-1}	
Temperature	Ambient	
Internal standard	Probenecid	
Injection size	10 μl	
Detection	UV, λ = 210 nm	
Recording	Waters M 730 data module	
Retention times, min	Internal standard	3.5
	Chlorthalidone	8.6

Ref.	Macgregor TR, Farina PR et al (1984) Ther. Drug Monit. 6:83

HPLC 5

Sample	Tablet formulations
Column	Reversed-phase; octylsilane 5–10 μm, 250 × 4.6 mm ID
Mobile phase	0.01 M Dibasic ammonium phosphate/methanol (3:2), pH 5.5 with phosphoric acid
Flow-rate	1 ml min^{-1}
Internal standard	2,7-Naphthalenediol
Injection size	Appr. 25 μl
Detection	UV, λ = 254 nm
Retention times	(relative) Chlorthalidone 0.8 Internal standard 1.0
Ref.	Anonym (1985): The United States Pharmacopeia XXI and The National Formulary XVI. United States Pharmacopeial Convention Inc., Rockville, p 210

---

45 Choline

CAS-No	62-49-7 (cation); 67-48-1 (chloride)		
Chemical name	2-Hydroxyethyltrimethylammonium hydroxide		
Empirical formula	Choline	Chloride	
	$C_5H_{15}NO_2$	$C_5H_{14}ClNO$	
Molecular mass	121.2	139.6	
Proprietary names	Colyne; Neurotropan; Wallachol		
Clinical use	Cholinergic		
Solubility,	Water	alcohol	ether
Choline	very soluble	very soluble	insoluble
Chloride	very soluble	very soluble	insoluble

HPLC 1

Sample	Plant materials or plant saps
Extraction	Propan-2-ol, and then with hyrochloric acid
Apparatus	Spectra-Physics 8700 LC with Rheodyne injector and Cecil CE 212A UV detector
Column	Reversed-phase; Dionex MPIC-NSI, 10 μm, 200 × 4 mm ID

Pre-column	Dionex MPIC-NGI mounted in a Dionex 2010i ion chromograph with cation fibre suppressor
Mobile phase	2 mM Methanesulfonic acid and acetonitrile Gradient: 0 to 10% acetonitrile in 20 min
Flow-rate	1 ml min^{-1}
Injection size	20 µl
Detection	UV, λ = 200 nm
Recording	Pye DP 88 integrator
Retention times, min	Choline 2.6 Acetyl-β-methylcholine 4.5 Chlorocholine 2.8 Butyrylcholine 11.5 Acetylcholine 3.1
Ref.	Gorham J (1986) J. Chromatogr. 362:243

HPLC 2

Sample	Plant materials and plant saps
Apparatus	Spectra-Physics 8700 HPLC system and Cecil CE 212A UV detector
Column	Reversed-phase; Dionex MPIC-NSI, 10 µm, 200 × 4 mm ID
Pre-column	Dionex MPIC-NGI mounted in a Dionex 2010i ion chromatograph with cation fibre suppressor
Mobile phase	2 mM Hexanesulfonic acid and acetonitrile Isocratic: A) 0.75% acetonitrile; B) 7% acetonitrile Gradient: 0 min, 0.6% acetonitrile; 9 min, 0.9% acetonitrile; 20 min, 3% acetonitrile; 30 min, 15% acetonitrile; 38 min, 30% acetonitrile
Flow-rate	1 ml min^{-1}
Internal standard	Trimethylaminopropan-2-ol
Injection size	20 µl
Detection	UV, λ = 200 nm
Recording	Pye DP 88 integrator

Retention times, min

System	isocratic A	isocratic B	gradient
Ammonium	7.0		6.0
Aminoethanol	8.0		10.0
Methylaminoethanol	9.5		
Dimethylaminoethanol	11.5		
Choline	12.5	4.5	16.8
Trimethylaminopropan-1-ol	14.5		
Acetylcholine		6.5	29.5
Triethylaminoethanol		7.5	
Phosphorylcholine		9.2	34.0
Butyrylcholine			41.0
Benzoylcholine			46.0
Trimethylaminopropan-2-ol			21.0

Ref.	Gorham J (1986) J. Chromatogr. 362:243

HPLC 3

Sample	Rat brain homogenate
Apparatus	HPLC with Biometra electrochemical detector EP 30
Column	Biometra-ACC (EP 17), 125 × 4 mm ID
Pre-column	Biosensor (EP 25) cartridge with acetyl-choline esterase and choline oxidase
Mobile phase	0.1 M Sodium phosphate buffer, pH 7.4 plus 10 mM tetramethylammonium chloride
Flow-rate	0.6 ml min^{-1}
Injection size	20 µl
Detection	EC, + 500 mV
Retention times, min	(estimated from drawing) Cholin 6.0 Acetylcholine 11.0
Ref.	Biometra HPLC-Gesamtkatalog (1989) HPLC-Gesamtkatalog, Goettingen, p 52

46 Cimetidine

CAS-No	51481-61-9
Chemical name	2-Cyano-1-methyl-3-[2-(5-methylimidazol-4-yl-methylthio)ethyl]-guanidine
Empirical formula	$C_{10}H_{16}N_6S$
Molecular mass	252.3
Proprietary names	Tagamet; Tametin; Ulcomet
Clinical use	Histamine H_2-Receptor Antagonist
Solubility	1 Part in appr. 88 parts water

HPLC 1

Sample	Serum and urine
Extraction	Methanol on Bond Elut C_{18} column
Apparatus	Waters M 510 LC pump, ultraviolet detector and WISP TM autosampler
Column	Reversed-phase; Resolve TM C_8, 10 µm, 100 × 5 mm ID in RCM-TM
Pre-column	Corasil C_{18}, 27–50 µm, 25 × 4 mm ID

Mobile phase	0.01 M Acetic acid/acetonitrile (84:16, v/v) plus triethylamine 0.12–0.16 g/l, pH 4.1
Flow-rate	2 ml min^{-1}; ca 623.5 psi
Internal standard	Ornidazole
Injection size	100 µl
Detection	UV, λ = 229 nm; 0.02-0.04 AUFS

Retention times, min

Theophylline	1.3	Ampicillin	4.0
Cefoperazone	1.5	Cimetidine-sulfoxide	5.2
β-Hydroxytheophylline	1.5	Hydroxylmethylcimetidine	6.0
Cephalothin	1.6	Internal standard	7.0
Theobromine	1.6	Tobramycin	7.2
Ceftazidime	2.0	Sulfamethoxazole	8.0
Metronidazole	2.0	Cimetidine	11.0
Caffeine	2.4	Cloxacillin	16.0
Imipenem	2.4	Trimethoprim	40.0

Ref. Strong HA and Spino M (1987) J. Chromatogr. Biomed. Appl. 422:301

HPLC 2

Sample	Serum and urine
Extraction	Methylene chloride
Apparatus	LDC Constametric III LC pump, Spectromonitor III UV detector and Rheodyne M 7125 injector
Column	Reversed-phase; Partisil ODS-3, 10 µm, 250 × 4.6 mm ID
Mobile phase	7% Acetonitrile in 10 mM potassium phosphate buffer, pH 4.8
Flow-rate	2.0 ml min^{-1}; 1000-1500 psi
Temperature	21 °C
Internal standard	Procaine hydrochloride
Injection size	20 µl
Detection	UV, λ = 228 nm; 0.01-2.0 AUFS
Recording	LDC M 308 computing integrator
Retention times, min	Cimetidine 6.2 Internal standard 12.9
Ref.	Guay DRP, Bockbrader HN and Matzke GR (1982) J. Chromatogr. Biomed. Appl. 228:398

HPLC 3

Sample	Blood plasma and gastric fluid
Extraction	Acetonitrile under alkaline, acidic and again under alkaline condition
Apparatus	HPLC System with M U6K injector from Waters and Kratos spectroflow monitor SF 770

Column	Reversed-phase; RCM-100
Mobile phase	340 g Methanol in 1 liter phosphate buffer (containing 8.7 mmol KH_2PO_4 and 3.04 mmol Na_2HPO_4, pH 7.41)
Flow-rate	3.0 ml min^{-1}; 6900 kPa
Internal standard	Guanidine derivative
Injection size	15 µl
Detection	UV, λ = 228 nm
Retention times, min	Internal standard 2.00 Cimetidine 2.78
Ref.	Abdel-Rahim M, Ezra D et al (1985) Clin. Chem. 31:621

HPLC 4

Sample	Blood plasma
Extraction	Methylene chloride/isopropanol (90:10) under alkaline condition
Apparatus	Varian 5000 HPLC with UV detector
Column	Alkyl-nitrile; Micropak CN-10, 10 µm, 3000 × 4 mm ID
Mobile phase	50% Acetonitrile and 50% 0.01 M sodium dihydrogen phosphate
Flow-rate	1.6 ml min^{-1}
Internal standard	Ranitidine
Injection size	10 µl
Detection	UV, λ = 228 nm
Retention times, min	Cimetidine 4.0 Ranitidine 5.0
Ref.	Fleitman J, Torosian G and Perrin JH (1982) J. Chromatogr. Biomed. Appl. 229:255

HPLC 5

Sample	Blood plasma
Extraction	Octanol at pH 9.0
Column	Ion-exchange; Nucleosil 10 SA, 150 × 4.6 mm ID
Mobile phase	85% Potassium sodium phosphate buffer (0.067 M), pH 6.0 containing potassium chloride 20 g l^{-1} and 15% methanol
Flow-rate	1.2 ml min^{-1}
Injection size	10 and 50 µl
Detection	UV, λ = 228 nm
Retention times, min	(estimated from drawing) Cimetidine sulfoxide 4.0 Cimetidine 8.0
Ref.	Larsson R, Erlanson P et al (1982) Eur. J. Clin. Pharmacol. 21:325

HPLC 6

Sample	Serum and urine
Extraction	Deproteinization with acetonitrile, and then with water-saturated isoamyl alcohol and methylene chloride
Apparatus	Perkin-Elmer Series 2 HPLC with Schoeffel M 770 UV detector
Column	Normal phase; Zorbax Sil, 250 × 4.6 mm ID
Pre-column	Whatman HC Pellosil
Mobile phase	Acetonitrile/methanol/water/ammonium hydroxide (1000:50:50:2, v/v), pH 10.5
Flow-rate	3.0 ml min^{-1}
Internal standard	SKF 92374
Injection size	25 µl containing at least 0.05 µg ml^{-1}
Detection	UV, λ = 228 nm; 0.01 AUFS
Retention times, min	

Cimetidine	3.6	Cimetidine sulfoxide	17.1
Internal standard	5.4	Guanyl urea cimetidine	25.8
Hydroxymethyl cimetidine	7.2	Creatinine	13.2

Ref. Ziemniak JA, Chiarmonte DA and Schentag JJ (1981) Clin. Chem. 27:272

HPLC 7

Sample	Whole blood and plasma
Extraction	Methanol
Apparatus	Waters M 6000 LC pump with U6K injector and Kratos M 773 UV detector
Column	Reversed-phase; styrene-divinyl benzene 10 µm, 60 × 4.6 mm ID
Mobile phase	12.5% Acetonitrile in 20 mM potassium phosphate, pH 9.5
Flow-rate	1.0 ml min^{-1}
Injection size	20 µl
Detection	UV, λ = 210 nm; 0.1–0.005 AUFS
Retention times, min	Cimetidine sulfoxide 2.6
	Hydroxymethyl cimetidine 3.1
	Cimetidine 6.0
Ref.	Rustum AM and Hoffman NE (1988) J. Assoc. Off. Anal. Chem. 71:519

47 Clavulanic Acid

CAS-No	58001-44-8
Chemical name	3-(2-Hydroxyethylidene)-7-oxo-4-oxa-1-azabicyclo [3.2.0]heptane-2-carboxylic acid

Empirical formula	$C_8H_9NO_5$
Molecular mass	199.2
Proprietary name	Augmentin
Clinical use	Amoxicillin potentiator
Light absorption	λ_{max} = 277 nm in methanol

HPLC 1

Sample	Serum and urine
Extraction	Centrifugation and filtration using 0.2 M phosphate buffer, pH 7.0
Apparatus	LDC Constametric pump IIG, Waters M 441 UV detector and Rheodyne M 7210 loop injector
Column	Reversed-phase; μ-Bondapak C_{18}, 300 × 3.9 mm ID
Pre-column	Reversed-phase; μ-Bondapak C_{18}, 30 × 4.6 mm ID
Mobile phase	Methanol/aqueous 0.03 M phosphate buffer, pH 7 (20:80, v/v)
Flow-rate	2.0 ml min^{-1}
Temperature	Ambient
Derivatization	Pre-column; with 2 M 1,2,4-triazole reagent
Injection size	50 μl, at least 1 μg ml^{-1}
Detection	UV, λ = 313 nm
Recording	Varian M 4290 computing integrator
Retention time, min	Clavulanate chromophore 7.5 (estimated from drawing)
Ref.	Martin J and Mendez R (1988) J. Liq. Chromatogr. 11:1697

HPLC 2

Sample	Blood plasma and urine
Extraction	Deproteinization with acetonitrile
Apparatus	Jasco LC pumps and Shimadzu M 240-UV detector
Column	Reversed-phase; Develosil ODS-10, 250 × 4.6 mm ID
Pre-column	Reversed-phase; Lichrosorb C_2, 50 × 4.6 mm ID
Mobile phase	A. 5 mM Tetrabutylammonium bromide + 0.1 mM Na_2HPO_4 + 0.1 mM NaH_2PO_4/methanol (3:1, v/v) for blood plasma B. 5 mM Tetrabutylammonium bromide + 1 mM Na_2HPO_4 + 1 mM NaH_2PO_4/methanol (5:1, v/v) for urine
Flow-rate	1.2 ml min^{-1}
Temperature	Ambient
Derivatization	Post-column; with NaOH saturated methanol
Injection size	5–20 μl
Detection	UV, λ = 270 nm; 0.01–0.04 AUFS

Retention times, min

(estimated from drawing)	plasma (A)	urine (B)
Clavulanic acid	12.5	18.0

Ref.	Haginaka J, Yasuda H et al (1983) Chem. Pharm. Bull. 31:4436

HPLC 3

Sample	Clavulanate potassium
Column	Reversed-phase; ODS 3–10 µm, 300 × 4 mm ID
Mobile phase	7.8 g Monobasic sodium phosphate in 1000 ml water, pH 4.4 and methanol (95:5)
Flow-rate	2.0 ml min^{-1}
Internal standard	Amoxicillin
Injection size	Appr. 20 µl
Detection	UV, λ = 210 nm
Retention times	(relative) Clavulanic acid 0.5 Internal standard 1.0
Ref.	Anonym (1985): The United States Pharmacopeia XXI and The National Formulary XVI, Supplement 4, 1986. United States Pharmacopeial Convention Inc., Rockville, p 2149

48 Clindamycin

CAS-No	18323-44-; 58207-19-5 (hydrochloride, monohydrate); 24729-96-2 (phosphate)			
Chemical name	(7*S*)-Chloro-7-deoxylincomycin			
Empirical formula	Clindamycin	Hydrochloride	Phosphate	
	$C_{18}H_{33}ClN_2O_5S$	$C_{18}H_{33}ClN_2O_5S$, HCl, H_2O	$C_{18}H_{34}ClN_2O_8PS$	
Molecular mass	425.0	479.5	505.0	
Proprietary names	Cleocin; Dalacin; Sobelin			
Clinical use	Antibiotic			
Solubility,	Water	alcohol	chloroform	methanol
Hydrochloride	1 in 2	1 in 200	very slightly sol	soluble
Phosphate	1 in 2.5		insoluble	

HPLC 1

Sample	Pharmaceutical formulations and raw materials
Apparatus	Altex M 110A LC pump, Waters R401 refractometer, LDC M 1203 UV detector and Rheodyne M 7120 injector
Column	Reversed-phase; µ-Bondapak C_{18}, 10 µm, 300 × 3.9 mm ID

Mobile phase	A. For RI: 0.005 M D,L-10-sodium camphor sulfonate and 0.035 M glacial acetic acid in methanol/water (60:40), pH 6.0 B. For UV: Methanol/water (60:40), 0.01 M phosphate buffer and 0.005 M sodium pentate-sulfonate
Flow-rate	1.0 ml min^{-1}; 900 psi
Internal standard	Phenethyl alcohol
Injection size	25 µl containing 375 µg
Detection	Refractive index detector; UV, $\lambda = 214$ nm
Retention times	(retention volume, in mobile phase A)

Lincomycin B	4.41	Clindamycin B	7.69
Lincomycin	4.50	7-Epiclindamycin	9.15
Phenethyl alcohol	5.90	Clindamycin	11.30

Ref. Landis JB, Grant ME and Nelson SA (1980) J. Chromatogr. 202:99

HPLC 2

Sample	Test substance
Apparatus	Du Pont 830 LC with differential refractometer detector
Column	Reversed-phase; µ-Bondapak C_{18}, 300 × 3.9 mm ID
Mobile phase	125 ml Water, 1 g dioctyl sodium sulfosuccinate, 1 ml formic acid and methanol to make 500 ml
Flow-rate	1 ml min^{-1}; 612 psi
Injection size	10 µl
Detection	RI
Retention times	(relative)

Clindamycin hydrochloride

Lincomycin B hydrochloride	0.49
Epilincomycin hydrochloride	0.55
Lincomycin hydrochloride	0.55
Clindamycin B hydrochloride	0.79
Clindamycin hydrochloride	1.00
Epiclindamycin hydrochloride	1.00
Testosterone propionate (IS)	1.25

Clindamycin palmitate

Clindamycin hydrochloride	0.23
Palmitic acid	0.44
Clindamycin 4-palmitate	0.59
Methylpalmitate (IS)	0.70
Clindamycin 3-palmitate	0.75
Clindamycin 2-palmitate	1.00

	Clindamycin phosphate	
	Lincomycin 2-phosphate	0.45
	Lincomycin hydrochloride	0.61
	Clindamycin 3-phosphate	0.90
	Clindamycin 4-phosphate	0.92
	Clindamycin 2-phosphate	1.00
	Propylparaben (IS)	1.29
	Clindamycin hydrochloride	1.94
Ref.	Brown LW (1978) J. Pharm. Sci. 67:1254	

HPLC 3

Sample	Clindamycin hydrochloride
Column	Reversed-phase; ODS 5–10 µm, 300 × 4 mm ID
Mobile phase	200 ml Water with 2 g dl-10 camphorsulfonic acid, 1 g ammonium acetate and 1 ml glacial acetic acid and methanol up to 500 ml, pH 6.0
Flow-rate	1 ml min^{-1}
Internal standard	Phenylethyl alcohol
Injection size	Appr. 25 µl
Detection	Refractive index detector
Retention times	(relative) Internal standard 0.60 Clindamycin 1.00
Ref.	Anonym (1985): The United States Pharmacopeia XXI and The National Formulary XVI, Supplement 4, 1986. United States Pharmacopeial Convention Inc., Rockville, p 2150

49 Clonazepam

CAS-No	1622-61-3		
Chemical name	5 (2 Chlorophenyl)-1,3-dihydro-7-nitro-2*H*-1,4-benzodiazepine-2-one		
Empirical formula	$C_{15}H_{10}ClN_3O_3$		
Molecular mass	315.7		
Proprietary names	Clonopin; Iktorivil; Rivotril		
Clinical use	Anticonvulsant		
Solubility	Water	alcohol and ether	chloroform
	insoluble	slightly soluble	sparingly soluble
Light absorption	λ_{max} = 273 nm in aqueous acid		

HPLC 1

Sample	Serum
Extraction	Toluene/heptane/isoamyl alcohol (76:20:4)
Apparatus	Waters M 6000A LC pump, U6K injector and 450 UV detector
Column	Reversed-phase; Supelcosil C_{18}, 5 μm, 250 × 4.5 mm ID
Pre-column	Perisorb RP-18
Mobile phase	10 mM Potassium phosphate buffer, pH 6.0 containing 2.6 g l^{-1} sodium pentanesulfonic acid/methanol/acetonitrile (50:35:15, v/v)
Flow-rate	1 ml min^{-1}
Internal standard	Flunitrazepam
Injection size	20 μl, 25 ng ml^{-1}
Detection	UV, λ = 313 nm; 0.01 AUFS
Retention times, min	(estimated from drawing) Clonazepam 12.86 Carbamazepine 11.14 Flunitrazepam 14.24
Ref.	Haver VM, Porter WH et al (1986) Ther. Drug Monit. 8:352

HPLC 2

Sample	Serum
Extraction	n-Butyl chloride at pH 9.2 and back extraction with 2 N HCl, and then with chloroform/ isopropanol (9:1, v/v) at pH 11.0
Column	Cyanopropyl; silane CN, 5–6 μm, 250 × 4.6 mm ID
Pre-column	Permaphase ETH, 30 μm, 50 × 4.6 mm ID
Mobile phase	0.5 N Acetic acid/acetonitrile/n-butylamine (59:41:0.005, v/v)
Flow-rate	1.2 ml min^{-1}
Internal standards	Flunitrazepam and desmethyldiazepam
Injection size	20 μl, 0–400 ng ml^{-1}
Detection	UV, λ = 245 nm; 0.01 AUFS
Retention times, min	(estimated from drawing) Carbamazepine 5.30 Flunitrazepam 7.09 Clonazepam 6.30 Desmethyldiazepam 8.26
Ref.	Doran TC (1988) Ther. Drug Monit. 10:474

HPLC 3

Sample	Serum
Extraction	Diethyl ether at pH 9.1
Apparatus	Hewlett-Packard 1084B HPLC with UV detector
Column	Reversed-phase; Lichrosorb C_8, 7 μm, 250 × 4 mm ID

Mobile phase	28% Acetonitrile in water Gradient: Acetonitrile 28% to 38% in 8 min; until start wash program with 80% acetonitrile
Flow-rate	1.9 ml min^{-1}
Temperature	35 °C
Injection size	40 µl, 200 nM
Detection	UV, λ = 320 nm
Retention times, min	(estimated from drawing) Carbamazepine 5.00 Clonazepam 6.50
Ref.	Wad N (1986) Ther. Drug Monit. 8:358

HPLC 4

Sample	Human serum
Extraction	n-Hexane/ethyl acetate (9:1)
Apparatus	Perkin-Elmer Series 2 LC pump and LC 75 UV detector
Column	Reversed-phase; C_8, 5 µm, 150×4.6 mm ID
Pre-column	Pelliguard LC-8, 40 µm, 20 mm
Mobile phase	Acetonitrile/1.75 mM hydrochloric acid/50 mM sodium acetate (36:10:54)
Flow-rate	1.5 ml min^{-1}
Temperature	Ambient
Internal standard	Flunitrazepam
Injection size	100 µl
Detection	UV, λ = 220 nm; 0.080 AUFS
Retention times, min	(estimated from drawing) Flunitrazepam 5.80 Nitrazepam 4.14 Clobazam 6.90 Clonazepam 4.55
Ref.	Zilli MA and Nisi G (1986) J. Chromatogr. Biomed. Appl. 378:492

HPLC 5

Sample	Tablet formulations
Column	Reversed-phase; ODS 5 to 10 µm, 300×4 mm ID
Mobile phase	Water/methanol/acetonitrile (4:3:3)
Internal standard	*o*-Dichlorobenzene
Injection size	Appr. 25 µl, 40 µg ml^{-1}
Detection	UV, λ = 254 nm
Retention times, min	Chlonazepam 6.0 Internal standard 14.5
Ref.	Anonym (1985): The United States Pharmacopeia XXI and The National Formulary XVI. The United States Pharmacopeial Convention Inc., Rockville, p 233

50 Clonidine

CAS-No	4205-90-7; 4205-91-8 (hydrochloride)			
Chemical name	2,6-Dichloro-*N*-(imidazolidin-2-ylidene)aniline			
Empirical formula	Clonidine	Hydrochloride		
	$C_9H_9Cl_2N_3$	$C_9H_9Cl_2N_3,HCl$		
Molecular mass	230.1	266.6		
Proprietary names	Catapres; Dixarit; Isoglaucon			
Clinical use	Antihypertensive			
Solubility,	Water	alcohol	chloroform	ether
Clonidine			1 in 8	1 in 50
Hydrochloride	1 in 13	1 in 25	1 in 250	insoluble
Light absorption	λ_{max} = 271 nm in aqueous acid			

HPLC 1

Sample	Tablets
Apparatus	Waters M 6000A LC pump, 440 Ultraviolet detector and 710 autosampler
Column	Reversed-phase; Zorbax TMS 5–6 μm, 250 × 4.6 mm ID
Mobile phase	Methanol/0.0022 M KH_2PO_4 with 0.016 M Na_2HPO_4, pH 7.9 (65:35)
Flow-rate	1 ml min^{-1}
Temperature	23 °C
Injection size	25 μl, 4 mg ml^{-1}
Detection	UV, λ = 254 nm; 0.02 AUFS
Retention times, min	(estimated from drawing) Clonidine 3.5 Chlorthalidone 7.6
Ref.	Walters SM and Stonys DB (1983) J. Chromatogr. Sci. 21:43

HPLC 2

Sample	Dosage forms
Apparatus	Pye-Unicam LC-XPD pump, LC-UV detector and Rheodyne M 7125 injector
Column	Reversed-phase; Nucleosil 5, C_8, 125 × 4.6 mm ID
Mobile phase	Methanol/water/triethylamine (80:20:0.005)
Flow-rate	1 ml min^{-1}
Injection size	20 μl, 2.15–44.8 μg ml^{-1}
Detection	UV, λ = 240 nm; 0.02–0.16 AUFS
Retention times, min	Clonidine 2.76 (estimated from drawing)
Ref.	Wilczynska-Wojtulewicz I and Sadlej-Sosnowska N (1986) J. Chromatogr. 367:434

HPLC 3

Sample	Tablet formulations
Column	Reversed-phase; octasilane 5–10 μm, 150 × 4 mm ID
Pre-column	Reversed-phase; octasilane 30–50 μm, 50 × 4.6 mm ID
Mobile phase	Triethylamine solution (containing 3200 ml water, 16 ml triethylamine and 6.4 ml phosphoric acid, pH 5.8)/acetonitrile/methanol (16:3:1)
Flow-rate	3 ml min^{-1}
Injection size	Appr. 50 μl, 1 μg ml^{-1}
Detection	UV, $\lambda = 220$ nm
Ref.	Anonym (1985): The United States Pharmacopeia XXI and The National Formulary XVI. United States Pharmacopeial Convention Inc., Rockville, p 234

51 Cloxacillin

CAS-No	61-72-3; 7081-44-9 (sodium salt, monohydrate)	
Chemically	Semisynthetic penicillin	
Empirical formula	Cloxacillin	Sodium Salt
	$C_{19}H_{18}ClN_3O_5S$	$C_{19}H_{17}ClN_3NaO_5S,H_2O$
Molecular mass	435.9	475.9
Proprietary names	Cloxapen; Orbenin; Solcillin	
Clinical use	Antibiotic	

Solubility,	Water	ethanol	chloroform
Sodium salt	1 in 2.5	1 in 30	1 in 500

Light absorption: $\lambda_{max} = 352$ nm in aqueous acid (sodium salt)

HPLC 1

Sample	Capsules and oral solutions
Apparatus	Spectra-Physics 8000 LC with 8300 fixed wavelength detector
Column	Reversed-phase; C_2, 10 μm, 250 × 4.6 mm ID
Mobile phase	0.05 M Potassium dihydrogen phosphate, pH 4.5/acetonitrile (800:200, v/v)
Flow-rate	1.5 ml min^{-1}
Temperature	30 °C
Internal standard	Dimethylphtalate
Injection size	10 μl
Detection	UV, $\lambda = 254$ nm; 0.04 AUFS

Retention times	(relative)			
	Cloxacilloic acid	0.38	Cloxacillin	1.00
	Cloxalloic acid	0.50	Ethylparaben	1.50
	Oxacillin	0.68	Internal standard	1.59
	Methylparaben	0.86	Dicloxacillin	1.82
Ref.	Lauriault G, Lebelle MJ and Vilim A (1982) J. Chromatogr. 246:157			

HPLC 2

Sample	Raw materials and blood plasma from rats
Apparatus	Lewa FLM/50 HPLC pump with pulse-dampening system and Perkin-Elmer LC 55 UV detector
Column	Reversed-phase; Nucleosil C_{18}, 5 µm, 250 × 3 mm ID
Mobile phase	0.05 M Phosphate buffer, pH 7/acetonitrile (7:2.5, v/v)
Flow-rate	1.8 ml min^{-1}
Temperature	Ambient
Injection size	10 µl, 1 mg ml^{-1}
Detection	UV, $\lambda = 220$ nm
Recording	Hewlett-Packard 3352B computer-system

Retention times, min	(estimated from drawing)		Cloxacillin	6.9
	Ampicillin	1.5	Flucloxacillin	9.0
	Oxacillin	4.5	Dicloxacillin	13.3
Ref.	Hartmann V and Roediger M (1976) Chromatographia 9:266			

HPLC 3

Sample	Raw materials and pharmaceutical dosage forms
Apparatus	Tracor Instruments M 950 LC pump, 970A UV detector and Rheodyne M 7125 loop injector
Column	Reversed-phase; Chromegabond C_{18}, 10 µm, 300 × 4.6 mm ID
Mobile phase	Acetonitrile/methanol/0.01 M KH_2PO_4, pH 4.7 (19:11:70, v/v)
Flow-rate	1 ml min^{-1}; 790 psi
Temperature	Ambient
Injection size	20 µl containing 0.1 µg ml^{-1}
Detection	UV, $\lambda = 225$ nm; 0.16 AUFS

Retention times, min	Amoxicillin	3.6	Oxacillin	10.0
	Ampicillin	4.0	Cloxacillin	14.4
	Methicillin	5.2	Nafcillin	21.0
	Penicillin G	6.2	Dicloxacillin	26.8
	Penicillin V	8.4		
Ref.	Briguglio GT and Lau-Cam CA (1984) J. Assoc. Off. Anal. Chem. 67:228			

HPLC 4

Sample	Milk
Extraction	Deproteinization with acetonitrile after enzymatic hydrolysis with penicillase and then with methylene chloride
Apparatus	Waters M 440 LC with 660 gradient controller and S.I./McPherson FL-749 fluorescence detector
Column	Reversed-phase; Lichrosorb C_{18}, 10 μm, 250 × 4 mm ID
Mobile phase	Acetonitrile/water (58:42)
Flow-rate	1 ml min^{-1}
Internal standard	Benzaldehyde
Derivatization	Pre-column; with dansyl hydrazine
Injection size	20 μl
Detection	Fluorescence, λ = 254/500 nm
Retention times, min	

Methicillin	5.23	Cloxacillin	8.18
Benzylpenicillin	5.74	Nafcillin	9.70
Penicillin V	6.73	Dicloxacillin	10.73
Oxacillin	7.21	Internal standard	12.17
Phenethicillin	7.68		

Ref. Munns RK, Shimoda W et al (1985) J. Assoc. Off. Anal. Chem. 68:968

HPLC 5

Sample	Body fluids
Extraction	Methylene chloride or methanol-perchloric acid
Apparatus	Waters M 6000A LC pump, U6K injector and Perkin-Elmer LC-55 spectrophotometer
Column	Reversed-phase; Lichrosorb C_8, 5 μm, 250 × 4.6 mm ID
Pre-column	Pellicullar ODS, 50 × 2.1 mm ID
Mobile phase	I. 0.02 M Sodium acetate, pH 5.5/methanol (10:8, v/v) II. 0.02 M Ammonium acetate, pH 6.6/acetonitrile (100:34, v/v)
Flow-rate	0.8 ml min^{-1}
Injection size	20 μl
Detection	UV, λ = 220 nm; 0.008–0.4 AUFS
Retention times, min	(estimated from drawing)

Mobile phase	I	II
Penicilloic acid	7.0	
5-Hydroxymethyl metabolite	9.0	6.0
Cloxacillin	13.5	9.2
Oxacillin		6.8
Flucloxacilin	15.0	11.8

Ref. Thijssen HHW (1980) J. Chromatogr. Biomed. Appl. 183:339

HPLC 6

Sample	Serum
Extraction	Centrifugation with methanol
Column	Reversed-phase; Lichrosorb C_{18}, 10 µm, 250 × 4.6 mm ID
Mobile phase	20% Acetonitrile in 0.03 M ammonium acetate buffer, pH 7.0
Flow-rate	1.6 ml min^{-1}
Internal standard	Probenecid
Injection size	30–35 µl
Detection	UV, λ = 230 nm
Recording	Hewlett-Packard 3392A integrator

Retention times, min				
	Benzylpenicillin	5.30	or	12.30
	Gentamicin	12.30	or	5.30
	Probenecid (IS)	13.80		
	Cloxacillin	16.65		

Ref. Lee HS, Ti TY and Khoo YM (1988) Ther. Drug Monit. 10:340

52 Cocaine

CAS-No	50-36-2; 53-21-4 (hydrochloride)			
Chemical name	(1*R*, 2*R*, 3*s*, 5*S*)-2-Methoxycarbonyltropan-3-yl benzoate			
Empirical formula	Cocaine	Hydrochloride		
	$C_{17}H_{21}NO_4$	$C_{17}H_{21}NO_4$,HCl		
Molecular mass	303.4	339.8		
Proprietary names	Bonain' solution; Cocaine Eye-drops; Cocaine solution			
Clinical use	Local Anaesthetic			
Solublity,	Water	ethanol	chloroform	ether
Cocaine	1 in 600	1 in 7	1 in 0.5	1 in 4
Hydrochloride	1 in 0.5	1 in 3.5–4.5	1 in 15–18	insoluble
Light absorption	λ_{max} = 233 nm in aqueous acid			

HPLC 1

Sample	Raw materials
Apparatus	Waters M 6000 LC pump and Cecil CE-212 variable-wavelength UV detector

Column	Reversed-phase; μ-Bondapak 10 μm, C_{18} ODS, 300 × 4 mm ID
Mobile phase	55% Methanol in 0.025 M phosphate buffer, pH 7.0
Flow-rate	2 ml min^{-1}
Temperature	Ambient
Injection size	2 μl containing 1–10 μg
Detection	UV, λ = 220 nm
Retention times, min	(estimated from drawing)

Caffeine	2.0	Cocaine	7.0
Ephedrine	3.4	Quinine	9.0
Methylamphetamine	4.0	Tubocurarine	12.3

Ref. Twitchett PJ and Moffat AC (1975) J. Chromatogr. 111:149

HPLC 2

Sample	Illicit cocaine and coca leaves
Apparatus	Spectra-Physics 8100 LC with autosampler and loop injector and Kratos M 773 UV detector
Column	Reversed-phase; Supelcosil 5 μm, LC-8, 250 × 4.6 mm ID
Mobile phase	Acetonitrile/tetrahydrofuran/0.1%, (v/v) triethylamine in water (40:10:50, v/v)
Flow-rate	1 ml min^{-1}
Temperature	Ambient
Injection size	10 μl
Detection	UV, λ = 254 nm
Recording	Spectra-Physics 4270 computing integrator
Retention times, min	(estimated from drawing)

Norcocaine	5.2
Cocaine	6.5
cis-Cinnamoylmethylecgonine	7.0
trans-Cinnamoylmethylecgonine	7.4

Ref. LeBella MJ, Callahan SA et al (1988) Analyst. 113:1213

HPLC 3

Sample	Test solution containing cocaine hydrochloride, phentermine, ephedrine and proptriptyline
Apparatus	HPLC equipment with ultraviolet detector
Column	Normal phase; Spherisorb S5W, 250 × 4.5–4.9 mm ID
Mobile phase	2700 ml Methanol and 300 ml aqueous ammonium nitrate buffer (containing 27 g ammonium nitrate and 90 ml concentrated ammonia, specific gravity 0.880 in 900 ml water), pH 9.3–9.5

Flow-rate	2 ml min^{-1}
Temperature	Ambient (19–33 °C)
Internal standard	Protriptyline
Injection size	5 µl
Detection	UV, λ = 254 nm
Recording	Electronic integrator
Retention times, min	Cocaine 1.7; Ephedrine 3.7; Phentermine 2.5; Protriptyline 4.7
Ref.	Gill R, Osselton MD et al (1987) J. Chromatogr. 386:65

HPLC 4

Sample	Cocaine mixture
Column	Reversed-phase; Nucleosil C_{18}, 5 µm
Mobile phase	Tetrahydrofuran/water (20:80) containing 5 mM n-heptanesulfonate sodium and 2% acetic acid
Flow-rate	1 ml min^{-1}
Temperature	15 °C
Injection size	0.370 µg
Detection	UV, λ = 235 nm; 0.005 A
Retention times, min	(estimated from drawing) Cocaine 9.4; Allopseudococaine 12.6; Allococaine 10.2; Pseudococaine 15.5
Ref.	Olieman C, Maat L et al (1980) Recl. Trav. Chim. Pays-Bas., p 501

HPLC 5

Sample	Brompton mixture
Apparatus	Spectra-Physics 3500B HPLC pump with 770 UV detector and Valco injector
Column	Reversed-phase; µ-Bondapak C_{18}, 10 µm, ODS
Mobile phase	0.015 M KH_2PO_4, pH 3.0 with 2 N phosphoric acid/acetonitrile (3:1, v/v), pH 3.3
Flow-rate	0.8 ml min^{-1}; 620 psi
Temperature	Ambient
Internal standard	4-Chloropyridine hydrochloride
Injection size	10 µl containing 3 nM
Detection	UV, λ = 235 nm; 0.02 AUFS

Retention times, min

Morphine	4.50	Diacetylmorphine	13.67
6-Aminoacetyl-morphine	7.00	Benzoic acid	16.20
Benzoylecgonine	8.33	Cocaine	19.00
4-Chloropyridine	10.80	2-Chloro-3-nitro-pyridine	23.17

Ref. Poochikian GK and Cradock JC (1980) J. Pharm. Sci. 69:637

53 Codeine

CAS-No	76-57-3 (anhydrous); 6059-47-8 (monohydrate); 6854-40-6 (sulfate, trihydrate)			
Chemical name	3-*O*-Methylmorphine monohydrate			
Empirical formula	Codeine	Sulfate		
	$C_{18}H_{21}NO_3,H_2O$	$(C_{18}H_{21}NO_3)_2$, $H_2SO_4,3H_2O$		
Molecular mass	317.4	750.9		
Proprietary names	Codicept; Kaodene; Paveral			
Clinical use	Analgesic			
Solubility,	Water	ethanol	chloroform	ether
Codeine	1 in 120	1 in 2	1 in 0.5	1 in 50
Sulfate	1 in 30	1 in 1300	insoluble	insoluble
Light absorption	λ_{max} = 285 nm in aqueous acid			

HPLC 1

Sample	Elixir
Apparatus	Dupont M LC-850 with autosampler, Waters M 440 UV detector and Hewlett-Packard 1040A photodiode-array detector
Column	Reversed-phase; μ-Bondapak TM C_{18}, 300 × 3.9 mm ID
Mobile phase	Buffer (containing 2.4 g 1-Butanesulfonic acid sodium, 2.04 g KH_2PO_4 and 2 ml triethylamine in 1 liter water, pH 4.8)/methanol (80:20, v/v)
Flow-rate	2 ml min^{-1}
Temperature	50 °C
Injection size	20 μl
Detection	UV, λ = 214 nm; 0.08 AUFS
Recording	Dupont M 4100 integrator
Retention times, min	*p*-Aminophenol 1.8; Acetaminophen 2.8; Sodium benzoate 3.9; Codeine phosphate 4.9; Codeine-N-oxide 6.4; Codeinone 8.8
Ref.	Sisco WR, Rittenhouse CT et al (1986) J. Chromatogr. 354:355

HPLC 2

Sample	Blood
Extraction	Methanol/1% ortho-H_3PO_4/acetonitrile (50:30:20) under alkaline condition on C_{18} extraction column

Apparatus	Spectra-Physics 8700 XR LC pump, Pye-Unicam 4020 variable-wavelength ultraviolet detector
Column	Reversed-phase; Spherisorb 5 µm ODS, 250 × 4.6 mm ID
Mobile phase	0.05 M Sodium pentanesulfonic acid/acetonitrile (70:30), pH 2.0 with H_3PO_4
Flow-rate	1.2 ml min^{-1}
Injection size	20 µl
Detection	UV, λ = 220 nm
Retention times, min	(estimated from drawing) Morphine 4.0 Codeine 6.2 Monoacetylmorphine 4.5
Ref.	Tebbett IR (1987) Chromatographia 23:377

HPLC 3

Sample	Test solution containing Papaverine, dipipanone hydrochloride, codeine phosphate, methdilazine hydrochloride and proptriptyline
Column	Normal phase; Spherisorb S5W, 250 × 4.5–4.9 mm ID
Mobile phase	2700 ml Methanol and 300 ml aqueous ammonium nitrate buffer (containing 27 g NH_4NO_3 and 90 ml concentrated ammonia, specific gravity 0.880 in 900 ml water), pH 9.3–9.5
Flow-rate	2.0 ml min^{-1}
Temperature	Ambient (19–33 °C)
Internal standard	Protriptyline
Injection size	5 µl
Detection	UV, λ = 254 nm
Retention times, min	Papaverine 1.6 Methdilazine 3.6 Dipipanone 2.2 Proptriptyline 4.7 Codeine 3.0
Ref.	Gill R, Osselton MD et al (1987) J. Chromatogr. 386:65

HPLC 4

Sample	Raw materials
Apparatus	Pye-Unicam LC-XPS HPLC pump, Altex M 153 UV detector and Rheodyne M 7125 injector
Column	Normal phase; Spherisorb S5W, 5 µm, 250 × 5 mm ID
Pre-column	Silica
Mobile phase	90 ml Methanol and 10 ml ethylenediamine-ammonium nitrate buffer, pH 10.2 (containing 10.5 g ammonium nitrate, 15 ml ethylenediamine and 200 ml water, and dilute 10-fold)
Flow-rate	2.0 ml min^{-1}

Temperature	30 °C			
Internal standard	Protriptyline hydrochloride, 0.15 mg ml^{-1}			
Injection size	5 µl, 1.07 mg ml^{-1}			
Detection	UV, λ = 254 nm			
Capacity factor, *k'*	Dipipanone	0.58	Phenylephrine	1.26
	Promazine	0.78	Ephedrine	1.50
	Codeine	0.92	Protriptyline	2.13
	Prolintane	1.11	Strychnine	2.77
Ref.	Smith RM and Rabuor JO (1989) J. Chromatogar. 464:117			

---

54 Colistin

CAS-No	1066-17-7; 1264-72-8 (sulfate); 8068-28-8 (sodium methansulfonate)			
Chemically	Mixture of polypeptides produced by Bacillus polymyxa var colistinus			
Empirical formula	Colistin A	Sulfate	Sodium Methansulfonate	
	$C_{53}H_{100}N_{16}O_{13}$		$C_{58}H_{105}N_{16}Na_5O_{28}S_5$	
Molecular mass	1168		1749 (calculated)	
Proprietary names	Belcomycine; Colimicina; Coly-Mycin M			
Clinical use	Antibiotic			
Solubility,	Water	alcohol	methanol	ether
Sulfate	1 in < 2	1 in 300	slightly soluble	insoluble
Sodium methansulfonate	1 in < 2	1 in 500	soluble	insoluble

HPLC 1

Sample	Raw materials	
Apparatus	Du Pont M 830 LC with 837 UV detector	
Column	Reversed-phase; Nucleosil 5 µm, C_{18}, 150 × 4.6 mm ID	
Mobile phase	22% Acetonitrile in 0.023 M H_3PO_4, 0.01 M CH_3COOH and 0.05 M sodium sulfate buffer, pH 2.5 with triethylamine	
Flow-rate	0.9 ml min^{-1}; 1204 psi	
Injection size	20 µl containing 8–100 µg	
Detection	UV, λ = 220 nm	
Recording	Hewlett-Packard 3380A integrator	
Retention times, min	(estimated from drawing)	
	Polymyxin E_2	8.0–8.5
	Polymyxin E_1	15.0–16.0

Ref.	Elverdam I, Larsen P and Lund E (1981) J. Chromatogr. 218:653

HPLC 2

Sample	Raw materials
Apparatus	LDC modular LC, high-pressure Milton Roy mini-pump with a pulse dampener, moving-wire flame ionization detector and ultragradient mixer
Column	Reversed-phase; μ-Bondapak C_{18}, 300 × 4.6 mm ID
Mobile phase	A. Acetonitrile/water/200 mM phosphate buffer, pH 2.0 (200:700:100) B. Acetonitrile/absolute methanol/water/200 mM phosphate buffer with Na_2HPO_4, pH 2.0 (500:200:290:10) Linear gradient: 20% Acetonitrile to 50% acetonitrile and 20% methanol both pH 2.0, in 16 min
Flow-rate	1 ml min^{-1}; 1000 psi
Temperature	Ambient
Injection size	8 μl
Detection	FID; 175 °C
Retention times, min	(estimated from drawing) Colistin B 8.5 Colistin A 10.3
Ref.	Tsuji K and Robertson JH (1975) J. Chromatogr. 112:663

HPLC 3

Sample	Raw materials of colistin S
Apparatus	Waters M 6000A LC pump, 660 solvent programmer and 440 UV detector
Column	Reversed-phase; Vydac C_{18}, 300 × 3.6 mm ID
Mobile phase	(1) 0.01 M Pi (monosodium phosphate), (2) 0.05 M Pi, and (3) 0.01 M HSA (1-heptane-sulfonic acid sodium salt), each with 80% MeOH (methanol), 60% AcCN (acetonitrile) and 40% THF (tetrahydrofuran) and pH each 2.0, 3.4 and 6.6
Flow-rate	1 ml min^{-1}
Temperature	Ambient
Injection size	1–10 μl
Detection	UV, λ = 254 nm; 0.2 A
Recording	Computing integrator (Supergrator-1, Columbia Scientific)
Retention times, min	Mobile phase

Buffer	Modifier	pH		
		2.0	3.4	6.6

	0.01 M Pi	AcCN	2.78	2.94	2.74
		THF	3.09	3.06	2.98
		MeOH	3.21	3.18	3.03
	0.05 M Pi	AcCN	3.26	3.27	3.47
	0.01 M HSA	AcCN	3.02	3.60	3.85
		THF	4.30	4.95	3.88
		MeOH	4.69	4.70	3.75

Ref. Aszalos A and Aquilar A (1984) J. Chromatogr. 290:83

HPLC 4

Sample	Raw materials
Apparatus	LDC Constametric IIG LC pump with Cecil CE 272 spectrophotometer
Column	Reversed-phase; Spherisorb ODS 5 μm, 150×4 mm ID
Mobile phase	0.005 M Tartrate buffer, pH 3.0 containing 0.005 M sodium butane-1-sulfonate, 0.2 M sodium sulfate and 18% (v/v) acetonitrile
Flow-rate	2.0 ml min^{-1}; 1800 psi
Injection size	7.5 μl, 10 mg ml^{-1}
Detection	UV, λ = 220 nm; 0.02 AUFS
Recording	Electronic integrator
Retention times, min	Colistin E_2 6.25
	Polymyxin B_2 8.50
	Colistin E_1 13.50
	Polymyxin B_1 18.50
Ref.	Thomas AH, Thomas JM and Holloway I (1980) Analyst 105:1068

55 Cortisone

CAS-No	53-06-5; 50-04-4 (acetate)		
Chemical name	17α,21-Dihydroxypregn-4-ene-3,11,20-trione		
Empirical formula	Cortisone	Acetate	
	$C_{21}H_{28}O_5$	$C_{23}H_{30}O_6$	
Molecular mass	360.4	402.5	
Proprietary names	Adreson; Corlin; Cortone		
Clinical use	Corticosteroid		
Solubility,	Water	alcohol	acetone
Cortisone	very slightly sol	soluble	
Acetate	1 in 5000	1 in 300–350	1 in 75
Light absorption	λ_{max} – 240 nm in ethanol		

HPLC 1

Sample	Swine blood plasma
Extraction	Ethyl acetate or with methyl tert.-butyl ether
Apparatus	Waters M 6000A HPLC pump with 710B automatic injector and Kratos 773 varibale-wavelength detector
Column	Reversed-phase; Brownlee Special C_{18}, 5 μm, 220 × 4.6 mm ID
Pre-column	Reversed-phase; Brownlee C_{18}, 5 μm, 30 × 4.6 mm ID
Mobile phase	Tetrahydrofuran/water (25:75, v/v)
Flow-rate	1 ml min^{-1}; 3002–3205 psi
Temperature	Ambient (25 °C)
Injection size	100 μl containing 10 ng
Detection	UV, λ = 240–242 nm; 0.004 AUFS
Recording	Waters M 730 data module

Retention times, min

Prednisone	9.80	Dexamethasone	24.30
Cortisone	10.30	Prednisolone acetate	35.70
Prednisolone	11.60	Cortisone acetate	37.10
Hydrocortisone	12.60	Hydrocortisone-acetate	40.00
Methylprednisolone	18.90		

Ref. Carlin AS, Prasad VK et al (1988) J. Chromatogr. Biomed. Appl. 425:162

HPLC 2

Sample	Dosage forms and bulk drug substances
Apparatus	HPLC system and ultraviolet detector
Column	Reversed-phase; ODS 10 μm
Mobile phase	I. Acetonitrile/water (50:50) II. Acetonitrile/0.025 M acetate buffer with 20 ml 1 N NaOH, 150 ml 0.5 N KCl and 50 ml 0.5 M sodium acetate in 1 liter water, pH 4.0 (50:50)
Flow-rate	2.0 ml min^{-1}
Temperature	Ambient
Detection	UV, λ = 254 nm

Retention times, min

Mobile phase	I.	II.
Cortisone alcohol	2.45	
Cortisone acetate	4.69–4.97	4.74–4.95

Ref. Ng LL (1987) J. Assoc. Off. Anal. Chem. 70:829

HPLC 3

Sample	Urine
Extraction	Acetonitrile/water (90:10, v/v) on Sep-pak C_{18} extraction column

Apparatus	Spectra-Physics 8700 LC with gradient system, Magnus M 7110 autosampler with Rheodyne injector and Pye-Unicam LC-UV detector
Column	Reversed-phase; Ultrasphere C_{18}, 5 μm, 250 × 4.6 mm ID
Pre-column	Reversed-phase; Hypersil ODS 5 μm, 50 × 5 mm ID
Mobile phase	Water/acetonitrile Gradient: Acetonitrile 0% to 50% in 25 min; 50% to 100% in 1 min; 100% to 0% in 7 min
Flow-rate	0.9 ml min^{-1}
Internal standard	6α-Methylprednisolone
Injection size	2000 μl
Detection	UV, λ = 260 nm
Recording	Spectra-Physics 4100 computing integrator
Retention times, min	Cortisol 25.2 Cortisone 25.8 Internal standard 27.4 (estimated from drawing)
Ref.	Jenner DA and Richards J (1985) J. Pharm. Biomed. Anal. 3:251

HPLC 4

Sample	Serum
Extraction	Methylene chloride
Apparatus	Shimadzu LC-3A pump and ultraviolet detector
Column	Reversed-phase; Zorbax ODS, 250 × 4.6 mm ID
Mobile phase	Methanol/isopropanol/water (44:10:46)
Flow-rate	0.7 ml min^{-1}
Temperature	55 °C
Injection size	25 μl, 5–500 μg ml^{-1}
Detection	UV, λ = 240 nm; 0.01 AUFS
Retention times, min	Cortisone 6.54 Cortisol 7.10 Corticosterone 10.11 11-Deoxycortisol 10.58 Androstenedione 17.09 11-Deoxycorticosterone 17.09 17α-Hydroxyprogesterone 19.54 Progesterone 45.09
Ref.	Ji-qing W, Xian-teng Z and Ji-lu W (1987) Clin Chem. 33:1354

HPLC 5

Sample	Serum
Extraction	Ethanol on Sep-Pak C_{18} cartridge and then with dichloromethane after derivatization on extraction column
Apparatus	Waters M 6000A HPLC pump, Hitachi M 650–10 LC fluorescence spectrophotometer and Waters U6K loop-valve injector

Column	Cosmosil 5SL, 150 × 4 mm ID and Lichrosorb RP 18, 125 × 4 mm ID
Mobile phase	Hexane/ethyl acetate (6:5)
Flow-rate	1 ml min^{-1}
Temperature	Ambient
Derivatization	Pre-column; with 9-anthroyl nitrile
Internal standard	Prednisolone
Injection size	10 μl containing 20 fmol
Detection	Fluorescence, λ = 360/460 nm
Retention times, min	Cortisol 7.0
	Prednisolone 9.5
	Cortisone 12.3 (estimated from drawing)
Ref.	Goto J, Shamsa F et al (1983) J. Pharm. Biomed. Anal. 1:83

56 Cyanocobalamin (Vitamin B_{12})

CAS-No	68-19-9			
Chemical name	Coα-[α-(5,6-Dimethylbenzimidazolyl)]-Coβ-cyanocobamide			
Empirical formula	$C_{63}H_{88}CoN_{14}O_{14}P$			
Molecular mass	1355.4			
Proprietary names	Berubi; Cytacon; Rubramin			
Clinical use	Vitamin			
Solubility	Water	alcohol	acetone	chloroform
	1 in 80	1 in 180	insoluble	insoluble
Light absorption	λ_{max} = 361 nm in water			

HPLC 1

Sample	Raw materials
Apparatus	Waters ALC M 202/401 LC with 660 solvent programmer and 440 ultraviolet detector
Column	Reversed-phase; μ-Bondapak C_{18}, 10 μm, 300 × 3.9 mm ID
Mobile phase	A. 0.05 M Sodium acetate buffer, pH 4.0 (containing 2.9 ml glacial acetic acid and 2.6 ml of 3.5 M NaOH in 1 liter water) B. 0.05 M Sodium acetate buffer (as in A.) in methanol, pH 4.0 (73:27, v/v) Concave gradient: Methanol 27% to 95% in 10 min; reverse in 5 min

Flow-rate	1.8 ml min^{-1}
Injection size	15 µl containing 9.8 µg
Detection	UV, λ = 254 nm; 0.5 AUFS
Ref.	Frenkel EP, Kitchens RL and Prough R (1979) J. Chromatogr. 174:393

HPLC 2

Sample	Raw materials
Apparatus	Du Pont M 850 LC with gradient controller, ultraviolet detector and temperature controller
Column	Reversed-phase; Zorbax ODS 6 µm, 250 × 4.6 mm ID
Mobile phase	A. 0.05 M Sodium dihydrogen phosphate B. Methanol Concave gradient: 23% B for 3 min; sample injection and hold 2 min; 23% B to 70% B in 10 min; hold 1 min; 70% B to 23% B in 3 min
Flow-rate	1.8 ml min^{-1}
Temperature	40 °C
Injection size	15 µl containing 9.8 µg
Detection	UV, λ = 254 nm; 0.5 AUFS
Retention times, min	(estimated from drawing) Hydroxycobalamin 2.8 Deoxyadenosylcobalamin 9.0 Cyanocobalamin 5.3 Methylcobalamin 10.6
Ref.	Frenkel EP, Kitchens RL and Prough R (1979) J. Chromatogr. 174:393

HPLC 3

Sample	Dosage forms
Apparatus	Knauer HPLC system and spectrophotometric detector
Column	Reversed-phase; Lichrosorb 5 µm, C_{18}, 250 × 4 mm ID
Mobile phase	I. Methanol/water (80:20) II. Methanol/water (50:50)
Flow-rate	2.0 ml min^{-1}; 2610 psi
Temperature	Ambient (22–25 °C)
Injection size	5–25 µl, 100–500 ng ml^{-1}
Detection	UV, λ = 254 nm; 1.28 AUFS
Retention times, min	see below
Ref.	Amin M and Reusch J (1987) J. Chromatogr. 390:448

Mobile phase	I.	II.
Vitamin B_1	1.0–1.1	0.9–1.0
Vitamin B_6	1.0–1.1	1.2–1.3
Vitamin B_{12}	1.5–1.6	1.5–1.6
Vitamin B_2	1.7–1.8	1.9–2.0

HPLC 4

Sample	Fat soluble vitamins and vitamin B_{12}
Apparatus	Azuma Electric MF-2 Microfeeder HPLC system with Jasco Uvidec 100 ultraviolet detector
Column	A. Vydac 201 TPB5, 500 × 0.53 mm ID B. Polymer-based octadecyl, 500 × 0.53 mm ID
Mobile phase	For A. Acetonitrile/ethanol/methanol (35:45:20) For B. Acetonitrile/methanol (75:25)
Flow-rate	4 μl min^{-1}
Temperature	20 °C
Detection	UV, $\lambda = 280$ nm
Retention times, min	(estimated from drawing)

Column	A.	B.
Vitamin K_3	20.8	21.0
Vitamin B_{12}	21.0	18.0
Vitamin A	23.5	24.8
Vitamin E	35.5	44.0
Vitamin D_2	36.5	39.8
Vitamin E-acetate	39.0	36.0
Vitamin D_3	40.0	41.0
Vitamin K_1	45.0	47.0

Ref. Wahyuni WT and Jinno K (1987) Chromatographia 23:320

HPLC 5

Sample	Multivitamin
Column	Reversed-phase; Vydac C_{18}, 5 μm, 250 × 4.6 mm ID
Mobile phase	A. 0.1 M Potassium acetate, pH 4.7 B. Acetonitrile/water (50:50) Gradient: B 5% to 100% in 15 min
Flow-rate	1.5 ml min^{-1}
Detection	UV
Retention times, min	(estimated from drawing)

Vitamin C	2.04	Nicotinamide	7.13
Nicotinic acid	3.85	Folic acid	8.60
Vitamin B_6	5.80	Vitamin B_{12}	10.51
Vitamin B_1	6.52	Vitamin B_2	11.09

Ref. Vydac Application Notes on HPLC Columns and Separation Materials (1987–1988) Vydac, California, p 17

HPLC 6

Sample	Blood plasma
Extraction	Methanol on reversed-phase Sep-Pak C_{18} column
Apparatus	Knauer 52.00 LC pump and 85.00 UV detector

Column	Normal phase; Lichrosorb Si 60, 5 μm, 250 × 4.6 mm ID
Mobile phase	Methanol/aqueous sodium cyanide, 0.1 g per liter (40:60, v/v)
Flow-rate	0.8 ml min^{-1}
Temperature	30 °C
Internal standard	Bromcresol green
Injection size	80 μl, 20–500 ng ml^{-1}
Detection	UV, λ = 368 nm; 0.03 AUFS
Retention times, min	(estimated from drawing) Dicyanocobalamin 5.0–5.4 Internal standard 6.4–6.8
Ref.	Butte W, Riemann H-H and Walle AJ (1982) Clin. Chem. 28:1778

57 Cyclosporin A

CAS-No	59865-13-3
Chemically	Cyclic polypeptide consisting of 11 amino acids
Empirical formula	$C_{62}H_{111}N_{11}O_{12}$
Molecular mass	1203
Clinical use	Immunosuppressant
Solubility	Water insoluble; alcohol, methanol and ether soluble

HPLC 1

Sample	Blood
Extraction	Methyl tert.-butyl ether
Apparatus	LKB M LC-2150 pump, Valco CV-6-UHPa-N60 injector and Pye-Unicam UV detector
Column	Reversed-phase; Hypersil-WP-300, C_4, 5 μm, 100 × 4.6 mm ID
Pre-column	RSIL C_{18}, 10 μm, 150 × 4.6 mm ID
Mobile phase	Acetonitrile/methanol/water (40:15:45, v/v)
Flow-rate	0.7 ml min^{-1}; 1015 psi
Temperature	70 °C
Injection size	50 μl, 30–1000 ng ml^{-1}
Detection	UV, λ = 214 nm
Recording	Philips PM 8251 recorder
Retention times, min	Cyclosporin A 5.0 Cyclosporin D 6.5

Ref.	Lambert WE, De Leenheer AP et al (1988) J. Chromatogr. Biomed. Appl. 427:195

HPLC 2

Sample	Blood
Extraction	Diethyl ether under acidic condition
Apparatus	Varian LC-5060 pump with gradient system and Perkin-Elmer LC-55 UV detector
Column	Reversed-phase; Varian MCH 5 μm, 150 × 4.6 mm ID (gradient) Reversed-phase; Supelcosil LC-18, 3 μm, 50 × 4.6 mm ID (isocratic)
Mobile phase	A. Water/methanol (90:10) B. Acetonitrile/methanol (90:10) I. Gradient: B 44% to 49% in 5 min; hold 5 min; B 49% to 76% in 23 min; hold 10 min; B 76% to 100% in short; hold 7 min II. Isocratic: Acetonitrile/methanol/water (46:20:34)
Flow-rate	1 ml min^{-1} (I); 2 ml min^{-1} (II)
Temperature	70 °C (B)
Injection size	80 μl
Detection	UV, λ = 210 nm; 0.002 AUFS
Retention times, min	(estimated from drawing) (gradient) Metabolites (1–4) 22–33 Cyclosporin A 32.5 Cyclosporin D 35.8
Ref.	Bowers LD and Singh J (1987) J. Liq. Chromatogr. 10:411

HPLC 3

Sample	Body fluids
Extraction	Water-saturated ether under acidic and alkaline conditions
Apparatus	Waters M 6000A LC pump and 481 ultraviolet detector
Column	Reversed-phase; μ-Bondapak C_{18}
Mobile phase	Acetonitrile/methanol/water (60:15:25, v/v)
Flow-rate	1 ml min^{-1}
Temperature	70 °C
Injection size	25 μl, 200 ng ml^{-1}
Detection	UV, λ = 214 nm
Retention times, min	Cyclosporin A 8.2 Cyclosporin D 10.9 (estimated from drawing)

Ref.	Buice RG, Stentz FB and Gurley BJ (1987) J. Liq. Chromatogr. 10:421

HPLC 4

Sample	Blood and urine
Extraction	On C_{18} Bond Elut cartridge with 25% isopropyl alcohol in ethyl acetate
Apparatus	Waters M LC-510 pump with Shimadzu SPD-6A ultraviolet detector
Column	A. Reversed-phase; 5 μm, C_{18}, 150 × 4.6 mm ID B. Cyano phase; Zorbax 5 μm, 250 × 4.6 mm ID
Mobile phase	For A. Acetonitrile/water (68.5:31.5, v/v) For B. Water/acetonitrile/methanol (54.5:31:14.5, v/v)
Flow-rate	1.4 ml min^{-1} (A); 0.7 ml min^{-1} (B)
Temperature	70 °C (A); 60 °C (B)
Injection size	90 μl
Detection	UV, λ = 202 nm (A) and 214 nm (B)
Recording	Spectra-Physics 4290 integrator

Retention times, min

Column	A.	B.
Cyclosporin A (CsA)	7.6	31.0
Cyclosporin D (CsD)	10.5	37.0
Metabolite-17 (CsA)		16.0
Metabolite-1 (CsA)		18.0
Metabolite-21 (CsA)		23.0

Ref.	Awni WM and Maloney JA (1988) J. Chromatogr. Biomed. Appl. 425:233

HPLC 5

Sample	Whole blood
Extraction	Ethanol on Bond-Elut TM C_{18} extraction column
Apparatus	Perkin-Elmer Serie 10, LC pump and LC 95 UV/VIS detector
Column	Reversed-phase; 3 μm-Pecosphere, C_{18}, 33 × 4.6 mm ID
Mobile phase	Acetonitrile/water (63:37)
Temperature	70 °C
Flow-rate	1 ml min^{-1}
Internal standard	Cyclosporin D
Injection size	10 μl
Detection	UV, λ = 210 nm; 0.016 AUFS
Recording	Perkin-Elmer LCI-100 integrator

Retention times, min

Cyclosporin	6.2
Internal standard	8.8

Ref.	Kabra PM and Wall JH (1987) J. Liq. Chromatogr. 10:477

58 Daunorubicin

CAS-No	20830-81-3; 23541-50-6 (hydrochloride)			
Chemically	Anthracycline antibiotic produced by Streptomyces coeruleorubidus or Streptomyces peucetius			
Empirical formula	Daunorubicin		Hydrochloride	
	$C_{27}H_{29}NO_{10}$		$C_{27}H_{29}NO_{10}$, HCl	
Molecular mass	527.5		564.0	
Proprietary names	Cerubidin; Daunoblastin			
Clinical use	Antineoplastic			
Solubility,	Water	methanol	chloroform	ether
Hydrochloride	soluble	soluble	insoluble	insoluble

HPLC 1

Sample	Pharmaceuticals and raw materials
Apparatus	Knauer HPLC pump with Hewlett-Packard 1040A multidiode array detector, or Chromatem M 380 pump with Spectromonitor LDC III UV detector
Column	Normal phase A. Lichrosorb Si 60, 5 μm, 250 × 4.6 mm ID B. Lichrosorb Si 60, 5 μm, 250 × 4.6 mm ID C. Lichrosorb Si 60, 5 μm, 100 × 4.6 mm ID D. Nucleosil 50, 5 μm, 100 × 4.6 mm ID E. Spherisorb, 5 μm, 100 × 4.6 mm ID
Mobile phase	Methanol/water/formic acid buffer, pH 2.5 (containing 30 ml water, 18 ml formic acid, 10 g NH_4Cl, and 9 g NH_4OH in 100 ml water)/ dichloromethane (130:6.5:0.75:863)
Flow-rate	A. 1.5 ml min^{-1}; B. 3.0 ml min^{-1}; C., D., and E. 1.0 ml min^{-1}
Injection size	20 μl, 1 mg ml^{-1}
Detection	UV, λ = 254 nm
Recording	Hewlett-Packard 3390A or Shimadzu C-R3A integrator

Capacity factor, *k'*

Column	B	C	D	E
Daunorubicin	3.14	3.27	4.49	3.96
Dexorubicin	5.17	5.85	8.46	7.25
Duborimycin	6.63	7.51	10.29	9.04

Ref.	Miller JH McB, Pascal C and Tissieres M (1987) J. Chromatogr. 392:361

HPLC 2

Sample	Pharmaceuticals and raw materials
Apparatus	Knauer HPLC pump with Hewlett-Packard 1040A multidiode array detector, or Chromatem M 380 pump with Spectromonitor LDC III UV detector
Column	Reversed-phase; 250 × 4.6 mm ID: Lichrosorb C_{18}, 5 μm; Partisil C_{18}, ODS-3; Spherisorb C_{18}, 5 μm; Nucleosil C_{18}, 5 μm; Hypersil ODS 5 μm and μ-Bondapak C_{18}, 10 μm
Mobile phase	Acetonitrile/water (50:50) containing 0.01 M sodium dodecyl sulfate and 0.02 M H_3PO_4
Flow-rate	2.0 ml min $^{-1}$
Injection size	20 μl, 1 mg ml^{-1}
Detection	UV, λ = 254 nm
Recording	Hewlett-Packard 3390A or Shimadzu C-R3A integrator
Retention times, min	(estimated from drawing) (in Licrosorb C_{18}, 5 μ)

Daunorubicin aglycone	4.6	Duborimycin	9.0
Doxorubicin	7.0	Daunorubicin	13.9

Ref.	Miller JH McB, Pascal C and Tissieres M (1987) J. Chromatogr. 392:361

HPLC 3

Sample	Blood plasma from dogs
Extraction	Chloroform/isopropanol (1:1, v/v)
Apparatus	Altex M 110 LC pump with Spectra-Physics fluorimeter
Column	Aminocyano; Partisil-10 PAC, 250 × 4.6 mm ID
Mobile phase	Chloroform/methanol/acetic acid/water (80:20:2:3)
Flow-rate	3.5 ml min $^{-1}$; 2700–4500 psi
Internal standard	Doxorubicin, 250 ng ml $^{-1}$
Injection size	20 μl
Detection	Fluorescence, λ = 380–480/560–570 nm
Retention times, min	

Adriamycin aglycone	0.5	Daunorubicin	2.2
Carminomycin aglycone	0.5	Rudolphomycin	2.2
Daunorubicin aglycone	0.5	Carubicin	2.3
Pyrromycinone	0.5	Proflavine	2.4
Quinine sulfate	0.8	Puromycin	2.5
Acridine	1.0	Musettamycin	2.6
Aclacinomycin A	1.1	Carminomycinol	3.3
Parromycin A	1.7	Doxorubicin	3.9
Bleomycin	2.0	Adramycinol	4.5
Marcellomycin	2.0	Quinidine sulfate	4.8

Ref.	Averbuch SD, Finkelstein TT et al (1981) J. Pharm. Sci. 70:265

HPLC 4

Sample	Test samples
Column	Aminocyano silica; Partisil PXS 10/25 PAC, 10 μm, 250 × 4.6 mm ID
Mobile phase	Hexane/ethylene chloride/methanol/acetic acid/water (40:40:10:4:1)
Flow-rate	3.0 ml min $^{-1}$
Injection size	20 μl
Detection	Fluorescence, λ = 380–480/560–570 nm
Retention times, min	Aclacinomycin A 1.3 Quinine sulfate 3.8
	Acridine 1.3 Carminomycinol 4.0
	Carubicin 3.0 Proflavine 4.0
	Quinidine sulfate 3.3 Daunorubicinol 4.8
	Daunorubicin 3.6 Doxorubicin 6.3
	Puromycin 3.6
Ref.	Reich SD, Martinez L et al (1981) J. Pharm. Sci. 70:268

HPLC 5

Sample	Pharmaceutical preparations
Apparatus	Waters M 6001A LC pump, U6K injector and 240 UV detector
Column	Reversed-phase; μ-Bondapak C_{18}, 10 μm, 250 × 4.6 mm ID
Mobile phase	A. 650 ml Methanol and 350 ml water containing 1.15 g monobasic ammonium phosphate dissolved in 5 ml acetic acid, pH 4.0 B. Methanol/0.005 M heptanesulfonic acid aqueous solution (62.5:37.5), pH 3.5
Flow-rate	2.0 ml min $^{-1}$; 1600 psi
Temperature	Ambient
Injection size	5 μl
Detection	UV, λ = 254 nm
Recording	Hewlett-Packard 3380A integrator
Retention times, min	Mobile phase A. B.
	Doxorubicin 4.00 6.28
	Doxorubicinone 5.14 5.07
	Daunorubicin 6.18 10.39
	Daunorubicinone 7.70 7.58
Ref.	Haneke AC, Crawford J and Aszalos A (1981) J. Pharm. Sci. 70:1112

59 Deoxycortone (Corticosterone)

CAS-No	64-85-7; 56-47-3 (acetate)	
Chemical name	21-Hydroxypregn-4-ene-3,20-dione	
Empirical formula	Deoxycortone	Acetate
	$C_{21}H_{30}O_3$	$C_{23}H_{32}O_4$
Molecular mass	330.5	372.5
Proprietary names	Corticosterone; Cortiron; Percorten	
Clinical use	Corticosteroid	

Solubility,	Water	ethanol	acetone	chloroform
Deoxycortone		freely soluble		
Acetate	insoluble	1 in 50	1 in 30	1 in 1.5

Light absorption: λ_{max} = 240 nm in ethanol

HPLC 1

Sample	Corticosteroids
Apparatus	Kronton M 410 HPLC pump, Pye-Unicam LC-UV spectrophotometer detector and Rheodyne M 7125 injector
Column	A. Normal phase; silica, 120 × 4.6 mm ID B. Reversed-phase; ODS-silica, 120 × 4.6 mm ID
Mobile phase	For A. Methanol/water/0.2 M phosphate buffer, pH 7.0 (50:45:5) containing 2.5 mM cetyltrimethylammonium bromide For B. Methanol/water (60:40)
Flow-rate	1 ml min $^{-1}$
Injection size	20 µl, 0.1 % solution or 10 µl, 0.05 % solution
Detection	UV, λ = 254 nm; 0.01 AU
Recording	Hewlett-Packard 3353A laboratory data system and Kipp & Zonen BD-8 recorder
Retention times	Separation factors between hydrocortisones

	1	2	3	4	5	6	7	8	9	10
A. Silica										
Lichrosorb Si 60	0.10	0.22	0.33	0.39	0.49	0.55	0.67	0.76	1.33	1.51
Nucleosil 50–5	0.08	0.21	0.32	0.38	0.48	0.54	0.65	0.76	1.36	1.50
Zorbax SIL	0.10	0.22	0.33	0.40	0.49	0.56	0.66	0.77		1.52
Partisil 5	0.09	0.22	0.32	0.39	0.49	0.55	0.66	0.76	1.35	1.51
Spherisorb S5 W	0.11	0.23		0.41	0.52	0.57	0.71	0.84	1.37	1.56
Lichrosorb Si 100	0.13	0.25		0.42		0.56		0.80		1.52
Nucleosil 100–5	0.11	0.24	0.33	0.42	0.51	0.55	0.69	0.80	1.31	1.53
Hypersil	0.13	0.23		0.42		0.54	0.72		1.32	1.54

B. ODS-Silica									
Lichrosorb RP-18	0.17	0.34	0.46	0.61		0.79	1.38	1.46	1.81
Nucleosil 5 C_{18}	0.17	0.33	0.46	0.61	0.72	0.81	1.40	1.47	1.85
Hypersil ODS	0.12	0.29	0.39	0.52	0.74				1.62
Zorbax ODS	0.12	0.29	0.41	0.57	0.69			1.62	2.27
Partisil ODS-3	0.16	0.32	0.45	0.61	0.69			1.43	1.92
Spherisorb S5 ODS	0.19	0.33	0.50	0.67			1.45	1.63	2.05

*Values ommitted imply that the corresponding peak coincided with the main peak or with closely adjacent peaks from other impurities due to a less effecient column and/or a less retentive column.

1. Cortisone acetate, 2. Hydrocortisone, 3. Hydrocortisone acetate, 4. Prednisolone, 5. Prednisolone acetate, 6. Corticosterone, 7. Triamcinolone acetonide, 8. Methylprednisolone, 9. Betamethasone, 10. Fluocinolone acetonide, 11. Dexamethasone, 12. Deoxycortone acetate

Ref. Helboe P (1986) J. Chromatogr. 366:191

HPLC 2

Sample	Blood plasma
Extraction	Diethyl ether/dichloromethane (60:40, v/v) under alkaline condition
Apparatus	Beckman M 344 gradient LC and 165 variable-wavelength detector
Column	Reversed-phase; Waters Novapak C_{18}, 4 µm, 150 × 3.9 mm ID
Pre-column	Whatman pellicular C_{18} sorbent
Mobile phase	Methanol/water (55:45, v/v)
Flow-rate	1 ml min $^{-1}$
Internal standard	19-Nortestosterone, 2.3 µg ml $^{-1}$
Injection size	20 µl, 0.5–5 µg ml $^{-1}$ eluent
Detection	UV, λ = 250 nm
Recording	Shimadzu Chromatopac C-RIB data processor
Retention times, min	Corticosterone 4.7 19-Nortestosterone 7.4
Ref.	Woodward CJH and Emery PW (1987) J. Chromatogr. Biomed. Appl. 419:280

HPLC 3

Sample	Raw materials
Column	Nitro-phase; Nucleosil NO_2 10 µm, 300 × 4 mm ID
Mobile phase	0.2% Water and 15% acetonitrile in carbon-tetrachloride
Flow-rate	0.9 ml min $^{-1}$; 1305 psi
Injection size	200 µl
Detection	UV, λ = 242 nm; fluorescence, λ = 366/450 nm

Retention times, min	(estimated from drawing)		
	Detection	UV	Fluorescence
	Androstenedione	4.44	6.67
	Impurity	11.11	13.33
	Corticosterone	14.44	16.30
	Cortisone	24.07	26.67
Ref.	Reh E and Schwedt G (1980) Fresenius Z. Anal. Chem. 303:117		

HPLC 4

Sample	Steroids, peptides and antibiotic mixture
Column	Reversed-phase; Nucleosil C_{18}, 5 μm, 250 × 3 mm ID
Mobile phase	Methanol/water/acetic acid (60:38:2)
Temperature	40 °C
Derivatization	Post-column; with *o*-phthalaldehyde
Injection size	7 μg
Detection	UV, λ = 254 nm
Retention times, min	(estimated from drawing) Corticosterone 4.75 Cortisone 2.5 Cortexone 6.25
Ref.	Von Arx E and Faupel M (1978) J. Chromatogr. 154:68

60 Dexamethasone

CAS-No	50-02-2; 1177-87-3 (acetate, anhydrous); 2392-39-4 (sodium phosphate)			
Chemical name	9α-Fluoro-16α-methylprednisolone			
Empirical formula	Dexamethasone	Acetate	Sodium Phosphate	
	$C_{22}H_{29}FO_5$	$C_{24}H_{31}FO_6$	$C_{22}H_{28}FNa_2O_8P$	
Molecular mass	392.5	434.5	516.4	
Proprietary names	Decadron; Fortecortin; Hexadrol			
Clinical use	Glucocorticoid			
Solubility,	Water	ethanol	chloroform	methanol
Dexamethasone	insoluble	1 in 42	1 in 165	springly sol
Acetate	insoluble	1 in 40	1 in 33	freely soluble
Sodium phosphate	1 in 2		insoluble	
Light absorption	λ_{max} = 240 nm in methanol			

HPLC 1

Sample	Mixture of diphenhydramine, dequalinium chloride and dexamethasone

Apparatus	Perkin-Elmer Series 2 HPLC and LC-55 spectrophotometric detector
Column	A. Reversed-phase; Lichrosorb select-B C_8, 7 μm, 125 × 4 mm ID; B. Reversed-phase; Nucleosil C_{18}, 7 μm, 250 × 4 mm ID
Mobile phase	For A: Water/2 M H_3PO_4 and KH_2PO_4 buffer/acetonitrile/tetrahydrofuran (1480:20:440:80, v/v) For B: 5.59 g Tetraethylammonium trifluoromethanesulfonate in 1060 ml water + 840 ml acetonitrile + 100 ml tetrahydrofuran
Flow-rate	1.5 ml min $^{-1}$ (A); 0.8 ml min $^{-1}$ (B)
Temperature	30 °C (B)
Injection size	50 μl
Detection	UV
Recording	Perkin-Elmer dual canal L 6512 recorder
Retention times, min	(estimated from drawing) A B Diphenhydramine 4.0 7.2 Dequalium chloride 5.0 9.0 Dexamethasone 9.5 5.8
Ref.	Surmann JP and Glienke M (1988) Fresenius Z. Anal. Chem. 331:630

HPLC 2

Sample	Commercial preparations
Apparatus	Perkin-Elmer Series 3B HPLC with two reciprocating pumps and controller, LC-75 UV spectrophotometric detector and Rheodyne M 7010 injector
Column	Normal phase; Alumina, 10 μm, 250 × 4.1 mm ID
Mobile phase	Methanol/dichloromethane (20:80) containing 0.4% ammonium hydroxide for dexamethasone and 0.2% for tramazoline
Flow-rate	1 ml min $^{-1}$
Injection size	20 μl
Detection	UV, λ = 240 nm; 0.02 AUFS
Recording	LKB M 2210 Bromme potentiometric recorder
Retention times, min	Dexamethasone 2.8 Tramazoline 5.4
Ref.	Kountourellis JE and Raptouli A (1988) Anal. Lett. 21:1361

HPLC 3

Sample	Human blood plasma
Extraction	Methylene chloride
Apparatus	Varian M 5000 HPLC system, Verichrom UV detector and Valco injection valve
Column	Reversed-phase; μ-Bondapak C_{18}, 10 μm, 300 × 3.9 mm ID

Mobile phase	0.002 M Acetate buffer, pH 4.8 with acetic acid/acetonitrile (58:42)
Flow-rate	1 ml min^{-1}
Internal standard	Equilenine
Injection size	20 µl, 1.25–20 ng ml^{-1}
Detection	UV, λ = 246 nm; 0.001 AUFS
Retention times, min	(estimated from drawing) Cortisol 5.0 Dexamethasone 5.5 Equilenine 12.0
Ref.	Lamiable D, Vistelle R et al (1986) J. Chromatogr. Biomed. Appl. 378:486

HPLC 4

Sample	Dexamethasone sodium phosphate formulations
Apparatus	Waters M 6000A LC pump, U6K injector and 440 UV detector
Column	Reversed-phase; µ-Bondapak C_{18}, 300 × 3.9 mm ID
Mobile phase	0.07 M Potassium dihydrogen phosphate/tetrahydrofuran/methanol (300:18:182)
Flow-rate	1.5 ml min^{-1}
Internal standard	Diethyl phthalate
Injection size	5 µl containing 0.15–1.5 µg
Detection	UV, λ = 254 nm
Recording	Hewlett-Packard 3370B integrator
Retention times, min	Xylocaine 3.2 Benzyl alcohol 3.8 Methylparaben 5.6 Ethylparaben 8.4 Dexamethasone sodium phosphate 10.5 Diethyl phthalate 13.6 Propylparaben 18.2 Dexamethasone 18.9
Ref.	Gagné D and Lodge BA (1980) J. Chromatogr. 193:160

HPLC 5

Sample	Raw materials
Apparatus	Waters HPLC pump and M 440 UV detector
Column	Reversed-phase; µ-Bondapak C_{18}, 300 × 3.9 mm ID
Mobile phase	Methanol/water (1:1, v/v) containing 0.01 M PIC-A reagent (tetrabutylammonium phosphate in phosphate buffer)
Flow-rate	1.5 ml min^{-1}
Temperature	24 °C
Injection size	5–20 µl containing 11 µg
Detection	UV, λ = 254 nm; 1 AUFS
Retention times, min	(estimated from drawing) Adduct formed 4.0 Dexamethasone sodium phosphate 11.0

Ref.	Dijkstra J and Dekker D (1982) J. Chromatogr. 238:247

61 Dextrose

CAS-No	50-99-7
Chemical name	D-(+)-Glucopyranose
Empirical formula	$C_6H_{12}O_6$
Molecular mass	180.2
Proprietary names	Emdex; Glucodin; Glutose
Clinical use	Sugar
Solubility	Water 1 in 1; alcohol 1 in 200; glycerol soluble; ether insoluble

HPLC 1

Sample	Sausage products
Extraction	First ethanol and then water on C_{18} Sep-Pak and on ion exchange resin columns
Apparatus	Waters M 244 LC, WISP 710B sample processor and R-401 refractometer detector
Column	Amino phase; Bio-sil 5S, 250 × 4.0 mm ID
Mobile phase	Acetonitrile/water (78:22)
Flow-rate	1.0 ml min^{-1}
Detection	RI
Recording	Waters M 730 data module
Retention times, min	Dextrose 14.25; Maltose 24.37; Sucrose 20.25; Lactose 28.10
Ref.	Ali MS (1988) J. Assoc. Off. Anal. Chem. 71:1097

HPLC 2

Sample	Mixture of reducing sugars
Apparatus	Hitachi M 635 HPLC with 204 spectrofluorophotometer
Column	Normal phase; Lichrosorb SI 100, 250 × 4 mm ID
Mobile phase	I. Ethanol/water/chloroform (8:0.5:93.5, v/v) II. Ethanol/water/chloroform (12:0.5:87.5, v/v) III. Ethanol/water/chloroform (20:0.6:79.4, v/v)
Flow-rate	1.5 ml min^{-1}; 1740 psi
Temperature	27 °C
Derivatization	Pre-column; with trichloroacetic acid (0.5%, w/v) in ethanol, and Dns-hydrazine solution (1.0%, w/v) in ethanol

Injection size	0.05–1.0 mmol
Detection	Fluorescence, λ = 350/500 nm
Retention times, min	

Mobile phase	I.	II.	III.
Glyceraldehyde	4.6	3.3	
Ribose	4.9	3.4	
Fucose	5.3	3.2	
Rhamnose	6.8	3.7	
Arabinose	7.2	4.2	
Xylose	7.8	4.1	
Glucose	14.5	6.8	3.6
Galactose	15.5	7.5	3.8
Mannose	19.0	8.2	3.8
N-Acetylglucosamine		10.2	4.6
N-Acetylgalactosamine		10.2	4.8
N-Acetylmannosamine		14.9	6.2
Maltose			7.4
Cellobiose			8.6
Gentiobiose			11.0
Lactose			12.2

Ref. Takeda M, Maeda M and Tsuji A (1982) J. Chromatogr. 244:347

HPLC 3

Sample	Raw materials
Apparatus	Shimadzu M LC-6A HPLC system and Waters R 401 differential refractometer detector
Column	α-Cyclodextrin; 5 μm, 250 × 4.6 mm ID
Mobile phase	Acetonitrile/water (78:22)
Flow-rate	1.5 ml min^{-1}
Temperature	22 °C
Injection size	40 μg
Detection	Refractive index
Recording	Shimadzu C-R3A Chromatopac
Retention times, min	(estimated from drawing)

Phenyl-β-D-glucopyranoside	2.5	Glucose	5.7
1-D-deoxy-1-nitro-D-sorbitol	3.0	Galactitol	6.0
D-Erythrose	3.2	Sucrose	7.5
2-Deoxy D ribose	3.6	Collobiose	7.9
Ribose	4.0	Lactose	8.9
Lyxose	4.1	Melibiose	9.7
Tagatose	4.5	Malezitose	11.0
Xylose	5.2	Raffinose	12.8
		Stachyose	21.9

Ref. Armstrong DW and Jin HL (1989) J. Chromatogr. 462:219

HPLC 4

Sample	Rat gastric mucous
Extraction	Trifluoroacetic acid
Apparatus	Varian M 5000 HPLC system with UV-100 detector
Column	Reversed-phase; Hibar 5 μm, C_8, 250 × 4.6 mm ID
Mobile phase	A. Acetonitrile B. 0.5% Aqueous acetic acid Gradient: B 95% to 75% in 35 min; hold 5 min; B 75% to 20% in 10 min; hold 5 min; B 20% to 95% in 15 min
Flow-rate	0.8 ml min^{-1}
Temperature	Ambient
Derivatization	Pre-column; with 2.5% dansyl hydrazine
Injection size	20 μl containing 0.1 μg ml^{-1}
Detection	UV, λ = 254 nm; 0.1 AUFS
Retention times, min	(estimated from drawing) Galactose 35.0 Xylose 40.0 Glucose 36.0 Ribose 41.0 Mannose 38.0 Fucose 42.5
Ref.	Corti F, Luzzani F and Ventura P (1988) J. Pharm. Biomed. Anal. 6:1049

HPLC 5

Sample	Fruit juice, wine and alcohol-free beer
Column	Reversed-phase; Biometra-RPE, 125 × 4 mm ID (EP 18)
Pre-column	Biosensor S (EP 45, EP 35) containing alcohol oxidase and glucose oxidase (Biometra)
Mobile phase	0.1 M Sodium phosphate buffer, pH 7.5
Flow-rate	0.6 ml min^{-1}
Injection size	20 μl
Detection	EC, + 500 mV
Retention times, min	(estimated from drawing) Glucose 2.0 Ethanol 3.5
Ref.	Biometra HPLC-Gesamtkatalog (1989) Biometra, Goettingen, p 54

---

62 Diazepam

CAS-No	439-14-5
Chemical name	7-Chloro-1,3-dihydro-1-methyl-5-phenyl-2*H*-1,4-benzodiazepin-2-one

Empirical formula	$C_{16}H_{13}ClN_2O$
Molecular mass	284.8
Proprietary names	Avex; Sedaril; Valium
Clinical use	Tranquillizer
Solubility	Water 1 in 333; alcohol 1 in 25; chloroform 1 in 2; ether 1 in 39
Light absorption	λ_{max} = 242 nm in aqueous acid

HPLC 1

Sample	Raw materials and their metabolites
Apparatus	Waters M 6000A LC pump, Perkin-Elmer LC 75 UV detector and Rheodyne 7120 injector
Column	Reversed-phase; ODS-Hypersil 5 µm, 200 × 5 mm ID
Mobile phase	Methanol/0.1 M phosphate buffer, pH 7.67/ water (70:20:10, v/v)
Flow-rate	1.5 ml min^{-1}
Injection size	20 µl
Detection	UV, λ = 240 nm
Retention times, min	(estimated from drawing)

Diazepem	4.2	Desoxychlordiazepoxide	7.0
Ketazolam	4.8	Prazepam	8.0
Flurazepam	6.0	Medazepam	11.5

Ref. Gill R, Law B and Gibbs JP (1986) J. Chromatogr. 356:37

HPLC 2

Sample	Body fluids
Extraction	Dichloromethane/pentane (1:1) under alkaline condition
Apparatus	Waters M 6000A LC pump, 440 UV detector and U6K injector
Column	Reversed-phase; µ-Bondapak 10 µm, C_{18}, 300 × 3.9 mm ID
Pre-column	ODS 37–53 µm, C_{18}, 70 × 2.1 mm ID
Mobile phase	Methanol/0.025 M sodium and potassium phosphate buffer, pH 6 (55:45, v/v)
Flow-rate	2 ml min^{-1}
Temperature	Ambient
Internal standard	Carbamazepine
Injection size	2.5 ng
Detection	UV, λ = 254 nm
Recording	Hewlett-Packard 3390A integrator
Retention times, min	

Internal standard	5.2	Desmethyldiazepam	11.0
Oxazepam	6.8	Diazepam	13.5
Temazepam	8.4		

Ref.	Klockowski PM and Levy G (1987) J. Chromatogr. Biomed. Appl. 422:334

HPLC 3

Sample	Dosage forms
Apparatus	Beckmann M 114 LC pump with 165 UV detector
Column	Reversed-phase; Altech RPP 6.5 μm, C_8, 250 × 4.6 mm ID
Mobile phase	Water/methanol/acetonitrile (10:9:1)
Flow-rate	1.2 ml min^{-1}
Temperature	Ambient
Internal standard	N-Desmethyldiazepam
Injection size	10 μl
Detection	UV, λ = 225 nm; 0.1 AUFS
Retention times, min	Oxazepam 7.0 N-Desmethyldiazepam 8.0–10.0 Diazepam 9.5 Degradation product 13.5–14.2
Ref.	Abdel-Hamid ME and Abuirjeie MA (1988) Analyst 113:1443

HPLC 4

Sample	Blood plasma and urine of humans and cats
Extraction	Benzene/methylene chloride (9:1), at pH 9.0
Apparatus	Waters ALC/GPC M 204/6000A HPLC with U6K injector and 440 UV detector
Column	Reversed-phase; Bondapak 10 μm, C_{18}, 300 × 3.95 mm ID
Mobile phase	Methanol 550 ml diluted to 1000 ml with water
Flow-rate	2.0 ml min^{-1}
Injection size	25 μl containing 25–500 ng
Detection	UV, λ = 254 nm; 0.02 AUFS
Capacity factor, *k'*	Oxazepam 7.2 Nordiazepam 11.0 Temazepam 8.3 Diazepam 13.8
Ref.	Cotler S, Puglisi CV and Gustafson JH (1981) J. Chromatogr. Biomed. Appl. 222:95

HPLC 5

Sample	Pharmaceutical preparations
Apparatus	Waters M 6000 HPLC, Rheodyne M 7125 injector and EG & G electrochemical detector and Perkin-Elmer LC 55 UV detector
Column	Reversed-phase; Nucleosil C_{18}, 5 μm, 250 × 4 mm ID
Mobile phase	500 ml Methanol + 60 ml ethyl acetate + 100 ml 1 M glacial acetic acid + 75 ml 1 M NaOH + 50 ml 1 M sodium nitrate, pH 6.0 with 1 M NaOH and make 1000 ml with water

Flow-rate	1.5 ml min^{-1}
Internal standard	Nitroglycerin, 0.2 mg ml^{-1}
Injection size	20 μl, 0.5 mg ml^{-1}
Detection	EC, − 1.15 V (− 0.4 V, − 0.75 V); UV, λ = 254 nm
Retention times, min	(estimated from drawing) Internal standard 8.0 Isosorbide 6.0 Diazepam 12.0
Ref.	Kral K and Kainz G (1983) Fresenius Z. Anal. Chem. 316:497

63 Dicloxacillin

CAS-No	3116-76-5; 13412-64-1 (sodium salt, monohydrate)		
Chemical name	3-(2,6-Dichlorophenyl)-5-methyl-4-isoxazolyl-penicillin		
Empirical formula	Dicloxacillin	Sodium Salt	
	$C_{19}H_{17}Cl_2N_3O_5S$	$C_{19}H_{16}Cl_2N_3NaO_5S,H_2O$	
Molecular mass	470.3	510.3	
Proprietary names	Diclocil; Diflor; Pathocil		
Clinical use	Antibiotic		
Solubility,	Water	methanol	ether
Sodium salt	freely soluble	soluble	insoluble

HPLC 1

Sample	Canine and equine sera
Extraction	Ethanol/acetonitrile/water (40:10:50)
Apparatus	Waters M 590 LC pump and 481 UV detector
Column	Phenyl phase; Sperisorb-5 MPLC, 220 × 4.6 mm ID
Mobile phase	10 ml Dodecanesulfonate (0.005 M) and water to 1 liter
Flow-rate	1.2–1.5 ml min^{-1}
Temperature	50 °C
Injection size	10–50 μl
Detection	UV, λ = 230 nm
Recording	Waters M 730 data module
Retention times, min	(estimated from drawing) Oxacillin 7.0 Cloxacillin 10.3 Dicloxacellin 15.9
Ref.	Tyczkowska K and Aronson AL (1988) J. Assoc. Off. Anal. Chem. 71:773

HPLC 2

Sample	Blood plasma
Extraction	Acetonitrile/water (35:65, v/v) using Bond-Elut C_{18} extraction column after deproteinization with acetonitrile
Apparatus	LKB M 2150 HPLC, Shimadzu SPD-6A variable-wavelength detector and Waters M 712 WISP autoinjector
Column	Reversed-phase; Hypersil ODS 5 μm, 100 × 2 mm ID
Mobile phase	Acetonitrile/water (40:60, v/v) containing 10 mM Na_2HPO_4, pH 2.0 with *o*-H_3PO_4
Flow-rate	0.5 ml min^{-1}
Injection size	20–25 μl, 10 μg ml^{-1}
Detection	UV, λ = 220 nm; 0.001 AU
Retention times, min	(estimated from drawing) Flucloxacillin 2.7; Cloxacillin 2.3; Dicloxacillin 3.5
Ref.	Hung CT, Lim JKC et al (1988) J. Chromatogr. Biomed. Appl. 425:331

HPLC 3

Sample	Raw materials and rat blood plasma
Apparatus	Lewa FLM M 50 HPLC pump with pulse-dampening system and Perkin-Elmer LC 55 UV detector
Column	A. Reversed-phase; Lichrosorb C_8, 10 μm, 250 × 3 mm ID B. Cyano phase; Nucleosil CN 5 μm, 250 × 3 mm ID
Mobile phase	For A: 0.05 M Phosphate buffer, pH 7.0/ methanol (5:1, v/v) For B: As for A but (7:3.6, v/v)
Flow-rate	3.2 ml min^{-1}
Temperature	Ambient
Injection size	10 μl, 1 mg ml^{-1}
Detection	UV, λ = 220 nm
Recording	Hewlett-Packard 3352B computer-system
Retention times, min	(estimated from drawing)

Column	A.	B.
Ampicillin	1.39	2.69
Epicillin		3.90
Oxacillin	6.16	6.29
Cloxacillin	8.95	9.00
Flucloxacillin	10.14	9.00
Dicloxacillin	16.90	11.50

Ref. Hartmann V and Roediger M (1976) Chromatographia 9:266

HPLC 4

Sample	Raw materials
Apparatus	Varian M 5000 LC, UV-50 detector and Valco automatic loop injector
Column	Reversed-phase; Micropak SPC-18, 4 μm, 150 × 4.6 mm ID
Pre-column	Special
Mobile phase	Gradient: 0.01 M ortho-Phosphoric acid/ acetonitrile (85:15 to 40:60 in 20 min)
Flow-rate	1 ml min^{-1}
Injection size	200 μl containing 2 μg
Detection	UV, λ = 220 nm; 0.2 AUFS
Retention times, min	(estimated from drawing)

Carbenicillin	11.5	Penicilin V	15.0
Methicillin	12.5	Cloxacillin	17.8
Penicillin G	13.6	Nafcillin	18.9
		Dicloxacillin	20.4

Ref.	Moats WA (1986) J. Chromatogr. 366:69

HPLC 5

Sample	Penicillins
Apparatus	HPLC low-pressure ternary gradient system
Column	Reversed-phase; Superspher 100, C_{18}, 125 × 4 mm ID
Mobile phase	Acetonitrile/1% triethylamine acetate, pH 4.0 (30:70, v/v)
Flow-rate	1.5 ml min^{-1}; 3000 psi
Injection size	10 μl containing 10 μg
Detection	UV, λ = 220 nm; 0.16 AUFS
Retention times, min	Ampicilin, oxacillin, colxacilin, and dicloxacillin, in ascending order (from drawing)
Ref.	Merck Application Manual on HPLC, TLC and HPTLC (1986). E. Merck, Darmstadt, p 80

64 Dicoumarol

CAS-No	66-76-2
Chemical name	3,3′ Methylenebis(4-hydroxycoumarin)
Empirical formula	$C_{19}H_{12}O_6$
Molecular mass	336.3
Proprietary names	Apekumarol; Dicumol; Dufalone
Clinical use	Anticoagulant

Solubility	Water and alcohol insoluble	chloroform slightly sol	aq NaOH freely soluble
Light absorption	λ_{max} = 314 nm in alkali hydroxides		

HPLC 1

Sample	Tablets
Apparatus	Du Pont LC-8800 and ultraviolet detector
Column	Reversed-phase; μ-Bondapak 10 μm, C_{18}, 300 × 3.9 mm ID
Mobile phase	Tetrahydrofuran/methanol/water/glacial acetic acid (35:10:65:0.1, v/v)
Flow-rate	1.5 ml min^{-1}
Temperature	Ambient
Injection size	20 μl, 0.05 mg ml^{-1}
Detection	UV, λ = 311 nm; 16 AUFS
Ref.	Moore ES (1987) J. Assoc. Off. Anal. Chem. 70:834

HPLC 2

Sample	Tablets			
Apparatus	Du Pont LC-8000, ultraviolet detector and M 7125 Rheodyne 20 μl sample loop			
Column	Reversed-phase; μ-Bondapak 10 μm, C_{18}, 300 × 3.9 mm ID			
Mobile phase	Tetrahydrofuran/methanol/water/glacial acetic acid (35:10:65:0.1, v/v)			
Flow-rate	1.5 ml min^{-1}			
Injection size	20 μl, 0.05 mg ml^{-1}			
Detection	UV, λ = 311 nm; 16 AUFS			
Retention times, min	Ethyl biscoumacetate	2.0	Acenocoumarol	14.5
	Dicoumarol	5.0	Phenprocoumon	17.0
	Warfarin sodium	9.0		
Ref.	Moore E and Lau-cam C (1986) J. Assoc. Off. Anal. Chem. 69:629			

65 Digitoxin

CAS-No	71-63-6
Chemically	Extract from the leaves of suitable Digitalis species
Empirical formula	$C_{14}H_{64}O_{13}$
Molecular mass	764.9
Proprietary names	Crystodigin; Digimerck; Purodigin
Clinical use	Cardiac Glycoside

Solublity	Water	alcohol	chloroform	methanol and ether
	insoluble	1 in 150	1 in 40	slightly soluble
Light absorption	λ_{max} = 216 nm in methanol			

HPLC 1

Sample	Urine
Extraction	Methylene chloride at pH 8.5
Apparatus	Spectra-Physics 8700 LC, Kratos ultraviolet detector and gradient controller
Column	Reversed-phase; Nucleosil 5 μm, C_{18}, 250 × 4.0 mm ID
Mobile phase	A. Methanol B. Water containing PIC B7 (Waters) Isocratic: A/B (68:32) Gradient: 66% A to 68% A in 5 min and to 75% A in 12 min and hold 8 min
Flow-rate	0.8 ml min^{-1}; 1668 psi
Injection size	100 μl containing 0.5 μg
Detection	UV, λ = 222 nm
Recording	Shimadzu C-R3A integrator

Retention times, min (estimated from drawing)

	I	II.
Monodigitoxide	5.0	5.0
Digoxin	8.0	8.0
Digitoxigenin	9.0	10.0
Bisdigitoxoside-digitoxigenin	16.0	18.0
Digitoxin	25.0	22.0

Ref. Santos SRCJ, Kirch W and Ohnhaus EE (1987) J. Chromatogr. Biomed. Appl. 419:155

HPLC 2

Sample	Digoxin, digitoxin and their cardioactive metabolites
Apparatus	Perkin-Elmer Series LC-3B pump, LC 85B UV/VIS detector and Rheodyne 7105 injector
Column	Reversed-phase; Nucleosil 10 μm, C_{18}, 250 × 4.0 mm ID × 2
Mobile phase	Acetonitrile/methanol/water (30:30:40)
Flow rate	1 ml min^{-1}
Injection size	40 μl containing 1 μg
Detection	UV, λ = 220 nm

Retention times, min

Digoxigenin	2.6	Digitoxigenin	6.5
Digoxigenin monodigitoxoside	2.8	Digitoxigenin monodigitoxoside	8.6
Digoxigenin bis-digitoxoside	3.4	Digitoxigenin bisdigitoxoside	12.1
Digoxin	4.2	Digitoxin	18.6

Ref.	Plum J and Daldrup T (1986) J. Chromatogr. Biomed. Appl. 377:221

HPLC 3

Sample	Digoxin, digitoxin and their cardioactive metabolites
Column	Reversed-phase; Nucleosil C_{18}, 10 µm 250 × 4 mm ID × 2
Mobile phase	A. Acetonitrile B. Acetonitrile/water (20:80, w/w) Gradient: A 0.1% for 3 min; A 0.1% to 15% in 1 min; A 15% to 20% in 12 min; A 20% to 40% in 12 min; A 40% to 0.1% in 10 min (percentages, v/v)
Flow-rate	1 ml min^{-1}; 1160 psi
Injection size	40 µl
Detection	UV, λ = 220 nm
Retention times, min	Digoxigenin 6.4 Digoxigenin monodigitoxoside 8.3 Digoxigenin bisdigitoxoside 10.5 Digoxin 11.5 β-Methyldigoxin 15.7 β-Acetlydigoxin 17.3 Digitoxigenin 18.9 Digitoxigenin monodigitoxoside 20.1 Digitoxigenin bisdigitoxoside 22.5 Digitoxin 24.1
Ref.	Plum J and Daldrup T (1986) J. Chromatogr. Biomed. Appl. 377:221

HPLC 4

Sample	Test substances
Apparatus	HPLC low-pressure ternary gradient system
Column	Reversed-phase; Lichrosorb RP-8, 250 × 4 mm ID
Mobile phase	Water/acetonitrile (60:40, v/v)
Flow-rate	2.5 ml min^{-1}; 2500 psi
Injection size	20 µl
Detection	UV, λ = 220 nm; 0.08 AUFS
Retention times, min	(estimated from drawing) Esculin 1.5 g-Strophanthin 3.7 Digitoxin 7.0
Ref.	Merck Application Manual on HPLC, TLC and HPTLC (1986). E. Merck, Darmstadt, p 31

HPLC 5

Sample	Raw materials and dosage forms
Apparatus	Hewlett-Packard 1084-LC with UV detector
Column	Reversed-phase; Nucleosil C_{18}, 10 µm, 250 × 4 mm ID
Mobile phase	Acetonitrile/0.05 M ortho-phosphoric acid (50:50)
Flow-rate	3.0 ml min^{-1}
Temperature	40 °C
Internal standard	Gitoxin

Injection size	5–30 μl
Detection	UV, λ = 220 nm
Recording	Hewlett-Packard LC-Terminal 79850 B integrator
Retention times, min	Gitoxin 1.62
	Digitoxigenin 2.15
	Digitoxin 2.74
Ref.	Rehm KD and Steinigen M (1981) Pharm. Ztg. 126:99

66 Digoxin

CAS-No	20830-75-5
Chemically	Extract from the leaves of Digitalis lanata
Empirical formula	$C_{41}H_{64}O_{14}$
Molecular mass	780.9
Proprietary names	Allocor; Cardiox; Novodigal
Clinical use	Cardiac Glycoside
Solubility	Water insoluble; alcohol 80% 1 in 122; pyridine 1 in 4; chloroform slightly sol
Light absorption	λ_{max} = 220 nm in ethanol

HPLC 1

Sample	Raw materials
Apparatus	Altex M LC-110A pump, loop injector and fluorescence and ultraviolet detectors
Column	Reversed-phase; ODS-5 μm, C_{18}, 250 × 4.6 mm ID
Mobile phase	Methanol/dichloromethane/water/isopropanol (41:3:50:6)
Flow-rate	0.4 ml min^{-1}
Derivatization	Post-column; with fluorigenic derivatization reagent
Injection size	20 μl
Detection	UV, λ = 254 nm; fluorescence, λ = 360/460 nm
Recording	Omniscribe B-5000 dual pen recorder
Retention times, min	(estimated from drawing)
	UV, λ = 254 nm; 0.02 AUFS
	Digoxigenin 8.9
	Digoxigenin monodigitoxoside 10.0
	Digoxigenin bisdigitoxoside 13.6
	Digoxin 19.7
	Follow fluorescence detection, λ = 360/460 nm
	Digoxigenin 24.7

	Digoxigenin monodigitoxoside	25.8
	Digoxigenin bisdigitoxoside	30.0
	Dihydrodigoxin	33.6
	Digoxin	36.4
Ref.	Desta B (1987) J. Chromatogr. Biomed. Appl. 421:381	

HPLC 2

Sample	Urine
Extraction	Dichloromethane on Extrelut column
Apparatus	Spectra-Physics LC-8700 and Pye-Unicam M 4020 UV detector
Column	Normal phase; Lichrosorb Si 60, 5 µm, 200 × 4 mm ID
Mobile phase	n-Hexane/dichloromethane/methanol (82.9:14.2:2.9)
Flow-rate	1.2 ml min^{-1}
Temperature	Ambient
Derivatization	A. Pre-column; with 4-nitrobenzyl chloride (4-NBCl) in dry pyridine B. Acid hydrolysis
Injection size	100 µl
Detection	UV, λ = 258 nm
Recording	Spectra-Physics 4270 integrator
Retention times, min	(estimated from drawing) derivatization

		A.	B.
	Digitoxigenin	8.00	9.50
	Digoxigenin	12.00	15.00
	(R)-Dihydrodigoxigenin	13.00	15.81
	(S)-Dihydrodigoxigenin	13.50	17.0

Ref.	Jakobsen P and Waldorff S (1986) J. Chromatogr. Biomed. Appl. 382:349

HPLC 3

Sample	Digoxin and digitoxin and their cardioactive metabolites
Apparatus	Perkin-Elmer Series 3B LC pump and LC-85B Variable-wavelength detector
Column	Reversed-phase; Nucleosil C_{18}, 10 µm, 250 × 4.0 mm ID × 2
Mobile phase	I. Acetonitrile/water (25:75) II. Acetonitrile/methanol/water (20:20:60)
Flow-rate	1 ml min^{-1}; 943–986 psi
Injection size	20 µl with 1 µg
Detection	UV, λ = 220 nm; 0.02 AU

Retention times, min	Mobile phase I. II. Digoxigenin 4.1 4.1 Digoxigenin monodigitoxoside 4.4 4.9 Digoxigenin bisdigitoxoside 6.3 7.7 Digoxin 8.8 12.0
Ref.	Plum J and Daldrup T (1986) J. Chromatogr. Biomed. Appl. 377:221

HPLC 4

Sample	Tablets
Apparatus	Familic-100 micro HPLC with Uvidec-100 UV detector (Japan Spectroscopic)
Column	Reversed-phase; ODS 5 μm (SC-01), 95 × 0.5 mm ID
Mobile phase	I. Acetonitrile/water (28:72) II. Acetonitrile/water (33:67) III. Methanol/water (1:1)
Flow-rate	8 μl min^{-1}
Internal standard	Dexamethasone
Injection size	0.2 μl
Detection	UV, λ = 220 nm
Retention times, min	(estimated from drawing)

Mobile phase	I.	II.	III.
Digoxigenin	5.5		5.5
Digoxigenin monodigitoxoside	6.0		6.0
Digoxigenin bisdigitoxoside	7.4		12.0
Digoxin	10.5	5.0	25.0
Dexamethasone (IS)	14.0		19.5
β-Methyldigoxin	25.0	11.0	
Dimethyldigoxin 1		22.5	
Dimethyldigoxin 2		27.0	

Ref. Fujii Y, Ikeda Y and Yamazaki M (1988) J. Chromatogr. 448:157

HPLC 5

Sample	Lanatoside C-group
Apparatus	HPLC system with UV and fluorescence detectors
Column	Reversed-phase; Nucleosil C_{18}, 10 μm, 250 × 4 mm ID
Mobile phase	Acetonitrile/dioxane (1:1) in water (40:60)
Flow-rate	0.4 ml min^{-1}
Injection size	20 μl
Detection	UV and fluorescence

Retention times, min	(estimated from drawing)		
	Detection	UV, λ = 220 nm	
	Digoxigenin	11.8	
	Desacetyl lanatoside C	14.0	
	Lanatoside C	21.0	
	Digoxin	24.0	
	Follow Fluorescence detection, λ = 350/485 nm		
	Digoxigenin	31.8	(11.8)
	Desacetyl lanatoside C	34.0	(14.0)
	Lanatoside C	41.4	(21.0)
	Digoxin	44.0	(24.0)
Ref.	Gfeller JC, Frey G and Frei RW (1977) J. Chromatogr. 142:271		

67 Diltiazem Hydrochloride

CAS-No	42399-41-7; 33286-22-5 (hydrochloride)		
Chemical name	(2*S*-cis)-3-(Acetyloxy)-5-[2-(dimethylamino)-ethyl]-2,3-dihydro-2-(4-methoxyphenyl)-1,5-benzothiazepin-4(5*H*)-one		
Empirical formula	Diltiazem	Hydrochloride	
	$C_{22}H_{26}N_2O_4S$	$C_{22}H_{26}N_2O_4S,HCl$	
Molecular mass	414.5	451.0	
Proprietary names	Herbesser; Masdil		
Clinical use	Vasodilator		
Solubility, Hydrochloride	Water, methanol and chloroform freely soluble		alcohol 100% slightly sol
Light absorption	λ_{max} = 237 nm in aqueous alkali		

HPLC 1

Sample	Blood plasma
Extraction	1% 2-Propanol in n-hexane under mild alkaline condition
Apparatus	Hewlett-Packard 1090-LC pump, variable-wavelength photodiode array and autosampler
Column	Cyanopropylsilane, 150 × 4.6 mm ID
Mobile phase	Acetonitrile/0.05 M sodium acetate, pH 4.0 (45:55)
Flow-rate	1.5 ml min^{-1}
Temperature	40 °C

Internal standard	Diazepam
Injection size	20 μl
Detection	UV, λ = 239 nm
Recording	Hewlett-Packard 3392A integrator
Retention times, min	Procainamide 2.29; Labetalol 2.34; Diazepam (IS) 2.50; Warfarin 2.54; Metoprolol 2.55; Pindolol 2.61; Isosorbide dinitrate 2.64; Dipyridamole 3.00; Disopyramide 3.39; Desacetyldiltiazem 3.82; Propranolol 3.82; Quinidine 3.87; Chlorothiazide 4.25; Diltiazem 4.90; Captopril 7.17
Ref.	Johnson SM and Wahba Khalil SK (1987) J. Liq. Chromatogr. 10:673

HPLC 2

Sample	Human blood plasma
Extraction	On-line clean-up method over pre-column
Apparatus	Kontron M 414 T LC pump and Pye-Unicam 4025 spectrophotometric LC detector
Column	Supelcosil LC-18-DB, 5 μm, 250 × 4.6 mm ID
Pre-column	Supelcosil LC-18-DB, 5 μm, 20 × 4.6 mm ID
Mobile phase	Acetonitrile/0.05 M phosphate buffer, pH 2.5 (65:35) with 0.25% (v/v) triethylamine
Flow-rate	1 ml min^{-1}
Temperature	Ambient
Internal standard	Benzothiazepine derivate
Injection size	200 μl; 10–1000 ng ml^{-1}
Detection	UV, λ = 238 nm
Recording	Spectra-Physics 4270 integrator
Retention times, min	Diltiazem 10.8; Internal standard 12.1
Ref.	Ascalone V and Dal Bo L (1987) J. Chromatogr. Biomed. Appl. 423:239

HPLC 3

Sample	Raw materials
Apparatus	Varian M 2010 LC, 2050 ultraviolet detector and Rheodyne M 7125 injector
Column	Reversed-phase; Micropak MCH-5, 150 × 4 mm ID
Mobile phase	Acetonitrile/water (48:52), pH 3.3 with H_3PO_4 and containing 0.01 M sodium n-octanesulfonate
Flow-rate	1 ml min^{-1}
Temperature	Ambient
Internal standard	Clonazepam
Injection size	40 μl, 1–10 μg ml^{-1}

Detection	UV, λ = 239 nm; 0.02 AUFS
Retention times, min	Internal standard 3.0 Diltiazem 9.0
Ref.	Abdel-Hamid ME, Suleiman MS et al (1988) Anal. Lett. 21:2263

HPLC 4

Sample	Blood plasma
Extraction	Ether/hexane (50:50)
Apparatus	Waters M 45 pump with WISP autosampler and Shimadzu SPD-6A UV detector
Column	Normal phase; Spherisorb S5W, 250 × 4.6 mm ID
Mobile phase	0.3 ml Ammonia, 30 ml methanol, 30 ml dichloromethane and 360 ml hexane
Internal standard	Carboxylic acid derivate
Flow-rate	3.0 ml min^{-1}; 1400 psi
Injection size	50 µl, 0–400 ng ml^{-1}
Detection	UV, λ = 240 nm
Recording	Shimadzu C-R3A computing integrator
Retention times, min	Internal standard 2.9 Diltiazem 3.7 Desacetyldiltiazem 4.4
Ref.	Kinney CD and Kelly JG (1986) J. Chromatogr. Biomed. Appl. 382:377

HPLC 5

Sample	Serum or plasma
Extraction	Methyl tert.-butyl ether
Apparatus	Applied Chromatography Systems M 750/04 reciprocating HPLC pump, Rheodyne M 7125 injection loop and Jasco Uvidec-100-V UV detector
Column	Normal phase; Spherisorb S5W 5 µm, 250 × 5 mm ID
Mobile phase	10 mM Ammonium perchlorate in methanol, pH 7.6 with 1 M methanolic sodium hydroxide
Flow-rate	2.0 ml min^{-1}
Temperature	Ambient
Internal standard	Fenethazine hydrochloride
Injection size	100 µl, 10–400 ng ml^{-1}
Detection	UV, λ = 240 nm
Recording	Hewlett-Packard 3392A integrator
Retention times, min	(estimated from drawing) Desacetyldiltiazem 3.8 Diltiazem 3.0 Internal standard 5.8
Ref.	Bhamra RK, Ward AE and Holt DW (1987) Biomed. Chromatogr. 2:180

68 Dimetridazole

CAS-No	551-92-8; 25332-20-1 (hydrochloride)			
Chemical name	1,2-Dimethyl-5-nitroimidazole			
Empirical formula	Dimetridazole	Hydrochloride		
	$C_5H_7N_3O_2$	$C_5H_7N_3O_2, HCl$		
Molecular mass	141.1	177.6		
Proprietary names	Emtryl; Emtrylvet			
Clinical use	Antiprotozoal (vet)			
Solubility,	Water	ethanol	chloroform	aq acid
Dimetridazole	slightly sol	1 in 30	1 in 5	soluble
Hydrochloride	freely sol	freely sol		
Light absorption	λ_{max} = 277 nm in aqueous acid			

HPLC 1

Sample	Swine feed
Apparatus	Hewlett-Packard 1048B LC with autosampler and UV detector
Column	Reversed-phase; Resolve TM C_{18}, 5 µm, 150 × 3.9 mm ID
Mobile phase	I. 5% Methanol in a mixture of acetonitrile/ 0.05 M sodium acetate (1:9), pH 6.0 with acetic acid; isocratic II. A. Acetonitrile/0.05 M sodium acetate (1:9), pH 6.0 with acetic acid II. B. Methanol Gradient: B 5% to 55% in 10 min; B 55% to 5% in 4 min
Flow-rate	1.5 ml min^{-1}; 2538 psi
Temperature	Ambient
Injection size	20 µl containing 22.4 ng
Detection	UV, λ = 320 and 550 nm for reference; 0.08 AUFS
Retention times, min	(estimated from drawing)

Mobile phase	I.	II.
Dimetridazole	4.35	3.73
Ipronidazole		7.02

Ref. Royal JE, Munns RK et al (1987) J. Assoc. Off. Anal. Chem. 70:626

HPLC 2

Sample	Chicken tissue
Extraction	Dichloromethane under acidic and alkaline conditions

Apparatus	Perkin-Elmer Series 2/2-LC pump and LC-75 UV detector
Column	Reversed-phase; Spherisorb 5 ODS, 250 × 4.5 mm ID
Pre-column	Brownlee RP-18
Mobile phase	1.36 g Potassium dihydrogen ortho-phosphate in 1 l water and mix with methanol in the ratio (7:3)
Flow-rate	1 ml min^{-1}
Temperature	Ambient
Internal standard	Metronidazole
Injection size	15 μl, 0.5–5.0 μg ml^{-1}
Detection	UV, λ = 315 nm
Retention times, min	Metronidazole 5.8 Dimetridazole 12.5
Ref.	Anonym (1985): Analytical Methods Committee, Royal Society of Chemistry, London Analyst 110:1391

HPLC 3

Sample	Poultry tissue
Extraction	Dichloromethane under acidic and alkaline conditions
Apparatus	Perkin-Elmer Series 2/2-LC pump with LC-75 UV detector
Column	Reversed-phase; Spherisorb 5 ODS, 250 × 4.5 mm ID
Pre-column	Brownless RP-18
Mobile phase	0.01 M Potassium dihydrogen ortho-phosphate/ methanol (7:3)
Flow-rate	1.0 ml min^{-1}
Temperature	22 °C
Internal standard	Metronidazole
Injection size	15 μl
Detection	UV, λ = 315 nm; 0.02 AUFS
Recording	Perkin-Elmer Sigma 15 data processor
Retention times, min	(estimated from drawing) Metronidazole 6.40 Dimetridazole 13.40
Ref.	Hobson-Frohock A and Reader JA (1983) Analyst 108:1091

HPLC 4

Sample	Pork tissue
Extraction	Blood: Deproteinization with acetonitrile Tissue: First ethyl alcohol, and then with hexane and mobile phase and chromatograph the lower phase

Apparatus	Kratos M 400 LC pumps, Spectra-Physics 8780 XR autosampler and Coulochem M 5100A electrochemical detector
Column	A. Reversed-phase; Spherisorb 10 C_{18}, 10 μm, 250 × 4.6 mm ID B. Reversed-phase; Ultrasphere ODS 5 μm, 150 × 4.6 mm ID
Mobile phase	Mixture of 3 ml triethylamine, 50 ml 1 M acetic and 800 ml water, pH 5.0 with diluted NH_4OH and add 150 ml acetonitrile
Flow-rate	1 ml min^{-1}
Internal standard	Nitrofurazone
Injection size	200 μl containing 1–10 ng
Detection	EC, − 1.20 V, − 0.9 V vs Ag/AgCl RE
Recording	Isaac M 42A integrator
Retention times, min	Major metabolite 18.1 Dimetridazole 29.3 Nitrofurazone 31.9
Ref.	Carignan G, Skakum W and Sved S (1988) J. Assoc. Off. Anal. Chem. 71:1141

69 Dopa (Levodopa)

CAS-No	59-92-7
Chemical name	Dihydroxyphenylalanine
Empirical formula	$C_9H_{11}NO_4$
Molecular mass	197.2
Proprietary names	Brocadopa; Dopasol; Larodopa
Clinical use	CNS-Neurotransmitter
Solubility	Water: 1 in 300; mineral acids and alkalis: soluble; alcohol: insoluble
Light absorption	λ_{max} = 280 nm in aqueous acid

HPLC 1

Sample	Enantiomers of norepinephrine, dopa and dopamine
Apparatus	Shimadzu M LC-3A pump with SGA-1A gradient programmer and RF-500 LC fluorescence detector
Column	Reversed-phase; Develosil ODS-5, 100 × 6.0 mm ID
Mobile phase	A. 50 mM Phosphate buffer, pH 6.5/methanol (55:45) B. 50 mM Phosphate buffer, pH 6.5/methanol (63:35)

Flow-rate	1 ml min^{-1}
Derivatization	Pre-column; with *o*-phthalaldehyde-N-acetyl-L-cysteine
Injection size	5 μl with 10 ng
Detection	Fluorescence, λ = 340/450 nm
Retention times, min	Mobile phase A. B. L-Dopa 9.0 8.0 D-Dopa 11.5 10.0
Ref.	Nimura N, Iwaki K and Kinoshita T (1987) J. Chromatogr. 402:387

HPLC 2

Sample	Blood plasma
Extraction	200 mM Aqueous acetic acid
Apparatus	Waters M 6000A HPLC, U6K injector and electrochemical detector
Column	Reversed-phase; Beckman CA-94710, 3 μm, ODS
Mobile phase	A. 6.8 g Sodium acetate, 1 g Waters PIC B 7 (low UV) heptanesulfonic acid and 100 mg EDTA in 1 liter water, pH 3.5 with H_3PO_4 B. As A plus 40 ml acetonitrile Gradient: B 0% to 100% in 6 min after sample injection
Flow-rate	1.5 ml min^{-1}
Internal standard	N-Methyldopamine
Injection size	90 μl containing 250 pg
Detection	EC, + 0.3, + 0.15 and − 0.35 V
Recording	Waters data module
Retention times, min	Dopa 3.90 Epinephrine 10.46 Dopac 5.45 Dopamine 19.63 Norepinephrine 6.16 N-Methyldopamine 21.43
Ref.	Goldstein DS, Stull R et al (1984) Clin. Chem. 30:815

HPLC 3

Sample	Catecholamines
Column	Reversed-phase; Lichrosorb C_{18}, 5 μm, 150 × 4.6 mm ID
Mobile phase	1 mM NaCl, 50 mM NaH_2PO_4, 0.1 M $NaClO_4$, 0.1% heptanesulfonic acid and 2% isopropyl alcohol in water, pH 4.2
Flow-rate	1 ml min^{-1}
Injection size	1 ng
Detection	Coulometric, + 0.75 V
Retention times, min	(estimated from drawing) Adrenaline 6.2 Dopa 3.0 Octopamine 7.1 Noradrenaline 5.0 Metanephrine 11.6

Ref.	Meyer V (1984): Praxis der Hochleistungs-fluessigchromatographie, 3rd edn. Diesterweg Salle and Sauerlaender, Frankfurt am Main and Salzburg, p 152

70 Dopamine

CAS-No	51-61-6; 62-31-7 (hydrochloride)	
Chemical name	4-(2-Aminoethyl)pyrocatechol	
Empirical formula	Dopamine	Hydrochloride
	$C_8H_{11}NO_2$	$C_8H_{11}NO_2$, HCl
Molecular mass	153.2	189.6
Proprietary names	Intropin; Revivan; Tensophoril	
Clinical use	Sympathomimetic	
Solubility,	Water	alcohol
Hydrochloride	soluble	soluble
Light absorption	λ_{max} = 280 nm in aqueous acid	

HPLC 1

Sample	Raw materials
Apparatus	Waters LC with M 45 pump and Uvikon 720 LC spectrophotometer (Kontron)
Column	Reversed-phase; Lichrosorb 10 µm, C_{18}, 250 × 4.6 mm ID
Mobile phase	20 g Citric acid + 18.6 ml *o*-phosphoric acid + 850 ml water, pH 4.5 with ammonium hydroxide + 80 ml methanol and fill up to 1000 ml
Flow-rate	1 ml min^{-1}; 1000 psi
Injection size	20 µl, 200 µg ml^{-1}
Detection	UV, λ = 280 nm
Recording	Shimadzu printer-plotter C-RIA
Retention times, min	Dopamine 6.48 Salsolinol 9.35
Ref.	Thomas H, Stammel W and Brossi A (1983) Chromatogr. Sci. 21:481

HPLC 2

Sample	Raw materials
Apparatus	Waters LC with M 45 pump and Uvikon 720 LC spectrophotometer (Kontron)
Column	Cation-exchange; Nucleosil 10SA 10 µm, 250 × 4.6 mm ID

Mobile phase	Citric acid 7.68 g + *o*-phosphoric acid 6.66 ml + water 850 ml, pH 4.5 with ammonium hydroxide + methanol 80 ml and fill up to 1000 ml
Flow-rate	1 ml min^{-1}; 1100 psi
Injection size	20 µl, 250 µg ml^{-1}
Detection	UV, λ = 280 nm
Recording	Shimadzu printer-plotter C-RIA
Ref.	Thomas H, Stammel W and Brossi A (1983) Chromatogr. Sci 21:481

HPLC 3

Sample	Raw materials
Apparatus	Spectra-Physics LC-8770 and electrochemical detector and Valco injector
Column	Reversed-phase; Chromspher C_8, 5 µm, 250 × 4.6 mm ID loaded with tributyl-n-phosphate
Mobile phase	0.05 M Sodium citrate, 0.2 M sodium perchlorate and 1.5 mM Na_2EDTA, pH 4.9 and saturated with tributyl-n-phosphate
Flow-rate	0.5 ml min^{-1} (A); 0.4 ml min^{-1} (B)
Temperature	21.2 °C
Injection size	500 µl
Detection	EC, 800 mv

Retention times, min

Flow-rate	(A)	(B)
Dihydroxyphenylacetic acid	48.84	64.49
Dopamine	67.24	82.87

Ref.	Van der Krogt JA, Van Valkenburg CFM and Belfroid RDM (1988) J. Chromatogr. Biomed. Appl. 427:9

HPLC 4

Sample	Solid dosage forms
Apparatus	Du Pont M 880 LC with variable-wavelength UV detector and Rheodyne M 7125 injector
Column	Reversed-phase; µ-Bondapak C_{18}, 10 µm, 300 × 3.9 mm ID
Mobile phase	3% Acetic acid
Flow-rate	1.5 ml min^{-1}
Temperature	Ambient
Internal standard	Acetaminophen and methyldopa
Injection size	20 µl containing 0.100 µg
Detection	UV, λ = 280 nm; 0.02–0.08 AUFS

Retention times, min (estimated from drawing)

		Methoxytyrosine	5.8
Levodopa	3.0	Acetaminophen	10.0
Methyldopa	4.3	3-0-Methyl-carbidopa	13.0
Carbidopa	5.0		

Ref.	Ting S (1986) J. Assoc. Off. Anal. Chem. 69:169

HPLC 5

Sample	Raw materials
Apparatus	Altex M 100A LC pump with Bioanalytical Systems LC-4 amperometric detector and Rheodyne M 7125 injector
Column	Reversed-phase; Spherisorb I ODS 5 μm, 250 × 4.6 mm ID
Pre-column	Whatman, Co-Pell ODS 30–38 μm, 50 × 4.6 mm ID
Mobile phase	0.1 mol/l Sodium acetate trihydrate/methanol (85:15, v/v), pH 4.2 with glacial acetic acid
Flow-rate	1.2 ml min^{-1}
Injection size	50 μl
Detection	EC, GC electrode, + 0.6 V vs Ag/AgCl RE

Capacity factor, *k'*		
	Norepinephrine	0.35
	Tryptamine	0.56
	Epinephrine	0.60
	Normetanephrine	0.77
	Dopamine	0.78
	5-Hydroxytryphtophan	0.87
	Vanillylmandelic acid	1.00
	Metanephrine	1.07
	3,4-Dihydroxyphenylacetic acid	1.26
	5-Hydroxytryptamine	2.22
	5-Hydroxyindoleacetic acid	2.61
	5-Hydroxytryptophol	2.77
	Homovanillic acid	3.15
	5-Methoxytryptamine	8.86

Ref. Tagari PC, Boullin DJ and Davies CL (1984) Clin. Chem. 30:131

HPLC 6

Sample	Blood plasma
Extraction	n-Heptane/chloroform (75:25, v/v)
Apparatus	Merck-Hitachi M 655A LC pump, Waters WISP 710B autosampler and ESA Coulochem M 5100A electrochemical detector
Column	Normal phase; Silicagel 10 μm
Mobile phase	A: 0.03 M ortho-Phosphate in 0.035 M citrate buffer, pH 4.0 B: Buffer as A, pH 6.4 C: 85% Buffer as A, pH 6.4 and 15% acetonitrile D. Buffer as A, pH 8.0
Flow-rate	1 ml min^{-1}
Temperature	40 °C
Injection size	50–60 μl
Detection	EC, + 0.3 V

Retention times, min	Mobile phase	A.	B.	C.	D.
	Epinephrine	4.6	6.5	6.1	8.0
	Norepinephrine	4.9	6.7	7.2	7.2
	Metanephrine	5.0	7.7	5.4	7.6
	Normetanephrine	5.0	6.9	6.4	7.6
	Dopamine	5.5	8.5	8.4	10.3
	Epinine	5.5	10.0	7.4	15.5
Ref.	Meineke I, Stuewe E et al (1989) J. Chromatogr. Biomed. Appl. 493:287				

71 Doxorubicin (Adriamycin)

CAS-No	23214-92-8; 25316-40-9 (hydrochloride)	
Chemically	Antracycline antibiotic produced by Streptomyces peucetius var caesius	
Empirical formula	Doxorubicin	Hydrochloride
	$C_{27}H_{29}NO_{11}$	$C_{27}H_{29}NO_{11}$, HCl
Molecular mass	543.5	580.0
Proprietary names	Adriamycin; Adriblastin; Farmiblastina	
Clinical use	Antineoplastic	
Solubility, Hydrochloride	Water soluble; methanol soluble; ethanol 1 in 75; chloroform insoluble	
Light absorption	λ_{max} = 233 nm in methanol	

HPLC 1

Sample	Raw materials
Apparatus	Haskel Engineering M 26980-4 LC pump, Altex M 153 ultraviolet detector and Rheodyne M 70–10 injector
Column	Reversed-phase; Lichrosorb 5 µm, C_8, 150 × 4.6 mm ID
Mobile phase	Acetonitrile/0.025 M camphorsulfonic acid (50:50), pH 3.8 with sodium hydroxide
Flow-rate	1.1 ml min^{-1}; 1842 psi
Temperature	Ambient
Injection size	25 µl containing 0.29 µg
Detection	UV, λ = 254 nm; 0.08 AUFS
Retention times, min	(estimated from drawing) Daunomycin 7.0; Adriamycin 4.0; Carminomycin 9.2
Ref.	White ER and Zarembo JE (1981) Antibiotics 34:836

HPLC 2

Sample	Blood
Extraction	Eluent over AASP using ethyl C_2 cassettes
Apparatus	Shimadzu LC-3A pump, Shimadzu RF 530 fluorescence detector and Varian advanced automated sample processor (AASP)
Column	Phenyl phase; μ-Bondapak 10 μm, 300 × 3.9 mm ID
Pre-column	C_2-Ethyl cassettes for AASP
Mobile phase	Acetonitrile/0.1 M ammonium formate buffer, pH 4 (31:69)
Flow-rate	1.5 ml min^{-1}
Internal standard	Daunorubicin 200 ng ml^{-1}
Detection	Fluorescence, λ = 470/565 nm
Recording	Shimadzu C-R1B integrator
Retention times, min	(estimated from drawing) Doxorubicin 3.0 Doxorubicinol 2.0 Internal standard 6.0
Ref.	Dobbs NA and James CA (1987) J. Chromatogr. Biomed. Appl. 420:184

HPLC 3

Sample	Blood plasma and urine
Extraction	Chloroform/isopropyl alcohol (4:1, v/v) (blood)
Apparatus	Orlita LC pump, Waters M 710B WISP automatic injector and Perkin-Elmer M 3000 fluorescence detector
Column	Reversed-phase; CP Microsphere C_{18}, 3 μm, 250 × 4.6 mm ID
Mobile phase	I. Blood: 0.06 M KH_2PO_4 buffer, pH 4.0/acetonitrile (70:32.5, v/v) II. Urine: 0.02 M KH_2PO_4, buffer, pH 4.0/acetonitrile (50:20, v/v)
Flow-rate	0.7 ml min^{-1}
Internal standard	Daunorubicin
Injection size	30 μl
Detection	Fluorescence, λ = 470/580 nm
Retention times, min	(estimated from drawing; mobile phase I.) 1,2,Doxorubicinol 6.5 4'-Epidoxorubicinol 7.5 Doxorubicinol aglycone 9.0 Doxorubicin 10.0, 23.0 4'- Epidoxorubicin 12.0 7-Deoxydoxorubicinol 15.0 Doxorubicinone 16.0 7-Deoxydoxorubicinone 29.0
Ref.	Weenen H, Osterop APRM et al (1986) J. Pharm. Sci. 75:1201

HPLC 4

Sample	Rat serum and tissues
Extraction	Serum: With acetonitrile Tissue: With normal saline and then isopropanol/acetonitrile (90:10, v/v)
Apparatus	Waters M 6000A pump and Rheodyne M 7125 injector, Schoeffel FS 970 fluorimeter
Column	Reversed-phase; Hypersil-ODS, 100 × 4.6 mm ID
Mobile phase	Acetonitrile/aqueous 80 mM sodium lauryl sulfate and 30 mM KH_2PO_4 (1:1, v/v), pH 2.0 with phosphoric acid
Flow-rate	2 ml min^{-1}
Internal standard	Daunomycin HCl
Injection size	20 and 100 µl, 10 ng ml^{-1}
Detection	Fluorescence, λ = 470/580 nm
Retention times, min	(estimated from drawing) Doxorubicinol 1.5 Daunomycin 3.0 Doxorubicin HCl 2.0
Ref.	Gallo JM, Hung CT and Perrier DG (1986) J. Pharm. Biomed. Anal. 4:483

HPLC 5

Sample	Serum and tissues
Extraction	Chloroform/methanol (4:1)
Apparatus	Hitachi M 635A HPLC with M 650-10S fluorescence detector
Column	Normal phase; Zorbax Sil 5 µm, 150 × 4.6 mm ID
Mobile phase	3.8% Sodium acetate, pH 4.5 in isopropanol
Flow-rate	1.0 ml min^{-1}
Internal standard	Daunomycin
Injection size	5 µl
Detection	Fluorescence; λ = 470/585 nm
Retention times, min	(estimated from drawing) Adriamycinone 4.0 Adriamycin 10.5 Daunomycin 8.0 Adriamycinol 13.5
Ref.	Shinozawa S, Mimaki Y et al (1980) J. Chromatogr. 196:463

HPLC 6

Sample	Blood plasma
Extraction	Chloroform/1-heptanol (8:2) on extraction column, pH 8.6 and then with phosphate buffer, pH 2.2
Apparatus	LDC M 711-LC pump, Spectromonitor-II UV and Schoeffel FS 970 fluorimetric detectors
Column	Reversed-phase; Lichrosorb C_2, 5 µm, 150 × 4 mm ID
Mobile phase	Acetonitrile/water/0.1 M phosphoric acid (20:70:10)

Flow-rate	1 ml min^{-1}
Temperature	25 °C
Injection size	100 μl, 10–300 ng ml^{-1}
Detection	Fluorescence, λ = 435/550 nm; 0.0004 AU
Retention times, min	(estimated from drawing) Adriamycinol 3.4 Adriamycin 5.7
Ref.	Eksborg S, Ehrsson H and Andersson I (1979) J. Chromatogr. Biomed. Appl. 164:479

HPLC 7

Sample	Anthracyclines in acidic solution
Apparatus	Waters M 45 LC pump, 440 UV detector and Perkin-Elmer M 650 fluorescence detector
Column	Reversed-phase; Lichrosorb C_8, 10 μm, 300 × 3.9 mm ID
Mobile phase	Acetonitrile/water (50:50, v/v), pH 2.0 with perchloric acid
Flow-rate	1.5 ml min^{-1}
Temperature	25 °C
Injection size	20 μl
Detection	UV, λ = 254 nm; 0.005 AUFS Fluorescence, λ = 465/550 nm
Retention times, min	(estimated from drawing) Doxorubicin 2.8 Doxorubicinone 5.1
Ref.	Beijnen JH, Wiese G and Underberg WJM (1985) Pharm. Weekbl. Sci. 7:109

HPLC 8

Sample	Blood plasma
Extraction	Chloroform/1-heptanol (9:1), at pH 8.4 and re-extraction with 0.3 M H_3PO_4
Apparatus	Spectra-Physics M 3500 HPLC, Rheodyne M 7120 injector and UV and fluorescence detectors
Column	Phenyl phase; μ-Bondapak 10 μm, 300 × 2 mm ID
Mobile phase	Acetonitrile/0.05 M KH_2PO_4 (35:65)
Flow-rate	0.4 ml min^{-1}
Temperature	23 °C + 0.5 °C
Injection size	170 μl, 40 ng ml^{-1}
Detection	UV, λ = 254 nm; fluorescence, λ = 254/550 nm
Recording	Spectra-Physics 4050 terminal printer and plotter
Retention times, min	(estimated from drawing) Doxorubicin 14.00 13-Dihydro-4-demethoxydaunorubicin 17.00 4-Demethoxydaunorubicin 24.00

Ref.	Moro E, Bellotti V et al (1983) J. Chromatogr. Biomed. Appl. 274:281

72 Doxycycline

CAS-No	564-25-0; 17086-28-1 (monohydrate); 24390-14-5 (hydrochloride, hemiethanolate hemihydrate)			
Chemical name	6-Deoxy-5β-hydroxytetracycline monohydrate			
Empirical formula	Dexycyclin monohydrate	Hydrochloride		
	$C_{22}H_{24}N_2O_8$, H_2O	$C_{22}H_{24}N_2O_8$, HCl, $1/2C_2H_5OH$, $1/2H_2O$		
Molecular mass	462.5	512.9		
Proprietary names	Doxiclin; Isodox; Vibramycine			
Clinical use	Antibiotic			
Solubility,	Water	alcohol	ether	aq acids
Doxycyclin	very slightly soluble	sparingly soluble	insoluble	freely soluble
Hydrochloride	1 in 3	1 in 60	insoluble	
Light absorption	$\lambda_{max} = 269$ nm in aqueous acid			

HPLC 1

Sample	Serum and urine
Extraction	0.01 M H_3PO_4 in acetonitrile using Bond-Elut C_{18} cartridge (serum)
Apparatus	Beckman M 114 pump, Kontron 660 MSI autosampler and Kratos spectroflow M 773 UV detector
Column	Reversed-phase; Novapak C_{18}, 4 μm, 150 × 3.9 mm ID
Pre-column	Reversed-phase; Lichrosorb C_{18}
Mobile phase	Acetonitrile/acetic acid/0.1 M KH_2PO_4 (75:150:125) and (65:150:125) for serum and urine, respectively
Flow-rate	1.0 ml min^{-1}
Temperature	Ambient
Injection size	100 μl, 0–500 ng ml^{-1}
Detection	UV, $\lambda = 340$ nm; 0.01 AUFS
Retention times, min	Doxycycline 3–3.8 (estimated from drawing)
Ref.	Sheridan ME and Clarke GS (1988) J. Chromatogr. Biomed. Appl. 434:253

HPLC 2

Sample	Human blood plasma and urine
Extraction	0.1 M Methanolic HCl using extraction column

Apparatus	Merck-Hitachi M 655A LC, Waters M 710B WISP autosampler and Kratos SP-757 variable wavelength detector
Column	Serum: Reversed-phase; Nucleosil 100 C_{18}, 5 μm, 125 × 4.0 mm ID Urine: Reversed-phase; Nucleosil 100 C_{18}, 5 μm, 200 × 4.0 mm ID
Mobile phase	1650 g Water, 550 g acetonitrile, 2.0 g NaH_2PO_4, 0.75 g EDTA and 11.5 g 60 % perchloric acid
Flow-rate	0.7 ml min^{-1} (serum); 0.8 ml (urine) column
Temperature	30 °C
Internal standard	Demeclocycline salt
Injection size	25–100 μl, 0.125–2.5 μg ml^{-1}
Detection	UV, λ = 350 nm; 0.1 AUFS
Retention times, min	Sample — serum — urine Doxycycline — 3.3 — 5.1 Demeclocycline — 5.7 — 8.3
Ref.	Nieder M and Jaeger H (1988) Chromatographia 25:526

HPLC 3

Sample	Raw materials
Mobile phase	0.1 M Phosphate buffer, pH 3.0/acetonitrile (60:40)
Detection	UV, λ = 350 nm
Column and retention	(relative retention)

Column	doxycline/demeclocycline
Hypersil ODS, 5 μm	1.3
Lichrosorb CH_{18}, 4 μm	1.3
μ-Bondapak C_{18}, 10 μm	1.5
Novapak C_{18}, 4 μm	1.5
Nucleosil C_{18}, 5 μm	1.6
Nucleosil C_{18}, 7 μm	1.6
Nucleosil C_{18}, 10 μm	1.6
Lichrosorb RP_{18}, 5 μm	1.9

Ref.	Nieder M and Jaeger H (1988) Chromatographia 25:526

HPLC 4

Sample	Pharmaceutical preparations (capsules)
Apparatus	Latek HPLC system with ultraviolet detector
Column	Reversed-phase; Nucleosil C_{18}, 7.5 μm, 250 × 4.2 mm ID
Mobile phase	Water/acetonitrile (70:30), pH 1.4 with 85% ortho-phosphoric acid
Flow-rate	1 ml min^{-1}; 1450 psi
Temperature	20 °C

Injection size	10 µl containing 4 µg
Detection	UV, λ = 270 nm
Ref.	Schaefer U, Oelschlaeger H and Stricker H (1988) Pharm. Ind. 50:588

HPLC 5

Sample	Serum
Extraction	Mobile phase and 2% phosphoric acid by micro-separation
Apparatus	Waters M 590 LC pump, 710B autosampler and 481 UV detector
Column	Phenyl phase; Spherisorb-5 MPLC, 220 × 4.6 mm ID in 220 mm MPLC holder
Mobile phase	100 ml Methanol, 15 ml acetonitrile, 20 ml 85% phosphoric acid, 5 ml triethylamine and 10 g disodium hydrogen phosphate in 1 liter water
Flow-rate	0.8 ml min^{-1}
Temperature	50 °C
Injection size	10 µl, 10 µg ml^{-1}
Detection	UV, λ = 267 nm; 0.02 AUFS
Recording	Waters M 730 data module
Retention times, min	(estimated from drawing)

Oxytetracycline	7.43	Tetracycline	12.50
Minocycline	8.40	Doxycycline	18.14

Ref.	Tyczkowska K and Aronson AL (1986) J. Assoc. Off. Anal. Chem. 69:760

HPLC 6

Sample	Biological materials from mouse
Extraction	Acetonitrile and 0.03 M H_3PO_4
Apparatus	Waters M 6000A LC pump and Schoeffel SF-770 UV detector
Column	Reversed-phase; C_8, 10 µm, 250 × 4 mm ID
Mobile phase	27% Acetonitrile in 0.01 M NaH_2PO_4, pH 2.4
Flow-rate	1.9 ml min^{-1}
Injection size	50 or 100 µl
Ref.	Boecker R and Weber A (1980) Zbl. Bakt. Hyg. I Abt. Orig. A 247:276

73 Ephedrine

CAS-No	50906-05-3; 134-72-5 (sulfate)
Chemical name	(1*R*,2*S*)-2-Methylamino-1-phenylpropan-1-ol hemihydrate

Empirical formula	Ephedrine		Sulfate	
	$C_{10}H_{15}NO, 1/2H_2O$		$(C_{10}H_{15}NO)_2, H_2SO_4$	
Molecular mass	174.2		428.5	
Proprietary names	Asmapax; Ectasule Minus; Franol			
Clinical use	Sympathominetic			
Solubility,	Water	alcohol	ether	olive oil
Ephedrine	1 in 20	1 in < 1	soluble	1 in 25
Sulfate	1 in 1.3	1 in 90		insoluble
Light absorption	λ_{max} = 251 nm in aqueous acid			

HPLC 1

Sample	Dosage forms (oily)
Apparatus	Waters M 6000A LC pump, 480 ultraviolet detector and U6K universal injector
Column	Silica gel 60, C_8, 300 × 4.6 mm ID 300 × 4.5 mm ID
Mobile phase	Acetonitrile/0.01 M KH_2PO_4 (10:90), pH 3.0 with 50% phosphoric acid
Flow-rate	1 ml min^{-1}
Internal standard	Procaine hydrochloride
Injection size	10 μl
Detection	UV, λ = 214 nm; 0.02 AUFS
Recording	Waters M 730 data module
Retention times, min	(estimated from drawing)
	Ephedrine 4.6
	Procaine 6.0
Ref.	Pietta P, Manera E and Ceva P (1986) J. Chromatogr. 367:228

HPLC 2

Sample	Raw materials
Apparatus	Spectra-Physics 3500B LC pump and 770 ultraviolet detector
Column	Reversed-phase; C_{18}, 10 μm, 300 × 3.9 mm ID
Mobile phase	Methanol/acetic acid/triethylamine/water (5.0:1.5:0.5:93.0)
Flow-rate	1.5 ml min^{-1}
Temperature	Ambient
Injection size	10 μl, 1 mg ml^{-1}
Detection	UV, λ = 254 nm; 0.16 AUFS
Recording	Hewlett-Packard 3380A integrator
Retention times, min	(estimated from drawing)
	Phenylpropanolamine 7.2
	Diastereoisomer ephedrine 10.0
	Pseudoephedrine 12.0
Ref.	Roos RW and Lau-Cam CA (1986) J. Chromatogr. 370:403

HPLC 3

Sample	Raw materials
Apparatus	Waters M 6000A LC pump, U6K injector and 440 UV detector
Column	Normal phase; μ-Porasil, 300 × 3.9 mm ID
Pre-column	Whatman HC Pellosil, 70 × 2.1 mm ID
Mobile phase	n-Hexane/methylene chloride/methanol/diethylamine (300:30:5:0.5)
Flow-rate	1.0 ml min^{-1}
Temperature	Ambient
Detection	UV, $\lambda = 254$ and 280 nm
Recording	Omniscribe dual pen recorder
Retention times, min	(estimated from drawing)
	Methamphetamine 8.0
	Phenmetrazine 8.5
	Phentermine 9.0
	Amphetamine 11.2
	Pseudoephedrine 16.5
	Ephedrine 22.5
	Phenylpropanolamine 34.8
Ref.	Noggle Jr FT and Clark CR (1984) J. Assoc. Off. Anal. Chem. 67:687

HPLC 4

Sample	Syrup preparations
Apparatus	Du Pont M 830 LC with UV photometer
Column	Cation exchange; Zipax SCX, 50 × 2.1 mm ID
Mobile phase	0.02 M Dibasic ammonium phosphate in 36% dioxane/water, pH 8.2
Flow-rate	1.1 ml min^{-1}; 100 psi
Temperature	Ambient
Internal standard	Phenylpropanolamine
Detection	UV, $\lambda = 254$; 0.004 AU
Recording	Autolab model system integrator
Retention times, min	(estimated from drawing) Pheniramine 8.2
	Internal standard 2.7 Pyrilamine 16.5
	Ephedrine 4.4
Ref.	Mollica JA and Stober HC (1978) in: Tsuji K (ed) GLC and HPLC Determination of Therapeutic Agents, Part 2. Marcel Dekker Inc., New York and Basel, p 691

HPLC 5

Sample	Amphetamine analogues
Column	Normal phase; Spherisorb S5W, 250 × 5 mm ID
Mobile phase	Methanol/aqueous ammonium nitrate buffer, pH 10.1 (9:1)

Flow-rate	2.0 ml min^{-1}
Detection	UV, λ = 254 nm
Retention times, min	(estimated from drawing)

Diethylpropion	1.50	4-Hydroxy-amphetamine	2.90
Methylphenidate	1.70	2-Phenethylamine	3.20
Phenylpropanol-amine	2.25	Ephedrine	3.75
Amphetamine	2.70	Methylamphetamine	4.00
		Mepentermine	4.80

Ref. Gilbert MT (1987): High Performance Liquid Chromatography. Wright, Bristol, p 148

HPLC 6

Sample	Human urine with amphetamine related compounds
Extraction	Diethyl ether under alkaline condition
Apparatus	Shimadzu M LC-6A pump, Rheodyne M 7125 injector, Shimadzu RF-530 fluorescence detector, Atto AC-2220 luminomonitor and Hitachi M 650-10S fluorimeter
Column	Intersil Reversed-phase; ODS-2, 250 × 4.6 mm ID
Pre-column	Reversed-phase; Spherisorb-5 C_{18}, 30 × 4.6 mm ID
Mobile phase	0.001 M Imidazole in acetonitrile/water, pH 7.0 with HNO_3. The ratios of acetonitrile/ waters (8:2, v/v) for CBI derivatives, (7:3) for Dns derivatives and (6:4) for NBD derivatives, pH 7.9 each
Flow-rate	1.0 ml min^{-1}
Temperature	Ambient
Derivatization	Post-column; with CBI (cyanobenz[f]isoindole); dansyl chloride (Dns-Cl) and 4-fluoro-7-nitro-benzoxadiazole (NBD-F)
Injection size	20 μl
Detection	Fluorescence, λ = 343/530 nm (Dns-derivatives), 418/483 nm (CBI-derivatives) and 470/530 (NBD-derivatives)
Retention times, min	(estimated from drawing)

Derivatization with	Dns	NBD	CBI
Phenylpropanolamine	7.0	7.5	6.5
Benzylamine	8.0	9.0	7.0
Ephedrine	8.8	8.0	
Phenylethylamine	9.5	12.0	8.5
Phenylpropylamine	11.5	15.0	11.0
Phenylbutylamine	14.0	19.5	13.0
N-Methylphenethylamine	15.0	15.0	
Methamphetamine	18.0	17.0	
N-Isopropylbenzylamine	20.5	19.0	

Ref.	Hayakawa K, Hasegawa K et al (1989) J. Chromatogr. 464:343

HPLC 7

Sample	Cough and cold products
Apparatus	IBM Instruments LC/9533 ternary gradient LC with UV detector
Column	Adamantyl phase; modified Si 100, 10 μm, 50 × 4.6 mm ID
Mobile phase	A. 0.01 M $NH_4H_2PO_4$, pH 2.2 B. 90% 0.01 M Phosphate buffer and 10% acetonitrile
Flow-rate	3.5 ml min^{-1} (A); 1.0 ml min^{-1} (B)
Temperature	Ambient
Injection size	20 μl
Detection	UV, $\lambda = 216$ nm
Recording	IBM Instruments M 9000 computer
Retention times, min	(estimated from drawing)

Column	A.	B.
Phenylpropanolamine-HCl	1.0	1.6
Ephedrine hydrochloride	2.3	2.2

Ref.	Yang SS and Gilpin RK (1988) J. Chromatogr. Sci. 26:416

74 Epinephrine (Adrenaline)

CAS-No	51-43-4; 51-42-3 (bitartrate)		
Chemical name	(*R*)-1-(3,4-Dihydroxyphenyl)-2-methylaminoethanol		
Empirical formula	Epinephrine	Bitartrate	
	$C_9H_{13}NO_3$	$C_9H_{13}NO_3$, $C_4H_6O_6$	
Molecular mass	183.2	333.3	
Proprietary names	Eppy; Epitrate; Suprarenin		
Clinical use	Sympathomimetic		
Solubility,	Water	alcohol	aq HCl, aq NaOH
Epinephrine	Sparingly soluble	insoluble	freely soluble
Bitartrate	1 in 3	1 in 520	
Light absorption	$\lambda_{max} = 280$ nm in aqueous acid		

HPLC 1

Sample	Urine
Extraction	100 mM Acetic acid over an affinity chromatography column (Glycotest 25), pH 8.00

Apparatus	Waters M 6000A LC pump and Bioanalytical Systems LC-4B amperometric detector
Column	Reversed-phase; 5 µm Hypersil ODS, 150 × 4.5 mm ID
Pre-column	10 µm ODS, 30 × 4.5 mm ID
Mobile phase	Methanol/buffer (containing 50 mM Na-formate, 0.15% formic acid-99%, 0.1% triethylamine, 0.25 mM 1-octanesulfonic acid and 0.2 mM disodium EDTA) (5:95, v/v), pH 4.0
Flow-rate	1 ml min^{-1}
Internal standard	3,4-Dihydroxybenzylamine (DHBA)
Injection size	50 µl, 12 nmol l^{-1}
Detection	EC, + 0.65 vs Ag/AgCl RE
Recording	Spectra-Physics 4270 computing integrator
Retention times, min	Norepinephrine 3.5 Internal standard 7.0 Epinephrine 4.8 Dopamine 12.0
Ref.	Peaston RT (1988) J. Chromatogr. Biomed. Appl. 424:263

HPLC 2

Sample	Mice brain tissue
Extraction	5% Perchloric acid
Apparatus	Waters M LCG-590 pump, WISP sample processor and Bioanalytical Systems LC-4B amperometric detector
Column	Reversed-phase; Biophase ODS 5 µm, C_{18}, 250 × 4.6 mm ID
Pre-column	Reversed-phase; Biophase ODS 5 µm, C_{18}, 30 × 4.6 mm ID
Mobile phase	0.1 M Citric acid, 0.06% diethylamine, 0.05 mM Na_2EDTA, 225 ml per liter heptanesulfonic acid and 3% acetonitrile, pH 2.55
Flow-rate	1.5 ml min^{-1}
Temperature	Ambient
Internal standard	Isoproterenol and L-cysteine
Injection size	50 µl, 20 ng ml^{-1}
Detection	EC, GC electrode, + 0.85 V
Recording	Shimadzu C-R3A integrator
Retention times, min	(estimated from drawing)

Norepinephrine	3.3
3-Methoxy-4-hydroxyphenylglycol	4.0
DOPA(L-β-3,4-dihydroxyphenylalanine)	4.8
DL-Normetanephrine hydrochloride	5.4
Dopamine	6.9
3,4-Dihydroxyphenylacetic acid	8.0
L-(−)Isoproterenol	11.4

	5-Hydroxy-L-tryptophan	13.4
	5-Hydroxyindole-3-acetic acid	15.0
	3-Methoxytyramine	18.2
	Serotonin	19.1
	Homovanillic acid	23.4
Ref.	Hadfield MG, Milio C and Narasimachari N (1986) J. Chromatogr. 369:449	

HPLC 3

Sample	Biological materials
Extraction	Deproteinization and dilution using 0.2 M perchloric acid, 3 M HCl and water
Apparatus	Waters M 6000A LC pump, WISP 710B autosampler and Bioanalytical Systems electro-chemical detector
Column	Reversed-phase; Supelcosil 5 µm, C_{18}, LC-18-DB
Mobile phase	100 mM NaH_2PO_4,H_2O/methanol (840:160) plus 0.0026 M octyl sodium sulfate, 0.0001 M disodium EDTA and 0.00025 triethylamine, pH 3.85–4.25
Flow-rate	1 ml min^{-1}
Temperature	35 °C
Injection size	100 µl, 100 ng ml^{-1}
Detection	EC, GC electrode vs Ag/AgCl RE
Recording	Hewlett-Packard HP 3385A data automation

Retention times, min	Vanillylmandelic acid	2.73
	L-DOPA (3,4-hydroxyphenylalanine)	3.19
	3-Methoxy-4-hydroxyphenylglycol	3.74
	3,4-Dihydroxyphenylacetic acid,	5.33
	Norepinephrine	6.93
	5-Hydroxyindolacetic acid	7.61
	Epinephrine	8.51
	Homovanillic acid	10.40
	Dopamine	17.05
	Serotonin	39.36
	3-Methoxytyramine	42.06
Ref.	Seegal RF, Brosch KO and Bush B (1986) J. Chromatogr. Biomed. Appl. 377:131	

HPLC 4

Sample	Blood plasma and urine
Apparatus	Waters M LC-6000A pump and Bioanalytical Systems LC-4B electrochemical detector
Column	Reversed-phase; 5 µm, C_{18}, 30 × 2.1 mm ID and 5 µm, C_8, 220 × 2.1 ID
Mobile phase	55 mM Citric acid/1 M sodium acetate (750:2), pH 2.35 containing 0.35 M Na_2EDTA
Flow-rate	0.16 ml min^{-1}

Injection size	10 µl, 2–5 µg ml^{-1}
Detection	EC, + 700 mv vs Ag/AgCl RE
Retention times, min	Norepinephrine 5.5 Dopamine 9.2 Epinephrine 6.4 Dihydroxyphenyl- DOPA 7.6 acetic acid 13.4
Ref.	Benedict CR (1987) J. Chromatogr. 385:369

HPLC 5

Sample	Blood plasma
Extraction	0.1 M Perchloric acid using alumina column for sample-clean-up
Apparatus	Beckman M 340 LC pump and Bioanalytical Systems LC-4B electrochemical detector
Column	Cation exchange; Bio-Rad, 150 × 4.6 mm ID
Pre-column	Cation exchange; Bio-Rad, 150 × 4.6 mm ID
Mobile phase	15% Acetonitrile in 0.07 M sodium citrate, pH 6.4 with concentrated H_3PO_4
Flow rate	0.55 ml min^{-1}
Temperature	Ambient
Internal standard	3,4-Dihydroxybenzylamine (DHBA)
Injection size	25 µl, 9 pg ml^{-1},
Detection	EC, + 0.4 V vs Ag/AgCl RE
Recording	Hewlett-Packard 3390A integrator
Retention times, min	Epinephrine 7.8 Dopamine 10.0 Norepinephrine 9.0 IS 11.0 (estimated from drawing)
Ref.	Koch DD and Polzin GL (1987) J. Chromatogr. 386:19

HPLC 6

Sample	Urine
Extraction	I. On alumina oxide and ion-pair extraction columns II. On cation exchange and alumina extraction columns with 0.7 M H_2SO_4 and 2 M $(NH_4)_2SO_4$
Apparatus	Waters M 45 LC pumps, U6K sample injector and ESA and Bioanalytical Systems electrochemical detectors
Column	Reversed-phase; µ-Bondapak C_{18}, 10 µm, 250 × 4.6 mm ID
Mobile phase	0.07 M NaH_2PO_4 Buffer containing 0.1 M Na_2EDTA, 1 mM n-heptanesulfonic acid and methanol (90:10, v/v), pH 3.9
Flow-rate	1 ml min^{-1}
Internal standards	Dihydroxybenzylamine (I), Epinine (II)
Detection	EC, GC electrode, + 0.25 V, − 0.15 V; + 0.7 V

Recording	Potentiometric recorder			
Retention times, min	(estimated from drawing)		Internal standard I	12.5
	Norepinephrine	6.5	Dopamine	20.0
	Epinephrine	9.0	Internal standard II	24.0
Ref.	Claeys M, Schepers A et al (1988) J. Pharm. Biomed. Anal. 6:895			

75 Ergometrine

CAS-No	60-79-7; 129-51-1 (maleate)			
Chemical name	*N*-[(*S*)-2-Hydroxy-1-methylethyl]-D-lysergamide			
Empirical formula	Ergometrine	Maleate		
	$C_{19}H_{23}N_3O_2$	$C_{19}H_{23}N_3O_2,C_4H_4O_4$		
Molecular mass	325.4	441.5		
Proprietary names	Ergonovine; Ergotrate; Ermetrine			
Clinical use	Ergot Alkaloid			
Solubility,	Water	ethanol	chloroform	ether
Ergometrine	slightly sol	soluble	sparingly soluble	
Maleate	1 in 40	1 in 100	insoluble	insoluble
Light absorption	λ_{max} = 310 nm in aqueous alkali			

HPLC 1

Sample	Formulation with oxytocin		
Apparatus	Altex M 110 or LDC minipump and Cecil M 272 or Hewlett-Packard 1030B UV detector or Perkin-Elmer M 2000 fluorescence detector		
Column	Reversed-phase; Nucleosil C_{18}, 5 μm, 100 × 4.6 mm or 250 × 5 mm ID		
Mobile phase	0.05% (w/v) Sodium tetradecyl sulfate in 0.83 mM H_3PO_4, and 35% (v/v) acetonitrile (A) and with 37% (v/v) acetonitrile (B), pH 5.0 with triethylamine		
Flow-rate	2.5 ml min^{-1} (A); 1 ml min^{-1} (B)		
Temperature	18 °C (mobile phase)		
Injection size	50–80 μl		
Detection	UV; 0.002–0.005 AU		
Retention times, min	(estimated from drawing)	Mobile	phase
		A.	B.
	Oxytocin	8.0	5.0
	Ephedrine	10.0	6.5
	Ergometrine	18.0	12.0
Ref.	Pask-Hughes RA, Corran PH and Calam DH (1981) J. Chromatogr. 214:307		

HPLC 2

Sample	Formulations with oxytocin
Apparatus	Altex M 110 or LDC minipump and Cecil M 272 or Hewlett-Packard 1030B UV detector or Perkin-Elmer M 2000 fluorescence detector
Column	Cation exchange; Partisil 5 SCX, 50 × 4.6 mm ID
Pre-column	Normal phase; Lichroprep Si 60, 5–25 μm, 50 × 4.6 mm ID
Mobile phase	8.3 mM H_3PO_4, pH 5.0 with NH_3/acetonitrile (88:12, v/v)
Flow-rate	1.0 ml min^{-1} (A); 2 ml min^{-1} (B)
Injection size	50–80 μl
Detection	UV; 0.005 AU
Retention times, min	(estimated from drawing)

	A.	B.
Ergometrine	9.0	20.8
Ephedrine	14.0	31.0
Oxytocin		10.6

Ref. Pask-Hughes RA, Corran PH and Calam DH (1981) J. Chromatogr. 214:307

HPLC 3

Sample	Pharmaceutical products (combinations)
Apparatus	Altex M 110 pump or Cecil M 301 pump, Cecil M 272 UV spectrophotometer and Perkin-Elmer M 2000 fluorescence spectrophotomer
Column	Reversed-phase; HSCP C_{18}, 100 × 5 mm ID
Mobile phase	8.3 mM H_3PO_4 and triethylamine, pH 5.0, containing 0.1% (w/v) sodium dodecylsulfate and 40% (v/v) acetonitrile
Injection size	20 μg
Detection	UV, λ = 210 nm; fluorescence, λ = 313/442 nm
Retention times, min	(estimated from drawing)

		Chlorobutanol	9.00
Lysergic acid	5.50	Ergometrine	13.00
Oxytocin	6.50	Ergometrinine	17.25

Ref. Pask-Hughes RA, Corran PH and Calam DH (1981) Anal. Proc. 18:247

HPLC 4

Sample	Tablets
Apparatus	Waters M 6000A HPLC and 770 Spectroflow spectrophotometer
Column	Reversed-phase; μ-Bondapak C_{18}, 300 × 4 mm ID
Mobile phase	20% (v/v) Acetonitrile and 1% (v/v) acetic acid in water
Flow-rate	1 ml min; 800 psi

Injection size	20 μl, 0.2 mg ml^{-1}
Detection	UV, λ = 312 nm; 0.4 AUFS
Recording	Hewlett-Packard 2100 RTE computing integrator
Retention times, min	Ergometrine (ergonovine) maleate, lysergic acid, ergometrine diastereomer and ergometrine nitrate, in ascending order (from drawing)
Ref.	Sondack DL (1978) J. Chromatogr. 166:615

76 Ergotamine

CAS-No	113-15-5; 379-79-3 (tartrate)			
Chemical name	12′-Hydroxy-2′-methyl-5′α-benzylergotaman-3′,6′,18-trione			
Empirical formula	Ergotamine $C_{33}H_{35}N_5O_5$	Tartrate $(C_{33}H_{35}N_5O_5)_2C_4H_6O_6$		
Molecular mass	581.7	1313.4		
Proprietary names	Ergomar; Gynergen; Lingraine			
Clinical use	Ergot Alkaloid			
Solubility,	Water	ethanol	chloroform	ether
Ergotamine	insoluble	slightly soluble	freely soluble	
Tartrate	1 in 500	1 in 500	insoluble	insoluble
Light absorption	λ_{max} = 310 nm in aqueous alkali			

HPLC 1

Sample	Pharmaceutical raw materials
Apparatus	Waters M 6000 or Altex 100 LC pump and Altex M 153 UV detector
Column	Reversed-phase; Nucleosil C_{18}, 5 μm, 150 × 3 mm ID
Mobile phase	A. 0.01 M Ammonium carbonate B. Acetonitrile Gradient: 8% B in A; B 8% to 15% in 1 min; B 15% to 30% in 1 min; B 30% to 40% in 3 min; B 40% to 50% in 2 min; B 50% to 60% in 4 min
Flow-rate	1.5 ml min^{-1}; 5075 psi
Injection size	30 μl, 0.6 mg ml^{-1}
Detection	UV, λ = 254 nm
Recording	Hewlett-Packard 3352B data system
Retention times, min	(estimated from drawing)

Lysergic acid	1.5	aci-Ergotamine	6.8
Isolysergic acid	3.7	aci-Ergotaminine	8.5
Lysergic acid amide	5.1	Ergotamine	9.1
Isolysergic acid amide	5.8	Ergotaminine	10.8

Ref.	Erni F, Frei RW and Lindner W (1976) J. Chromatogr. 125:265

HPLC 2

Sample	Dosage forms (tablets)
Apparatus	Autoanalyser with Hewlett-Packard 1010 HPLC, Perkin-Elmer LC 55 UV detector and 3000 fluorimetric detector
Column	Reversed-phase; Hibar C_{18}, 10 μm, 250 × 4 mm ID
Mobile phase	Acetonitrile/0.01 M ammonium carbonate (47:53, v/v)
Flow-rate	3 ml min^{-1}
Injection size	30 μl
Detection	Fluorescence, λ = 320/420 nm; VIS, λ = 550 nm
Retention times, min	(estimated from drawing) Ergotamine 4.2 Isomers and hydrolytic products 1.0, 2.0 Ergotaminine 11.0
Ref.	Kubin H, Brehm M and Ulmen J (1982) Pharm. Ind. 44:1269

HPLC 3

Sample	Tablets
Apparatus	Waters M 6000A LC pump, Perkin-Elmer LS 5 fluorescence spectrophotometer and Rheodyne M 7125 injector
Column	Reversed-phase; Alltech C_{18}, 250 × 4.6 mm ID
Mobile phase	Acetonitrile/water/triethylamine (700:300:0.5)
Flow-rate	1.0 ml min^{-1}
Injection size	20 μl
Detection	Fluorescence; λ = 250/430 nm
Retention times, min	Ergonovine maleate 2.5 Ergotamine tartarate 4.5 Ergotaminine 5.5
Ref.	Cieri UR (1987) J. Assoc. Off. Anal. Chem. 70:538

HPLC 4

Sample	Pharmaceutical tablets
Apparatus	Du Pont M 8800 pump, UV absorbance detector and Rheodyne M 7125 injector
Column	Reversed-phase; Polymeric 5 μm, C_{18}, 250 × 4.6 mm ID
Mobile phase	0.01 M Ammonium acetate in acetonitrile/water/triethylamine (35:64:1), pH 3.7 with glacial acetic acid
Flow-rate	1.5 ml min^{-1}
Internal standard	Ethylparaben, 5.6 μg ml^{-1} methanol

Injection size	10 μl, 16–48 μg ml^{-1}
Detection	UV, λ = 254 nm; 0.08 AUFS
Recording	Spectra-Physics 4100 integrator
Retention times, min	Ergotamine tartarate 3.46 Internal standard 4.14 Cyclizine hydrochloride 4.87
Ref.	Jalal IM, Sa'sa SI and Yasin TA (1988) Anal. Lett. 21:1561

HPLC 5

Sample	Blood plasma
Extraction	Cyclohexane/1-butanol (9:1)
Apparatus	Waters M 6000A LC pumps with U6K injector and Schoeffel FS 970 fluorescence detector
Column	Reversed-phase; Hypersil-ODS 5 μm, 250 × 4.6 mm ID
Mobile phase	Acetonitrile/0.01 M ammonium carbonate (50:50)
Flow-rate	1.5 ml min^{-1}
Internal standard	Ergocristine, 100 ng ml^{-1}
Injection size	50–150 μl, 0–75 ng ml^{-1}
Detection	Fluorescence, λ = 328/389 nm
Retention times, min	(estimated from drawing) Ergotaminine 6.2 Ergotamine 3.8 Ergocristine 8.8
Ref.	Edlund PO (1981) J. Chromatogr. Biomed. Appl. 226:107

77 Erythromycin

CAS-No	114-07-8 (erythromycin A); 3521-62-8 (estolate)	
Chemically	Macrolide antibiotic produced by Streptomyces erythreus	
Empirical formula	Erythromycin	Estolate
	$C_{37}H_{67}NO_{13}$	$C_{40}H_{71}NO_{14},C_{12}H_{26}O_4S$
Molecular mass	733.9	1056.4
Proprietary names	Erycinum; Ilotycin; Mistral	
Clinical use	Antibiotic	

Solubility,	Water	alcohol	chloroform	methanol	aq HCl
Erythromycin	1 in 1000	1 in 5	1 in 6	soluble	soluble
Estolate	insoluble	1 in 2	1 in 10		insoluble

Light absorption λ_{max} = 280 nm at pH 6.3

HPLC 1

Sample	Body fluids
Extraction	Acetonitrile/0.05 M phosphate buffer (3:2) using Bond Elut C_{18} extraction column
Apparatus	Waters M 6000A LC pump and ESA Coulochem M 5100A electrochemical detector
Column	Reversed phase; 10 µm, C_{18}, 250 × 3.9 mm ID
Mobile phase	Acetonitrile/0.05 M phosphate buffer (65:35, v/v), pH 7.0
Flow-rate	1.3 ml min^{-1}; 1494 psi
Temperature	35 °C
Internal standard	Oleandomycin phosphate, 20 µg ml^{-1}
Injection size	5–20 µl, 0.25–4 µg ml^{-1}
Detection	EC, + 0.7 V, + 0.9 V and guard cell, + 1.00 V
Retention times, min	Internal standard 4.8 Erythromycin 6.0 Propionyl erythromycin 9.0
Ref.	Stubbs C and Kanfer I (1988) J. Chromatogr. Biomed. Appl. 427: 93

HPLC 2

Sample	Tablets and capsules
Apparatus	Perkin-Elmer Series 4 LC pump, LC-95 UV detector and LC-420 autosampler
Column	Reversed-phase; Partisil 5-ODS3, RAC II, C_{18}, 100 × 4.6 mm ID
Mobile phase	Acetonitrile/methanol/N,N-dimethylcyclohexamine/ 0.012 M $(NH_4)_2HPO_4$ (350:100:1.5:550), pH 5.3 with acetic acid
Flow-rate	0.5 ml min⁻1
Injection size	20 µl, 0.4–2.8 mg ml^{-1}
Detection	UV, λ = 215 and 288 nm
Recording	Spectra-Physics computing integrator SP 4270
Retention times, min	Erythromycin C 3.70 Erythromycin A 4.65 Erythronolide B 6.35 Erythronolide B 7.28 3-Mycarosyl erythronolide B 11.98
Ref.	Geria T, Hong W-H and Daly RE (1987) J. Chromatogr. 396: 191

HPLC 3

Sample	Body fluids
Extraction	Diethyl ether
Apparatus	Waters M 510 LC pump, 460 electrochemical detector and WISP 710B autosampler
Column	Reversed-phase; Novapak C_{18}

Mobile phase	56 mM Sodium acetate buffer/acetonitrile/ methanol (56:50:4), pH 7.0 with acetic acid
Flow-rate	1.1 ml min^{-1}; 769 psi
Internal standard	Roxithromycin
Injection size	40 µl
Detection	EC, + 0.9 vs Ag/AgCl RE
Recording	Hewlett-Packard 3390 data station
Retention times, min	CP-62993 (Pfizer) 3.8 Erythromycin base 6.0 Erythrolosamine 6.9 Erythromycin B 7.1 4″-Acetylerythromycin 7.4 6-O-Methylerythromycin 10.3 Roxithromycin 14.7 Erythromycin estolate 34.5 Erythromycin ethylsuccinate 35.5
Ref.	Croteau D, Vallée F et al (1987) J. Chromatogr. Biomed. Appl. 419:205

HPLC 4

Sample	Human blood plasma
Extraction	Methyl tert.-butyl ether after slight alkalizing with saturated K_2CO_3 solution, pH 9.8
Apparatus	Spectra-Physics 8700 XR LC pump and LC 4 amperometric detector (Bioanalytical Systems)
Column	PRP-1; 5 µm, 150 × 4.1 mm ID
Mobile phase	Potassium phosphate (50 mM, pH 9.5)/acetonitrile (60:40, v/v)
Flow-rate	0.7 ml min^{-1}; 2248 psi
Temperature	25 °C
Internal standard	Oleandomycin
Injection size	25 µl, 2.5 nmol
Detection	EC, 1.1 V vs Ag/AgCl RE
Recording	Spectra-Physics 4270 integrator
Retention times, min	(estimated from drawing) Internal standard 7.5 Erythromycin 11.6
Ref.	Nilsson L-G, Walldorf B and Paulsen O (1987) J. Chromatogr. Biomed. Appl. 423:189

HPLC 5

Sample	Serum
Apparatus	LDC M 19-60066-002 high-pressure minipump
Column	Reversed-phase; Lichrosorb C_{18}, 250 × 4.6 mm ID
Pre-column	µ-Bondapak C_{18}, 50 × 2.1 mm ID
Mobile phase	Acetonitrile/0.2 M ammonium acetate/water (65–60:10:25–30), pH 7.0

Flow-rate	0.9 ml min^{-1}
Temperature	70 °C ± 2 °C
Derivatization	Post-column
Detection	Fluorescence
Retention times, min	(estimated from drawing) Erythromycin 6.0 4″-Acetylerythromycin or Erythrolosamine 7.5 Anhydroerythromycin 8.3 Erythromycin ethylsuccinate 12.1
Ref.	Tsuji K (1978) J. Chromatogr. 158:337

78 Estradiol

CAS-No	50-28-2; 50-50-0 (benzoate)
Chemical name	Estra-1,3,5(10)-triene-3,17β-diol
Empirical formula	Estradiol $C_{18}H_{24}O_2$; Benzoate $C_{25}H_{28}O_3$
Molecular mass	272.4; 376.5
Proprietary names	Gynoestryl; Ovocyclin; Progynon
Clinical use	Sex Hormone

Solubility,	Water	alcohol	acetone	chloroform
Estradiol	insoluble	1 in 28	1 in 17	1 in 435
Benzoate	insoluble	slightly sol	1 in 50	1 in 5

Light absorption λ_{max} = 231 nm in ethanol (benzoate)

HPLC 1

Sample	Human blood plasma
Extraction	Centrifugation
Apparatus	Yanaco M L-100 and Yanaco L-4000W as pretreatment and analytical pumps respectively, manual injector, and electrochemical detector and UV detector
Column	Reversed-phase; Yanapak ODS-A, 250 × 4.6 mm ID
Pre-column	Yanapak C_8, 30 × 4 mm ID for pretreatment
Mobile phase	For pretreatment: Water, methanol and 50 mM potassium phosphate buffer, pH 3.1/acetonitrile (8:2, v/v) For analysis: 50 mM Potassium phosphate buffer, pH 3.1/acetonitrile/methanol (10:7:3, v/v)
Flow-rate	1 ml min^{-1}
Temperature	25 °C
Internal standard	4′-Methoxyoctananilide

Injection size	9.6 ng			
Detection	EC, + 1.0 V vs. Ag/AgCl RE; UV, λ = 245 nm			
Recording	Chromatogramm data processor (System Instruments)			
Retention times, min	Estradiol	15.8		
	Estrone	21.6		
	Internal standard	46.0		
Ref.	Suzuki Y, Hayashi N and Sekiba K (1988) J. Chromatogr. Biomed. Appl. 426:33			

HPLC 2

Sample	Equine estrogens			
Apparatus	Beckmann M 100A and 110A LC pumps with 420 system controller, 210 sample injection valve and 155 UV detector			
Column	Reversed-phase; Ultrasphere C_{18}, 5 μm, 250 × 4.6 mm ID			
Mobile phase	I. Methanol/dichloromethane/water/isopropanol (40:4:48:8)			
	II. Methanol/dichloromethane/water/isopropanol (41:3:49:7)			
	III. Methanol/dichloromethane/water/isopropanol (43:2:52:3)			
Flow-rate	1.2 ml min^{-1} (I, II); 1.3 ml min^{-1} (III)			
Injection size	20 μl			
Detection	UV, λ = 220 nm; 0.02 AUFS			
Recording	Shimadzu C-RIA electronic data processor			
Retention times, min	Mobile phase	I.	II.	III.
	17β-Dihydroequilenin	5.5	7.5	14.2
	17α-Dihydroequilenin	6.2	8.5	16.3
	17β-Dihydroequilin	7.3	8.8	18.1
	17α-Dihydroequilin	7.5	10.3	21.0
	17α-Ethynylestradiol	8.3		
	17β-Estradiol	8.7	10.8	26.8
	17α-Estradiol	11.5	12.6	22.7
	Equilenin	12.0	16.0	24.7
	Equilin	13.5	18.5	29.3
	Estrone	15.6	21.6	34.0
Ref.	Desta B (1988) J. Chromatogr. 435:385			

HPLC 3

Sample	Pharmaceutical ointments
Apparatus	Spectra-Physics 8700 LC pump, Rheodyne M 7125 injector and amperometric detector and Pye- Unicam LC 871 UV/VIS detector

Column	Reversed-phase; Lichrosorb C_{18}, 7 µm, 250 × 4.00 mm ID
Pre-column	Reversed-phase; Lichrosorb C_{18}, 40 µm, 30 × 4.00 mm ID
Mobile phase	I. Methanol/water (90:10, v/v) for estradiol benzoate II. Methanol/water/acetic acid (60:35:5, v/v) with lithium perchlorate 1 g l^{-1} for estradiol
Flow-rate	1.5 ml min^{-1} (I) and 1.2 ml min^{-1} (II)
Injection size	20 µl containing 20 ng and 2 ng
Detection	UV, λ = 230 nm; EC, + 0.85 V, GC electrode
Retention times, min	

Mobile phase/detection	I/UV	II/EC
Estradiol		4.0
Estradiol benzoate	6.2	

Ref.	Leroy P, Benoit E and Nicolas A (1986) J. Chromatogr. 367:428

HPLC 4

Sample	Iodinated estradiol and estriol preparations
Apparatus	Waters M 6000A HPLC, U6K injector and 441 UV detector
Column	Reversed-phase; Radial-Pak C_{18}, 10 µm, 100 × 5 mm ID
Mobile phase	I. Methanol/water/acetic acid (50:50:2) II. Methanol/water/acetic acid (70:30:2)
Flow-rate	1 ml min^{-1}
Temperature	Ambient
Injection size	5000 cpm estradiol and 6000 cpm estriol
Detection	UV, λ = 280 nm; 0.1 AUFS
Recording	Zinsser Hydrogramma 16-γ-counter
Capacity factor, *k'*	

Mobile phase	I.	II.
Estradiol		2.3
2-Iodo-estradiol		4.5
4-Iodo-estradiol		6.7
2,4-diiodo-estradiol		14.0
Estriol	3.8	
2-Iodo-estriol	11.3	
4-Iodo-estriol	13.2	
2,4-diiodo-estriol	45.0	

Ref.	Dibbelt L, Gloning K and Kuss E (1986) J. Chromatogr. 358:268

HPLC 5

Sample	Iodinated estradiol and estriol preparations
Column	Normal phase; SI Radial-Pak 10 µm, 100 × 5 mm ID

Mobile phase	I. Chloroform/methanol (99:1)		
	II. Chloroform/methanol (96:4)		
Flow-rate	1 ml min^{-1}		
Temperature	Ambient		
Injection size	5000 cpm Estradiol and 6000 cpm estriol		
Detection	UV, λ = 280 nm; 0.1 AUFS		
Recording	Zinsser Analytic Hydrogramma 16-γ-counter		
Capacity factor, k'	Mobile phase	I.	II.
	2,4-Diiodo-estriol		5.3
	2-Iodo-estriol		6.5
	4-Iodo-estriol		6.5
	Estriol		7.8
	2,4-Diiodo-estradiol	4.4	
	2-Iodo-estradiol	5.4	
	4-Iodo-estradiol	6.2	
	Estradiol	10.8	
Ref.	Dibbelt L, Gloning K and Kuss E (1986) J. Chromatogr. 358:268		

HPLC 6

Sample	Estradiol benzoate injection
Apparatus	HPLC with UV detector
Column	Normal phase; silicagel 10 μm, 300 × 4 mm ID
Mobile phase	Cyclohexane/1,4-dioxane (9:1)
Flow-rate	2.0 ml min^{-1}
Internal standard	4-Hydroxybenzaldehyde
Detection	UV, λ = 254 nm
Ref.	Anonym (1980): British Pharmacopoeia. London Her Majesty's Stationary Office, p 644

HPLC 7

Sample	Conjugated estrogens in tablets
Extraction	Centrifugation with water, 0.1 M acetate buffer, pH 3.0 and then with tetrabutylammonium chloride and chloroform for estrogen sulfate
Apparatus	HPLC with UV detector
Column	Reversed-phase; Zorbax ODS 6 μm, 150 × 4.6 mm ID
Mobile phase	Acetonitrile/methanol/water/tetrabutylammonium hydroxide (400:350:1000:13, v/v/v/w)
Flow-rate	1 ml min^{-1}
Internal standard	Prednisolone hemisuccinate, 100 μg ml^{-1}
Injection size	20 μl, 1–500 μg ml^{-1}
Detection	UV, λ = 214 nm
Ref.	Townsend RW, Keuth V et al (1988) J. Chromatogr. 450:414

79 Fenbendazole

CAS-No	43210-67-9
Chemical name	Methyl 5-phenylthio-1*H*-benzimidazol-2- ylcarbamate
Empirical formula	$C_{15}H_{13}N_3O_2S$
Molecular mass	299.4
Proprietary names	HOE 881 V; Panacur (vet)
Clinical use	Anthelmintic
Solubility	Water: insoluble; dimethyl sulfoxide: freely soluble
Light absorption	λ_{max} = 238 nm in aqueous alkali

HPLC 1

Sample	Body fluids
Extraction	Ether
Apparatus	Altex M 110 LC pump and Cecil M 2012 UV detector
Column	Reversed-phase; ODS-Hypersil 10 μm, 100 × 8 mm ID
Mobile phase	Methanol/ammonium carbonate (0.05 M) (65:35)
Flow-rate	1.5 ml min^{-1}
Injection size	5 μl, 0.05–10 μg ml^{-1}
Detection	UV, λ = 292 nm
Retention times, min	(estimated from drawing) Oxibendazole 5.2 Fenbendazole 10.1
Ref.	Borgan JA and Marriner S (1980) J. Pharm. Sci. 69:422

HPLC 2

Sample	Blood plasma
Extraction	Ethyl acetate at pH 8.0
Apparatus	Waters M 45 LC pump and Kratos M 757 UV detector
Column	Reversed-phase; Hypersil SAS (C_2, C_4, C_6) 5 μm, 125 × 4.6 mm ID
Mobile phase	TAAP (tetraethylammonium phosphate) buffer consisting 0.005 M H_3PO_4 (pH 5.9 with tetraethylammonium hydroxide)/acetonitrile (70:30)
Flow-rate	2 ml min^{-1}
Temperature	Ambient (18–22 °C)
Internal standard	Mebendazole
Injection size	100 μl containing 1 and 10 μg ml^{-1}
Detection	UV, λ = 300 nm

Retention times, min	Metabolite II	1.5	Triclabendazole sulfoxide	10.4
	Metabolite III	2.5		
	Internal standard	3.2	Triclabendazole sulfone	13.0
	Fenbendazole	8.6		
Ref.	Lehr KH and Damm P (1986) J. Chromatogr. Biomed. Appl. 382:355			

HPLC 3

Sample	Blood plasma from cattle and sheep
Extraction	Multistep, using water, methanol and acetonitrile over Sep-pak column after deproteinization with acetone
Apparatus	Waters M 6000A LC pump, WISP 710B sample processor and 481 LC spectrophotometer detector
Column	Reversed-phase; μ-Bondapak 10 μm, C_{18}, 300 × 3.9 mm ID
Mobile phase	Acetonitrile/0.04 M $(NH_4)_2HPO_4$ (41:59, v/v), pH 7.5–7.6 with 20% phosphoric acid
Flow-rate	1.5 ml min^{-1}
Injection size	100 μl, 0.05–5 μg ml^{-1}
Detection	UV, λ = 300 nm
Recording	Hewlett-Packard 3390A integrator
Retention times, min	Fenbendazole sulfoxide 3.2 Fenbendazole sulfone 4.5 Fenbendazole 11.3 Triclabendazole 40.0
Ref.	Bull MS and Shume GRE (1987) J. Pharm. Biomed. Anal. 5:501

HPLC 4

Sample	Mixture of seven benzimidazole anthelmintics
Apparatus	Hewlett-Packard 1090 HPLC with photodiode array detector
Column	Reversed-phase; MicroPak-10 μm, ODS, 300 × 4 mm ID
Mobile phase	0.017 M H_3PO_4/acetonitrile (60:40, v/v)
Flow-rate	1 ml min^{-1}
Temperature	45 °C
Internal standard	Mebendazole
Injection size	20 μl, 1000 ng ml^{-1}
Detection	UV, λ = 290 nm
Retention times, min	(estimated from drawing)

Thiabendazole	4.3	Fenbendazole sulfone	6.3
Oxfendazole	4.8	Mebendazole	7.3
p-Hydroxy-oxyfenbendazole	5.7	Albendazole	8.4
		Fenbendazole	12.1

Ref.	Long AR, Hsieh LC et al (1989) J. Chromatogr. 475:404

80 Fentanyl

CAS-No	437-38-7; 990-73-8 (citrate)			
Chemical name	*N*-(1-Phenethyl-4-piperidyl)propionanilide			
Empirical formula	Fentanyl	Citrate		
	$C_{22}H_{28}N_2O$	$C_{22}H_{28}N_2O,C_6H_8O_7$		
Molecular mass	336.5	528.6		
Proprietary names	Leptanal; Sublimaze; Thalamonal			
Clinical use	Narcotic Analgesic			
Solubility,	Water	chloroform	methanol	ethanol
Fentanyl	sparingly sol			
Citrate	1 in 40	1 in 350	1 in 10	1 in 140
Light absorption	λ_{max} = 258 nm in methanol			

HPLC 1

Sample	Dosage forms
Apparatus	Waters M LC-510 pump with 712 WISP autosampler and 820 data station and Kratos M 757 UV/VIS detector
Column	Reversed-phase; Whatman Partisil ODS-3, 10 µm, 250 × 4.6 mm ID
Mobile phase	Water/methanol/85% H_3PO_4 (500:500:4)
Temperature	Ambient
Flow-rate	1.0 ml min^{-1}
Injection size	20 µl, 0.05 mg ml^{-1}
Detection	UV, λ = 229 nm; 0.05 AUFS
Recording	Recordall M 5000
Retention time, min	Fentanyl 8.0 (estimated from drawing)
Ref.	Wilson TD, Maloney T and Amsden WB (1988) J. Chromatogr. 445:299

HPLC 2

Sample	Blood plasma
Extraction	Heptane under alkaline condition
Apparatus	LDC Constametric III pump, Waters M U6K universal injector and Lambdamax 481 LC spectrophotometer
Column	Nitrile-phase; Spherisorb 5 µm, S5CN, 250 × 4.6 mm ID
Mobile phase	Acetonitrile/0.01 M trisodium orthophosphate, pH 2.6 with 1 M orthophosphoric acid (25:75, v/v)
Flow-rate	1.8 ml^{-1}

Temperature	18–22 °C
Injection size	100 µl, 0–100 ng ml^{-1}
Detection	UV, λ = 195 nm
Retention times, min	Alfentanil 6.2
	Fentanyl 7.2
Ref.	Kumar K, Morgan DJ and Crankshaw DP (1987) J. Chromatogr. Biomed. Appl. 419:464

81 Flumequine

CAS-No	42835-25-6		
Chemical name	9-Fluoro-6,7-dihydro-5-methyl-1-oxo-1*H*,5*H*-benzo[*ij*]quinolizine-2-carboxylic acid		
Empirical formula	$C_{14}H_{12}FNO_3$		
Molecular mass	261.3		
Proprietary name	Apurone		
Clinical use	Chemotherapeutic		
Solubility	Water	alcohol	aq NaOH
	insoluble	soluble	soluble

HPLC 1

Sample	Blood plasma
Extraction	Chloroform under acidic and alkaline conditions
Apparatus	Merck-Hitachi M 655A-11 HPLC pump, Merck LMC system SM-25 UV detector and Rheodyne M 7125 injector
Column	Reversed-phase; Lichrosorb C_{18}, 5 µm, 250 × 4 mm ID
Mobile phase	0.01 M Potassium dihydrogen phosphate/methanol (45:55, v/v), pH 3.0 with H_3PO_4
Flow-rate	1.2 ml min^{-1}
Temperature	Ambient
Internal standard	Quinaldic acid
Injection size	20 µl, 2.5–50 µg ml^{-1}
Detection	UV, λ = 254 nm; 0.08–0.16 AUFS
Retention times, min	(estimated from drawing)
	Internal standard 3.0
	7-Hydroxyflumequine 4.0
	Flumequine 7.5
Ref.	Decolin D, Nicolas A and Siest G (1987) J. Chromatogr. Biomed. Appl. 414:499

HPLC 2

Sample	Premixes, milk-replacer and feeds
Extraction	Acetonitrile/chloroform (1:1)
Apparatus	Varian M 5000 LC pump with UV-100 detector and Merck-Hitachi D-2000 CI
Column	Reversed-phase; Nucleosil C_{18}, 5 µm, 250 × 4 mm ID
Mobile phase	45 ml Acetonitrile and 55 ml from a solution of 99 ml 0.05 M phosphoric acid and 1 ml dioxane
Flow-rate	0.8 ml min^{-1}
Injection size	10 µl
Detection	UV, λ = 323 nm
Recording	Canon A-1210 Plotter and IBM PC-XT
Retention times, min	Flumequine 9.9
Ref.	Backe W (1987) Arch. Pharm. (Weinheim) 320:1093

82 5-Fluorouracil

CAS-No	51-21-8		
Chemical name	5-Fluoropyrimidine-2,4(1*H*,3*H*)-dione		
Empirical formula	$C_4H_3FN_2O_2$		
Molecular mass	130.1		
Proprietary names	Adrucil; Efudix; Fluoroplex		
Clinical use	Antineoplastic		
Solubility	Water	alcohol	ether,chloroform
	sparingly sol	slightly sol	insoluble
Light absorption	λ_{max} = 266 nm in aqueous acid		

HPLC 1

Sample	Blood plasma
Extraction	Acetonitrile and zinc sulfate
Apparatus	Waters M 6000 LC pump, Kratos M 773 ultraviolet detector and Rheodyne M 7125 injector
Column	Hamilton PRP-1; 10 µm, 100 × 4.6 mm ID
Mobile phase	10 mM Tetrabutylammonium hydroxide/methanol (74:26, v/v), pH 11.0
Flow-rate	1 ml min^{-1}
Injection size	5 µl
Detection	UV, λ = 266 nm; 0.002–0.5 AUFS
Recording	Woburn TM 4500 microscribe recorder
Retention times, min	5 Fluorouracil 8.00
Ref.	Rustum AM and Hoffman NE (1988) J. Chromatogr. Biomed. Appl. 426:121

HPLC 2

Sample	Blood plasma
Extraction	Deproteinization with 4.4 M perchloric acid
Apparatus	Waters M 5000A LC pump, U6K injector and 440 UV detector
Column	Reversed-phase; μ-Bondapak C_{18}, 250 × 2.9 mm ID
Mobile phase	50 mM Potassium dihydrogen phosphate, pH 4.6
Flow-rate	1.5 ml min^{-1}
Temperature	25 °C
Injection size	20 μl containing 10 μmol l^{-1}
Detection	UV, λ = 254 and 280 nm; 0.01 AUFS
Recording	Waters M 730 data module integrator
Retention times, min	5-Fluorouracil 4.52; Xanthine 8.60 Uric acid 6.05; Oxipurinol 10.78 Hypoxanthine 7.25; Allopurinol 12.18 Uridine 7.90
Ref.	Wung WE and Howell SB (1980) Clin. Chem. 26:1704

HPLC 3

Sample	Blood plasma
Extraction	Ether/n-propanol (80:20) and then re-extraction on Shephadex LH-20 extraction column with 0.005 M K_2HPO_4, pH 9.8
Apparatus	Waters ALC M 202 LC with 440 UV detector and U6K injector
Column	Anion-exchange; Aminex A-25, 17.5 μm, 300 × 2.0 mm ID
Mobile phase	0.3 M Acetate buffer, pH 4.5
Flow-rate	0.7 ml min^{-1}
Temperature	30 °C
Internal standard	5-Chlorouracil
Injection size	5–25 μl, 100–300 ng ml^{-1}
Detection	UV, λ = 254 nm
Retention times, min	5-Fluorouracil 8.0 5-Chlorouracil 17.0
Ref.	Cohen JL and Brown RE (1978) J. Chromatogr. 151:237

HPLC 4

Sample	Mixture of 5-Fluorouracil derivatives
Apparatus	Waters M 6000A LC, 440 UV detector and U6K injector
Column	A. Reversed-phase; μ-Bondapak C_{18}, 10 μm, 300 × 4 mm ID B. Reversed-phase; RCM-C_{18}, 10 μm, 100 × 8 mm ID

	C. Reversed-phase; Zorbax TMS 7–8 μm, 250 × 4.6 mm ID
	D. Reversed-phase; Zorbax C_8, 7–8 μm, 250 × 4.6 mm ID
Mobile phase	For A: 0.5 mM Tetrabutylammonium hydrogen sulfate C_{16} ions and 20 mM sodium acetate, pH 4.0
	For B: 0.1 mM Tetrabutylammonium hydrogen sulfate C_{16} ions and 10 mM sodium acetate, pH 6.3
	For C: 2.5 mM Triethylammonium hydroxide C_6, pH 4.3
	For D: 50 mM Triethylammonium hydroxide C_6 ions, 10 mM tetrabutylammonium hydrogen sulfate C_{16} ions, 10 mM sodium acetate and 1 mM Na_3PO_4, pH 6.2
Detection	UV, λ = 254 and 280 nm
Capacity factor, *k′*	

Column	A	B	C	D
5-Fluorouracil	1.2	2.2	1.8	1.7
5-Fluorouridine	2.4	5.0	1.8	3.2
5-Fluorodeoxyuridine	3.3	6.4	1.8	3.6

Ref. AU JL-S, Wientjes MG et al (1982) J. Chromatogr. Biomed. Appl. 228:245

HPLC 5

Sample	5-Fluorouracil derivatives
Apparatus	LKB M 2150 HPLC pumps with 2152 HPLC controller, connected with column switching system, Perkin- Elmer M 650-10S fluorescence spectrophotometer and LKB M 2151 UV detector
Column	A. Cyanopropyl phase; CPS-Hypersil, 150 × 4.6 mm ID
	B. Reversed-phase; ODS-Hypersil, 150 × 4.6 mm ID
Mobile phase	For A. Methanol/water (1:1, v/v)
	For B. Methanol/water (6:4, v/v)
Flow-rate	1 ml min^{-1}
Temperature	35 °C (A); ambient (B)
Derivatization	Pre-column; with 4-bromomethyl-7-methoxy-coumarin
Injection size	100 μl
Detection	Fluorescence, λ = 325/395 nm
	UV, λ = 325 nm
Retention times, min	(estimated from drawing)

Mobile phase	A.	B.
5-Fluorouracil derivative	10.5	6.8

Ref. Kindberg CG, Slavik M et al (1989) J. Pharm. Biomed. Anal. 7:459

HPLC 6

Sample	Bile, liver homogenate, plasma and urine from pigs
Extraction	Deproteinization with acetonitrile
Apparatus	Orlita M 1044 LC pump, Rheodyne M 7125 injector and Kratos M 757 UV detector
Column	Hamilton PRP-1; 10 µm, 100 × 3 mm ID
Mobile phase	50 mM Tris (tris-(hydroxymethyl)-aminomethane) and 5 mM cetyltrimethylammonium bromide, pH 8.2 with concentrated HCl
Flow-rate	1 ml min^{-1}
Temperature	Ambient
Injection size	100 µl containing 5 ng ml^{-1}
Detection	UV, λ = 268 nm; 0.0001 AU
Retention times, min	5-Fluorouracil 4.0 (estimated from drawing)
Ref.	Bierman JAC, Lingeman H et al (1989) J. Pharm. Biomed. Anal. 7:611

83 Flurbiprofen

CAS-No	5104-49-4	
Chemical name	2-(2-Fluorobiphenyl-4-yl)propionic acid	
Empirical formula	$C_{15}H_{13}FO_2$	
Molecular mass	244.3	
Proprietary names	Cebutid; Froben	
Clinical use	Antiphlogistic/Analgesic	
Solubility	Water	most organic solvents
	slightly soluble	freely soluble
Light absorption	λ_{max} = 247 nm in aqueous acid	

HPLC 1

Sample	Dosage forms and biological materials
Extraction	Etyhl acetate under alkaline condition
Apparatus	Waters M 6000A LC pump with 660 solvent programmer and 420E fluorescence detector
Column	Phenyl phase; µ-Bondapak, 300 × 3.9 mm ID
Mobile phase	Acetonitrile/water (35:65)
Flow-rate	1.5 ml min^{-1}; 1500 psi
Injection size	15 µl, 0.5–20 µg ml^{-1}
Detection	Fluorescence
Recording	Philips PM 8251 single-pen recorder
Retention times, min	Flurbiprofen 6.5
Ref.	Babhair SA (1988) J. Liq. Chromatogr. 11:463

HPLC 2

Sample	Blood plasma and breast milk
Extraction	Deproteinization with 30% perchloric acid
Apparatus	Waters M 6000A LC pump, WISP 710A injector and 440 UV detector
Column	Reversed-phase; μ-Bondapak C_8, Radial Pak, 10 μm, 100 × 8 mm ID
Pre-column	Reversed-phase; C_{18}
Mobile phase	Acetonitrile/methanol/1% acetic acid (30:30:40)
Flow-rate	2 ml min^{-1}
Temperature	Ambient (23 °C)
Internal standard	Propionic acid derivative
Injection size	50 μl, 0.05–2.0 μg ml^{-1}
Detection	UV, λ = 254 nm
Recording	Water M 730 data module
Retention times, min	Flurbiprofen 6.41 Internal standard 7.28
Ref.	Johnson VA and Wilson JT (1986) J. Chromatogr. Biomed. Appl. 382:367

84 Folic Acid

CAS-No	59-30-3			
Chemical name	Pteroylglutamic acid			
Empirical formula	$C_{19}H_{19}N_7O_6$			
Molecular mass	441.4			
Proprietary names	Folacid; Foldine; Folvite			
Clinical use	Vitamin			
Solubility	Water	alcohol	aq NaOH	aq HCl
	insoluble	insoluble	soluble	soluble
Light absorption	λ_{max} = 256 nm in aqueous alkali			

HPLC 1

Sample	Raw materials
Apparatus	Waters M 510 HPLC pumps, U6K injection valve, 481 variable-wavelength detector and Bioanalytical Systems TL-5A electrochemical detector
Column	Reversed-phase; Zorbax ODS 5 μm, 150 × 4.6 mm ID
Mobile phase	Methanol/0.1 M phosphate buffer, pH 6.0 (1:3, v/v)
Flow-rate	1 ml min^{-1}
Temperature	Ambient
Injection size	20 μl containing 0.5 ppm

Detection	UV, $\lambda = 263$ nm; EC, GC electrode + 0.6 + 1, 1 V vs Ag/AgCl RE
Retention times, min	Ascorbic acid 1.65 Folic acid 3.10 *p*-Aminobenzoic acid 2.20 Nicotinamide 3.80
Ref.	Wang E and Hou W (1988) J. Chromatogr. 447:256

HPLC 2

Sample	Raw materials
Apparatus	Du Pont M 820 HPLC system and ultraviolet detector
Column	Reversed-phase; μ-Bondapak C_{18}, 10 μm, 300 × 4.0 mm ID
Mobile phase	35.1 g Sodium perchlorate monohydrate, 1.36 g monobasic potassium phosphate, 6.94 ml of 1 N potassium hydroxide and 40 ml methanol and water up to 1 liter, pH 7.2 with 1 N KOH
Flow-rate	1.0–5.0 ml min^{-1}
Temperature	Ambient and 35 °C
Injection size	3–6 μl containing 30 μg
Detection	UV, $\lambda = 254$ nm; 0.08 AUFS

Retention times, min

Temperature	Ambient	35 °C
p-Aminobenzoic acid	2.9	2.7
N-(*p*-aminobenzoyl)-L glutamic acid	2.9	2.7
Pteridinecarboxylic acid derivate	8.2	5.7
2-Amino-4(1H)-pteridione	13.9	
Folic acid	19.4	12.1

Ref.	Reif VD, Reamer JT and Grady LT (1977) J. Pharm. Sci. 66:1112

HPLC 3

Sample	Multivitamin preparation
Column	Reversed-phase; Micropak MCH
Mobile phase	A. Water containing 0.01 M pentanesulfonic acid and 0.1 N citric acid B. Methanol/water (90:10) containing 0.01 M pentanesulfonic acid and 0.1 N citric acid Gradient: B 0% to 50% at 2% min^{-1}
Injection size	1 μg
Detection	Fluorescence, λ = 7–54 + 7–60/4–76 + 3–70 UV, $\lambda = 254$ nm; 0.2 AUFS
Retention times, min	(estimated from drawing)

Ascorbic acid	1.4	Thiamine	11.2
Niacin	2.5	Folic acid	18.4
Niacinamide	3.2	Riboflavine	20.2
Pyridoxine	6.0	Vitamin B_{12}	23.0

Ref.	Wehr CT: Varian Liquid chromatography at work 61. Varian, California

HPLC 4

Sample	Raw materials
Column	Reversed-phase; ODS 5–10 µm, 300 × 4 mm ID
Mobile phase	35.1 g Sodium perchlorate, 1.40 g monobasic potassium phosphate, 7.0 ml of 1 N potassium hydroxide and 40 ml methanol and make 1 liter with water, pH 7.2 with 1 N KOH
Flow-rate	According to suitability
Temperature	Ambient
Injection size	25 µl containing 5 to 20 µg
Detection	UV, $\lambda = 254$ nm
Ref.	Anonym (1985): The United States Pharmacopeia XXI and The National Formulary XVI. United States Pharmacopeial Convention Inc., Rockville, p 449

HPLC 5

Sample	Blood (with 5-methyltetrahydrofolate)
Extraction	1% (w/v) Ascorbate solution on extraction columns
Column	Reversed-phase; µ-Bondapak C_{18}, 10 µm, 300 × 3.9 mm ID
Mobile phase	A. 0.1 M Acetate buffer, pH 5.5 B. Acetonitrile Linear gradient: B 0% to 34% in 30 min
Flow-rate	0.5 ml min^{-1}
Injection size	100–200 µl
Detection	UV, $\lambda = 280$ nm
Retention times, min	5-Methyltetrahydrofolate 24.0 (estimated from drawing)
Ref.	Shearer MJ (1986) in: Lim CK (ed) HPLC of small molecules, a practical approach. IRL Press, Oxford, Washington DC, p 157

85 Fructose

CAS-No	57-48-7
Chemical name	D-(−)-Fructopyranose
Empirical formula	$C_6H_{12}O_6$
Molecular mass	180.2
Proprietary names	Fructal; Laevosan; Laevuflex

Clinical use	Sugar
Solubility	Water 1 in 0.3; alcohol 1 in 15; methanol 1 in 14; ether, chloroform insoluble

---

HPLC 1

Sample	Sugars
Apparatus	Applied Chromatography Systems M 750/03 HPLC and 750/13 differential refractometer
Column	Normal phase; Lichrosorb Si 60, 5 μm, 220 × 4 or 7 mm ID
Mobile phase	73% (v/v) Acetonitrile in water and with 0.01% amine modifier I (from Natec Hamburg)
Flow-rate	2.2 ml min^{-1}
Injection size	20 μl
Detection	RI
Retention times, min	(estimated from drawing) Glycerol 7.0; Fructose 12.0; Glucose 14.8; Sucrose 19.5; Maltose 24.0
Ref.	Johncock SIM and Wagstaffe PJ (1980) Analyst 105:581

HPLC 2

Sample	Sugar solution
Apparatus	HPLC System with refractive index detector
Column	Aminex (Q-150-S(K^+); Biorad, 600 × 6 mm ID
Mobile phase	Water
Temperature	60 °C
Injection size	5–40 μl, 1–50 mg ml^{-1}
Detection	RI
Retention times, min	(estimated from drawing)

Flow-rate, ml min^{-1}	1.85	0.85
Sucrose	6.50	17.00
Glucose	8.00	21.50
Fructose	9.00	22.80

Ref.	Palmer JK (1979) in: Tsuji K (ed) GLC and HPLC Determination of Therapeutic Agents, Part 3. Marcel Dekker Inc., New York and Basel, p 1317

HPLC 3

Sample	Sugars and sweet wood extract
Apparatus	HPLC with electrochemical, UV and refractive index detectors
Column	Reversed-phase
Mobile phase	Acetonitrile/water (75:25)
Flow-rate	Adjusted to retention times

Detection	EC; UV, λ = 200 nm; RI
Retention times, min	(estimated from drawing) Fructose 6.5 Sucrose 12.8 Glucose 8.0 Maltose 16.5
Ref.	Meyer V (1984): Praxis der Hochleistungsfluessig-chromatographie, 3. edn. Diesterweg Salle Sauerlaender, Frankfurt and Salzburg, p 66

HPLC 4

Sample	Monosaccharides
Apparatus	Altex HPLC with Knauer 103.07 refractive index detector and Spectra-Physics SF 770 UV detector
Column	Amino phase; Polygosil 60–5 NH_2, 250 × 4.6 mm ID
Mobile phase	Acetonitrile/water (75 : 25)
Flow-rate	2.0 ml min^{-1}
Detection	UV, λ = 188 nm
Retention times, min	(estimated from drawing) Fructose 9.0 Rhamnose 5.7 Mannose 10.0 Xylose 6.4 Glucose 11.2 Arabinose 7.5 Galactose 12.4
Ref.	Binder H (1980) J. Chromatogr. 189 : 414

86 Furazolidone

CAS-No	67-45-8
Chemical name	3-(5-Nitrofurfurylideneamino)-2-oxazolidone
Empirical formula	$C_8H_7N_3O_5$
Molecular mass	225.2
Proprietary names	Furoxone; Giardil; Tricofuron
Clinical use	Antimicrobial
Solubility	Water and ethanol: Very slightly soluble; chloroform: Slightly soluble; ether: insoluble
Light absorption	λ_{max} = 259 nm and 367 nm in dimethylformamide and dilution with water

HPLC 1

Sample	Swine blood plasma and tissues
Extraction	Ethyl acetate on Merck Extrelut column
Apparatus	Waters M 6000A LC pumps with WISP 710B autosampler and Kratos M 773 spectrophotometric detector

Column	Reversed-phase; Shandon Hypersil ODS 5, 250 × 4.6 mm ID
Pre-column	Brownlee Labs OD-GU RP18, 30 × 4.6 mm ID
Mobile phase	Water/acetonitrile/acetate buffer (containing 41 g sodium acetate and 30 g acetic acid in 1 liter water (675 : 250 : 75)
Flow-rate	1 ml min^{-1}
Injection size	50–100 µl, 0–1000 ng ml^{-1}
Detection	UV, λ = 362 nm; 0.005 AUFS
Retention times, min	Furazolidone 8.5–10.0 (estimated from drawing)
Ref.	Vroomen LHM, Berghmans MCJ and Van der Struijs TDB (1986) J. Chromatogr. 362 : 141

HPLC 2

Sample	Swine liver and kidneys
Extraction	First ethyl acetate, and then with acetonitrile
Apparatus	Hewlett-Packard 1084B LC pump and HP 79875A variable wavelength UV detector
Column	Reversed-phase; Hypersil SAS, 250 × 4.6 mm ID
Pre-column	Reversed-phase; C_8, 30–40 µm, 100 × 2.1 mm ID
Mobile phase	Water/acetonitrile (75 : 25)
Flow-rate	1.75 ml min^{-1}
Temperature	Ambient
Injection size	20 µl containing 10–100 ng
Detection	UV, λ = 360 nm
Retention time, min	Furazolidone 4.8 (estimated from drawing)
Ref.	Ernst GF and Van der Kaaden A (1980) J. Chromatogr. 198 : 526

HPLC 3

Sample	Pork tissues
Extraction	First absolute ethanol, and then with ethyl acetate on alumina column
Apparatus	Waters ALC/GPC M 204 with U6K injector and M 6000A pump and Perkin-Elmer LC-75 variable-wavelength detector
Column	Reversed-phase; Brownlee C_8, 10 µm, 250 × 4.6 mm ID
Pre-column	Brownlee RP-GU MPLC C_8, 10 µm
Mobile phase	Acetonitrile/0.01 M ammonium acetate/ethanol (25 : 70 : 5), pH 6.8
Flow-rate	1 ml min^{-1}
Injection size	4–100 µl containing 2–40 ng
Detection	UV, λ = 350
Recording	Hewlett-Packard 3385A integrator

Retention times, min	Quinoxaline-2-carboxylic acid 3.49	Nitrofurazone 6.55
	Carbadox 4.98	Furazolidone 8.66
		Deoxycarbadox 10.37
Ref.	Macintosh AI and Neville GA (1984) J. Assoc. Off. Anal. Chem. 67 : 958	

HPLC 4

Sample	Fish
Extraction	95% Acetonitrile on alumina column
Column	Reversed-phase; Nucleosil C_{18}, 5 µm, 250 × 4 mm ID
Mobile phase	Acetonitrile/acetic acid/water (70 : 0.1 : 30)
Flow-rate	0.8 ml min^{-1}
Injection size	10 µl, 2.5 µg ml^{-1}
Detection	UV, λ = 380 nm; 0.04 AUFS
Retention times, min	(estimated from drawing) Furazolidone 2.7 Difurazone 7.9
Ref.	Hori Yoshihiro (1984) J. Fd. Hyg. Soc. Jap. 25 : 158

HPLC 5

Sample	Swine blood plasma
Extraction	Ethyl acetate
Apparatus	Altex M 100A LC pump with 210 injector, Bioanalytical Systems electrochemical and Schoeffel SF 770 UV detectors
Column	Reversed-phase; Partisil PXS 10/25 ODS
Mobile phase	Methanol/water (30 : 70), pH 4.0 with 16.5 mM dibasic sodium phosphate and 13.1 mM sodium citrate
Flow-rate	1.1 ml min^{-1}
Injection size	70 µl containing 47 ng
Detection	UV, λ = 362 nm; EC, − 0.75 V vs Ag/AgCl RE
Retention time, min	Furazolidone 4.35 (estimated from drawing)
Ref.	Veale HS and Harrington GW (1982) J. Chromatogr. 240 : 230

87 Furosemide

CAS-No	54-31-9
Chemical name	4-Chloro-*N*-furfuryl-5-sulfamoylanthranilic acid
Empirical formula	$C_{12}H_{11}ClN_2O_5S$
Molecular mass	330.8
Proprietary names	Frusid, Lasix; Uritol

Clinical use	Diuretic
Solubility	Water insoluble; alcohol 1 in 75; methanol soluble; aq NaOH freely soluble
Light absorption	λ_{max} = 235 nm in aqueous acid

HPLC 1

Sample	Urine
Extraction	Centrifugation
Apparatus	Altex M 110A LC pump, LDC-Spectromonitor III UV detector, Perkin-Elmer LC 1000 fluorescence detector and Rheodyne M 70–10 injector
Column	Reversed-phase; Lichrosorb C_8, 5 or 10 μm, 150 × 4.5 mm ID
Mobile phase	33% Acetonitrile in 0.1 M phosphoric acid
Flow-rate	1.0 ml min^{-1}
Temperature	Ambient
Injection size	20 μl, 0.2 μmol l^{-1}
Detection	UV, λ = 274 nm Fluorescence, λ = 270/389 nm
Retention time, min	Furosemide 7.0 (estimated from drawing)
Ref.	Lagerstroem P-O (1981) J. Chromatogr. Biomed. Appl. 225 : 476

HPLC 2

Sample	Horse blood plasma
Extraction	Filtration
Column	ISRP (internal surface reversed-phase) 5 μm; 250 × 6.0 mm ID
Mobile phase	I. 0.1 M KH_2PO_4/isopropanol/tetrahydrofuran (84 : 10 : 6), pH 6.8 II. 0.1 M KH_2PO_4/tetrahydrofuran (90 : 10), pH 7.5
Flow-rate	0.6 ml min^{-1}
Temperature	Ambient
Injection size	10 μl, 20 μg ml^{-1}
Detection	UV, λ = 241 nm; 0.05 AUFS

Retention times, min

Mobile phase	Oxyphen-butazone	Phenylbutazone	Furosemide
I.	8.0	11.0	13.0
II.	9.0	12.0	17.0

(estimated from drawing)

Ref.	Pinkerton TC, Perry JA and Rateike JD (1986) J. Chromatogr. 367 : 412

HPLC 3

Sample	Blood plasma and urine
Extraction	Diethyl ether under acidic, and then under alkaline conditions

Apparatus	Varian HPLC system with Rheodyne injection loop and Schoeffel FS 970 fluorescence monitor
Column	Reversed-phase; Varian CH–10
Mobile phase	0.5% Acetic acid in acetonitrile/water (30 : 70)
Flow-rate	1 ml min^{-1}; 2498 psi
Injection size	21 µl, 0.2–500 µg ml^{-1}
Detection	Fluorescence, $\lambda = 350/389$ nm
Retention time, min	Furosemide 3.0
Ref.	Swezey SE, Meffin PJ and Blaschke TF (1979) J. Chromatogr. 174 : 469

HPLC 4

Sample	Blood plasma
Extraction	Deproteinization with acetonitrile
Apparatus	Waters M 6000A HPLC with U6K injection loop and Schoeffel FS 970 fluorescence detector
Column	Reversed-phase; µ-Bondapak C_{18}, 10 µm, 300 × 3.9 mm ID
Mobile phase	Acetonitrile/0.05% phosphoric acid (30 : 70)
Flow-rate	2 ml min^{-1}; 1500 psi
Temperature	Ambient (24 °C)
Injection size	10 µl, 1.27 µg ml^{-1}
Detection	Fluorescence, $\lambda = 225$ nm (excitation)
Retention time, min	Furosemide 8.0 (estimated from drawing)
Ref.	Nation RL, Peng GW and Chiou WL (1979) J. Chromatogr. Biomed. Appl. 162 : 88

HPLC 5

Sample	Blood plasma and urine
Extraction	Plasma: Deproteinization with methanol (Method I) and with acetonitrile (Method II); Urine: Direct injection
Apparatus	Waters ALC/GPC M LC-244, UV detector and U6K injector
Column	Reversed-phase; µ-Bondapak C_{18}, 10 µm, 300 × 3.9 mm ID
Mobile phase	Method I: Methanol/0.01 M sodium acetate, pH 5.0 (35 : 65) (prepared with 350 ml methanol, 650 ml water and 0.6 ml CH_3COOH, pH 5 with 4 M NaOH) Method II: Acetonitrile/0.01 M sodium acetate, pH 5.0 (25 : 75) (prepared with 250 ml acetonitrile, 750 ml water, 0.6 ml glacial acetic acid, pH 5.0 with 4 M NaOH)
Flow-rate	2.0 ml min^{-1}, 6090 psi
Temperature	Ambient

Internal standards	Sodium cephalothin (Method I); sodium phenobarbital (Method II)
Injection size	15–20 μl
Detection	UV; λ = 280 nm; 254 nm (phenobarbital); 0.01 AUFS
Retention times, min	Method I II Cephalothin 6.0 Furosemide 8.0 5.0 Pentobarbital 7.0
Ref.	Lin ET, Smith DE et al (1979) J. Chromatogr. Biomed. Appl. 163 : 315

HPLC 6

Sample	Blood plasma and urine
Extraction	Diethyl ether/n-hexane (65:35) under acidic condition
Apparatus	Hewlett-Packard 1081B HPLC with Perkin-Elmer M 3000 fluorescence spectrometer
Column	Reversed-phase; Lichrosorb C_8, 5 μm, 150 × 4.6 mm ID
Mobile phase	Methanol/0.02 M phosphate buffer, pH 3.0 (1:1)
Flow-rate	1.0 ml min^{-1}; 3480 psi
Temperature	35 °C
Internal standard	Desmethylnaproxen
Injection size	10 μl
Detection	Fluorescence, λ = 275/410 nm (plasma) and 235/400 nm (urine)
Retention times, min	Furosemide 4.2 Degradation products Internal standard 6.0 Furosemide 1 1.93 Furosemide 2 2.85 Furosemide 3 5.43
Ref.	Kerremans ALM, Tan Y et al (1982) J. Chromatogr. Biomed. Appl. 229:129

88 Gentamicin

CAS-No	1403-66-3; 1405-41-0 (sulfate)
Chemically	Aminoglycoside antibiotic containing gentamicin C_1, C_{1A}, C_2 and C_{2A} and C_{2B} (sagamicin) produced by Micromonospora purpurea
Empirical formula	Gentamicin C_1 Gentamicin C_2 Sulfate $C_{21}H_{43}N_5O_7$ $C_{20}H_{41}N_5O_7$
Molecular mass	477.6 463.6
Proprietary names	Biogen; Cidomycin; Garamycin
Clinical use	Antibiotic

Solubility,	Water	alcohol, chloroform and ether
Gentamicin	freely soluble	sparingly soluble
Sulfate	soluble	insoluble
Light absorption	λ_{max} = 251 nm in aqueous alkali	

HPLC 1

Sample	Raw materials
Apparatus	Beckmann M LC-114 pumps, 420 gradient controller and 165 UV/VIS detector
Column	Reversed-phase; Ultrasphere ODS, 150 × 4.6 mm ID
Pre-column	Reversd-phase; Vydac, 45 × 4.6 mm ID
Mobile phase	A. 5 g Heptanesulfonic acid sodium salt in acetic acid/water/methanol (50:250:700) B. Methanol Linear gradient: In 2 min after injection of the sample, 100% A to A-B (75:25) in 3 min at a rate of 8.33% B min^{-1}
Flow-rate	1.5 ml min^{-1}
Derivatization	Pre-column; with *o*-phthalaldehyde
Injection size	20 µl, 400 µg ml^{-1}
Detection	UV, λ = 330 nm; 0.2 AUFS
Recording	Shimadzu C-R1B recording integrator
Retention times, min	Gentamicin C_1 5.07 Gentamicin C_{2a} 12.67 Gentamicin C_{1a} 11.06 Gentamicin C_2 13.77
Ref.	Albracht JH and De Wit MS (1987) J. Chromatogr. 389:306

HPLC 2

Sample	Serum
Extraction	Methanol after derivatization on a silicic acid column
Apparatus	Water M 6000A LC pump and Schoeffel FS 9701C fluorescence detector
Column	Reversed-phase; µ-Bondapak C_{18}, 10 µm, 300 × 3.9 mm ID
Pre-column	Reversed-phase; µ-Bondapak C_{18}/Porasil B, 23 × 3.9 mm ID
Mobile phase	1% Triethylamine, pH 6.2 with H_3PO_4/methanol (79:21)
Flow-rate	2 ml min^{-1}
Derivatization	Pre-column; with *o*-phthalaldehyde
Injection size	20 µl
Detection	Fluorescence, λ = 260/418 nm
Retention times, min	Gentamicin C_1 4.8 Gentamicin C_2 8.6 Gentamicin C_{1a} 6.8

Ref.	Rumble RH and Roberts MS (1987) J. Chromatogr. Biomed. Appl. 419:408

HPLC 3

Sample	Raw materials
Apparatus	Dosapro Milton Roy LC minipump with pulse dampener, Rheodyne M 7125 injector and Waters M 420 fluorescence detector
Column	Cation-exchange; 10 μm, 250 × 4.6 mm ID × 2
Mobile phase	1.1 M KCl, pH 3.0 with HCl
Derivatization	Post-column; with *o*-phthalaldehyde
Flow-rate	2 ml min^{-1}
Temperature	Ambient
Injection size	20 μl containing 800 ng
Detection	Fluorescence, λ = 338/455 nm
Recording	Hewlett-Packard 3390 integrator
Retention times, min	(estimated from drawing) Gentamicin C_{2a} 10.0 Gentamicin C_{1a} 6.4 Gentamicin C_{2b} 11.6 Gentamicin C_2 7.8 Gentamicin C_1 14.0
Ref.	Seidl G and Nerad HP (1988) Chromatographia 25:169

HPLC 4

Sample	Injectable dosage forms
Apparatus	Varian M 5000 LC pump with Spark SPH 125 autosampler, Rheodyne M 7010 injection valve and Perkin-Elmer LC-55 UV/VIS detector
Column	Reversed-phase; Hypersil ODS-5 μm, 160 × 4.6 mm ID
Mobile phase	Methanol/water (72:28, v/v) containing 5% glacial acetic acid and 0.5% w/v 1-heptanesulfonic acid, sodium salt, pH 3.00
Flow-rate	1.3 ml min^{-1}
Temperature	30 °C
Derivatization	Pre-column; with *o*-phthalaldehyde
Injection size	20 μl
Detection	UV, λ = 330 nm
Recording	Spectra-Physics 4270 computing integrator
Retention times, min	Gentamicin C_1 4.9 Gentamicin C_{2a} 17.5 Gentamicin C_{1a} 12.5 Gentamicin C_2 20.9 (estimated from drawing)
Ref.	Van Duijn JW, Barends DM et al (1988) Pharm. Weekbl. Sci. 10:267

HPLC 5

Sample	Guineapig and human serum
Extraction	Deproteinization with acetonitrile

Apparatus	Perkin-Elmer Series 410-LC pump and LC-75 UV/VIS detector
Column	Reversed-phase; μ-Bondapak C_{18}, 10 μm, 300 × 3.9 mm ID
Pre-column	Reversed-phase; C_{18}, 5 μm, 33 × 4.6 mm ID
Mobile phase	300 ml Water with 700 ml acetonitrile and 1 ml acetic acid
Flow-rate	2.2 ml min^{-1}; 11.8 MPa
Temperature	Ambient
Derivatization	Pre-column; with 1-fluoro-2,4-dinitrobenzene
Injection size	50 μl
Detection	UV, $\lambda = 365$ nm
Recording	Perkin-Elmer computing integrator ICI-100
Retention times, min	Gentamicin C_{1a} 11.0 Netilmicin (IS) 13.5
Ref.	Dionisotti S, Bamonte F et al (1988) J. Chromatogr. Biomed. Appl.434:169

HPLC 6

Sample	Serum
Extraction	Methylene chloride after deproteinization with acetonitrile, pH 7.4
Apparatus	Pye-Unicam LC-3 HPLC and UV detector LC-3
Column	Reversed-phase; Lichrosorb C_{18}, 10 μm, 250 × 4.6 mm ID
Mobile phase	Acetonitrile/methylene chloride/water/methanol (80:10:8:4)
Flow-rate	4.0 ml min^{-1}
Derivatization	Pre-column; with benzene sulfonyl chloride (BSC)
Internal standard	Netilmicin
Injection size	20 μl
Detection	UV, $\lambda = 230$ nm
Retention times, min	Gentamicin 3.5 Netilmicin 5.5
Ref.	Larsen N-E and Marinelli K (1980) J. Chromatogr. Biomed. Appl. 221:182

89 Globulins

Chemically	Protein
Proprietary names	Gamma Globulin Kabi; Gammar; Immu-G
Clinical use	Antiinfective

HPLC 1

Sample	Skim milk
Extraction	Precipitation of casein at pH 4.6
Column	TSK 3000 SW; silica gel 10 μm, 600 × 7.5 mm ID
Mobile phase	Buffer with 0.1 M NaH_2PO_4, 0.05 M NaCl and 0.02% $NaNO_3$, pH 6.8
Flow-rate	0.5 ml min^{-1}
Injection size	100 μl
Detection	UV, λ = 280 nm
Retention times, min	(estimated from drawing)

Proteins of higher molecular weight		18.0
γ-Globulin	M = 150.000	26.0
Bovine serum albumin	69.000	28.0
β-Lactoglobulin	35.000	33.0
α-Lactoglobulin	16.500	37.0

Ref.	Meyer V (1984): Praxis der Hochleistungsfluessigchromatographie, 3rd edn. Diesterweg Salle and Sauerlaender, Frankfurt am Main and Salzburg, p 171

HPLC 2

Sample	Intravenous Immunoglobulin-G Albumin preparations
Apparatus	Tosoh CCPM LC pump, Rheodyne M 7125 injector and Tosoh 8000-UV detector
Column	TSK-gel; G-3000SWXL, 300 × 7.8 mm ID
Mobile phase	A. 50 mM Sodium phosphate, pH 6.7 containing 0.3 M NaCl B. 50 mM Sodium phosphate, pH 5.0 containing 0.1 M Na_2SO_4
Flow-rate	1.0 ml min^{-1}
Temperature	25 °C
Injection size	100 μl containing 100 μg IgG and albumin
Detection	UV, λ = 280 nm; 0.64 AUFS
Retention times, min	(estimated from drawing)

Immuno Globulin G	7.50 (A and B)
Albumin	8.75 (B)

Ref.	Yamasaki Y and Kato Y (1989) J. Chromatogr. 467:436

HPLC 3

Sample	Intravenous Immunoglobulin-G Albumin formulations
Apparatus	Du Pont HPLC with M 8800 gradient controller, Kratos M 783 UV detector and Waters WISP 710B autosampler
Column	Anion-exchange; Waters Protein Pak DEAE 5 PW, 75 × 7.5 mm ID

Mobile phase	A. 20 mM Tris, pH 8.5 B. 20 mM Tris, pH 7.0 containing 0.3 M sodium chloride Linear gradient: 100% A to 100% B in 30 min; 100% B to 100% A in 5 min
Flow-rate	1.0 ml min^{-1}
Temperature	Ambient
Injection size	200 µl
Detection	UV, λ = 280 nm; 1.0 AUFS
Recording	Hewlett-Packard 3357 LAS computer with Omniscribe recorder
Retention times, min	(estimated from drawing) Immunoglobulin G 14.0 Sodium acetyl-tryptophanate 21.6 Albumin 23.4
Ref.	Lee JK, Deluccia FJ et al (1988) J. Chromatogr. 444:141

HPLC 4

Sample	Intravenous Immunoglobulin G-Albumin formulations
Apparatus	Waters M 6000A LC pump, 440 UV detector and WISP 710B autosampler
Column	Beckman Spherogel TM; TSK 3000 SW, 600 × 7.5 mm ID
Mobile phase	0.2 M Dibasic potassium phosphate, pH 7.0 containing 0.02% sodium azide
Flow-rate	0.5 ml min^{-1}
Temperature	Ambient
Injection size	200 µl
Detection	UV, λ = 280 nm; 0.5 AUFS
Recording	Hewlett-Packard 3357 LAS computer with Omniscribe recorder
Retention times, min	Immunoglobulin G: Aggregate 18.2, Dimer 21.6, Monomer 25.8 Albumin: Aggregate 18.4, Dimer 25.5, Monomer 30.1
Ref.	Lee JK, Deluccia FJ et al (1988) J. Chromatogr. 444:141

90 Griseofulvin

CAS-No	126-07-8
Chemical name	7-Chloro-2′,4,6-trimethoxy-6′-methylspiro-[benzofuran-2(3*H*),1′-[2]cyclohexene]-3′,4′-dione produced by Penicillium griseofulvum

Empirical formula	$C_{17}H_{17}ClO_6$
Molecular mass	352.8
Proprietary names	Fulcin; Fulvicin; Likuden
Clinical use	Antifungal
Solubility	Water: very slightly soluble; alcohol 100%: 1 in 300; acetone: 1 in 20; chloroform: 1 in 25
Light absorption	λ_{max} = 236 nm and 291 nm in dehydrated alcohol

---

HPLC 1

Sample	Blood plasma
Extraction	Diethyl ether on Merck Extrelut extraction column or deproteinization with ethanol
Apparatus	Orlita DMPAE M 10.4 pump, Perkin-Elmer LC 1000 fluorescence detector and Hewlett-Packard 7671 autosampler
Column	Reversed-phase; Lichrosorb C_8, 5 μm, 100 × 3 mm ID
Mobile phase	Acetonitrile/water (40:60)
Flow-rate	1.2 ml min^{-1}; 2146 psi
Temperature	20 °C
Injection size	25 μl, 1–10 μg ml^{-1}
Detection	Fluorescence, λ = 297/428 nm
Retention times, min	Griseofulvin 5.0
Ref.	Kreuzig F (1979) J. Chromatogr. Biomed. Appl. 163:322

HPLC 2

Sample	Bulk drug and dosage forms
Apparatus	Waters M 6000A LC pumps, 440 UV detector and U6K injector
Column	Cyano phase; Zorbax CN, 10 μm, 300 × 4.6 mm ID
Mobile phase	Methanol/water (3:2, v/v)
Flow-rate	1 or 2 ml min^{-1}
Internal standard	*m*-Phenylphenol
Injection size	20 μl
Detection	UV, λ = 254 nm
Retention times, min	(estimated from drawing)

Griseofulvic acid	2.0	Isogriseofulvin	6.9
Dechlorogriseofulvin	5.0	Tetrahydro-griseofulvin	7.8
Griseofulvin	6.5	Internal standard	8.8

Ref.	Townley E and Roden P (1980) J. Pharm. Sci. 69:523

HPLC 3

Apparatus	Waters M 6000A LC pumps, U6K injector and Schoeffel FS 970 fluorescene detector

Column	Reversed-phase; μ-Bondapak, 300 mm microparticulate
Mobile phase	Water/acetonitrile (1:1)
Flow-rate	2 ml min^{-1}
Temperature	Ambient
Detection	Fluorescence, λ = 260 nn (ex)
Recording	Laboratory data control potentiometer
Retention time, min	Griseofulvin 3.8
Ref.	Nation RL, Peng GW et al (1978) J. Pharm. Sci. 67:805

HPLC 4

Apparatus	Waters HPLC pumps with M 450 UV detector and U6K loop injector
Column	Reversed-phase; μ-Bondapak C_{18}, 300 × 4 mm ID
Mobile phase	45% Acetonitrile in 45 mM potassium phosphate, pH 3.0 with phosphoric acid
Flow-rate	2.5 ml min^{-1}
Detection	UV, λ = 295 nm
Retention time, min	Griseofulvin 4.8
Ref.	Hackett LP and Dusci LJ (1978) J. Chromatogr. 155:206

HPLC 5

Sample	Raw materials
Column	Reversed-phase; ODS-Sil-X-II, 1000 × 2.1 ID
Mobile phase	Methanol/0.05 M ammonium carbonate (30:70)
Flow-rate	0.6 ml min^{-1}; 1000 psi
Detection	UV, λ = 254 nm; 0.16 AUFS
Retention time, min	Griseofulvin 12.0 (estimated from drawing)
Ref.	White ER, Carroll MA et al (1975) J. Antibiotics 28:205

91 Haloperidol

CAS-No	52-86-8
Chemical name	4-[4-(4-Chlorophenyl)-4-hydroxy-1-piperidinyl]-1-(4-fluorophenyl)-1-butanone
Empirical formula	$C_{21}H_{23}ClFNO_2$
Molecular mass	375.9
Proprietary names	Haldol; Brotopon; Sigaperidol
Clinical use	Tranquillizer

Solubility	Water	ethanol	chloroform	methanol
	insoluble	1 in 50–60	1 in 15–20	1 in 55
Light absorption	λ_{max} = 245 nm in methanolic acid			

HPLC 1

Sample	Blood
Extraction	Hexane containing 2% isoamyl alcohol under alkaline condition
Apparatus	Waters M 6000A LC pump, IBM 9523 UV detector and Rheodyne M 7126 injector
Column	Reversed-phase; octadecyl 10 μm, C_{18}, 250 × 4.6 mm ID
Mobile phase	Acetonitrile/100 mM KH_2PO_4 (55:45), pH 3.8–4.0 with H_3PO_4
Flow-rate	2 ml min^{-1}
Internal standard	Chlorohaloperidol
Injection size	90 μl, 2–100 ng ml^{-1}
Detection	UV, λ = 196 nm
Recording	IBM laboratory computer CS-9000
Retention times, min	Reduced haloperidol 4.6 Haloperidol 6.7 Internal standard 9.0
Ref.	Miller RL and Devane CL (1986) J. Chromatogr. Biomed. Appl. 374:405

HPLC 2

Sample	Blood plasma
Extraction	Diethyl ether under acidic condition and then with chloroform under alkaline condition
Apparatus	Hitachi M 635A LC, 635-0900 multi-wavelength UV detector and high pressure sample valve
Column	Reversed-phase; Nucelosil C_{18}, 5 μm, 250 × 4 mm ID
Mobile phase	Methanol/water (63:37) containing 0.2 M ammonium acetate
Flow-rate	0.5 ml min^{-1}
Temperature	55 °C
Internal standard	Diphenylamine, 1.0 μg ml^{-1}
Injection size	30–40 μl, 10–80 ng ml^{-1}
Detection	UV, λ = 250 nm; 0.005 AUFS
Recording	Takeda digital integrator (Tr 2221 A)
Retention times, min	Spiroperidol 10.9 Haloperidol 12.5 Methylperidol 11.6 Diphenylamine 20.0
Ref.	Miyazaki K and Arita T (1981) J. Chromatogr. Biomed. Appl. 223:449

HPLC 3

Sample	Human blood plasma
Extraction	Hexane containing 2% isoamly alcohol under alkaline condition
Apparatus	Perkin-Elmer Series 2 HPLC with M 65T UV detector and Rheodyne M 7120 injector
Column	Reversed-phase; Biosil ODS, 10 μm, 250 nm
Mobile phase	40% (v/v) Acetonitrile in phosphate buffer, pH 3.8 (6.5 g KH_2PO_4 in 1 l water, pH 3.8–4.0 with ortho-phosphoric acid)
Temperature	50 °C
Internal standards	Desipramine and chlorohaloperidol
Injection size	100 μl, 9.4 ng ml^{-1}
Detection	UV, λ = 195 nm; 0.01 AUFS
Retention times, min	(estimated from drawing) Desipramine 5.3 Haloperidol 4.6 Chlorohaloperidol 6.1
Ref.	Jatlow PI, Miller R and Swigar M (1982) J. Chromatogr. Biomed. Appl. 227:233

HPLC 4

Sample	Blood plasma
Extraction	Heptane/isoamyl alcohol (98.5:1.5, v/v), and then with diluted H_2SO_4
Apparatus	Beckman M 331 pump or Hewlett-Packard 5880 LC, Altex injection valve and Kratos M 773 UV detector
Column	Reversed-phase; Ultrasphere C_8, 5 μm, 250 × 4.6 mm ID
Mobile phase	300 ml 16.3 mM Na_2HPO_4, pH 7.0, 450 ml acetonitrile and 250 ml methanol
Internal standard	Chlorohaloperidol, 50 ng ml^{-1}
Injection size	400–500 μl, 2.5–50 ng ml^{-1}
Detection	UV, λ = 254 nm; 0.001 AUFS
Recording	Hewlett-Packard 3390 integrator
Retention times, min	(estimated from drawing) Haloperidol 6.8 Chlorohaloperidol 7.5
Ref.	Dhar AK and Kutt H (1984) Clin. Chem. 30:1228

HPLC 5

Sample	Serum and plasma
Extraction	Hexane with 1.5% (v/v) isoamyl alcohol under alkaline and acidic conditions
Apparatus	Waters M 6000A HPLC pump, Valco injection valve and Bioanalytical Systems LC-4A amperometric detector

Column	Cyano phase; μ-Bondapak CN, 10 μm, 300 × 3.9 mm ID
Mobile phase	40 mM Potassium phosphate buffer, pH 6.8 with 5 M KOH/acetonitrile (55:45, v/v)
Temperature	30 °C
Internal standard	Chlorohaloperidol
Injection size	100 μl, 2–10 μg ml^{-1}
Detection	EC, GC electrode, + 0.9 V vs Ag/AgCl RE
Retention times, min	(estimated from drawing) Reduced Haloperidol 11.0 Haloperidol 12.5 Chlorohaloperidol 15.0
Ref.	Korpi ER, Phelps BH et al (1983) Clin. Chem. 29:624

92 Heroin (Diamorphine)

CAS-No	561-27-3; 1502-95-0 (hydrochloride, anhydrous)			
Chemical name	3,6-*O*-Diacetylmorphine			
Empirical formula	Heroin	Hydrochloride		
	$C_{21}H_{23}NO_5$	$C_{21}H_{23}NO_5$, HCl, H_2O		
Molecular mass	369.4	423.9		
Proprietary name	Diamorphine			
Clinical use	Analgesic			
Solubility,	Water	alcohol	chloroform	ether
Heroin	1 in 1700	1 in 31	1 in 1.5	1 in 100
Hydrochloride	1 in 1.6	1 in 12	1 in 1.6	insoluble
Light absorption	λ_{max} = 299 nm in aqueous alkali			

HPLC 1

Sample	Adulterated raw materials
Apparatus	Perkin-Elmer Series 4 LC pump with gradient system, Hewlett-Packard 1040A photodiode array and LC-4B electrochemical detector (Bioanalytical Systems)
Column	Reversed-phase; HS-5, C_{18}, 125 × 4.6 mm ID
Mobile phase	Start: Methanol/phosphate buffer, pH 2.2 with 0.023 M Hexylamine (5:95) Gradient: Methanol 5% to 30% in 20 min; hold 6 min; Methanol 30% to 60% in 10 min; hold 9 min
Flow-rate	1.5 ml min^{-1}

Internal standard	Propanophenone, 0.01–0.04 mg ml^{-1}
Injection size	50 µl, 0.01-0.04 mg ml^{-1}
Detection	UV, λ = 218–240 nm; EC, GC electrode, + 1.0 V and 1.1 V vs Ag/AgCl RE
Recording	Perkin-Elmer computing integrator LCI-100 and printer
Retention times, min	(estimated from drawing) Procaine 3.4 Heroin 17.5 Nalorphine 4.7 Noscapine 22.6 Quinine 9.7 Papaverine 26.2 Acetylcodeine 16.3 Propanophenone 32.0
Ref.	Lurie IS and McGuinness K (1987) J. Liq. Chromatogr. 10:2189

HPLC 2

Sample	Injectable solution
Apparatus	Pye-Unicam LC-3 and ultraviolet detector
Column	Reversed-phase; Hypersil-ODS 5 µm, 100 mm
Mobile phase	0.01 M Sodium pentanesulfonate/acetonitrile/ortho-phosphoric acid (69.5:30:0.5), pH 2.0
Flow-rate	2.0 ml min^{-1}
Temperature	Ambient
Injection size	20 µl, 7.8 mg ml^{-1}
Detection	UV, λ = 284 nm; 0.32 AUFS
Retention times, min	Morphine 0.7 Heroin 1.5 6-Mono-acetyl morphine 1.0 (estimated from drawing)
Ref.	Beaumont IM (1982) Anal. Proc. 19:128

HPLC 3

Sample	Illicit heroin
Apparatus	Waters M 6000A LC pump, 440 UV detector, Rheodyne M 7125 injector, Hewlett-Packard 1040A photodiode array and Perkin-Elmer M 3000 fluorescence detector
Column	Alumina; 250 × 4.6 mm ID with Spherisorb A5Y and Shandon C_{18}, 45 × 2.1 mm ID
Mobile phase	Methanol/acetonitrile/citrate-0.01 M TMS (tetramethylammonium hydroxide) buffer, pH 6.0 (28:17:55)
Flow-rate	1 ml min^{-1}
Injection size	20 µl
Detection	UV, λ = 260 nm; fluorescence, λ = 260/400 nm
Recording	Hewlett-Packard 85 desktop computer system

Retention times, min

Meconic acid	2.05	O^3-Monoacetylmorphine	10.39
Dipyrone	2.30	Cryptopine	10.83
Caffeine	2.96	O^6-Monoacetylmorphine	11.00
Meconine	3.15	Procaine	11.24
Dipyrone (2nd peak)	3.20	Thebaine	12.16
		Codeine	12.23
Papaverine	3.61	Strychnine	13.31
Noscapine	5.96	Morphine	14.35
Acetylthebaol	7.44	3,6-Dimethoxyphenantrene-4,5-epoxide	15.73
Alpha-thebaol	7.57		
Heroin	8.72		
Acetylcodeine	10.01		

Ref. Billiet HAH, Wolters R and De Galan L (1986) J. Chromatogr. 368:351

HPLC 4

Sample: Chinese heroin sample
Column: Normal phase; Partisil 6 μm, 250 × 4.6 mm ID
Mobile phase: Methanol/2 N ammonia solution/1 N ammonium nitrate solution (27:2:1, v/v)
Flow-rate: 1 ml min^{-1}; 1470 psi
Temperature: Ambient
Injection size: 1–5 μl
Detection: UV, λ = 278 nm
Retention times, min: (estimated from drawing)

Caffeine	4.8	Monoacetylmorphine	6.5
Heroin	6.0	Morphine	9.0
		Strychnine	14.0

Ref. Cornish DW, Jacobs AL et al (1978) in: Tsuji K (ed) GLC and HPLC Determination of Therapeutic Agents, Part 2. Marcel Dekker Inc., New York and Basel, p 519

HPLC 5

Sample: Blood
Extraction: Multistep, liquid-liquid extraction
Column: Normal phase
Mobile phase: Acetonitrile/methanol (75:25 v/v), pH 7.0 with ammonium hydroxide and acetic acid
Flow-rate: 1.33 ml min^{-1}
Temperature: 30 °C
Internal standard: l-α-Acetyl-methadol
Injection size: 200 μl containing at least 12.5 ng ml^{-1}
Detection: UV, λ = 218 nm
Retention times, min

Heroin	5.7	Morphine	10.0
6-Acetylmorphine	8.1	Internal standard	13.8

Ref.	Umans JG, Chiu TSK et al (1982) J. Chromatogr. Biomed. Appl. 233:213

93 Hexobarbital

CAS-No	56-29-1; 50-09-0 (sodium salt)				
Chemical name	5-(Cyclohex-1-enyl)-1,5-dimethylbarbituric acid				
Empirical formula	Hexobarbital	Sodium Salt			
	$C_{12}H_{16}N_2O_3$	$C_{12}H_{15}N_2NaO_3$			
Molecular mass	236.3	258.3			
Proprietary names	Evidorm; Evipan; Sombulex				
Clinical use	Hypnotic/Sedativ				
Solubility,	Water	ethanol	chloroform	methanol	aq NaOH
Hexobarbital	1 in 3000	1 in 45	1 in 4	soluble	soluble
Sodium salt	very sol	very sol	very sol		
Light absorption	λ_{max} = 243 nm in 0.05 M borax-buffer, pH 9.2				

HPLC 1

Sample	Rat blood	
Extraction	Methylene chloride under acidic condition	
Apparatus	Altex LC pump, Beckmann M 160 UV detector and Rheodyne injector	
Column	β-Cyclodextrin; 10 μm, 250 × 4.6 mm ID	
Pre-column	β-Cyclodextrin; C_{18}, 30–38 μm, 50 × 4.1 mm ID	
Mobile phase	Methanol/water (27:73, v/v)	
Flow-rate	1.3 ml min^{-1}	
Temperature	Ambient	
Internal standard	Alphenal, 6 μg ml^{-1}	
Injection size	20 μl, 1.0 μg ml^{-1}	
Detection	UV, λ = 214 nm	
Retention times, min	D-Hexobarbital	10.0
	L-Hexobarbital	11.1
	Alphenal	17.2
Ref.	Chandler MHH, Guttendorf RJ et al (1987) J. Chromatogr. Biomed. Appl. 419:426	

HPLC 2

Sample	Urine
Extraction	On-line sample clean up on 10 μm styrene-divinyl benzene copolymer and anion-exchange resin columns using buffer, water and methanol

Apparatus	Kontron M LC-410 pumps and Kratos M 575 ultraviolet detector
Column	Reversed-phase; Chromspher C_{18}, 5 µm, 100 × 3.0 mm ID
Pre-column	PRP-1, 20 × 4.0 mm and Aminex A-28, 50 × 3.0 mm ID
Mobile phase	0.1 M Acetate buffer, pH 5.0/methanol (50:50)
Flow-rate	2 ml min^{-1}
Injection size	25 µl, 1 µg ml^{-1}
Detection	UV, 220 nm; 0.01 AUFS
Retention times, min	(estimated from drawing) Amobarbital 7.9 Butobarbital 4.8 Secobarbital 9.1 Hexobarbital 5.9
Ref.	De Jong J, Nielen MWF et al (1986) J. Chromatogr. Biomed. Appl. 381:431

HPLC 3

Sample	Mixture of barbiturates
Apparatus	HPLC low-pressure ternary gradient system
Column	Reversed-phase; Superspher 100 C_{18}, 250 × 4 mm ID
Mobile phase	Methanol/water (60:40, v/v)
Flow-rate	0.5 ml min; 3000 psi
Injection size	25 µl containing 100 µg ml^{-1}
Detection	UV, $\lambda = 230$ nm; 0.02 AUFS
Retention times, min	(estimated from drawing) Hexobarbital 11.7 Barbital 5.6 Amobarbital 13.2 Phenobarbital 7.3 Secobarbital 16.1 Butabarbital 9.2
Ref.	Merck Application Manual on HPLC, TLC and HPTLC (1986). E. Merck, Darmstadt, p 63

HPLC 4

Sample	Raw materials and dosage forms
Apparatus	Hewlett-Packard 1084-LC with UV detector
Column	Reversed-phase; Nucleosil C_{18}, 10 µm, 250 × 4 mm ID
Mobile phase	Acetonitrile/0.05 M ortho-phosphoric acid (30:70)
Flow-rate	2 ml min^{-1}
Temperature	40 °C
Internal standard	Allobarbital
Injection size	5–30 µl
Detection	UV, $\lambda = 220$ nm
Recording	Hewlett-Packard LC-Terminal 79850B integrator

Retention times, min	Barbital	2.13	Propallylonal	4.71
	Allobarbital	3.14	Phenallymal	5.36
	Aprobarbital	3.67	Hexobarbital	8.03
	Phenobarbital	4.15		
Ref.	Rehm KD and Steinigen M (1981) Pharm. Ztg. 126:99			

94 Histamine

CAS-No	51-45-6; 51-74-1 (acid phosphate)		
Chemical name	2-(Imidazol-4-yl)ethylamine		
Empirical formula	Histamine	Acid Phosphate	
	$C_5H_9N_3$	$C_5H_9N_3, 2H_3PO_4$	
Molecular mass	111.2	307.1	
Proprietary names	Akrotherm; Alergenol		
Clinical use	Diagnostic Agent		
Solubility,	Water	alcohol	ether
Histamine	freely sol	freely sol	sparingly soluble
Acid phosphate	1 in 4	slightly sol	insoluble

HPLC 1

Sample	Blood plasma
Extraction	Butanol, in three steps under alkaline, acidic and alkaline conditions
Apparatus	LKB M 2156 LC pump and Shimadzu RF-530 fluorescence detector
Column	Reversed-phase; Nucleosil C_{18}, 5 μm, 250 × 4 mm ID
Mobile phase	Methanol/sodium chloride solution (containing 45% 0.2 N NaCl and 55% methanol, pH 3.0 with 1 N HCl) (5:95)
Flow-rate	0.5 ml min^{-1}
Derivatization	Pre-column; with *o*-phthalaldehyde (OPA)
Internal standard	Ranitidine
Injection size	200 μl, 0.05 ng ml^{-1}
Detection	Fluorescence, λ = 350/450 nm
Recording	Shimadzu C-R3A integrator
Retention times, min	Internal standard 5.8 Histamine 7.1
Ref.	Czerwonka R, Tsikas D and Brunner G (1988) Chromatographia 25:219

HPLC 2

Sample	Tears
Apparatus	Perkin-Elmer Series 3B LC with Rheodyne M 7125 injector and LS-4 fluorimetric detector
Column	Reversed-phase; Nucleosil C_8, 10 µm
Pre-column	Reversed-phase; Lichroprep C_8, 20–40 µm
Mobile phase	Acetonitrile/0.004 M KH_2PO_4, pH 3.5 (65:35)
Flow-rate	0.5 ml min^{-1}
Derivatization	Pre-column; with fluorescamine
Injection size	20 µl, 20 ng ml^{-1}
Detection	Fluorescence, λ = 390/480 nm
Recording	Perkin-Elmer Sigma 15 data station
Retention time, min	Histamine 4.2
Ref.	Bettero A, Angi MR et al (1984) J. Chromatogr. Biomed. Appl. 310:390

HPLC 3

Sample	Biological fluids and tissues
Extraction	Ethyl acetate (tissues) after derivatization
Apparatus	Waters M 6000A LC pump, U6K injector and J4-7461 Aminco fluoromonitor (American Instruments)
Column	Cyano phase; µ-Bondapak CN
Mobile phase	Methanol/0.08 M acetic acid (52:48, v/v), pH 2.9
Flow-rate	1.2 ml min^{-1}
Temperature	23 °C
Internal standard	3,4-Dihydroxybenzylamine
Derivatization	Pre-column; with *o*-phthalaldehyde
Injection size	20 µl
Detection	Fluorescence, λ = ex 340 nm
Recording	Supergrator-3 computing integrator (Columbia Scientific)
Retention times, min	(estimated from drawing) Internal standard 8.5

Taurine	2.7	Tyramine	10.8
Norepinephrine	6.0	Histamine	14.2
Octopamine	6.9		

Ref.	Mell, Jr. LD, Hawkins RN and Thompson RS (1979) J. Liq. Chromatogr. 2:1393

95 Hydrocortisone

CAS-No	50-23-7; 50-03-3 (acetate); 6000-74-4 (sodium phosphate)
Chemical name	11β,17α,21-Trihydroxypregn-4-ene-3,20-dione

Empirical formula	Hydrocortisone $C_{21}H_{30}O_5$	Acetate $C_{23}H_{32}O_6$		Sodium Phosphate $C_{21}H_{29}Na_2O_8P$
Molecular mass	362.5	404.5		486.4
Proprietary names	Cortril; Hydroderm; Ulcort			
Clinical use	Corticosteroid			
Solubility,	Water	alcohol	acetone	chloroform
Hydrocortisone	insoluble	1 in 40	1 in 80	slightly soluble
Acetate	insoluble	1 in 230		1 in 150–200
Sodium phosphate	1 in 4	slightly sol		insoluble
Light absorption	λ_{max} = 240 nm in ethanol			

HPLC 1

Sample	Bulk drugs and tablets
Apparatus	LC system and ultraviolet detector
Column	Normal phase; 5–6 µm diamporous silica, 250 × 4.6 mm ID
Mobile phase	Water/methanol/glacial acetic acid (2.75:52.25:1) and dilute to 1 liter with water-washed 1,2-dichloroethane
Flow-rate	1.5 ml min^{-1}
Temperature	Ambient
Internal standard	Acetaminophen
Injection size	10 µl
Detection	UV, λ = 254 nm
Retention times, min	(estimated from drawing) Hydrocortisone 8.4 Acetaminophen (IS) 10.5
Ref.	Walters MJ (1984) J. Assoc. Off. Anal. Chem. 67:218

HPLC 2

Sample	Mixture of 12 corticosteroids
Apparatus	Kontron M 410 LC pump, Pye-Unicam LC UV spectrophotometer detector and Rheodyne M 7125 injection valve
Column	Normal phase; Lichrosorb Si 60, 120 × 4.6 mm ID
Mobile phase	2.5 mM Cetyltrimethylammonium bromide and 5% phosphate buffer, pH 7.0 with organic modifiers as prescribed
Flow-rate	1 ml min^{-1}
Injection size	10 µl, 0.05% solution or 20 µl, 0.1% solution
Detection	UV, λ = 254 nm; 0.01 AU
Recording	Hewlett-Packard 3353A laboratory data system and Kipp & Zonen BD-8 recorder

Capacity factor, k' of 12 corticosteroids chromatographed on dynamically modified silica with various organic modifiers in the eluents

	k'			
Organic modifiers	50% methanol	32% acetonitrile	26% THF	25% MeOH + 13% THF
1. Cortisone acetate	2.32	3.25	4.51	3.19
2. Hydrocortisone	2.39	2.07	2.49	2.68
3. Hydrocortisone acetate	2.46	2.99	4.55	3.40
4. Prednisolone	2.46	2.08	2.29	2.60
5. Prednisolone acetate	2.52	2.92	3.99	3.28
6. Corticosterone	4.13	4.22	3.99	4.58
7. Triamcinolone acetonide	4.19	3.59	3.52	4.05
8. Methylprednisolone	4.30	3.58	3.66	4.42
9. Betamethasone	4.90	3.44	3.83	4.65
10. Fluocinolone acetonide	5.73	4.12	6.29	6.55
11. Dexamethasone	6.19	4.12	4.65	5.99
12. Deoxycortone acetate	6.52	11.54	10.16	8.18
Mean retention	4.01	3.99	4.49	4.46

(THF = Tetrahydrofuran; MeOH = Methanol)

Ref. Helboe P (1986) J. Chromatogr. 366:191

HPLC 3

Sample: Human blood plasma
Extraction: Dichloromethane
Apparatus: Varian M 8500 LC with stop-flow injector and LDC M 1203 UV detector
Column: Normal phase; Chrompack SI-10, 10 μm, 250 × 3 mm ID
Mobile phase: Chloroform/isooctane/methanol/water
I. (48.5:48.5:2.9:0.12, v/v) and
II. (71:25:3.75:0.25, v/v)
Flow-rate: 1.08 ml min^{-1}
Internal standards: Prednisone and prednisolone
Injection size: 20 μl containing 4.5 pmol
Detection: UV, λ = 254 nm; 0.008 AUFS
Retention times, min:

Mobile phase	I.	II.
11-Deoxycortisol	5.2	
Cortisol		6.0
Prednisone	10.5–11.0	
Prednisolone		7.8

Ref. De Vries CP, Lomecky-Janousek M and Popp-Snijders C (1980) J. Chromatogr. Biomed. Appl. 183:87

HPLC 4

Sample: Aqueous solutions and biological fluids
Extraction: Ethyl acetate at pH 6–7 or at 1.0

Apparatus	Altex M 100 HPLC pump, Rheodyne M 7105 injector and Cecil M 2012 variable-wavelength detector
Column	Reversed-phase; ODS 10 μm
Mobile phase	I. Acetonitrile/85% phosphoric acid/water (28:0.15:72, v/v) II. Acetonitrile/water/85% phosphoric acid (28:72:2, v/v)
Flow-rate	1.0 ml min^{-1} (I); 1.5 ml min^{-1} (II)
Temperature	25 °C
Internal standard	6α-Methylprednisolone
Injection size	20 μl
Detection	UV, λ = 254 nm
Recording	Hewlett-Packard 3390A integrator
Retention times, min	(estimated from drawing)

Mobile phase	I.	II.
Dexamethasone phosphate	11.5	
Hydrocortisone	13.5	8.0
6α-Methylprednisolone	21.0	13.5
Dexamethasone	24.0	
Triamcinolone		5.0
Triamcinolone acetonide		21.5

Ref.	Derendorf H, Rohdewald P et al (1986) J. Pharm. Biomed. Anal. 4:197

HPLC 5

Sample	Blood plasma or serum
Extraction	Methanol on C_{18} extraction column
Apparatus	Waters HPLC system with M 440 solvent delivery programmer, U6K injector and ultraviolet detector
Column	Reversed-phase; μ-Bondapak C_{18}, 300 × 3.9 mm ID and a C_{18} RCM Z-Module, 10 μm, 100 × 8 mm ID
Mobile phase	Methanol/water (60:40, v/v)
Flow-rate	1.0 ml min^{-1}
Temperature	Ambient
Internal standard	Corticosterone
Injection size	10–40 μl, 50–300 ng ml^{-1}
Detection	UV, λ = 254 nm; 0.005 A
Retention times, min	(estimated from drawing)
	Cortisol 6.7
	Corticosterone 10.0
Ref.	Hofreiter BT, Mizera AC et al (1983) Clin. Chem. 29:1808

HPLC 6

Sample	Blood plasma from asthmatic children
Extraction	Methanol

Apparatus	Waters M 6000A LC pump with 440 UV detector and U6K injector
Column	Reversed-phase; μ-Bondapak C_{18}, 300 × 4 mm ID
Mobile phase	0.05 M Sodium acetate buffer (pH 2.8 with acetic acid and containing 0.1 M sodium chloride)/acetonitrile (77:23, v/v)
Flow-rate	2.5 ml min^{-1}
Temperature	Ambient
Internal standard	*p*-Hydroxybenzoate-N-propyl
Injection size	50 μl, 0.5–10 μg ml^{-1}
Detection	UV, λ = 254 nm; 0.01 FS
Retention times, min	(estimated from drawing)

Orciprenaline	2.0	Hydrocortisone	4.8
Salbutamol	2.0	Tipepidine hibenzate	6.0
Terbutaline	2.0	Internal standard	7.2
Theophylline	2.0	Hydrocortisone succinate	8.4
Noscapine	3.0	Bromhexine	10.8

Ref.	Iwasaki E (1987) Clin. Chem. 33:1412

HPLC 7

Sample	Raw materials
Column	Normal phase, 5–10 μm, 300 × 4 mm ID
Mobile phase	Butyl chloride/water-saturated butyl chloride/tetrahydrofuran/methanol/glacial acetic acid (95:95:14:7:6)
Flow-rate	Adjusted to 1000 psi
Internal standard	Prednisone, 0.06 mg ml^{-1}
Injection size	Suitable volumes
Detection	UV, λ = 254 nm
Ref.	Anonym (1985): The United States Pharmacopeia XXI and The National Formulary XVI. United States Pharmacopeial Convention Inc., Rockville, p 499

96 Ibuprofen

CAS-No	15687-27-1
Chemical name	2-(4-Isobutylphenyl)propionic acid
Empirical formula	$C_{13}H_{18}O_2$
Molecular mass	206.3
Proprietary names	Amersol; Brufen; Relcofen
Clinical use	Analgesic

Solubility	Water	ethanol	chloroform	ether	aq NaOH
	insoluble	1 in 1.5	1 in 1	1 in 2	soluble
Light absorption	λ_{max} = 265 nm in aqueous alkali				

HPLC 1

Sample	Blood plasma and urine
Extraction	Blood: With methanol Urine: After treating with NaOH and HCl
Apparatus	Perkin-Elmer Series 10 LC with LC-75 UV detector and Rheodyne injector
Column	Reversed phase; Lichrosorb 10 μm, C_{18}, 250 × 4.5 mm ID
Mobile phase	Acetonitrile/0.1 M sodium acetate (35:65), pH 6.2 with acetic acid
Flow-rate	3 ml min^{-1}
Temperature	Ambient
Internal standard	Indomethacin
Injection size	20 μl
Detection	UV, λ = 220 nm
Recording	Perkin-Elmer R-100 chart-recorder
Retention times, min	Internal standard 3.8 Ibuprofen 5.2
Ref.	Avgerinos A and Hutt AJ (1986) J. Chromatogr. Biomed. Appl. 380:468

HPLC 2

Sample	Human blood plasma
Extraction	Hexane/2-propanol (85:15, v/v) under acidic condition
Apparatus	Waters M 6000A LC pump with WISP 710B automatic sampler and 441 UV detector
Column	Reversed-phase; Radial-Pak C_{18}, 10 μm, 100 × 5 mm ID
Pre-column	Co:Pell ODS, 50 × 4 mm ID
Mobile phase	Phosporic acid 85%/water/acetonitrile (4:600:400)
Flow-rate	3 ml min^{-1}
Internal standard	Ibufenac, 1–50 μg ml^{-1}
Injection size	50 μl; 5–50 μg ml^{-1}
Detection	UV, λ = 214 nm
Recording	Hewlett-Packard 3354C laboratory automation system
Retention times, min	Ibufenac 3.98 Ibuprofen 5.78
Ref.	Minkler PE, Hoppel CL (1988) J. Chromatogr. Biomed. Appl. 428:388

HPLC 3

Sample	Animal dosing formulations (liquid)
Apparatus	HPLC with autosampler and variable-wavelength UV detector
Column	Reversed-phase; Spherisorb ODS-2, 3 μm, 100 × 4.6 mm ID
Mobile phase	400 ml Water containing 1.0 g NaH_2PO_4, H_2O, 0.5 g H_3PO_4 (85%) and 3.0 g sodium perchlorate, H_2O plus 600 ml acetonitrile
Flow-rate	1.5 ml min^{-1}
Temperature	Ambient
Injection size	20 μl
Detection	UV, λ = 210 nm; 0.32 AUFS
Recording	Laboratory data system with recorder
Retention times, min	Pseudoephedrine 1.1 Ibuprofen 2.1 Terfenadine 4.7 (estimated from drawing)
Ref.	George RC and Contario JJ (1988) J. Liq. Chromatogr. 11:475

HPLC 4

Sample	Human urine
Extraction	2-Propanol 5% in methylene chloride under acidic condition
Apparatus	Shimadzu LC-6A pump with SCL-6A system controller, SPD-6A UV spectrophotometric detector and SIL-6A autoinjector
Column	α-Aminopropyl-phenylglycine; 5 μm spherical, 250 × 4.6 mm ID
Mobile phase	I. Hexane/chloroform/2-propanol (18:2:1) II. Hexane/chloroform/2-propanol (6:1:1)
Flow-rate	2 ml min^{-1}
Derivatization	Pre-column; with 4-methoxyaniline and ethyl chloroformate
Internal standard	α-Phenylpropionic acid
Injection size	25 μl, 12.5–200 μg ml^{-1}
Detection	UV, λ = 254 nm
Recording	Shimadzu C-R3A Chromatopac integrator
Retention times, min	(estimated from drawing)

Mobile phase	I.	II.
(S)–Ibuprofen	4.2	2.6
(R)–Ibuprofen	4.8	2.6
(S)–Internal standard	7.5	3.0
(R)–Internal standard	7.8	3.2
(S)–OH–Ibuprofen		5.8
(R)–OH–Ibuprofen		6.2

	(S)–CO_2H–Ibuprofen	10.5
	(R)–CO_2H–Ibuprofen	14.5
Ref.	Nicoll-Griffith DA, Inaba T et al (1988) J. Chromatogr. Biomed. Appl. 428:103	

97 Idoxuridine

CAS-No	54-42-2			
Chemical name	2′-Deoxy-5-iodouridine			
Empirical formula	$C_9H_{11}IN_2O_5$			
Molecular mass	354.1			
Proprietary names	Herpid; Iduviran; Kerecid			
Clinical use	Antiviral			
Solubility	Water	ethanol	methanol	chloroform
	1 in 500	1 in 400	1 in 230	insoluble
Light absorption	λ_{max} = 288 nm in aqueous acid			

HPLC 1

Sample	Raw materials
Apparatus	Altex M 110 LC pump, Cecil M 2012 spectrophotometric detector and Rheodyne M 7120 injector
Column	Reversed-phase; Hypersil ODS, C_{18}, 5 μm, 150 × 4.6 mm ID
Mobile phase	Acetonitrile/water (5:95)
Flow-rate	1.0 ml min^{-1}
Temperature	Ambient
Internal standard	Sulfanilamide
Detection	UV, λ = 254 nm; 0.02 AU
Recording	Hewlett-Packard 3353A laboratory data system
Retention times, min	(estimated from drawing)

2′-Deoxyuridine	3.8		5-Bromo-2′-deoxyuridine	7.7
Sulfanilamide	4.5		Idoxuridine	11.0
5-Iodouracil	6.6			

Ref.	Honoré-Hansen S and Thomsen M (1981) J. Chromatogr. 209:77

HPLC 2

Sample	Raw materials
Column	Reversed-phase; ODS C_{18}, 10 μm, 300 × 4 mm ID
Mobile phase	Methanol/water (13:87, v/v)
Flow-rate	1.7 ml min^{-1}
Internal standard	Sulfathiazole
Injection size	10 μl

Detection	UV, λ = 254 nm
Ref.	Anonym (1980) British Pharmacopoeia. London Her Majesty's Stationary Office, p 237

98 Imipramine

CAS-No	50-49-7; 113-52-0 (hydrochloride)				
Chemical name	*N*-(γ-Dimethylaminopropyl)iminodibenzyl				
Empirical formula	Imipramine	Hydrochloride			
	$C_{19}H_{24}N_2$	$C_{19}H_{24}N_2$, HCl			
Molecular mass	280.4	316.9			
Proprietary names	Antipress; Imidol; Tofranil				
Clinical use	Antidepressant				
Solubility,	Water	ethanol	chloroform	acetone	ether
Hydrochloride	1 in 2	1 in 1.5	1 in 1.5	1 in 15	insoluble
Light absorption	λ_{max} = 251 nm in aqueous acid				

HPLC 1

Sample	Biological samples
Apparatus	HPLC pump with electrochemical detector
Column	Phenyl phase; Spherisorb S5, 125 × 5 mm ID
Mobile phase	I. Methanolic ammonium perchlorate (10 mM) + 1 ml/l methanolic NaOH (0.1 M), pH 6.7 II. Methanolic ammonium acetate (10 mM) + 6 ml/l 10% (v/v) glacial acetic acid in methanol, pH 6.7
Flow-rate	2 ml min^{-1}
Injection size	20 μl; 10 μg ml^{-1}
Detection	EC, + 1.2 V vs Ag/AgCl

Retention times, min

Mobile phase	I.	II.
Nortriptyline	2.0	2.2
Meptazinol	2.2	2.7
Amitriptyline	3.0	3.0
Imipramine	3.5	3.9
Methdilazine	4.8	4.8

(estimated from drawing)

Ref.	Flanagan RJ, Bhamra RK et al (1988) J. Liq. Chromatogr. 11:1015

HPLC 2

Sample	Blood plasma
Extraction	Either on AASP (advanced automated sample processor) as pre-column or with hexane under alkaline condition

Apparatus	Varian M 2010 LC pump with 2050 spectrophotometer
Column	Alkyl-nitrile phase; Spherisorb CN 5 μm, 150 × 4.6 mm ID
Pre-column	AASP containing n-octyl-bonded silica (Varian) cassettes
Mobile phase	Acetonitrile/methanol/0.005 M phosphate buffer, pH 7.2 (50:25:30, v/v)
Flow-rate	1.5 ml min^{-1}
Internal standard	Clomipramin, 2 μg ml^{-1}
Injection size	50 μl, 2 μg ml^{-1}
Detection	UV, $\lambda = 252$ nm; 0.01 AUFS
Recording	Shimadzu C-R3A integrator
Retentiom times, min	(estimated from drawing) Clomipramine 4.6 Demethyl-clomipramine 7.0 Imipramine 5.4 Desipramine 8.2
Ref.	Ni P, Guyon F et al (1988) J. Liq. Chromatogr. 11:1087

HPLC 3

Sample	Blood plasma
Extraction	Diethyl ether under alkaline condition
Apparatus	Waters M 45 LC pump, U6K injector and 441 UV detector
Column	Reversed-phase; μ-Bondapak C_{18}, 10 μm, 300 × 3.9 mm ID
Mobile phase	Acetonitrile/0.025 M potassium dihydrogen Phosphate/water (45:50:5)
Flow-rate	1 ml min^{-1}
Temperature	Ambient
Internal standard	Clomipramine
Injection size	10–50 μl, 5–300 ng ml^{-1}
Detection	UV, $\lambda = 254$ nm
Recording	Spectra-Physics 4270 electronic integrator
Retention times, min	Desipramine 7.20 Imipramine 8.90 Monodesmethyl-trimipramine 8.60 Trimipramine 10.10 Clomipramine 13.00
Ref	Rop PP, Conquy T et al (1986) J. Chromatogr. Biomed. Appl. 375:339

HPLC 4

Sample	Mixtures of tricyclic antidepressant and related drugs
Apparatus	Waters M 6000 LC pump, Cecil M 272 UV detector and Rheodyne M 7120 injection valve
Column	Reversed-phase; Hypersil ODS 5 μm, 160 × 5 mm ID

Mobile phase	Acetonitrile/phosphate buffer (containing 0.6 ml n-nonylamine in 1000 ml 0.01 M NaH_2PO_4), pH 3.0 with H_3PO_4 (30:70)
Flow-rate	2 ml min^{-1}
Injection size	20 µl
Detection	UV, λ = 230 nm; 0.02 A
Retention times, min	(estimated from drawing)

Zimeldine	0.5	Imipramine	1.7
Noxiptyline	0.8	Amitriptyline	1.9
Doxepin	1.0	Trimipramine	2.2
Protriptyline	1.5	Butriptyline	2.6
		Clomipramine	3.4

Ref. Gill R and Wanogho SO (1987) J. Chromatogr. 391:461

HPLC 5

Sample	Blood or blood plasma
Extraction	Diethyl ether under alkaline and then under acidic conditions
Apparatus	Altex M 110A and Waters M 6000A pumps, Altex M 160 ultraviolet detector and Rheodyne M 7125 injector
Column	Reversed-phase; ODS 5 µm, 100 × 4.6 or (70 × 4.6) mm ID
Mobile phase	Acetonitrile/aqueous 10 mM Na_2HPO_4, pH 2 containing 80 mM sodium lauryl sulfate and 5 mM tetrabutylammonium bromide (50:50, v/v)
Flow-rate	2 ml min^{-1}
Internal standards	Doxepin, 100 ng ml^{-1} and nomifensine, 500 ng ml^{-1}
Injection size	100 µl, 50–500 ng ml^{-1}
Detection	UV, λ = 254 and 230 nm; 0.01–0.005 AUFS
Recording	Potentiometric recorder
Retention times, min	(estimated from drawing)

λ =	254 nm	230 nm
Doxepin	5.0	5.0
Imipramine	7.0	
Amitriptyline	8.0	
Desipramine	8.0	
Nortriptyline	9.0	
Clomipramine	10.0	
Nomifensine	17.0	17.0
Butriptyline		9.0
Maprotyline		10.0

Ref. Hung CT, Taylor RB and Paterson N (1983) J. Pharm. Biomed. Anal. 1:73

HPLC 6

Sample	Rat brain
Extraction	Diethyl ether/ethyl acetate (3:1) under alkaline condition
Apparatus	Toyo Soda HPLC, Rheodyne injector and Kyoto M 315 UV detector
Column	Reversed-phase; Cosmosil 5, C_{18}, 250 × 4.6 mm ID
Mobile phase	1% Triethylamine (pH 3.0 with H_3PO_4)/methanol/tetrahydrofuran (88:45:17)
Flow-rate	1 ml min^{-1}
Temperature	40 °C
Injection size	100 µl
Detection	UV, λ = 254 nm; 0.005 AUFS
Retention times, min	2-Hydroxyimipramine 5.2 Desipramine 12.3 2-Hydroxydesipramine 6.2 Nortriptyline 14.2 Imipramine 11.2
Ref.	Sugita S, Kobayashi A (1987) J. Chromatogr. Biomed. Appl. 421:412

HPLC 7

Sample	Five tricyclic drugs
Apparatus	Altex M 110 HPLC pump, Waters 440 UV detector and Rheodyne loop injector
Column	Cyanopropyl-phase; Zorbax CN, 5 µm, 150 × 4.6 mm ID
Mobile phase	0.025 M Sodium acetate and acetic acid buffer, pH 4.8/acetonitrile/methanol (20:35:45, v/v)
Flow-rate	2.5 ml min^{-1}
Temperature	Ambient
Injection size	10 µl
Detection	UV, λ = 254 nm; 0.05 AUFS
Recording	Hewlett-Packard 3390A integrator
Retention times, min	Norephedrine 2.7 Chlorpromazine 5.1 Imipramine 3.8 Trifluoperazine 5.4 Amitriptyline 4.0 Thioridazine 5.7
Ref.	Lovering EG, Beaulieu N et al (1985) J. Assoc. Off. Anal. Chem. 68:168

99 Indomethacin

CAS-No	53-86-1
Chemical name	[1-(4-Chlorobenzoyl)-5-methoxy-2 methylindol-3-yl]acetic acid
Empirical formula	$C_{19}H_{16}ClNO_4$

Molecular mass		357.8			
Proprietary names		Amuno; Rheumacin; Salinac			
Clinical use		Analgesic			
Solubility	Water	alcohol	chloroform	acetone	ether
	insoluble	1 in 50	1 in 30	soluble	1 in 40–45
Light absorption		λ_{max} = 230 nm and 279 nm in aqueous alkali			

HPLC 1

Sample	Raw materials and serum
Apparatus	Waters M 6000 LC pump, Pye-Unicam LC-UV detector and Rheodyne M 7125 injection valve
Column	Reversed-phase; Hypersil-ODS 5 μm, 160 × 5 mm
Mobile phase	Isopropanol/0.1 M potassium dihydrogen phosphate/formic acid (540:1000:1, v/v)
Flow-rate	1.5 ml min^{-1}
Injection size	10 μl
Detection	UV, λ = 240 nm
Retention times, min	(estimated from drawing)

Paracetamol	1.2	Salsalate	4.5
Salicylamide	1.6	Diflunisal	5.3
Phenacetin	1.8	Phenylbutazone	7.2
Propyphenazone	2.2	Indomethacin	8.1
Oxyphenbutazone	3.0	Diclofenac	11.9
Alclofenac	3.9	Phenylsalicylate	16.0

Ref. Stevens HM and Gill R (1986) J. Chromatogr. 370:39

HPLC 2

Sample	Blood plasma
Extraction	Methylene chloride under acidic condition
Apparatus	Altex M 334 HPLC with gradient system and Hitachi M 11– 30 variable wavelength-detector
Column	Reversed-phase; Altex ODS 5 μm, 150 × 4.6 mm ID
Mobile phase	Acetonitrile/water/glacial acetic acid (450:550:3.2, v/v)
Flow-rate	2.0 ml min^{-1}
Internal standard	Phenolphthalein, 62.5 μg ml^{-1}
Injection size	20 μl, 0.5–20 μg ml^{-1}
Detection	UV, λ = 240 nm; 0.05 AUFS
Recording	Hewlett-Packard 3385 integrator
Retention times, min	

Phenolphthelin	2.0	Fenprofen	5.4
Tolmetin	2.8	Indomethacin	7.5
Naproxen	3.4	Ibuprofen	8.0

Ref. Levine B and Caplan YH (1985) Clin. Chem. 31:346

HPLC 3

Sample	Blood from new borns
Extraction	Chloroform or diethyl ether at pH 5.0

Apparatus	Beckman M 110A HPLC pump with Rheodyne M 7125 injector and Waters M 440 ultraviolet detector
Column	Reversed-phase; Novapak-Radial C_{18}, 5 μm, 100 × 8 mm ID
Mobile phase	700 ml Methanol and 300 ml sodium acetate buffer, 10 mM, pH 3.6 with HCl
Flow-rate	4 ml min^{-1}
Temperature	Ambient
Internal standard	Flufenamic acid, 10.0 μg ml^{-1}
Injection size	20 μl containing 5 ng
Detection	UV, λ = 254 nm; 0.005 AUFS
Retention times, min	(estimated from drawing) Desbenzoyl indomethacin 1.0 Indomethacin 4.1 *o*-Desmethylindomethacin 2.0 Flufenamic acid 5.9
Ref.	Ou C-N and Frawley VL (1984) Clin. Chem. 30:898

HPLC 4

Sample	Neonates blood plasma
Extraction	Diethyl ether
Apparatus	Altex M 110A LC pump, Waters M 450 UV detector and U6K injector
Column	Reversed-phase; Radial-Pak, C_{18}, 10 μm, 100 × 8 mm ID
Pre-column	RCM-100
Mobile phase	Acetonitrile/water (45:55, v/v) containing 0.6% (v/v) acetic acid
Flow-rate	1.8 ml min^{-1}
Temperature	Ambient
Internal standard	Phenylbutazone
Injection size	100 μl, 0.25–1.0 μg ml^{-1}
Detection	UV, λ = 260 nm; 0.02 AUFS
Retention times, min	Indomethacin 6.0 Internal standard 8.0
Ref.	Mehta AC and Calvert RT (1983) Ther. Drug Monit. 5:143

100 Insulin

CAS-No	9004-10-8
Chemically	Mixture of peptides A and B containing 21 and 30 amino acids, respectively

Proprietary names	Monotard MC; Optisulin; Regular Iletin		
Clinical use	Antidiabetic		
Solubility	Water	alcohol	aq HCl and aq NaOH
	slightly soluble	insoluble	soluble
Light absorption	λ_{max} = 276 nm in aqueous acid		

HPLC 1

Sample	Adult and foetal rat insulin
Extraction	Thin-layer chromatography
Apparatus	Spectra-Physics LC-8700 pump with gradient controller Kratos M 773 UV detector and WISP M 710A autosampler
Column	Reversed-phase; Nucleosil 10 μm ODS, 250 × 4.6 mm ID
Mobile phase	A: 0.1 M NaH_2PO_4, pH 2.1 with H_3PO_4 B: Acetonitrile Start: A-B (75:25) Gradient: B 25% to 40% in 10 min; run 20 min; B 40% to 25% in 2 min
Flow-rate	1.4 ml min^{-1}
Detection	UV, λ = 220 nm
Recording	Hewlett-Packard 3390A integrator
Retention times, min	Insulin I 13.92–14.34 Insulin II 14.79–15.19
Ref.	De Gasparo M and Faupel M (1986) J. Chromatogr. 357:139

HPLC 2

Sample	Pharmaceutical dosage forms
Apparatus	ACS LC M 740 pump, Perkin-Elmer LC-75 UV detector and Rheodyne M 7125 injector
Column	A. Reversed-phase; 3 μm ODS, 30 × 5.2 mm ID B. Reversed-phase; ODS 5 μm, 250 mm
Mobile phase	For A: I. Human and porcine insulin: Buffer containing 0.1 M $(NH_4)_2SO_4$ and 0.005 M tartaric acid, pH 3.5 with H_2SO_4/acetonitrile (78:27.5) plus 0.002% cetrimide For A: II. Bovine insulin: Buffer as for porcine insulin/acetonitrile (78:22) plus cetrimide 0.002% For B: As in A: I
Flow-rate	2 ml min^{-1}
Injection size	20 μl, 0–80 IU ml^{-1}
Detection	UV, λ = 215 nm; 0.1 AUFS
Recording	Hewlett-Packard 3900A integrator

Retention times, min

	A.I.	A.II.	B.
Methylhydroxybenzoate	2.0		13.0
Human insulin	4.5		
Porcine insulin	6.0	20.0	21.5
Bovine insulin		5.0	
Desamidoinsulin	12.8		

Ref. Adams PS and Haines-Nutt RF (1986) J. Chromatogr. 351:574

HPLC 3

Sample: Soluble acid formulations
Apparatus: Altex M 110 or LDC 720 LC pump with Cecil M 212 UV detector
Column: Reversed-phase; Hypersil ODS 5 µm or, Nucleosil 5 µm, C_{18} or, Lichrosorb 5 µm, C_{18}, 100 or 150 × 5 mm ID
Mobile phase: 0.1 M Ammonium sulfate containing 5 mM (L +) tartaric acid, pH 3.0/acetonitrile (75:25) and 14 µM cetrimide
Flow-rate: 1 ml min^{-1}
Injection size: 10 µl
Detection: UV, λ = 225 or 280 nm
Retention times, min: (estimated from drawing)

Methylparaben	6.5
Insulin: Bovine native	10.4
Porcine native	15.3
Bovine monodesamido	18.0
Porcine monodesamidoinsulin	29.5

Ref. Lloyd LF (1982) Anal. Proc. 19:131

HPLC 4

Sample: Pharmaceutical dosage froms
Apparatus: ACS M LC-740 pump, Perkin-Elmer LC-75 UV detector and Rheodyne M 7125 injector
Column: Nitrile phase; 5 µm Spherisorb CN, 250 × 4 mm ID
Mobile phase: Water/acetonitrile/trifluoroacetic acid/sodium octanesulfonic acid (30:70:0.1:0.1)
Flow-rate: 2 ml min^{-1}
Injection size: 100 µl, 10–100 units l^{-1}
Detection: UV, λ = 215 nm
Recording: Hewlett-Packard 3900A integrator
Retention times, min

Methylhydroxybenzoate	2.4
Porcine insulin	5.0

Ref. Adams PS and Haines-Nutt RF (1986) J. Chromatogr. 351:574

HPLC 5

Sample	Insulin biosynthetic products in isolated rat and mouse islets
Extraction	3 M Acetic acid containing 0.1% human serum albumin
Apparatus	Waters M 6000A pumps, WISP 710A autosampler, 660 solvent programmer and Pye-Unicam LC-UV detector
Column	Reversed-phase; Lichrosorb C_{18}, 5 μm; 250 × 4.0 mm ID
Mobile phase	Gradient: A. Acetonitrile 25% to 30% in 0.125 M triethylammonium phosphate, pH 4.0 in 30 min B. Acetonitrile 30% to 36% in 0.1% trifluoroacetic acid in 60 min
Flow-rate	1 ml min^{-1}
Temperature	Ambient
Detection	UV, λ = 210 nm; 0.08–0.32 AUFS
Retention times, min	(estimated from drawing)

Mobile phase	A.	B.
C-Peptide II	21.5	26.0
C-Peptide I	23.5	28.0
Insulin I	27.5	42.0
Insulin II	29.6	47.5
Glucagon(porcine)	40.0	50.0
Proinsulin-like		52.0
Proinsulin-like		56.5
Proinsulin I		60.0
Proinsulin II		62.0

Ref. Linde S, Nielsen JH et al (1989) J. Chromatogr. 462:243

101 Isoniazid

CAS-No	54-85-3; 2066-89-9 (aminosalicylate)			
Chemical name	Isonicotinohydrazide			
Empirical formula	Isoniazid	Aminosalicylate		
	$C_6H_7N_3O$	$C_6H_7N_3O\ C_7H_7NO_3$		
Molecular mass	137.2	290.3		
Proprietary names	Isozid; Paraniazide; Rimifon			
Clinical use	Antituberculous			
Solubility,	Water	alcohol	chloroform	methanol
Isoniazid	1 in 8	1 in 45–50	1 in 1000	
Aminosalicylate	soluble			soluble
Light absorption	λ_{max} = 266 nm in aqueous acid			

HPLC 1

Sample	Tablet formulations
Apparatus	Altex M 110A LC pump, Valco CV-6-UHPa-N60 injector and Waters M 440 UV detector
Column	Cyanopropyl phase; Spherisorb CN, 5 μm, 250 × 3.2 mm ID
Mobile phase	0.01 M Acetate buffer, pH 3.5/acetonitrile (95:5, v/v)
Flow-rate	1.5 ml min^{-1}; 1160 psi
Injection size	10 μl, 0.1–1.0 mg ml^{-1}
Detection	UV, λ = 254 nm; 0.02–0.002 AUFS
Recording	Spectra-Physics electronic computing integrator
Retention times, min	Isonicotinic acid 1.25 4-Aminosalicylic acid 1.32 Isonicotinamide 3.07 Isoniazid 4.30 Pyridoxine hydrochloride > 15.00
Ref.	Butterfield AG, Lovering EG and Sears RW (1980) J. Pharm. Sci. 69:222

HPLC 2

Sample	Serum, polymorphonuclear cells and alveolar macrophages
Extraction	First acetone, and then re-extraction with butanol/chloroform (30:70)
Apparatus	Waters M 6000A HPLC, U6K injector and 440 UV detector
Column	Reversed-phase; μ-Bondapak C_{18}, 10 μm, 300 × 3.9 mm ID
Mobile phase	Methanol/water (5:95) containing 5 mM n-heptanesulfonic acid (Pic B_7, Waters)
Flow-rate	2 ml min^{-1}
Injection size	90 μl containing at least 95 ng ml^{-1}
Detection	UV, λ = 280 nm
Retention times, min	(estimated from drawing) Acetylisoniazid 5.5–6.0 Isoniazid 6.5–7.0
Ref.	Guillaumont M, Leclercq M et al (1982) J. Chromatogr. Biomed. Appl. 232:369

HPLC 3

Sample	Serum
Extraction	Chloroform under alkaline condition
Apparatus	Waters ALC/GPL M 244 LC with 440 UV detector
Column	Reversed-phase; μ-Bondapak C_{18}, 10 μm, 300 × 3.9 mm ID

Mobile phase	5% Methanol in 95% 0.1 M KH_2PO_4, pH 6.9
Flow-rate	2 ml min^{-1}
Internal standard	Nicotiniamide
Injection size	30 µl
Detection	UV, λ = 254 nm
Retention times, min	Acetyl isoniazid 3.3 Isoniazid 3.9 Internal standard 5.8
Ref.	Moulin MA, Albessard F et al (1981) J. Chromatogr. Biomed. Appl. 226:250

HPLC 4

Sample	Raw materials
Column	Reversed-phase; ODS 5 to 10 µm, 250 × 4.6 mm ID
Mobile phase	4.4 g Docusate sodium in 600 ml methanol and 400 ml water, pH 2.5 with 2 N sulfuric acid
Flow-rate	1.5 ml min^{-1}
Injection size	Appr. 10 µl, 0.32 mg ml^{-1}
Detection	UV, λ = 254 nm
Ref.	Anonym (1985): The United States Pharmacopeia XXI and The National Formulary XVI, Supplement 4, 1986. United States Pharmacopeial Convention Inc., Rockville, p 2181

HPLC 6

Sample	Rabbit blood plasma
Extraction	Acetonitrile under acidic condition after deproteinization with ammonium sulfate
Apparatus	Waters M 45 LC pump and M 45 UV detector
Column	Phenyl phase; 10 µm, 300 × 3.9 mm ID
Mobile phase	10 mM Phosphate buffer containing 0.25 mM tetrabutylammonium phosphate, pH 4.1
Flow-rate	1.0 ml min^{-1}
Temperature	Ambient
Internal standard	Niacinamide
Injection size	50 µl
Detection	UV, λ = 280 nm
Retention times, min	(estimated from drawing) Acetylisoniazid 8.0 Niacinamide 13.0 Isoniazid 9.0 Isoniazid methane-sulfonate 18.0
Ref.	Hsu K-Y and Ho Y (1989) J. Chromatogr. Biomed. Appl. 493:305

102 Isoprenaline

CAS-No	7683-59-2; 51-30-9 (hydrochloride); 6700-39-6 (sulfate, dihydrate)			
Chemical name	1-(3,4-Dihydroxyphenyl)-2-isopropylaminoethanol			
Empirical formula	Isoprenoline	Hydrochloride	Sulfate	
	$C_{11}H_{17}NO_3$	$C_{11}H_{17}NO_3$, HCl	$(C_{11}H_{17}NO_3)_2$, H_2SO_4, $2H_2O$	
Molecular mass	211.3	247.7	556.6	
Proprietary names	Aludrin; Isomenyl; Proternol			
Clinical use	Sympathomimetic			
Solubility,	Water	alcohol	chloroform	ether
Hydrochloride	1 in > 1	1 in 50	insoluble	insoluble
Sulfate	1 in 4	slightly soluble	insoluble	insoluble
Light absorption	λ_{max} = 297 nm in aqueous alkali			

HPLC 1

Sample	Methylated metabolites in physiological salt solution	
Apparatus	Waters M 510 LC pump, U6K injector and 410 amperometric electrochemical detector	
Column	Reversed-phase; Novapak C_{18}, 5 µm, 150 × 3.9 mm ID	
Pre-column	C_{18} oder CN Guard-Pak module	
Mobile phase	70 mM Na_2HPO_4, 5 mM 1-heptane-sulfonic acid sodium, 0.1 mM Na_2 EDTA and 9% (v/v) acetonitrile, pH 3.6 with 8.7 M H_3PO_4	
Flow-rate	1 ml min^{-1}	
Injection size	0.1–240 pmol	
Detection	EC, GC electrode vs Ag/AgCl RE	
Retention times, min	Isoprenaline	4.2
	o-Methylisoprenaline	8.0
Ref.	Bryan LJ and O'Donnell SR (1989) J. Chromatogr. Biomed. Appl. 487:29	

HPLC 2

Sample	Raw materials
Apparatus	Waters M 6000A LC pump, 440 UV detector and Valco sample loop
Column	Normal phase; Nucleosil 100, 10 µm, 200 × 4 mm ID
Mobile phase	0.01 M Camphor-10-sulfonic acid, pH 2.1
Flow-rate	1.7 mm sec^{-1}; 7.7 MPa
Temperature	25 °C
Injection size	10 µl
Detection	UV, λ = 254 nm; 0.0005 A

Retention times, min	(estimated from drawing) Norepinephrine bitartrate 2.5 Epinephrine bitartrate 2.8 Phenylephrine hydrochloride 3.5 Isoproterenol sulfate 4.3 Phenylpropanolamine-HCl 5.2
Ref.	Crommen J (1979) J. Chromatogr. 186:705

HPLC 3

Sample	Raw materials
Column	Reversed-phase; ODS 5–10 μm, 300 × 4 mm ID
Mobile phase	0.17 N Acetic acid
Flow-rate	1.5 ml min^{-1}
Injection size	10 μl
Detection	UV, $\lambda = 278$ nm
Ref.	Anonym (1985): The United States Pharmacopeia XXI and The National Formulary XVI. United States Pharmacopeial Convention Inc., Rockville, p 569

HPLC 4

Sample	Raw materials
Apparatus	LDC/Milton Roy reciprocating pump with pulse dampener and Cecil M 212 UV detector
Column	Reversed-phase; ODS-Hypersil-5 μm, 100 × 4.5 mm ID
Mobile phase	Water/methanol/acetic acid/sodium lauryl sulfate (70:30:2:0.002)
Flow-rate	1.4 ml min^{-1}; 1100 psi
Internal standard	Adrenaline acid tartrate
Injection size	5–20 μl, 0.1–0.8 mg ml^{-1}
Detection	UV, $\lambda = 280$ nm; 0.01 AU
Recording	Potentiometric recorder with integrator
Retention times, min	Adrenaline acid tartrate 6.5 Isoprenaline hydrochloride 9.5 Isoproteronone 10.5
Ref.	Clements JA, Hasson K and Smith G (1980) J. Chromatogr. 189:272

103 Ivermectin

CAS-No	70288-86-7 (mixture); 70161-11-4 (compound B_{1a}) 70209-81-3 (compound B_{1b})

Chemically	Mixture of 22,23-dihydroavermectin B_{1a} (80%) and B_{1b} (20%)
Proprietary name	Ivomec (vet)
Clinical use	Anthelmintic
Solubility	Soluble in methanol
Light absorption	λ_{max} = 245 nm in methanol

HPLC 1

Sample	Human blood plasma and milk
Extraction	Methanol on C_2 extraction cartridge
Apparatus	Varian M 5000 LC with gradient controller and Perkin-Elmer M 650S fluorescence detector
Column	Reversed-phase; Sepralyte C_{18}, 3 μm, 50 × 4.6 mm ID
Mobile phase	Water/methanol (30:70) Gradient: Methanol 70% to 96% in 4 min; run 8 min; methanol 96% to 100% in 0.1 min; run 4.4 min; methanol 100% to 70% in 0.1 min
Flow-rate	1.5 ml min^{-1}
Temperature	50 °C
Internal standard	Ivermectin monosaccharide, 0.1 μg ml^{-1}
Derivatization	Pre-column; with 1-methylimidazole/acetic anhydride/N,N-dimethylformamide (3:7:2)
Injection size	30–50 μl
Detection	Fluorescence, λ = 364/470 nm
Recording	Spectra-Physics computing integrator SP 4270
Retention times, min	Internal standard 6.0 Ivermectin B_{1a} (H_2B_{1a}) 7.6
Ref.	Chiou R, Stubbs RJ and Bayne WF (1987) J. Chromatogr. Biomed. Appl. 416:196

HPLC 2

Sample	Cattle blood plasma
Extraction	2-Step process: Extraction and clean-up solvent extraction
Apparatus	Water M 6000A LC pump, U6K injector and 440 UV detector
Column	Normal phase, μ-Bondapak NH_2, 300 × 3.9 mm ID
Mobile phase	n-Hexane/tetrahydrofuran/methanol/dimethyl sulfoxide (75:25:4:2); n-hexane is mixed at the end
Flow-rate	1 ml min^{-1}; 800–2000 psi
Temperature	25 °C
Injection size	100 μl containing 4–5 ng ml^{-1}
Detection	UV, λ = 254 nm; 0.01 AUFS

Retention times, min	Dihydroavermectin B_{1a} (H_2B_{1a}) 10–13.2
Ref.	Schnitzerling HJ and Nolan J (1985) J. Assoc. Off. Anal. Chem. 68:36

HPLC 3

Sample	Cattle and sheep tissues
Extraction	Isooctane, and then with hexane
Apparatus	Du Pont M 830 LC, Rheodyne sample loop and Schoeffel FS 970 fluorescence detector
Column	Reversed-phase; Zorbax ODS, 150 × 4.6 mm ID
Pre-column	Co: Pell ODS (Reeve Angel)
Mobile phase	Methanol/water (95:5)
Flow-rate	1.8 ml min^{-1}
Temperature	Ambient
Injection size	50 µl
Detection	Fluorescence, $\lambda = 364/440$ nm
Retention time, min	Dihydroavermectin B_{1a} 14.0
Ref.	Tway PC, Wood Jr JS and Downing GV (1981) J. Agric. Food Chem. 29:1059

HPLC 4

Sample	Raw materials
Apparatus	Spectra-Physics 8700 XR ternary solvent pump with Rheodyne M 7125 injector, LDC spectromonitor III UV and Perkin-Elmer LC-1000 fluorescence detectors
Column	Reversed-phase; Zorbax C_8, 150 × 4.6 mm ID
Mobile phase	Methanol/acetonitrile/water (40:48:12)
Flow-rate	1 ml min^{-1}
Derivatization	Pre-column; with ammonium acetate in ethanol
Injection size	10 µl, 0.01–10 µg ml^{-1}
Detection	UV, $\lambda = 240$ nm Fluorescence, $\lambda = 361/445$ nm
Retention times	Reaction intermediates, H_2B_{1a}-5-one and several fluorescence derivatives at different reaction times (0–300 min)
Ref.	Strong JD (1987) Anal. Chem. 59:266

HPLC 5

Sample	Blood plasma
Extraction	Ethyl acetate
Apparatus	Spectra-Physics 740 pump with 740C pump controller and 714 pressure monitor and Perkin-Elmer LC-1000 fluorimeter
Column	Reversed-phase; Zorbax ODS, 250 × 4.6 mm ID
Pre-column	Du Pont Permaphase ODS, 100 mm
Mobile phase	Acetonitrile/methanol/water (100:15:1)

Flow-rate	2.4 ml min^{-1}
Temperature	Ambient
Derivatization	Pre-column; with acetic anhydride-pyridine
Injection size	100 µl
Detection	Fluorescence, λ = 364/390 and 480 nm
Retention times, min	(estimated from drawing)
	Avermectin B_{2a} 5.0
	Avermectin B_{1a} 6.5
	Avermectin B_{1a}-4″-acetate 6.5
	Avermectin dihydro-B_{1b} 8.5
	Avermectin dihydro-B_{1a} 10.0
Ref.	Tolan JW, Eskola P et al (1980) J. Chromatogr. 190:367

---

104 Kanamycin

CAS-No	59-01-8; 25389-94-0 (sulfate)			
Chemically	Aminoglycoside anitbiotic produced by Streptomyces kanamyceticus			
Empirical formula	Kanamycin	Sulfate		
	$C_{18}H_{36}N_4O_{11}$	$C_{18}H_{36}N_4O_{11}$, H_2SO_4		
Molecular mass	484.5	582.6		
Proprietary names	Kannasyn; Kantrex; Klebcil			
Clinical use	Antibiotic			
Solubility,	Water	alcohol	acetone	ether
Sulfate	1 in 8	insoluble	insoluble	insoluble
Light absorption	λ_{max} = 249 nm in aqueous acid			

HPLC 1

Sample	Raw materials
Apparatus	Perkin-Elmer Series 4 and Waters M 6000A LC pumps or LDC minipump, Waters M 710B injector and LDC ultraviolet detector
Column	Normal phase; Lichrosorb Si 60, 250 × 4.6 mm ID
Mobile phase	Chloroform/methanol/acetic acid (560:400:40, v/v)
Flow-rate	1.7 ml min^{-1}
Temperature	Ambient
Derivatization	Pre-column; with 2-naphthalenesulfonyl chloride followed by extraction
Detection	UV, λ = 254 and 227 nm
Recording	Hewlett Packard 3390A integrator

Retention time, min	Kanamycin 10.0
Ref.	Tsuji K and Jenkins KM (1986) J. Chromatogr. 369:105

HPLC 2

Sample	Raw materials
Apparatus	Varian Aerograph M 4200 LC with RI detector M-4200
Column	Cation-exchange; Aminex A-27-OH, 1000 × 2 mm ID
Mobile phase	Water
Flow-rate	0.2 ml min^{-1}
Temperature	40 °C
Injection size	100 µl and 2 µl containing 1.2–2.1 mg
Detection	RI
Retention times, min	(estimated from drawing) Kanamycin B 6.4 Deoxykanamycin 5.1 Kanamycin 14.6
Ref.	Ottake T and Yaguchi M: Varian Liquid Chromatography at Work 5. Varian, California

HPLC 3

Sample	Bulk drugs and pharmaceutical formulations
Apparatus	Waters M 6000A pump, WISP 710B autosampler and 480 spectrophotometric detector
Column	Reversed-phase; C_8, 5 µm, 250 × 4.6 mm ID
Mobile phase	0.02 M Potassium phosphate, pH 7.5 with 40% (w/v) potassium hydroxide/acetonitrile/methanol (400:460:150, v/v), (400:493:120) for gentamicin and (400:506:160) for sisomicin
Flow-rate	1.5 ml min^{-1}
Temperature	45 °C
Derivatization	Pre-column; with 2,4,6-trinitrobenzenesulfonic acid, and pyridine (for gentamicin and sisomicin)
Injection size	25 µl; 35 µl (gentamicin)
Detection	UV, λ = 350 nm
Recording	Waters M 730 printer-plotter
Retention times, min	(estimated from drawing)

Kanamycin derivative A	5.79	Sisomicin	10.20
Kanamycin, probably C	7.45	Gentamicin C_{2a}	11.00
Gentamicin C_1	7.50	Gentamicin C_2	13.00
Gentamicin C_{1a}	9.00	Tobramycin	13.00
Kanamycin derivative B	9.72		

Ref.	Gambardella P and Punziano R et al (1985) J. Chromatogr. 348:229

HPLC 4

Sample	Urine and bile from cattle
Extraction	0.07% NaOH on SP-Sephadex C-25 mini-extraction column (SO_3H-form)

Column	Reversed-phase; TSK 80 TM-ODS, 250 × 4 mm ID
Mobile phase	Acetonitrile/water (60:40) containing 0.1% acetic acid
Flow-rate	1.0 ml min^{-1}
Temperature	50 °C
Derivatization	Pre-column; with 2,4-dinitrofluorobenzene
Injection size	10 µl, 0.5–25.0 µg ml^{-1}
Detection	UV, λ = 365 nm
Retention times, min	(estimated from drawing) Kanamycin 6.00– 7.00 Fradiomycin 19.00–20.00
Ref.	Saito Tomio, Kaneshima H et al (1987) J. Fd. Hyg. Soc. Jap. 28:317

HPLC 5

Sample	Cattle muscles and kidneys
Extraction	Trichloroacetic acid, and on Amberlite CG-50 (NH_4-form) column
Apparatus	Japan Spectroscopic Trirotar-V with spectrofluorimeter FP 210
Column	TSK gel; ODS-120T, 150 × 4.0 mm ID
Mobile phase	Methanol/water/acetonitrile (65:30:5)
Flow-rate	0.8 ml min^{-1}
Derivatization	Pre-column; with *o*-phthalaldehyde
Injection size	5 µl
Detection	Fluorimetric, λ = 335/440 nm
Retention time, min	Kanamycin 15.0 (estimated from drawing)
Ref.	Nakaya K, Sugitani A and Yamada F (1986) J. Fd. Hyg. Soc. Jap. 27:258

HPLC 6

Sample	Raw materials
Apparatus	HPLC system with refractive index detector
Column	Anion-exchange; Bio-Rad, Aminex A-27-OH, 1000 × 2 mm ID
Mobile phase	Water
Flow-rate	0.33 ml min^{-1}; 2000 psi
Temperature	40 °C
Injection size	4 µl containing 172 µg kanamycin A and 164 µg kanamycin B
Detection	RI
Retention times	(relative) Kanamycin B 0.45 Kanamycin C 0.58 Kanamycin A 1.00
Ref.	Tsuji K (1978) in: Tsuji K (ed) GLC and HPLC Determination of Therapeutic Agents, Part 2. Marcel Dekker Inc., New York and Basel, p 739

105 Ketamine

CAS-No	6740-88-1; 1867-66-9 (hydrochloride)			
Chemical name	(±)-2-(2-Chlorophenyl)-2-(methylamino)cyclohexanone			
Empirical formula	Ketamine		Hydrochloride	
	$C_{13}H_{16}ClNO$		$C_{13}H_{16}$ ClNO, HCl	
Molecular mass	237.7		274.2	
Proprietary names	Ketaject; Ketalar; Ketanest			
Clinical use	General Anaesthetic			
Solubility,	Water	alcohol	methanol	ether
Hydrochloride	1 in 4	1 in 14	1 in 6	insoluble
Light absorption	λ_{max} = 269 nm in aqueous acid			

HPLC 1

Sample	Blood plasma
Apparatus	Hitachi M 635 LC, 638 UV/VIS detector and 833 microprocessor with column-switching valve-system, Waters M 710B autosampler and Shimadzu 6A LC pump
Column	A. TSK-gel; G-2000SW, 300 × 7.5 mm ID and B. Reversed-phase; ERC-ODS 1272, 200 × 6.0 mm ID
Mobile phase	For A: 0.05 M Potassium dihydrogen phosphate For B: 0.05 M Potassium dihydrogen phosphate/acetonitrile (3:1) Programme: Introduction from column 1 to column 2 at 16–19.5 min and stop at 30 min
Flow-rate	1 ml min^{-1}
Detection	UV, λ = 210 nm
Retention time, min	Ketamine 22.0 (estimated from drawing)
Ref.	Takahagi H, Inoue K and Horiguchi M (1986) J. Chromatogr. 352:369

HPLC 2

Sample	Rabbit serum
Extraction	Methanol
Apparatus	Jasco Trirotar-VI HPLC with Multi-320 UV detector
Column	Reversed-phase; Jasco Fine Pak Sil C_{18}, 125 × 4.6 mm ID
Mobile phase	Acetonitrile/water (30:70) containing 60 mM perchloric acid
Injection size	50 µl containing 2 µg
Detection	UV, λ = 215, 220 and 275 nm; less than 0.005 AU

Recording	Jasco DS-L 800 data processor
Retention times, min	Nor-ketamine 5.03
	Ketamine 5.86
Ref.	Hoshino T, Hondo T et al (1985) J. Chromatogr. 332:139

106 Ketoconazole

CAS-No	65277-42-1
Chemical name	cis-1-Acetyl-4[4-[[2-(2,4-dichlorophenyl)-2-(1*H*-imidazol-1-ylmethyl)-1,3-dioxolan-4-yl]-methoxy]-phenyl]piperazine
Empirical formula	$C_{26}H_{28}Cl_2N_4O_4$
Molecular mass	531.4
Proprietary names	Fungarest; Nizoral; Panfungol
Clinical use	Antifungal
Solubility	Water insoluble; ethanol 1 in 54; chloroform 1 in appr. 2; methanol 1 in 9
Light absorption	λ_{max} = 244 nm in methanol

HPLC 1

Sample	Blood plasma
Extraction	Diethyl ether under alkaline condition
Apparatus	Varian M 8500 HPLC, Schoeffel FS 970 fluorescence detector and Rheodyne loop injector
Column	Reversed-phase; µ-Bondapak C_{18}, 10 µm, 300 × 3.9 mm ID
Mobile phase	60% Acetonitrile and 40% 0.05 M Sorensen's phosphate buffer (pH 6.6 with hydrochloric acid)
Flow-rate	1.5 ml min^{-1}
Internal standard	R 41300 (Janssen, Belgium)
Injection size	25 µl containing 0.25 µg
Detection	Fluorescence, λ = 206/370 nm
Retention times, min	Ketoconazole 5.2
	Internal standard 9.7
Ref.	Swezey SF, Giacomini KM et al (1982) J. Chromatogr. Biomed. Appl. 227:510

HPLC 2

Sample	Human blood plasma
Extraction	First ethyl acetate, then with 1 N HCL and again with ethyl acetate

Apparatus	Waters M 6000A LC pump, U6K injector and Perkin-Elmer LC-55B UV detector
Column	Reversed-phase; μ-Bondapak CN, 300 × 3.9 mm ID
Mobile phase	0.05 M KH_2PO_4 and NaOH buffer, pH 6.0/ acetonitrile (65:35, v/v)
Flow-rate	2.0 ml min^{-1}
Temperature	Ambient
Injection size	5–20 μl containing 100 ng
Detection	UV, λ = 205 nm; 0.01 AUFS
Recording	Hewlett-Packard 3354B computing integrator and 18652A converter
Retention times, min	(estimated from drawing) Ketoconazole 6.74 Miconazole 14.16
Ref.	Alton KB (1980) J. Chromatogr. Biomed. Appl. 221:337

HPLC 3

Sample	Rat blood plasma and tissues
Extraction	Methanol on a reversed-phase C_{18} extraction cartridge
Apparatus	Beckman M 110A LC pump, 153 UV detector and 210 injector
Column	Reversed-phase; Novapak C_{18}, 5 μm, 100 × 3.9 mm ID
Mobile phase	A. Methanol/0.02 M KH_2PO_4, pH 6.8 with NaOH (70:30) B. Methanol/0.02 M KH_2PO_4, pH 6.8 with NaOH/ acetonitrile (35:35:30)
Flow-rate	2.0 ml min^{-1}
Temperature	Ambient
Internal standard	Clotrimazole
Injection size	50 μl
Detection	UV, λ = 254 nm; 0.008 AUFS
Retention times, min	Mobile phase A. B. Ketoconazole 4.5 3.8–4.2 Clotrimazole 6.8 6.0–6.8
Ref.	Riley CM and James MO (1986) J. Chromatogr. Biomed. Appl. 377:287

HPLC 4

Sample	Blood plasma and saliva
Extraction	Hexane/ethyl acetate (1:1) under alkaline condition
Apparatus	Waters M 510 HPLC pump, Bioanalytical Systems LC-4B electrochemical detector and Kratos M 757 spectrophotometric detector

Column	Reversed-phase; μ-Bondapak ODS 10 μm, 300 × 3.9 mm ID
Pre-column	Corasil dry pack, ODS
Mobile phase	0.15 M Formic acid and 0.01 M dibutylamine in 50% methanol, pH 3.0
Flow-rate	2.0 ml min^{-1}
Temperature	40 °C
Internal standard	Perphenazine
Injection size	3–500 pmol in 20 μl
Detection	EC, GC electrode, + 0.2–1.2 V UV, λ = 231–254 nm
Recording	Beckman M 427 integrator
Retention times, min	Ketoconazole 6.0 Perphenazine 9.3
Ref.	Hoffman DW, Jones-King KL et al (1988) Anal. Biochem. 172:495

HPLC 5

Sample	Biological fluids
Extraction	Methanol on reversed-phase C_{18} cartridge under alkaline condition
Apparatus	Spectra-Physics 3500B LC, Schoeffel GM 770 UV detector and Schoeffel FS 970 spectrofluorimeter
Column	Reversed-phase; ODS 5 μm, 250 × 4.6 mm ID
Pre-column	Whatman C_{18}, 30–38 μm, 70 mm
Mobile phase	Methanol/0.02 M NaH_2PO_4 (75:25, v/v), pH 6.8 with NaOH
Flow-rate	1 ml min^{-1}
Internal standard	Phenothiazine
Injection size	50 μl, 0.02–1.0 μg ml^{-1}
Detection	UV, λ = 231 nm Fluorescence, λ = 206/370 nm; 0.04 AUFS
Retention times, min	Phenothiazine 7.4 Ketoconazole 9.6
Ref.	Pascucci VL, Bennett J et al (1983) J. Pharm. Sci. 72:1467

107 Ketoprofen

CAS-No	22071-15-4
Chemical name	2-(3-Benzoylphenyl)propionic acid
Empirical formula	$C_{16}H_{14}O_3$
Molecular mass	254.3

Proprietary names	Alreumun; Ketalgin; Orudis
Clinical use	Analgesic/Anitphlogistic
Solubility	Water insoluble; alcohol, ether freely soluble; methanol soluble
Light absorption	λ_{max} = 260 nm in aqueous acid

HPLC 1

Sample	Raw materials
Apparatus	Waters M 590 LC pump, 440 UV detector and U6K universal injector
Column	Reversed-phase; Hibar Lichrocart C_{18}, 125 × 4 mm ID
Pre-column	Lichrocart PVDF, 40 × 4 mm ID
Mobile phase	Acetonitrile/water (37:63), pH 2.7 with 50% ortho-phosphoric acid
Flow-rate	2 ml min^{-1}
Internal standard	Ethyl-4-hydroxybenzoate, 8 μg ml^{-1}
Injection size	10–200 ng
Detection	UV, λ = 254 nm; 0.05 AUFS
Retention times, min	Internal standard 1.8 Ketoprofen 4.3 (estimated from drawing)
Ref.	Pietta P, Manera E and Ceva P (1987) J. Chromatogr. 387:525

HPLC 2

Sample	Human blood plasma and urine
Extraction	5% Isopropanol in isooctane and with water and chloroform, respectively
Apparatus	Waters M 590 LC pump, WISP 712 autosampler and M481 UV detector
Column	Reversed-phase; Partisil-5 ODS-3, 100 mm
Pre-column	Reversed-phase; 37–53 μm, C_{18}, 50 mm
Mobile phase	0.06 M KH_2PO_4/acetonitrile/triethylamine (64:36:0.02)
Flow-rate	1 ml min^{-1}
Temperature	Ambient
Internal standard	Racemic fenoprofen calcium
Injection size	10–40 μl
Detection	UV, λ = 275 nm
Recording	Hewlett-Packard 3390A integrator
Retention times, min	R(−)-Ketoprofen 9.8 R(−)-Fenoprofen 17.7 S(+)-Ketoprofen 11.3 S(+)-Fenoprofen 19.9
Ref.	Foster RT and Jamali F (1987) J. Chromatogr. Biomed. Appl. 416:388

HPLC 3

Sample	Human serum
Extraction	Chloroform/acetonitrile (60:40) or hexane/ether (1:1)
Apparatus	Spectra-Physics 8700 XR LC pump, Pye-Unicam 4020 UV detector and Negretti & Zambra injection system
Column	Reversed-phase; Spherisorb 5 μm ODS, 250 × 4.5 mm ID
Mobile phase	Water-ortho-phosphoric acid to pH 3.2/acetonitrile/methanol (52:35:13)
Flow-rate	1 ml min^{-1}
Injection size	20 μl containing 4 ng
Detection	UV, λ = 250 nm
Retention times, min	Acetylsalicylic acid 2.5, Salicylic acid 3.0, Ketoprofen 4.5, Naproxen 4.7, Fenoprofen 6.8, Flurbiprofen 8.4, Diflunisal 9.0, Ibuprofen 10.0, Indomethacin 11.8, Mefenamic acid 15.7 (estimated from drawing)
Ref.	Omile CI, Tebbett IR and Danesh B (1986) Chromatographia 22:187

HPLC 4

Sample	Blood plasma and urine
Extraction	Diethyl ether
Column	Reversed-phase; Spherisorb 5 μm octadecyl, 40 × 4.6 mm ID
Pre-column	Reversed-phase; Vydac 30–44 μm, 40 × 3.2 mm ID
Mobile phase	0.05 M Phosphate (Na_2HPO_4-KH_2PO_4) buffer, pH 7.0 containing 6–8% acetonitrile
Flow-rate	2 ml min^{-1}; 1000 psi
Injection size	5–200 μl
Detection	UV, λ = 262 nm; 0.01 AUFS
Retention times, min	Naproxen 10.0, Ketoprofen 16.0, Probenecid 18.0, Fenoprofen 20.0, Ibuprofen 24.0
Ref.	Upton RA, Buskin JN et al (1980) J. Chromatogr. 190:119

HPLC 5

Sample	Biological fluids
Extraction	Diethyl ether
Column	A. Reversed-phase; Lichrosorb C_{18}, 5 μm, 100 × 4.6 mm ID

	B. Reversed-phase; Lichrosorb C_8, 5 μm, 100 × 4.6 mm ID
Mobile phase	Acetonitrile/0.02 M phosphate buffer, pH 3.0 (45:55)
Flow-rate	1 ml min^{-1}; 1100 psi
Temperature	Ambient
Injection size	20 μl
Detection	UV, λ = 254 nm
Retention times, min	Column A. B.
	Ketoprofen 5.2 4.2
Ref.	Bannier A, Brazier JL et al (1980) J. Pharm. Sci. 69:763

HPLC 6

Sample	Blood plasma and urine
Extraction	Diethyl ether
Column	Reversed-phase; Spherisorb-5, ODS 5 μm, 50 × 4.6 mm ID
Mobile phase	35% Methanol in water, pH 3.5 with acetic acid
Flow-rate	2.0 ml min^{-1}
Internal standard	Oxyphenbutazone
Injection size	5 μl, 1.5–25 μg ml^{-1}
Detection	UV, λ = 260 nm; 0.05–0.02 AUFS
Capacity factor, *k'*	Ketoprofen 6.0
	Internal standard 8.0
Ref.	Jefferies TM, Thomas WOA and Parfitt RT (1979) J. Chromatogr. Biomed. Appl. 162:122

HPLC 7

Sample	Human blood plasma
Extraction	Methanol on Bond-Elut, ODS C_{18} extraction column
Apparatus	Waters M 6000A LC pump, U6K injector and 440 UV detector
Column	Reversed-phase; Ultrasphere 5 μm, 250 × 4.6 mm ID
Pre-column	Supelguard LC-8, 5 μm, 20 × 4.6 mm ID
Mobile phase	Methanol/water (80 : 20, v/v)
Flow-rate	1.2 ml min^{-1}
Temperature	20–22 °C
Injection size	25 μl, 0.5 μg ml^{-1}
Detection	UV, λ = 280 nm; 0.001 AUFS
Recording	Waters 740 data module integrator
Retention times, min	Furprofen 3.02
	Ketoprofen 3.95
Ref.	Palumbo G and Pantaleoni GC (1988) J. Chromatogr. Biomed. Appl. 424 : 201

108 Labetalol

CAS-No	36894-69-6; 32780-64-6 (hydrochloride)			
Chemical name	5-[1-Hydroxy-2-(1-methyl-3-phenylpropylamino)-ethyl]salicylamide			
Empirical formula	Labetalol		Hydrochloride	
	$C_{19}H_{24}N_2O_3$		$C_{19}H_{24}N_2O_3$, HCl	
Molecular mass	328.4		364.9	
Proprietary names	Trandate, Vescal			
Clinical use	Antihypertensive			
Solubility,	Water	alcohol	chloroform	ether
Hydrochloride	soluble	soluble	insoluble	insoluble
Light absorption	λ_{max} = 246 nm in aqueous alkali			

HPLC 1

Sample	Serum or plasma
Extraction	Diethyl ether, and then with chloroform
Apparatus	Waters M 6000A pump, U6K injector and 450 UV detector
Column	Reversed-phase; μ-Bondapak C_{18}, 300 × 4 mm ID
Mobile phase	Acetonitrile/phosphate buffer, pH 3.0 (32.6 : 67.5, v/v)
Flow-rate	1 ml min^{-1}; 1000 psi
Internal standard	Propericiazine, 2 μg μl^{-1}
Injection size	15 and 50 μl
Detection	UV, λ = 207 nm
Retention times, min	(estimated from drawing)
	Labetalol 6.5
	Internal standard 13.0
Ref.	Woodman TF and Johnson B (1981) Ther. Drug Monit. 3 : 371

HPLC 2

Sample	Human blood plasma
Extraction	Diethyl ether under alkaline condition
Apparatus	Waters HPLC with M 450 UV detector
Column	Reversed-phase; μ-Bondapak C_{18}, 300 × 4 mm ID
Mobile phase	Acetonitrile/45 mM potassium dihydrogen phosphate (35 : 65, v/v) pH 3.0 with H_3PO_4
Flow-rate	1 ml min^{-1}
Internal standard	Pericyazine
Detection	UV, λ = 233 nm

Retention times, min	Labetalol 3.9
	Internal standard 7.3
Ref.	Dusci LJ and Hackett LP (1979) J. Chromatogr. 175 : 208

HPLC 3

Sample	Human blood plasma
Extraction	Ethylene dichloride/diethyl ether/isopropanol (45 : 45 : 10)
Apparatus	Kipp M 9208 HPLC pump and Schoeffel FS 970 LC fluorimeter
Column	Reversed-phase; C_{18}, 10 µm, 250 × 4.6 mm ID
Mobile phase	Water/acetonitrile (68 : 32) containing 40 g $NaClO_4$, 40 g triethylammonium chloride and 4 g sodium acetate per liter, pH 4.5
	Post-column addition of borate buffer, pH 9.3
Flow-rate	1 ml min^{-1}
Temperature	Ambient
Internal standard	Chloroquine sulfate
Injection size	50 µl, 12.5–250 ng ml^{-1}
Detection	Fluorescence, λ = 335/370 nm
Retention times, min	(estimated from drawing)
	Labetalol 8.3
	Internal standard 11.25
Ref.	Oosterhuis B, Van den Berg M and Van Boxtel CJ (1981) J. Chromatogr. Biomed. Appl. 226 : 259

HPLC 4

Sample	Watery solution and blood plasma
Extraction	Methanol/phosphate buffer, pH 3.0 (50 : 50)
Apparatus	Varian M 5000 LC pump with Perkin-Elmer LS 4 fluorescence detector and Perkin-Elmer LC 90 UV detector
Column	Cyanopropyl-bonded phase; Lichrosorb 5 µm, 125 × 4 mm ID
Pre-column	Lichrosorb CN, 10 µm
Mobile phase	Acetonitrile/phosphate buffer, pH 3.0 (20 : 80)
Injection size	100 µl
Detection	UV, λ = 220 and 305 nm
	Fluorescence, λ = 335/370 nm
Recording	Shimadzu IC-R3A integrator
Capacity factor, *k′*	Mepindolol 1.86
	Labetalol 2.88
	Propranolol 3.47
Ref.	Musch G, Buelens Y and Massart DL (1989) J. Pharm. Biomed. Anal. 7 : 483

109 Lactose

CAS-No	63-42-3 (anhydrous)
Chemical name	4-*O*-β-D-Galactopyranosyl-α-D-glucopyranose
Empirical formula	$C_{12}H_{22}O_{11}$
Molecular mass	342.3
Clinical use	Laxative
Solubility	Water 1 in 6; alcohol insoluble; chloroform insoluble

HPLC 1

Sample	Raw materials
Apparatus	Shimadzu M LC-6A LC system and Waters R401 refractometer detector
Column	α-Cyclodextrin; 5 μm, 250 × 4.5 mm ID
Mobile phase	I. Acetonitrile/water (80 : 20, v/v) II. Acetone/water (85 : 15, v/v)
Flow-rate	1.5 ml min
Temperature	22 °C
Injection size	40 μg
Detection	Refractive index
Recording	Shimadzu AC-R3A or CR-2AX Chromatopac

Retention times, min

Mobile phase	I.	II.
Erythrose	5.82	4.53
Rhamnose		4.81
Glyceraldehyde		4.83
2-Deoxy-D-ribose	6.23	4.86
Glycerine	6.47	5.09
Ribose	6.74	4.98
6-Deoxy-D-glucose	6.79	5.08
3-O-Methyl-D-glucose	7.05	
2-Deoxy-D-glucose	7.19	5.19
Lyxose	7.24	5.21
Xylose	7.32	5.33
2-Deoxy-D-galactose		5.40
Erythritol	7.47	5.54
Talose	7.51	5.13
6-Deoxy-D-galactose		5.71
Arabinose	7.87	5.73
Tagatose	8.01	5.51
Ribitol	8.30	5.72
Fructose	8.32	5.84

Mobile phase	I.	II.
Sorbose	8.38	5.62
Arabitol	8.43	5.91
Mannose	8.47	5.79
Allose		6.00
Xylitol	8.48	6.01
Glucose	9.04	6.20
Glactose	9.45	
D-gluco-D-gulo-Heptose		6.38
Sorbitol	9.56	6.43
Mannitol	9.62	
Galactinol	9.74	6.56
Sucrose	11.31	6.63
Turanose	11.61	6.98
Cellobiose	12.27	7.85
Maltose	12.33	7.30
myo-Inositol	13.04	9.70
Lactose	13.23	8.22
Maltitol	13.23	7.03
Melibiose	14.40	8.83
Melezitose	15.62	8.53
Maltotriose		8.80
Raffinose	17.58	9.25
α-Cyclodextrin		12.97
Stachyose	27.87	13.88
β-Cyclodextrin		15.60
γ-Cyclodextrin		18.93

Ref. Armstrong DW and Jin HL (1989) J. Chromatogr. 462 : 219

HPLC 2

Sample: Raw materials

Apparatus: Shimadzu HPLC system with Hitachi U-2000 spectrophotometer

Column: β-Cyclodextrin; 5 μm, 250 × 4.6 mm ID

Mobile phase: Acetonitrile/water
Gradient: Water 8% to 30% in 30 min; hold 15 min

Flow-rate: 1 ml min^{-1}

Temperature: 22 °C

Injection size: 31 μg

Detection: UV, λ = 195 nm

Recording: Shimadzu C-R3A Chromatopac

Retention times, min: (estimated from drawing)

Phenyl-β-D-glucopyranoside	4.0	Maltose	23.0
		Lactose	24.0
2-Deoxy-D-ribose	4.5	Melibiose	27.0

Ribose	5.8	Melezitose	31.0
Xylose	6.0	Maltotriose	32.5
Talose	6.5	Stachyose	38.0
Sorbose	8.0	α-Cyclodextrin	38.5
Glucose	10.0	β-Cyclodextrin	41.5
Sorbitol	12.0	γ-Cyclodextrin	43.2
Sucrose	20.0		
Turanose	21.5		

Ref. Armstrong DW and Jin HL (1989) J. Chromatogr. 462 : 219

HPLC 3

Sample: Sugar mixture
Column: Amino phase; Lichrosorb NH_2 5 μm, 200 × 4.6 mm ID
Mobile phase: Acetonitrile/water (75:25)
Flow-rate: 1 ml min^{-1}
Detection: UV, λ = 190 nm
Retention times, min: (estimated from drawing)

Rhamnose	4.8	Saccharose	12.5
Xylose	5.5	Maltose	14.5
Fructose	7.0	Trehalose	16.0
Glucose	8.5	Lactose	18.5
Galactose	9.0	Melibiose	20.0
		Raffinose	28.5

Ref. Meyer V (1984): Praxis der Hochleistungsfluessig-chromatographie, 3rd edn. Diesterweg Salle and Sauerlaender, Frankfurt am Main and Salzburg, p 134

HPLC 4

Sample: Sugar solution
HPLC with refractive index detector
Column: μ-Bondapak-Carbohydrate; 300 × 4 mm ID
Mobile phase: Water/acetonitrile (20:80)
Flow-rate: 1.5 ml min^{-1}
Temperature: 25 °C
Injection size: 5–40 μl, 1–50 mg ml^{-1}
Detection: RI
Retention times, min: (estimated from drawing)

Glucose	6.8	Maltose	11.5
Sucrose	9.4	Lactose	12.8

Ref. Palmer JK (1979) in: Tsuji K (ed) GLC and HPLC Determination of Therapeutic Agents, Part 3. Marcel Dekker Inc., New York and Basel, p 1317

HPLC 5

Sample	Sugars
Column	Amino phase; Nucleosil 5 NH_2, 250 × 4 mm ID
Mobile phase	Acetonitrile/water (75:25)
Injection size	20 µl containing 0.2 µg
Detection	RI; UV
Retention times	Methylglycopyranoside, fructose, glycose, saccharose, lactose and maltose, in ascending order (from drawing)
Ref.	Busssemas HH (1984) LABO 15:788

110 Levamisole

CAS-No	14769-73-4; 16595-80-5 (hydrochloride)				
Chemical name	(*S*)-2,3,5,6-Tetrahydro-6-phenylimidazo [2,1-*b*]thiazole				
Empirical formula	Levamisole	Hydrochloride			
	$C_{11}H_{12}N_2S$	$C_{11}H_{12}N_2S$, HCl			
Molecular mass	204.3	240.8			
Proprietary names	Ketrax; Nilverm (vet); Solaskil				
Clinical use	Anthelmintic/Immunostimulant				
Solubility,	Water	methanol	ether	chloroform	aq CH_3COOH
Levamisole				soluble	soluble
Hydrochloride	1 in 2	1 in 5	insol		

HPLC 1

Sample	Biological fluids
Extraction	Chloroform under alkaline condition
Apparatus	Waters M 6000A LC pump and Schoeffel GM 770 variable-wavelength UV detector
Column	Reversed-phase; µ-Bondapak C_{18}, 10 µm, 250 × 4.6 mm ID
Mobile phase	0.2% Acetic acid in water/methanol/heptane-sulfonic acid (55:45:2)
Flow-rate	2 ml min^{-1}; 2900 psi
Temperature	Ambient (18–20 °C)
Injection size	10–100 µl containing 0.1–1.0 µg ml^{-1}
Detection	UV, λ = 225 nm; 0.01 AUFS
Retention time, min	Levamisole 3.8
Ref.	Alvinerie M, Galtier P and Escoula G (1981) J. Chromatogr. Biomed. Appl. 223:445

HPLC 2

Sample	Milk
Extraction	Dichloromethane using Extrelut 20 (Merck) extraction column
Apparatus	LDC Constametric III LC pump, Spectro-Monitor III with variable-wavelength UV detector and Valco N 60 injector
Column	TSK-Gel; LS 410 ODS SIL (Toyo Soda), 5 µm, C_{18}, 250 × 4.6 mm ID
Mobile phase	0.05 M Ammonium carbonate/methanol (35:65, v/v)
Flow-rate	1 ml min^{-1}
Temperature	Ambient
Injection size	22 µl, 1.0–40 µg ml^{-1}
Detection	UV, $\lambda = 220$ nm
Retention time, min	Levamisole 6.00–7.00 (estimated from drawing)
Ref.	Oesterdahl B-G, Johnsson H and Nordlander I (1985) J. Chromatogr. Biomed. Appl. 337:151

HPLC 3

Sample	Blood plasma and gastro-intestinal fluids
Extraction	Diethyl ether under alkaline condition
Apparatus	Altex M 110 LC pump and Cecil CE 2012 ultraviolet detector
Column	Reversed-phase; ODS Hypersil, 100 × 5 mm ID
Mobile phase	Methanol/0.05 M ammonium carbonate solution (65:35)
Flow-rate	1.0 ml min^{-1}
Injection size	5 µl, 5–20 µg ml^{-1}
Detection	UV, $\lambda = 220$ nm; 0.1 AUFS
Retention time, min	Levamisole 2.7
Ref.	Marriner S, Galbraith EA and Bogan JA (1980) Analyst 105:993

111 Lidocaine (Lignocaine)

CAS-No	137-58-6; 6108-05-0 (hydrochloride, monohydrate)	
Chemical name	2-Diethylaminoaceto-2′,6′-xylidide	
Empirical formula	Lidocaine	Hydrochloride
	$C_{14}H_{22}N_2O$	$C_{14}H_{22}N_2O$, HCl, H_2O
Molecular mass	234.3	288.8
Proprietary names	Ultracaine; Xylocaine; Xylocard	
Clinical use	Local Anaesthetic	

Solubility,	Water	ethanol, chloroform	ether
Lidocaine	insoluble	very soluble	freely soluble
Hydrochloride	1 in 0.7	1 in 1.5 1 in 40	insoluble
Light absorption	λ_{max} = 263 nm in aqueous acid		

HPLC 1

Sample	Blood plasma
Extraction	Ethyl acetate under alkaline condition and then with diluted H_2SO_4
Apparatus	Waters M 6000A LC pump with Perkin-Elmer LC-55 UV detector
Column	Alkyl-phenyl reversed-phase; μ-Bondapak, 10 μm, 300 × 3.9 mm ID
Mobile phase	Acetonitrile/0.006% phosphoric acid (30:70)
Flow-rate	2 ml min^{-1}; 1800 psi
Temperature	Appr. 24 °C
Internal standard	Ethylmethylglycinexylidide-HCl
Injection size	20–30 μl
Detection	UV, λ = 200 nm
Retention times, min	Glycinexylidide 5.9; Monoethylglycine-xylidide 7.4; Internal standard 8.3; Lidocaine 10.0
Ref.	Nation RL, Peng GW and Chiou WL (1979) J. Chromatogr. Biomed. Appl. 162:466

HPLC 2

Sample	Biological materials from rats
Extraction	Ethyl acetate under alkaline condition and then with diluted H_2SO_4
Apparatus	Shimadzu LC-3A pump and SPD-2A ultraviolet detector
Column	Reversed-phase; μ-Bondapak C_{18}, 10 μm, 300 × 4 mm ID
Mobile phase	Acetonitrile/0.1 M potassium phosphate buffer, pH 3 (1:9)
Flow-rate	2 ml min^{-1}
Temperature	40 °C ± 2 °C
Internal standard	Procaine hydrochloride
Injection size	5 and 50 μl, 11.8 nmol ml^{-1}
Detection	UV, λ = 210 nm; 0.02 AUFS
Retention times, min	Metabolite-3 2.6; Metabolite-2 3.5; Metabolite-4 4.2; Internal standard 5.0; Metabolite-1 5.7; Lidocaine 9.0

Ref.	Kawai R, Fujita S and Suzuki T (1985) J. Pharm. Sci. 74:1219

HPLC 3

Sample	Raw materials
Apparatus	Spectra-Physics 3500B LC pump and M 770 ultraviolet detector
Column	Reversed-phase; μ-Bondapak C_{18}, 10 μm, 300 × 3.9 mm ID
Mobile phase	Methanol/acetic acid/triethylamine/water (50:1.5:0.5:48)
Flow-rate	1.5 ml min^{-1}
Temperature	Ambient
Injection size	10 μl, 0.25 mg ml^{-1}
Detection	UV, λ = 254 nm; 0.16 AUFS
Recording	Hewlett-Packard 3380A integrator
Retention times, min	(estimated from drawing) Bupivacaine 10.0
	Lidocaine 6.0 Benzocaine 10.3
	Butacaine 7.7 Tetracaine 13.0
Ref.	Roos RW and Lau-Cam CA (1986) J. Chromatogr. 370:403

HPLC 4

Sample	Human blood plasma and urine
Extraction	Ethyl acetate under alkaline condition
Apparatus	Waters M 45 HPLC pump, WISP 721 autosampler and M 480 ultraviolet detector
Column	Reversed-phase; Novopak 5-μm, C_{18}, 115 × 8 mm ID
Mobile phase	Water/acetonitrile/phosphoric acid/ triethylamine (87.75:12:0.1:0.15), pH 3
Flow-rate	2 ml min^{-1}
Derivatization	Pre-column; with acetic anhydride
Internal standard	Procaine hydrochloride, 0.2–500 μg ml^{-1}
Injection size	50–200 μl
Detection	UV, λ = 200 nm
Recording	Waters M 480 data processing station
Retention times, min	(estimated from drawing)
	Procaine 8.0
	Metabolites (9) 10–65
	Lidocaine 14.6
Ref.	Tam YK, Tawfik SR and Coutts RT (1987) J. Chromatogr. Biomed. Appl. 423:199

HPLC 5

Sample	Blood plasma
Extraction	Centrifugation

Apparatus	LDC solvent delivery system 711–47 (Milton Roy minipump with pulse dampener) and Hitachi M 655A UV detector
Column	Reversed-phase; Lichrosorb 7 µm, 125 × 4.0 mm ID
Pre-column	Reversed-phase; Lichrosorb 10 µm, 10 × 3.2 mm ID
Mobile phase	1-Propanol/phosphate buffer, (containing 1.0 M NaH_2PO_4, 0.5 M Na_2HPO_4 or 1.0 M H_3PO_4, pH 3.0) (13.5:86.5, v/v) containing 1.0 mM decanesulfonate
Flow-rate	0.9 ml min^{-1}
Injection size	100 µl containing 1.05 µg ml^{-1}
Detection	UV, λ = 210 nm
Retention times, min	(estimated from drawing) Glycylxylidide 23.0 Monoethylglycylxylidide 28.0 Lidocaine 31.0
Ref.	Daoud N, Arvidsson T and Wahlund K-G (1987) J. Pharm. Biomed. Anal. 5:533

HPLC 6

Sample	Pharmaceutical formulations
Apparatus	Altex M 100A LC pump and 155 UV detector with Hitachi M 100-30 spectrophotometer
Column	Reversed-phase; Lichrosorb C_2, 10 µm, 250 × 3.2 mm ID
Mobile phase	Methanol/0.1% dibasic ammonium phosphate in water (50:50), pH 7.90
Flow-rate	1.5 ml min^{-1}; 2700 psi
Temperature	Ambient
Injection size	20 µl, 0.5–1.5 mg ml^{-1}
Detection	UV, λ = 263 and 273 nm
Retention times, min	(estimated from drawing)

1-Norepinephrine	1.0	Chlorocresol	2.8
m-Cresol	1.7	Propylparaben	3.0
Methylparaben	1.8	Lidocaine-HCl	5.0

Ref. Leung CP (1979) J. Chromatogr. 178:579

112 Lorazepam

CAS-No	846-49-1
Chemical name	7-Chloro-5-(2-chlorophenyl)-1,3-dihydro-3-hydroxy-2*H*-1,4-benzodiazepin-2-one
Empirical formula	$C_{15}H_{10}Cl_2N_2O_2$
Molecular mass	321.2

Proprietary names	Ativan; Lorans; Tavor		
Clinical use	Tranquillizer		
Solubility	Water insoluble	ethanol, acetone soluble	chloroform, ether slightly soluble
Light absorption	λ_{max} = 230 nm in ethanol		

HPLC 1

Sample: Benzodiazepines and their metabolites

Apparatus: Waters M 6000A LC pump and Perkin-Elmer LC-75 UV detector

Column: Normal phase; Spherisorb S5W, 5 μm, 250 × 5 mm ID

Mobile phase: Methanol/water/trifluoroacetic acid (997:2:1, v/v/v)

Flow-rate: 2 ml min^{-1}

Injection size: 20 μl

Detection: UV, λ = 240 nm

Capacity factor, *k'*:

Ketazolam	0	Nitrazepam	1.60
7-Aminonitrazepam	0	Prazepam	2.42
Desmethylclobazam	0	Alprazolam	2.79
Clobazam	0.02	Clorazepic acid	3.12
Lormetazepam	0.02	Nordazepam	3.12
Lorazepam	0.05	Diazepam	3.35
Demoxepam	0.07	7-Acetamido-nitrazepam	3.58
Clonazepam	0.13	Bromazepam	4.03
Flunitrazepam	0.21	Desmethylchlor-diazepoxide	4.65
Temazepam	0.30	Chlordiazepoxide	5.55
Oxazepam	0.36	Desoxychlor-diazepoxide	7.02
N-1-Hydroxyethyl-flurazepam	0.98	Medazepam	8.04
N-1-Desalkylflur-azepam	1.26	Flurazepam	10.86
Triazolam	1.37	Midazolam	10.86

Ref.: Gill R, Law B and Gibbs JP (1986) J. Chromatogr. 356:37

HPLC 2

Sample: Monkey blood plasma

Extraction: Diethyl ether at pH 9.0 with borate buffer

Apparatus: Waters M 6000A pump, Pye-Unicam LC 3 multi-wavelength detector and 710A autosampler

Column: Reversed-phase; μ-Bondapak C_{18}, 10 μm, 300 × 4 mm ID

Pre-column: Whatman-ODS, 27–37 μm, 70 × 0.2 mm ID

Mobile phase	Acetonitrile/0.1% aqueous NaH_2PO_4 (40:60), pH 3.0 with phosphoric acid
Flow-rate	2 ml min^{-1}
Internal standard	Diazepam, 10 µg ml^{-1}
Injection size	50 µl, 0–100 ng ml^{-1}
Detection	UV, λ = 230 nm
Recording	Trilap or HP 3380A computing integrator
Retention times, min	Lorazepam 4.3 Diazepam 8.8
Ref.	Walmsley LM and Chasseaud LF (1981) J. Chromatogr. Biomed. Appl. 226:155

HPLC 3

Sample	Blood plasma
Extraction	Methyl tert.-butyl ether
Apparatus	Altex M 110A LC pumps with 421 controller, and Schoeffel SF 770 UV detector and Rheodyne M 7125 injector
Column	Reversed-phase; C_{18}-microsorb, 5 µm, 150 × 4.6 mm ID
Mobile phase	Methanol/water (70:30, v/v)
Flow-rate	1.2 ml min^{-1}
Internal standard	Prazepam
Injection size	20 µl, 50–1000 ng ml^{-1}
Detection	UV, λ = 230 nm
Recording	Shimadzu C-RIA integrator
Retention times, min	5-Hydroxyphenyl-phenytoin 1.56 Pimidone 1.65 Phenobarbital 1.82 Carbamazepine 10,11-epoxide 1.84 Phenytoin 2.17 Carbamazepine 2.55 Lorazepam 3.05 Diazepam 5.04 Prazepam 8.72
Ref.	Gunawan S and Treiman DM (1988) Ther. Drug Monit. 10:172

HPLC 4

Sample	Blood plasma
Extraction	Micro column clean-up with water/acetonitrile (1:2, v/v)
Column	Reversed-phase; ODS-Hypersil 3 µm, 150 × 4.5 mm ID
Mobile phase	Methanol/1-propanol/0.02 M phosphate buffer, pH 6.0 (100:8:80, v/v)
Temperature	40 °C
Internal standard	6-Nitroquinoline
Injection size	10 µl containing 10 ng

Detection	EC, Mercury drop electrode, − 1.2 V vs Ag/AgCl RE			
Retention times, min	(estimated from drawing)			
	Aminonitrazepam	2.00	Oxazepam	5.00
	7-Acetamidonitrazepam	2.25	Loprazolam	5.40
	6-Nitroquinoline	3.10	Temazepam	5.80
	Demoxepam	3.45	Lormetazepam	5.94
	Nitrazepam	3.80	Chlordiazapoxide	6.52
	Triazolam	4.70	Desmethyldiazepam	7.20
	Lorazepam	4.90	Diazepam	8.30
Ref.	Lloyd JBF and Parry DA (1989) J. Anal. Toxicol. 13:163			

HPLC 5

Sample	Injectable formulations
Extraction	Mobile phase
Column	Reversed-phase; ODS 5–10 μm, 100–150 × 4.6 mm ID
Mobile phase	Methanol/0.05 M monobasic ammonium phosphate (50:50), pH 6.5 with ammonium hydroxide
Flow-rate	2.0 ml min^{-1}
Injection size	Appr. 20 μl
Detection	UV, λ = 240 nm
Ref.	Anonym (1985): The United States Pharmacopeia XXI and The National Formulary XVI, Supplement 6, 1987. United States Pharmacopeial Convention Inc., Rockville, p 2572

---

113 Lysergide (LSD)

CAS-No	50-37-3
Chemical name	(+)-*NN*-Diethyl-D-lysergamide
Empirical formula	$C_{20}H_{25}N_3O$
Molecular mass	323.4
Clinical use	Hallucinogen
Solubility	Soluble in water as tartrate salt
Light absorption	λ_{max} = 310 nm in aqueous alkali

HPLC 1

Sample	Human body fluids
Extraction	n-Heptane containing 2% isopentanol at pH 8.5
Apparatus	Waters M 6000 LC pump, U6K injector and Perkin-Elmer MPF-2A spectrofluorimeter

Column	A. Reversed-phase; Spherisorb ODS 5 μm, 100 × 4.6 mm ID B. Normal phase; Spherisorb S5W, 150 × 4.6 mm ID
Mobile phase	I. For A. Methanol/0.025 M disodium hydrogen phosphate (65:35), pH 8.0 with 10% orthophosphoric acid II. For B. Methanol/0.2 M ammonium nitrate (60:40)
Flow-rate	1 ml min^{-1}; 5.5 MPa (I) and 4.1 MPa (II)
Injection size	Less than 100 pg
Detection	Fluorescence, λ = 320/400 nm
Retention time, min	Lysergide 6.8 (A) (estimated from drawing)
Ref.	Twitchett PJ, Fletcher SM et al (1978) J. Chromatogr. 150:73

HPLC 2

Sample	Pharmaceutical raw materials
Apparatus	Waters M 6000 or Altex M 100 LC pump and 153 UV detector
Column	Reversed-phase; Nucleosil C_{18}, 5 μm, 75 × 3 mm ID
Mobile phase	A. 0.01 M Ammonium carbonate B. Acetonitrile Gradient: Start 8% B in A; B 8% to 15% in 32 sec; B 15% to 30% in 32 sec; B 30% to 40% in 32 sec; B 40% to 50% in 32 sec and B 50% to 60% in 62 sec
Flow-rate	2.5 ml min^{-1}; 3045 psi
Injection size	30 μl
Detection	UV, λ = 254 nm
Recording	Hewlett-Packard 3352B data system
Retention times, min	(estimated from drawing)

Lysergic acid	0.85
Isolysergic acid	1.85
Lysergic acid amide	2.5
Isolysergic acid amide	2.7
aci-Ergotamine	3.1
aci-Ergotaminine	3.5
Ergotamine	3.6
Ergotaminine	4.2

Ref.	Erni F, Frei RW and Lindner W (1976) J. Chromatogr. 125:265

HPLC 3

Sample	Illicit LSD sample
Extraction	Methanol

Apparatus	HPLC with UV and fluorescence detectors
Column	Reversed-phase; Micropak MCH-10, 250 × 2.1 mm ID
Mobile phase	Acetonitrile/0.1 M aqueous ammonium carbonate (50:50)
Flow-rate	1 ml min^{-1}
Injection size	130 pg
Detection	UV, λ = 254 nm Fluorescence, λ = 330/ > 420 nm
Retention time, min	Lysergide (LSD) 2.4 (estimated from drawing)
Ref.	Abbott SR and Berg JR (1977): Varian Liquid Chromatography at Work 32: Varian, California

HPLC 4

Sample	Lysergic acid derivatives and tablets
Extraction	Methanol
Apparatus	HPLC system with UV detector
Column	Normal phase; silica 6 μm, 250 × 4.6 mm ID
Mobile phase	Methanol/0.2 N ammonium nitrate solution (3:2)
Flow-rate	1.0 ml min^{-1}; 2700 psi
Temperature	Ambient
Injection size	1 μl
Detection	UV, λ = 320 nm
Retention times, min	(estimated from drawing) Lysergic acid 5.0 LSD 7.8 Lysergamide 6.0 iso-LSD 11.5
Ref.	McChesney JD (1978) in: Tsuji K (ed) GLC and HPLC Determination of Therapeutic Agents, Part 2. Marcel Dekker Inc., New York and Basel, p 417

114 Mebendazole

CAS-No	31431-39-7
Chemical name	Methyl 5-benzoyl-1*H*-benzimidazol-2-ylcarbamate
Empirical formula	$C_{16}H_{13}N_3O_3$
Molecular mass	295.3
Proprietary names	Mebutar; Telnin (vet); Vermox
Clinical use	Anthelmintic
Solubility	Water, ethanol, insoluble; methanol, ether insoluble; formic acid freely soluble
Light absorption	λ_{max} = 234 nm in acid isopropyl alcohol

HPLC 1

Sample	Tablets
Apparatus	Waters M 440 HPLC and UV detector
Column	Reversed-phase; μ-Bondapak C_{18}
Mobile phase	Tetrahydrofuran/0.5% formic acid (30:60)
Flow-rate	1.9 ml min^{-1}; 1500 psi
Temperature	25–28 °C
Internal standard	Salicylamide
Injection size	5 μl, 0.2-0.8 mg ml^{-1}
Detection	UV, $\lambda = 254$ nm; 1.0 AUFS
Retention times	Salicylamide and mebendazole, in ascending order (from drawing)
Ref.	Wang D-P and Lo H-S (1984) Analyst 109:669

HPLC 2

Sample	Human blood plasma
Extraction	Methanol on Sep Pak C_{18} extraction column
Apparatus	Altex M 322 MP HPLC with ultraviolet detector and Rheodyne M 7120 injector
Column	Reversed-phase; Lichrosorb C_8, 10 μm, 250 × 4.6 mm ID
Pre-column	Corasil C_{18}, 35 × 3.2 mm ID
Mobile phase	A. Methanol/water (55:45) B. Methanol/aqueous ammonium phosphate 0.05 M, pH 5.5 (55:45) Isocratic: 67% A and 33% B Gradient: A 100% for 7 min; B 0% to 45% in 6.2 min; hold 6.8 min
Flow-rate	1.7 ml min^{-1}; 1200–1300 psi
Temperature	Appr. 20 °C
Internal standard	5-Benzoylbenzimidazole carbamate
Injection size	20 μl, 20 ng ml^{-1}
Detection	UV, $\lambda = 254$ nm; 0.005 AUFS
Recording	Hewlett-Packard 3380A integrator

Retention times, min

System	isocratic	gradient
Metabolite III	3.30	15.40
Metabolite II	4.30	4.80
Metabolite I	5.00	16.80
Mebendazole	7.89	7.88
Internal standard	11.50	11.48

Ref. Allan RJ, Goodman HT and Watson TR (1980) J. Chromatogr. Biomed. Appl. 183:311

HPLC 3

Sample	Raw materials
Column	RSIL; C_{18} HL, 10 μm, 250 × 4.6 mm ID

Mobile phase	Methanol/formic acid (95:5)
Flow-rate	1.5 ml min^{-1}
Detection	UV
Retention time, min	Mebendazole 5.0 (estimated from drawing)
Ref.	Alltech Associates Inc. (1984) Chromatography, Catalog No 60, p 254

115 Meclizine

CAS-No	569-65-3; 1104-22-9 (hydrochloride, anhydrous)			
Chemical name	1-(4-Chlorobenzhydryl)-4-(3-methylbenzyl)-piperazine			
Empirical formula	Meclizine	Hydrochloride		
	$C_{25}H_{27}ClN_2$	$C_{25}H_{27}ClN_2$, 2HCl		
Molecular mass	391.0	463.9		
Proprietary names	Antivert; Bonamine; Postafen			
Clinical use	Antihistaminic			
Solubility,	Water	alcohol	chloroform	ether
Hydrochloride	1 in 1000	1 in 25	1 in 5	insoluble
Light absorption	λ_{max} = 230 nm in ethanol			

HPLC 1

Sample	Biological materials
Extraction	n-Hexane under alkaline condition
Apparatus	Altex M LC-110A pump, LDC ultraviolet detector and Rheodyne M 7010 injector
Column	Reversed-phase; Ultrasphere octyl 5 µm, 150 × 4.6 mm ID
Mobile phase	Acetonitrile/water (30:70, v/v)
Flow-rate	2.5 ml min^{-1}
Temperature	Ambient
Internal standard	Hydroxyzine hydrochloride
Injection size	20 µl, 0.02–2.0 µg ml^{-1}
Detection	UV, λ = 230 nm; 0.02 AUFS
Retention times, min	Hydroxyzine 8.0
	Meclizine 13.0
Ref.	Jun HW, Foda NH and McCall JW (1987) J. Pharm. Biomed. Anal. 5:295

HPLC 2

Sample	Tablet formulations
Apparatus	HPLC system with ultraviolet detector

Column	Reversed-phase; octyl 5 µm-ultrasphere, 150 × 4.5 mm ID
Mobile phase	Acetonitrile/water (7:3, v/v)
Flow-rate	2.5 ml min^{-1}
Temperature	Ambient
Internal standard	Hydroxyzine hydrochloride
Detection	UV, λ = 230 nm
Retention times, min	Hydroxyzine 8.0 Meclizine 13.0
Ref.	Foda NH, Jun HW and McCall JW (1988), Anal. Lett. 21:1177

HPLC 3

Sample	Meclizine hydrochloride tablets
Column	Cation-exchange; 10 µm, 250 × 4.6 mm ID
Pre-column	Silica gel; 30–50 µm, 40 × 4.6 mm ID
Mobile phase	Water/methanol (55:45) containing 0.69% NaH_2PO_4, pH 4.0 with H_3PO_4
Flow-rate	2.0 ml min^{-1}
Injection size	100 µl
Detection	UV, λ = 230 nm
Ref.	Anonym (1985): The United States Pharmacopeia XXI and The National Formulary XVI. The United States Pharmacopeial Convention Inc., Rockville, p 625

116 Medazepam

CAS-No	2898-12-6			
Chemical name	7-Chloro-2,3-dihydro-1-methyl-5-phenyl-1*H* 1,4-benzodiazepine			
Empirical formula	$C_{16}H_{15}ClN_2$			
Molecular mass	270.8			
Proprietary names	Benson; Nobrium; Tranquilax			
Clinical use	Tranquillizer			
Solubility	Water	alcohol	chloroform	ether
	insoluble	1 in 8	1 in 1	1 in 5
Light absorption	λ_{max} = 253 nm in aqueous acid			

HPLC 1

Sample	Raw materials and metabolites
Apparatus	Waters M 6000A LC pump and Perkin-Elmer 75-LC UV detector

Column	Reversed-phase; Hypersil ODS 5 µm, 200 × 5 mm ID
Mobile phase	Methanol/water/0.1 M phosphate buffer, pH 7.25 (55:25:20, v/v)
Flow-rate	1.5 ml min^{-1}
Injection size	20 µl
Detection	UV, λ = 240 nm
Capacity factor, *k'*	

7-Amino-nitrazepam	0.46	N-1-Desalkyl-flurazepam	5.19
7-Acetamido-nitrazepam	0.68	Temazepam	5.68
Chlorazepic acid	1.17	Lormetazepam	6.32
Bromazepam	2.32	Chlordiazepoxide	6.41
Demoxepam	2.42	Nordazepam	8.00
Clonazepam	2.85	Diazepam	9.47
Nitrazepam	2.96	Midazolam	9.75
Desmethyl-clobazam	3.06	Ketazolam	12.81
Flunitrazepam	3.15		
Clobazam	3.91		
N-1-Hydroxyethl-flurazepam	4.27		
Triazolam	4.38		
Desmethylchlor-diazepoxide	4.47		
Lorazepam	4.60		
Oxazepam	4.62		
Alprazolam	4.70		

Ref. Gill R, Law B and Gibbs JP (1986) J. Chromatogr. 356:37

HPLC 2

Sample	Raw materials and their metabolites
Apparatus	Waters M 6000A LC pump and Perkin-Elmer 75-LC UV detector
Column	Reversed-phase; µm Spherisorb S5W, 5 µm, 250 × 5 mm ID
Mobile phase	Methanol/perchloric acid 72% (1000:0.1, v/v)
Flow rate	2 ml min^{-1}
Injection size	20 µl
Detection	UV, λ = 240 nm
Retention times, min	(estimated from drawing)

Demoxepam	2.0	Nordazepam	5.0
Clonazepam	2.5	Diazepam	6.0
Temazepam	2.7	Chlordiazepoxide	6.8
Oxazepam	3.0	Medazepam	9.0
Nitrazepam	4.0		

Ref.	Gill R, Law B and Gibbs JP (1986) J. Chromatogr. 356:37

HPLC 3

Sample	Raw materials and dosage forms
Apparatus	Hewlett Packard 1084 LC and UV detector
Column	Reversed-phase; Nucleosil C_{18}, 10 μm, 250×4 mm ID
Mobile phase	Acetonitrile/0.05 M ortho-phosphoric acid (40:60)
Flow-rate	2 ml min^{-1}
Temperature	40 °C
Internal standard	Oxazepam
Injection size	5–30 μl
Detection	UV, $\lambda = 231$ nm
Recording	Hewlett-Packard 79850B LC-Terminal detector
Retention times, min	Chlordiazepoxide 1.55 Oxazepam 3.28 Medazepam 2.47 Diazepam 5.00
Ref.	Rehm KD and Steinigen M (1981) Pharm. Ztg. 126:99

117 Medroxyprogesterone Acetate

CAS-No	520-85-4 (medroxyprogesterone); 71-58-9 (acetate)
Chemical name	6α-Methyl-3,20-dioxopregn-4-en-17α-yl acetate
Empirical formula	$C_{24}H_{34}O_4$
Molecular mass	386.5
Proprietary names	Amen; Depo-Provera; Lutoral
Clinical use	Progestational Steroid
Solubility	Water insoluble; alcohol 1 in 800; acetone 1 in 50; chloroform 1 in 10
Light absorption	$\lambda_{max} = 241$ nm in ethanol

HPLC 1

Sample	Blood plasma
Extraction	Methanol on Sep-Pak extraction cartridge
Apparatus	Waters M 6000A pump, U6K injector and 440 UV detector
Column	Reversed-phase; Bondapak C_{18}, 5 μm, RCM, 100×8 mm ID
Mobile phase	Methanol/water (75:100, v/v), pH 4.00 with acetic acid
Flow-rate	2.5 ml min^{-1}; 1160 psi

Internal standard	19-Norprogesterone
Injection size	50 and 75 µl, 20–500 ng ml^{-1}
Detection	UV, λ = 254 nm
Recording	Waters data module integrator
Retention times, min	Medroxyprogesterone 8.10 Internal standard 9.85
Ref.	Milano G, Carle G et al (1982) J. Chromatogr. Biomed. Appl. 232:413

HPLC 2

Sample	Raw materials
Column	Normal phase; 5 µm, 250 × 2 mm ID
Mobile phase	700 ml Butyl chloride, 300 ml hexane, both previously saturated with water, and 80 ml acetonitrile
Flow-rate	For required resolution
Temperature	Ambient
Internal standard	Progesterone, 0.25 mg ml^{-1}
Injection size	Appr. 10 µl
Detection	UV, λ = 254 nm
Retention times, min	Progesterone appr. 12.0 Medroxyprogesterone acetate appr. 15.0
Ref.	Anonym (1985): The United States Pharmacopeia XXI and The National Formulary XVI. United States Pharmacopeial Convention Inc., Rockville, p 627

HPLC 3

Sample	Tablets
Apparatus	Waters M 510 LC pump, 481 UV detector and WISP 710B autosampler
Column	Reversed-phase; Novapak C_{18}, 5 µm, 150 × 3.9 mm ID
Mobile phase	Methanol/0.01 M $(NH_4)_2HPO_4$ (80:20, v/v) pH 7.2 with 85% phosphoric acid
Flow-rate	1.0 ml min^{-1}
Temperature	Ambient
Injection size	20 µl, 50–150 µg ml^{-1}
Detection	UV, λ = 254 nm
Recording	Hewlett-Packard 3390A integrator
Retention times, min	Medroxyprogesterone acetate 7.00
Ref.	Fatmi AA, Williams GV and Hickson EA (1988) J. Assoc. Off. Anal. Chem. 71:528

HPLC 4

Sample	Human blood plasma
Extraction	Hexane

Apparatus	Beckman M 110A LC pump and Pye-Unicam LC-UV detector	
Column	Reversed-phase; Spherisorb 5-ODS2, 5 μm, 250 × 4.6 mm ID	
Mobile phase	Methanol/0.02 M acetate buffer, pH 4.0 (79:21)	
Flow-rate	1.5 ml min^{-1}	
Internal standard	16β-Methylprogesterone	
Injection size	150 μl, 5–250 ng ml^{-1}	
Detection	UV, λ = 240 nm; 0.05 AUFS	
Retention times, min	Aldosterone	2.4
	Cortisone	2.5
	Corticosterone	3.0
	Estradiol	3.5
	Androstenedione	3.8
	17-α-Hydroxyprogesterone	4.0
	Testosterone	4.0
	Medroxyprogesterone acetate	5.3
	Progesterone	6.8
	Internal standard	9.0
Ref.	Read J, Mould G and Stevenson D (1985) J. Chromatogr. Biomed. Appl. 341:437	

HPLC 5

Sample	Blood plasma from cancer patient
Extraction	n-Hexane
Column	Reversed-phase; ODS-2,5 μm, 250 × 4.6 mm ID
Mobile phase	Methanol/0.02 M acetate buffer, pH 4.0 (79:21)
Flow-rate	1.5 ml min^{-1}
Internal standard	16β-Methylprednisolone
Injection size	150 μl, containing 4 ng ml^{-1}
Detection	UV, λ = 240 nm
Ref.	Mould GP, Read J et al (1989) J. Pharm. Biomed. Anal. 7:119

---

118 Melphalan

CAS-No	148-82-3
Chemical name	4-Bis(2-chloroethyl)amino-L-phenylalanine
Empirical formula	$C_{13}H_{18}Cl_2N_2O_2$
Molecular mass	305.2
Proprietary names	Alkeran; Alkerana
Clinical use	Antineoplastic

Solubility	Water insoluble	alcohol slightly soluble	ether insoluble	methanol 1 in 150	aq acid soluble
Light absorption	$\lambda_{max} = 260$ nm in methanol				

HPLC 1

Sample	Blood plasma
Apparatus	Pye-Unicam PU 4000 LC and PU 4020 ultraviolet detector
Column	Reversed-phase; Spherisorb ODS 5 µm, 250 × 4.6 mm ID
Pre-column	Spherisorb ODS 5 µm
Mobile phase	80% Methanol, 20% water and 0.0135% w/v sodium dodecyl sulfate, pH 3.11 with H_2SO_4
Flow-rate	1.3 ml min^{-1}
Temperature	40 °C
Injection size	200 µl, 100 µg ml^{-1}
Detection	UV, $\lambda = 260$ nm; EC, + 0.92 V
Recording	Pye-Unicam computing integrator CDP4
Retention time, min	Melphalan 9.5
Ref.	Adair CG, Thorburn Burns D and Harriot M (1986) Anal. Proc. 23:30

HPLC 2

Sample	Blood plasma
Extraction	Deproteinization, and then with methanol on a XAD-2 resin column
Apparatus	LDC LC pump, UV III monitor and Rheodyne injector
Column	Reversed-phase; ODS 5 µm, 250 × 4.6 mm ID
Mobile phase	Sodium dodecyl sulfate solution (0.675 g l^{-1}) /methanol (1:4, v/v), pH 3.00 with concentrated H_2SO_4
Temperature	40 °C ± 1 °C
Internal standard	Dns-arginine
Injection size	200 µl, 0.4 µg ml^{-1}
Detection	UV, $\lambda = 254$ and 263 nm; 0.002 AUFS
Retention times, min	(estimated from drawing) Melphalan 8.8; Monohydroxy-melphalan 6.8; Internal standard 11.5; Dihydroxy-melphalan 8.0
Ref.	Bosanquet AG and Gilby ED (1982) J. Chromatogr. Biomed. Appl. 232:345

HPLC 3

Sample	Human and animal blood
Extraction	Deproteinization with 5% trichloroacetic acid

Apparatus	Perkin-Elmer M 601 HPLC with gradient programmer and LC-55 UV detector
Column	Reversed-phase; HCODS/SIL X-C_{18}, 10 µm, 250 × 2.6 mm ID
Mobile phase	A. Acetonitrile B. 0.0175 M Acetic acid Concave gradient: Acetonitrile 12% to 80% in 14 min
Flow-rate	1.5 ml min^{-1}
Temperature	50 °C
Injection size	10 µl
Detection	UV, λ = 263 nm; 0.002 to 0.2 AUFS
Recording	Hewlett-Packard 3352-C laboratory integrator
Retention times, min	Hydroxy-melphalan (L-DOH) 6.5 Melphalan (L-PAM) 12.0
Ref.	Ahmed AE and Hsu T-F (1981) J. Chromatogr. Biomed. Appl. 222:453

HPLC 4

Sample	Mixture of hydrolytic products of melphalan
Apparatus	Spectra-Physics 3500B LC and 8200 UV detector
Column	Reversed-phase; ODS 10 µm, 250 × 3 mm ID
Mobile phase	Methanol/0.01 M sodium dihydrogen phosphate (1:1), pH 3.0 with 0.66 M H_3PO_4
Flow-rate	1 ml min^{-1}; 1400 psi
Temperature	Ambient
Internal standard	Benzophenone
Injection size	10 µl, 1–40 µg ml^{-1}
Detection	UV, λ = 254 nm; 0.04 AUFS
Retention times, min	(estimated from drawing) Dihydroxy-Melphalan 2.0 Mephalan 6.8 Monohydroxy-Melphalan 2.8 Internal standard 12.8
Ref.	Flora KP, Smith SL and Cradock JC (1979) J. Chromatogr. 177:91

HPLC 5

Sample	Tablets
Extraction	Alcohol/glacial acetic acid (75:2)
Column	Reversed-phase; octylsilane 5–10 µm, 250 × 4.2 mm ID
Mobile phase	Methanol/water/glacial acetic acid (50:49:1)
Injection size	10–20 µl
Detection	UV, λ = 254 nm
Retention time, min	Melphalan appr. 10.0

Ref.	Anonym (1985): The United States Pharmacopeia XXI and The National Formulary XVI, Supplement 1, 1984. United States Pharmacopeial Convention Inc., Rockville, p 1742

119 Mepyramine (Pyrilamine)

CAS-No	91-84-9; 59-33-6 (maleate)			
Chemical name	*N*-*p*-Anisyl-*N'N'*-dimethyl-*N*-(2-pyridyl) ethylenediamine			
Empirical formula	Mepyramine $C_{17}H_{23}N_3O$		Maleate $C_{17}H_{23}N_3O,C_4H_4O_4$	
Molecular mass	285.4		401.5	
Proprietary names	Anthisan; Flavelix; Pymal			
Clinical use	Antihistaminic			
Solubility,	Water	ethanol	chloroform	ether
Maleate	1 in 0.5	1 in 2.5	1 in 1.5	slightly sol
Light absorption	λ_{max} = 239 nm in aqueous acid			

HPLC 1

Sample	Dosage forms (capsules, tablets, solutions and sprays)
Apparatus	Kratos spectroflow M 400 LC pump and Milton Roy LDC spectromonitor D
Column	β-Cyclodextrin; 100 × 4.6 mm ID
Mobile phase	0.05 M Sodium acid phosphate, pH 7.0 with 0.1 M sodium hydroxide/methanol (70:30)
Injection size	50 µl, 5–35 µg ml^{-1}
Detection	UV, λ = 254 nm
Recording	Hewlett-Packard 3392A integrator
Retention times, min	

(estimated from drawing)		Pheniramine maleate	10.0
Phenylpropanolamine-hydrochloride	2.0	Tripolidine hydrochloride	12.7
Thenyldiamine hydrochloride	8.8	Pyrilamine maleate	14.0

Ref.	El Gezawi S, Omar N et al (1988) J. Pharm. Biomed. Anal. 6:393

HPLC 2

Sample	Tablets and syrups
Column	Cation exchange; Zipax SCX

Mobile phase	0.02 M Dibasic ammonium phosphate in 36% dioxane/water, pH 8.2		
Flow-rate	1.1 ml min^{-1}; 100 psi		
Temperature	Ambient		
Detection	UV, λ = 254 nm; 0.004 AU		
Retention times, min	(estimated from drawing)	Pheniramine	8.2
	Phenylpropanolamine 2.7	Pyrilamine	14.5
	Ephedrine 4.4		
Ref.	Mollica JA and Stober HC (1978) in: Tsuji K (ed) GLC and HPLC Determination of Therapeutic Agents, Part 2. Marcel Dekker Inc., New York and Basel, p 691		

HPLC 3

Sample	Nasal spray
Apparatus	Waters ALC/GPC M 204 HPLC system with UV detector
Column	Reversed-phase; μ-Bondapak C_{18}, 300 × 4 mm ID
Mobile phase	Methanol/water (65:35) containing 0.005 M heptane sulfonate (Waters PIC reagent B-7)
Flow-rate	2.0 ml min^{-1}
Internal standard	Naphazoline
Injection size	10 μl, 0.3–0.7 mg ml^{-1}
Detection	UV, λ = 254 nm; 0.2 AUFS
Retention times, min	(estimated from drawing) Naphazoline 3.0 Pyrilamine 4.5
Ref.	Wittmer DP and Pidacks C (1979) in: Tsuji K (ed) GLC and HPLC Determination of Therapeutic Agents, Part 3. Marcel Dekker Inc., New York and Basel, p 1211

120 Methadone

CAS-No	76-99-3; 297-88-1(±); 1095-90-5 (hydrochloride) and 125-56-4(±) (hydrochloride)	
Chemical name	(±)-6-Dimethylamino-4,4-diphenylheptan-3-one	
Empirical formula	Methadone	Hydrochloride
	$C_{21}H_{27}NO$	$C_{21}H_{27}NO,HCl$
Molecular mass	309.5	345.9
Proprietary names	Dolophine hydrochloride; Physeptone; L-Polamidon	
Clinical use	Narcotic Analgesic	

Solubility,	Water	ethanol	chloroform	ether
Hydrochloride	1 in 12	1 in 7	1 in 3	insoluble
Light absorption	λ_{max} = 292 nm in aqueous acid			

HPLC 1

Sample	Sustained release formulations
Apparatus	Waters M 6000A LC pump, U6K injector and UV detector
Column	Reversed-phase; μ-Bondapak C_{18}
Mobile phase	0.961 g Sodium 1-pentanesulfonic acid in 250 ml water and 750 ml methanol, pH 3.5 with glacial acetic acid
Flow-rate	1.0 ml min^{-1}
Temperature	Ambient
Internal standard	Anthracene, 0.2 μg ml^{-1}
Injection size	10 μl, 5–50 μg ml^{-1}
Detection	UV, λ = 230 nm; 0.01 AUFS
Retention times, min	Methadone 5.1
	Internal standard 8.1
Ref.	Hsieh J-W, Ma JKH et al (1978) J. Chromatogr. 161:366

HPLC 2

Sample	Mixture of opium alkaloids
Apparatus	Varian M 5020 HPLC with UV detector
Column	Reversed-phase; Micropak MCH-10, 300 × 4 mm ID
Mobile phase	A. 0.005 M Heptanesulfonic acid and 0.01 N citric acid in water/acetonitrile (80:20) B. Acetonitrile Gradient: B 20% to 40% in A over 10 min
Flow-rate	2.5 ml min^{-1}
Detection	UV, λ = 254 nm
Retention times, min	(estimated from drawing) Codeine 8.91
	Meperidine 4.38 Ethylmorphine 14.84
	Morphine 7.03 Methadone 18.20
Ref.	Wehr CT and Abbott S: Varian Liquid Chromatography at Work 66. Varian, California

HPLC 3

Sample	Opiates
Apparatus	HPLC low-pressure ternary gradient system
Column	Reversed-phase; Lichrosorb C_{18}, 125 × 4 mm ID
Mobile phase	Acetonitrile/0.1% triethylamine acetate, pH 4.0 (50:50, v/v)
Flow-rate	1.5 ml min^{-1}; 2400 psi

Injection size	10 µl containing 10 µg
Detection	UV, λ = 254 nm; 0.16 AUFS
Retention times	Codeine, morphine and methadone, in ascending order (from drawing)
Ref.	Merck Application Manual on HPLC, TLC and HPTLC (1986). E. Merck, Darmstadt, p 74

HPLC 4

Sample	Blood plasma from dogs
Extraction	Hexane, heptane, methylene chloride and chloroform, pH above 7.0
Apparatus	Perkin-Elmer Series 3B microcomputer controlled pump, 420B autosampler, M 650 fluorescence spectrophotometer and LC-75 UV detector
Column	Cyano-phase; CN-µ-Bondapak
Mobile phase	0.025 M Acetate buffer, pH 3.6/acetonitrile (80:20) containing 30 mg 9,10-dimethoxyanthracene-2-sulfonate per milliliter
Flow-rate	2 or 3 ml min^{-1}
Temperature	44 °C
Internal standard	1-(1-Phenylcyclohexyl)-4-hydroxypiperidine
Injection size	50 µl, 100–200 ng ml^{-1}
Detection	Fluorescence, λ = 380/445 nm; UV, λ = 210 nm
Recording	Perkin-Elmer Sigma 15 data station
Retention times, min	(estimated from drawing)

Detection	fluorescence	UV
Methadone	6.5	4.0, 5.0 (urine)
Internal standard	12.0	6.5

Ref.	Garrett ER, Derendorf H and Mattha AG (1985) J. Pharm. Sci.74:1203

HPLC 5

Sample	Methadone hydrochloride oral concentrate
Apparatus	HPLC system with double-beam photometric detector
Column	Reversed-phase; ODS 5–10 µm, 250 × 3.2 mm ID
Mobile phase	A. Freshly prepared formic acid/water/ ammonium hydroxide (100:1897.5:2.5, v/v) B. Acetonitrile Gradient: B 0% to 80% in about 10 to 20 min
Flow-rate	Suitable for 1500 psi
Temperature	Ambient
Injection size	Appr. 50 µl
Detection	UV, λ = 280 nm
Ref.	Anonym (1985): The United States Pharmacopeia XXI and The National Formulary XVI. United States Pharmacopeial Convention Inc., Rockville, p 649

HPLC 6

Sample	Blood plasma
Extraction	Hexane under alkaline condition
Apparatus	Perkin-Elmer Series 3B microcomputer controlled pumps, 420B autosampler and M 650 fluorimeter
Column	Cyano phase; μ-Bondapak CN
Mobile phase	0.025 M Acetate buffer, pH 3.6/acetonitrile (80:20) containing 30 mg per liter, 9,10-dimethoxyanthracene-2-sulfonic acid
Flow-rate	2 ml min^{-1}; 10 MPa
Temperature	50 °C and then mixed with chloroform from other pump
Internal standards	Cocaine, phencyclidine and methadone metabolites
Injection size	100 μl, 100 ng ml^{-1}
Detection	Fluorescence, λ = 380/445 nm

Retention times, min				
	Cocaine	6.0	Phencyclidine	10.0
	Phencyclidine metabolites	5–7.0	Methadone metabolite	14.0
			Methadone	17.0

Ref.	Derendorf H and Garrett ER (1983) J. Pharm. Sci. 72:630

---

121 Methylparaben

CAS-No	99-76-3; 5026-62-0 (sodium salt)			
Chemical name	Methyl parahydroxybenzoate			
Empirical formula	Methylparaben	Sodium Salt		
	$C_8H_8O_3$	$C_8H_7NaO_3$		
Molecular mass	152.1	174.1		
Proprietary names	Nipagin M; Nipagin M Sodium			
Clinical use	Preservative			
Solubility,	Water	alcohol	acetone	chloroform
Methylparaben	1 in 400–500	1 in 3–3.5	1 in 3	1 in 40
Sodium salt	1 in 2	1 in 50		

HPLC 1

Sample	Bulk drugs and elixirs
Apparatus	Spectra-Physics 8100 HPLC systems with Hitachi M 100-10 ultraviolet detector and Valco Instruments injection valve
Column	Normal phase; Zorbax-Sil 5 μm, 250 × 4.6 mm ID
Pre-column	Normal phase; silica 10 μm, 30 × 4.6 mm ID
Mobile phase	Glacial acetic acid/methanol/methelene chloride (1:45:954, v/v)

Flow-rate	1.2 ml min^{-1}
Internal standard	Cortisone
Injection size	15–30 μl
Detection	UV, λ = 254 nm
Recording	Shimadzu C-RIA integrator and strip-chart recorder
Retention times, min	Propylparaben 2.96 Cortisone 4.96 Methylparaben 3.19 Dexamethasone 6.84
Ref.	Bunch EA (1987) J. Assoc. Off. Anal. Chem. 70:967

HPLC 2

Sample	Food preservatives
Column	Reversed-phase; Nucleosil C_{18}, 10 μm, 250 × 4 mm ID
Mobile phase	Water/methanol (2:3)
Flow-rate	0.6 cm sec^{-1}; 1450 psi
Temperature	Ambient
Injection size	2 μl
Detection	UV, λ = 254 nm
Retention times, min	(estimated from drawing) *p*-Hydroxybenzoic acid 0.8 Ethylparaben 1.4 Methylparaben 1.2 Propylparaben 1.9
Ref.	Sebestian I (1987): High Performance Liquid Chromatography: Application No. 409 Macherey-Nagel, Dueren, p 263

HPLC 3

Sample	Urine of preterm infants
Extraction	Ether at pH 9.0, then with 0.1 M NaOH and re-extraction with ether at pH 2.0
Apparatus	Varian M 8500 LC and Waters M 440 UV detector
Column	Reversed-phase; μ-Bondapak, ODS C_{18}, 300 × 3.9 mm ID
Mobile phase	Acetonitrile/0.05% phosphoric acid (40:60)
Flow-rate	1.5 ml min^{-1}; 900 psi
Detection	UV, λ = 254 nm
Retention times, min	(estimated from drawing) Propylparaben 5.0–6.0 Methylparaben 2.5–3.7 Butylparaben 9.0
Ref.	Hindmarsh KW, John E et al (1983) J. Pharm. Sci. 72:1039

HPLC 4

Sample	Pharmaceutical formulations
Apparatus	Waters M 6000A pump with Milton Roy minipump and Waters M 440 and LDC M 1203 UV detectors
Column	Reversed-phase; μ-Bondapak C_{18}

Mobile phase	A. Water/acetonitrile/glacial acetic acid (58:40:2) B. Water/acetonitrile (70:30) containing 0.2 M ammonium acetate, pH 7.2 C. Water/acetonitrile (65:35) containing 0.01 M ammonium hydrogen phosphate, pH 2.2 with H_3PO_4
Flow-rate	2 ml min^{-1}
Internal standard	Butylparaben and ethylparaben
Injection size	10 µl, 0.07 mg ml^{-1}
Detection	UV
Retention times, min	(estimated from drawing/using mobile phase A.) Methylparaben 3.6 Propylparaben 6.4 Ethylparaben 4.6 Butylparaben 9.4
Ref.	Radus TP and Gyr G (1983) J. Pharm. Sci. 72:221

122 Metronidazole

CAS-No	443-48-1; 69198-10-3(hydrochloride)			
Chemical name	2-(2-Methyl-5-nitroimidazol-1-yl)ethanol			
Empirical formula	Metronidazole	Hydrochloride		
	$C_6H_9N_3O_3$	$C_6H_{10}ClN_3O_3$		
Molecular mass	171.2	207.7		
Proprietary names	Clont; Elyzol; Flagyl			
Clinical use	Antimicrobial/Antiprotozoal			
Solubility,	Water	ethanol	chloroform	ether
Metronidazole	1 in 100	1 in 200	1 in 250	slightly soluble
Light absorption	λ_{max} = 319 nm in aqueous alkali			

HPLC 1

Sample	Dosage forms (liquid)
Apparatus	Varian M 5000 LC, ultraviolet detector and 10 µl manual loop injector
Column	Reversed-phase; Micropak MCH-10, 300 × 4 mm ID
Mobile phase	0.02 M Ammonium acetate in acetonitrile/water (47:53), pH 4.5 with acetic acid
Flow-rate	2 ml min^{-1}
Internal standard	Etyhlparaben, 0.04 mg ml^{-1}
Injection size	10 µl, 0.1–0.75 mg ml^{-1} (benzoate)
Detection	UV, λ = 270 nm; 0.16 AUFS
Recording	Spectra-Physics 4100 digital integrator

Retention times, min	Benzoic acid	1.95	Metronidazole	2.90
	Methylparaben	2.16	Misonidazole	3.69
	2-methyl-5-nitro-Imidazole	2.39	Propylparaben	3.84
	Ethylparaben	2.78	Metroniazole benzoate	5.42
			Tinidazole	5.88

Ref. Sasa SI, Khalil HS and Jalal IM (1986) J. Liq. Chromatogr. 9:3617

HPLC 2

Sample	Suspension dosage forms
Apparatus	Perkin-Elmer Series 4-LC, LC-85B UV/VIS detector and Rheodyne M 7125-075 injector
Column	Reversed-phase; Perkin-Elmer C_{18}, 10 µm, 250 × 4 mm ID
Pre-column	Whatman-ODS, 30–38 µm Co:Pell, 50 × 2 mm ID
Mobile phase	0.005 M Acetate buffer and 0.05 M potassium nitrate in methanol/water (60:40), pH 5.2
Flow-rate	1 ml min^{-1}
Injection size	10 µl
Detection	UV, $\lambda = 254$ nm
Recording	Shimadzu C-R3A integrator
Retention times, min	Metronidazole 4.8; Benzoic acid 6.0; Methylparaben 7.2; Metronidazole benzoate 12.0; Propylparaben 14.0 (estimated from drawing)
Ref.	Pashankov PP and Kostova LL (1987) J. Chromatogr. 394:382

HPLC 3

Sample	Human blood plasma and faeces
Extraction	Ether/methylene chloride (1.5:1) at pH 9.0
Apparatus	Waters M 600 LC pump, U6K injector and Spectra-Physics 770 ultraviolet detector
Column	Reversed-phase; Sperisorb S5 ODS 5 µm, 150 × 4.6 mm ID
Mobile phase	0.01 M Phosphate buffer, pH 5.5/acetonitrile (85:15)
Flow-rate	1 ml min^{-1}
Temperature	Ambient (20–22 °C)
Injection size	20 µl, 0–16 µg ml^{-1}
Detection	UV, $\lambda = 320$ nm
Retention times, min	(estimated from drawing) Metronidazole 5.0 Tinidazole 9.2
Ref.	Lanbeck K and Lindström B (1979) J Chromatogr. Biomed. Appl. 162:117

HPLC 4

Sample	Blood plasma
Extraction	Deproteinization with trichloroacetic acid
Apparatus	Waters M 6000A HPLC with 440 absorbance monitor
Column	Reversed-phase; μ-Bondapak C_{18}, 300 × 3.9 mm ID
Mobile phase	Acetonitrile/5 mM potassium dihydrogen phosphate, (8:92, v/v), pH 3.0
Flow-rate	3.4 ml min^{-1}
Injection size	10–20 μl, 25–500 ng ml^{-1}
Detection	UV, λ = 313 nm; 0.02 AUFS
Retention times, min	Acid metabolite 1.7
	Hydroxy metabolite 2.2
	Metronidazole 3.6
Ref.	Gibson RA, Lattanzio L and McGee H (1984) Clin. Chem. 30:784

123 Mezlocillin

CAS-No	51481-65-3; 42057-22-7 (sodium salt, anhydrous)	
Chemically	Semisynthetic penicillin	
Empirical formula	Mezlocillin	Sodium Salt
	$C_{21}H_{25}N_5O_8S_2$	$C_{21}H_{24}N_5NaO_8S_2,H_2O$
Molecular mass	539.6	579.6
Proprietary name	Baypen	
Clinical use	Antibiotic	
Solubility,	Water	alcohol and acetone
Sodium salt	readily soluble	slightly soluble
Light absorption	λ_{max} = 259 nm in methanol	

HPLC 1

Sample	Biological fluids
Extraction	Diluting with Soerensen buffer, pH 7.4
Apparatus	LDC Constametric III-G LC pump with ultraviolet detector and Spectra-Physics 8780 XR autosampler
Column	Reversed-phase; Nucleosil 5 μm, C_{18}, 200 × 4 mm ID
Mobile phase	Acetonitrile/57.4 mM K_2HPO_4 (24:76), pH 4.0 with H_3PO_4
Flow-rate	1 ml min^{-1}
Temperature	Ambient
Injection size	20 μl, 5–250 μg ml^{-1}
Detection	UV, λ = 220 nm

Recording	Computing integrator (LDC-301)
Retention times, min	Penicillioate 6.0 Penilloate 10.0 Mezlocillin 15.9 (estimated from drawing)
Ref.	Knoeller J, Koenig W et al (1988) J. Chromatogr. Biomed. Appl. 427:257

HPLC 2

Sample	Biological fluids stored at different temperatures
Extration	Diluting with Sorensen buffer
Apparatus	LDC/Milton Roy Constametric III-G pump, variable-wavelength detector Spectromonitor D and Waters WISP M 710B sample processor
Column	Reversed-phase; C_{18}, 5 μm, 200 × 4 mm ID
Mobile phase	Acetonitrile/43 mM K_2HPO_4, pH 4.0 with phosphoric acid (24:76)
Flow-rate	1 ml min^{-1}
Temperature	Ambient
Injection size	20 μl, 1–250 μg ml^{-1}
Detection	UV, λ = 220 nm; 0.2 AUFS
Recording	Milton Roy/LDC computing integrator M 301
Retention times, min	Penicilloate 6.3 Mezlocillin 16.0 Penilloate 11.0
Ref.	Knoeller J, Bremm KD et al (1986) Antimicrob. Ag. Chemotherap. 29:527

HPLC 3

Sample	Serum
Extraction	Methylene chloride under acidic condition, and then with 0.1 M sodium acetate
Column	Reversed-phase; Spherisorb S5-ODS, 260 × 3.2 mm ID
Mobile phase	Acetonitrile/0.05 M phosphate buffer, pH 5.4 (22:78)
Flow-rate	1.5 ml min^{-1}
Internal standard	Primidone
Injection size	100 μl, 2.0–400 μg ml^{-1}
Detection	UV, λ = 214 nm; 0.01 AUFS
Ref.	Karnes HT, Beightol LA and Farthing D (1987) Ther. Drug Monit. 9:456

HPLC 4

Sample	Human and rat biological materials
Extraction	1,2-Dichloromethane
Apparatus	Waters M 45 LC pump, 710B sample processor, and 441 UV detector

Column	Reversed-phase; ODS C_{18}, 5 µm, 250 × 4.6 mm ID
Mobile phase	23% (v/v) Acetonitrile in 0.1 M phosphate buffer (3.5 g KH_2PO_4 and 11.05 g Na_2HPO_4 per liter), pH 7.0 with concentrated HCl
Flow-rate	1.3 ml min^{-1}
Internal standard	Piperacillin
Injection size	10–50 µl
Detection	UV, $\lambda = 214$ nm; 0.01 AU
Recording	Hewlett-Packard 3392A integrator
Retention times, min	Piperacillin 5.6
	Mezlocillin 7.8
Ref.	Jungbluth GL, Janicke DM and Jusko WJ (1989) J. Chromatogr. Biomed. Appl. 494:376

124 Miconazole

CAS-No	22916-47-8; 22832-87-7 (nitrate)		
Chemical name	1-[2,4-Dichloro-β-(2,4-dichlorobenzyloxy) phenethyl]imidazole		
Empirical formula	Miconazole	Nitrate	
	$C_{18}H_{14}Cl_4N_2O$	$C_{18}H_{14}Cl_4N_2O,HNO_3$	
Molecular mass	416.1	479.1	
Proprietary names	Daktarin; Dermonistat; Monistat		
Clinical use	Antifungal		
Solubility,	Water	chloroform	ethanol
Miconazole	very slightly soluble	very soluble	soluble
Nitrate	very slightly soluble	slightly sol	1 in 140
Light absorption	$\lambda_{max} = 264$ nm in methanol		

HPLC 1

Sample	Human serum
Extraction	Protein precipitation by acetonitrile (1:1, v/v)
Apparatus	Shimadzu LC-6A HPLC pump, SPD-6A UV detector and SIL-1A sample injector
Column	Reversed-phase; Yamamura Labs AM-312, ODS 5 µm, 150 × 6 mm ID
Mobile phase	0.05 M Sodium acetate, pH 7.4/acetonitrile (20:80, v/v)
Flow-rate	1.5 ml min^{-1}; 1000 psi
Temperature	Ambient
Injection size	20 µl, 0.1–5.0 µg ml^{-1}

Detection	UV, λ = 220 nm
Recording	Shimadzu C-R4A data module
Retention time, min	Miconazole 6.3
Ref.	Hosotsubo H (1988) Chromatographia 25:717

HPLC 2

Sample	Vaginal fluids
Extraction	Chloroform
Apparatus	Waters M 6000 LC pump, 481 UV detector and WISP 710B autosampler
Column	Reveresed-phase; Spherisorb C_8, 5 μm, 50 × 4.6 mm ID
Mobile phase	600 ml Acetonitrile and 10 ml 1 M sodium phosphate, pH 2.5 with 85% H_3PO_4 and make 1000 ml with water
Flow-rate	1.5 ml min^{-1}; 1.4 MPa
Temperature	Ambient
Internal standard	Hydrocortisone 21-caprylate
Injection size	10 μl, 0.2–200.0 μg ml^{-1}
Detection	UV, λ = 205 nm; 0.02 AUFS
Recording	Hewlett-Packard 3357 laboratory automation system
Retention times, min	Internal standard 2.2 Miconazole 6.5
Ref.	Selinger K, Matheou D and Hill HM (1988) J. Chromatogr. Biomed. Appl. 434:259

HPLC 3

Sample	Human saliva
Extraction	Acetonitrile
Apparatus	Waters M 6000A LC pump, U6K injector and M 450 UV detector
Column	RCM 100, Radial-Pak C_8, 5 mm ID
Pre-column	Bondapak C_{18}, Corasil
Mobile phase	77% Methanol in 0.01 M EDTA with 0.005 M n-nonylamine
Flow-rate	1.5 ml min^{-1}
Injection size	20 μl, 0.5–100 μg ml^{-1}
Detection	UV, λ = 230 nm
Recording	Waters M 730 data module plotter/intregrator
Retention time, min	Miconazole 5.05
Ref.	Turner A and Warnock DW (1982) J. Chromatogr. Biomed. Appl. 227:229

HPLC 4

Sample	Creams and suppositories
Extraction	Ethanol, and then acetonitrile/water (1:1)

Apparatus	Waters M 6000A LC pumps, WISP autosampler and 441 UV detector and Hewlett-Packard 1040A diode array
Column	Reversed-phase; μ-Bondapak C_{18}, 150 × 4.6 mm ID
Pre-column	C_{18} Guard-Pak
Mobile phase	550 ml Acetonitrile and 10 ml buffer (containing 100 ml triethylamine and 80 ml 85% phosphoric acid per liter water) and 300 ml water containing 1.0 g 1-octanesulfonic acid sodium salt
Flow-rate	2 ml min^{-1}
Injection size	20 μl, 8 μg ml^{-1}
Detection	UV, λ = 214 nm; 0.05 AUFS
Recording	Waters M 730 data module
Retention times, min	(estimated from drawing) Miconazole 3.1 Butylhydroxyanisol 2.2 Butylhydroxytoluol 10.0
Ref.	Tyler TA and Genzale JA (1989) J. Assoc. Off. Anal. Chem. 72:442

125 Minoxidil

CAS-No	38304-91-5			
Chemical name	2,6-Diamino-4-piperidinopyrimidine 1-oxide			
Empirical formula	$C_9H_{15}N_5O$			
Molecular mass	209.3			
Proprietary name	Alostil; Loniten; Tricoxidil			
Clinical use	Antihypertensive			
Solubility	Water	alcohol	chloroform	acetone
	1 in 500	readily soluble	insoluble	insoluble
Light absorption	λ_{max} = 230 nm and 281 nm in aqueous acid			

HPLC 1

Sample	Raw materials
Column	Reversed-phase; ODS 5–10 μm, 250 × 4 mm ID
Mobile phase	Methanol/water/glacial acetic acid (700:300:10) containing 3.0 g docusate sodium per liter, pH 3.0 with perchloric acid
Flow-rate	1 ml min^{-1}
Internal standard	Medroxyprogesterone acetate
Injection size	Appr. 10 μl
Detection	UV, λ = 254 nm

Retention times	(relative)
	Internal standard 0.8
	Minoxidil 1.0
Ref.	Anonym (1985): The United States Pharmacopeia XXI and The National Formulary XVI, Supplement 2. United States Pharmacopeial Convention Inc., Rockville, p 1867

HPLC 2

Sample	Bulk drugs and tablets
Apparatus	Altex M 110A LC pump, LDC UV-III detector and Rheodyne M 7125 injector
Column	Reversed-phase; ODS C_{18}, 10 μm, 300 × 3.9 mm ID
Mobile phase	6.7 mM Sodium dioctylsulfonsuccinate and 175 mM acetic acid in 70% methanol, pH 3.0
Flow-rate	1 ml min^{-1}; 1500 psi
Temperature	Ambient
Internal standard	Medroxyprogesterone acetate
Injection size	10 μl containing 0.25 mg ml^{-1}
Detection	UV, $\lambda = 254$ nm; 0.512 AUFS
Retention times, min	Internal standard 8.57
	Minoxidil 9.87
Ref.	Asmus PA, Landis JB et al (1984) J. Pharm. Sci. 73:1290

126 Morphine

CAS-No	6009-81-0 (monohydrate); 6211-15-0 (sulfate, pentahydrate)			
Chemical name	(5α,6α)-7,8-Didehydro-4,5-epoxy-17-methyl-morphinan-3,6-diol			
Empirical formula	Morphine	Sulfate		
	$C_{17}H_{19}NO_3,H_2O$	$(C_{17}H_{19}NO_3)_2,H_2SO_4,5H_2O$		
Molecular mass	303.4	758.8		
Proprietary names	Duromorph; MST-Continus; Nepenthe			
Clinical use	Narcotic Analgesic			
Solubility,	Water	ethanol	chloroform	ether
Morphine	1 in 5000	1 in 250	1 in 1500	insoluble
Sulfate	1 in 21	1 in 1000	insoluble	insoluble
Light absorption	$\lambda_{max} = 298$ nm in aqueous alkali			

HPLC 1

Sample	Injectable solution
Apparatus	Pye-Unicam LC3 or Altex M 110A pump and Pye-Unicam LC3 ultraviolet detector
Column	A. Reversed-phase; Hypersil ODS 5 μm, 100 × 5 mm ID B. Reversed-phase; Ultrasphere ODS 5 μm, 250 × 4.6 mm ID
Mobile phase	For A: 0.01 M Sodium pentenesulfonate/acetonitrile/ortho-phosphoric acid (69.5:30:0.5), pH 2.0 For B: As for A but (69.95:30:0.05)
Flow-rate	2.0 ml min^{-1} (A); 1.0 ml min^{-1} (B)
Temperature	Ambient
Injection size	20 μl, 0.1–2.0 mg ml^{-1}
Detection size	UV, $\lambda = 284$ nm; 0.32 AUFS (A) and 281 nm; 0.16 AUFS (B)
Recording	Bryans M 28000 chart-recorder

Retention times, min

Column	A.	B.
Morphine	0.62	3.7
6-Monoacetylmorphine	0.97	
Diamorphine	1.85	
Pseudomorphine		4.5
Morphine-N-oxide		5.0

Ref. Beaumont I and Deeks T (1982) J. Chromatogr. 238:520

HPLC 2

Sample	Pharmaceutical preparations
Apparatus	Waters LC pump, M 990 photodiode array and 440 UV detector
Column	Reversed-phase; μ-Bondapak C_{18}, 300 × 3.9 mm ID
Mobile phase	Methanol/50 mM ammonium phosphate (40:60, v/v), pH 6.22 with trifluoroacetic acid
Flow-rate	1 ml min^{-1}
Detection	UV, $\lambda = 280$ and 254 nm
Retention times, min	Morphine 5.4 Pseudomorphine 6.4
Ref.	Caute B, Monsarrat B et al (1988) J. Pharm. Pharmacol. 40:644

HPLC 3

Sample	Injectable formulations
Apparatus	Waters M 6000A LC pump, 441 UV detector and Rheodyne M 7125 injector

Column	Reversed-phase; Hypersil 5 µm ODS, 100 × 4.6 mm ID
Mobile phase	Acetonitrile/water (50:50, v/v) with 100 mM sodium lauryl sulfate, 40 mM sodium hydrogen phosphate and pH 3.0 with H_3PO_4
Flow-rate	2 ml min^{-1}
Injection size	20 µl, 100 µg ml^{-1}
Detection	UV, λ = 254 nm
Retention times, min	Morphine 1.80 Apomorphine 4.60 Codeine 3.15 Pseudomorphine 5.55
Ref.	Hung CT, Young M and Gupta PK (1988) J. Pharm. Sci. 77:719

HPLC 4

Sample	Guinea-pig serum
Extraction	Chloroform at pH 9.0
Apparatus	Kortec K25M HPLC pump with Biomedical Engineering M 1162 electrochemical detector and Valco C6U injector
Column	Reversed-phase; Ultrasphere ODS, 150 × 4.6 mm ID
Pre-column	Lichropep RP-18, 30 × 2 mm ID
Mobile phase	0.1 M NaH_2PO_4, 0.002 M octanesulfonic acid, 0.001 M disodium-EDTA and 25% (v/v) methanol
Flow-rate	1 ml min^{-1}
Temperature	25 °C
Internal standard	Naloxone
Injection size	25 µl containing 17 pg
Detection	EC, + 0.9 V vs Ag/AgCl RE
Retention times, min	Morphine 4.7 Naloxone 8.5
Ref.	Turner CA and Murphy R (1988) J. Chromatogr. Biomed. Appl. 428:383

HPLC 5

Sample	Biological fluids
Extraction	Benzene/1-butanol (85:15) under alkaline condition, pH 8.9
Apparatus	Milton Roy/LDC, Constametric III G high-pressure pump, Negretti M 190 six port injector and ESA Coulochem M 5100A coulometric detector
Column	Cyano phase; Zorbax 5 µm, 150 × 4.6 mm ID
Mobile phase	0.05 M KH_2PO_4/acetonitrile (94:6) containing 0.637 g tetrabutylammonium sulfate and 1 g 1-octane-sulfonic acid per liter
Flow-rate	1.1 ml min^{-1}; 1650 psi
Internal standard	Naloxone

Injection size	20 µl, 0–500 ng ml^{-1}
Detection	EC, + 0.25 V to + 0.45 V and 0.70 V
Retention times, min	Morphine 3.6
	Naloxone 5.6
Ref.	Derendorf H and Kaltenbach M (1986) J. Pharm. Sci. 75:1198

HPLC 6

Sample	Biological materials
Extraction	Methanol at pH 7.4
Apparatus	Waters M 510 LC pump, 440 ultraviolet detector with extended wavelength module
Column	Reversed-phase; YMC Al-312, ODS 5 µm, 150 × 6 mm ID
Pre-column	Reversed-phase; Lichrosorb 10 µm, C_{18}, 100 × 4 mm ID
Mobile phase	10 mM Sodium phosphate buffer, pH 6.8/acetonitrile (3:2, v/v)
Flow-rate	2 ml min^{-1}
Temperature	Ambient
Injection size	2–40 µl, 10 µg ml^{-1}
Detection	UV, λ = 214 nm; 0.02 AUFS
Recording	Shimadzu C-RIB integrator
Retention times, min	(estimated from drawing)

Normorphine	4.0	Morphine	8.4
Morphinone-2-mercapto-ethanol adduct	5.5	Dihydromorphine	12.0
		Dihydromorphinone	15.5

Ref.	Kumagai Y, Ishida T and Toki S (1987) J. Chromatogr. Biomed. Appl. 421:155

---

127 Nadolol

CAS-No	42200-33-9			
Chemical name	5-[3-[(1,1-Dimethylethyl)amino]-2-hydroxy-propoxy] 1,2,3,4-tetrahydro-2,3-naphthalenediol			
Empirical formula	$C_{17}H_{27}NO_4$			
Molecular mass	309.4			
Proprietary names	Corgard; Corgaretic 40; Solgol			
Clinical use	Beta-Blocker			
Solubility	Water	ethanol	chloroform	ether
	soluble	freely sol	slightly sol	insoluble
Light absorption	λ_{max} = 278 nm in methanol			

HPLC 1

Sample	Biological samples
Extraction	Diethyl ether/dichloromethane (5:2, v/v)
Apparatus	Altex M 110A LC pump, Perkin-Elmer 204 spectrofluorimeter and Rheodyne M 7125 injector
Column	Reversed-phase; ODS C_{18}, 5 µm, 200 × 4.6 mm ID
Mobile phase	Water/acetonitrile (4:1, v/v) containing triethylamine 1 ml per liter, pH 3.0 with H_3PO_4
Flow-rate	1.0 ml min^{-1}
Injection size	50–100 µl, appr. 200 ng ml^{-1}
Detection	Fluorimetric, $\lambda = 265/295$ nm
Retention time, min	Nadolol 5.0
Ref.	Liu LK and Robinson ML (1985) J. Pharm. Biomed. Anal. 3:351

HPLC 2

Sample	Blood plasma
Extraction	Deproteinization with acetonitrile, and then on extraction cartridge, with methanol + 0.1% propylamine or methanol/phosphate buffer, pH 3.0 (50:50)
Apparatus	Varian M 5000 LC pump, Perkin-Elmer LS 4 fluorescence detector and LC 90 UV detector and Rheodyne injector
Column	Cyanopropyl phase; Lichrosorb 5 µm, 125 × 4 mm ID
Pre-column	Lichrosorb CN 10 µm
Mobile phase	Acetonitrile/phosphate buffer, pH 3.0 (3:97)
Internal standard	Viloxazine
Injection size	100 µl, 300 ng ml^{-1}
Detection	UV, $\lambda = 220$ and 279 nm Fluorescence, $\lambda = 230/300$ nm
Recording	Shimadzu IC-R3A integrator
Retention times, min	(estimated from drawing) Nadolol 4.00 Internal standard 6.00
Ref.	Musch G, Buelens Y and Massart DL (1989) J. Pharm. Biomed. Anal. 7:483

HPLC 3

Sample	Tablets
Column	Dimethylsilane 5–10 µm, 250 × 4.6 mm ID
Mobile phase	700 ml Methanol and 1300 ml water containing 5.84 g sodium chloride and 1 ml 0.1 N HCl
Flow-rate	1.0 ml min^{-1}
Injection size	Appr. 20 µl
Detection	UV, $\lambda = 220$ nm

Ref.	Anonym (1985): The United States Pharmacopeia XXI and The National Formulary XVI, Supplement 4, 1986. United States Pharmacopeial Convention Inc., Rockville, p 2194

HPLC 4

Sample	Human blood plasma
Extraction	Diethyl ether under alkaline condition
Apparatus	Shimadzu M LC-4A pump, SPD-2AS UV detector and Rheodyne M 7125 injector or Shimadzu SIL 2AS autosampler
Column	Reversed-phase; Hypersil ODS 5 μm, 100 mm
Mobile phase	Methanol/water (57:43) with 0.1% octane sulfonic acid
Flow-rate	1.0 ml min^{-1}
Internal standard	Acebutolol
Injection size	10 μl
Detection	UV, λ = 220 nm; 0.01 AUFS
Retention times, min	(estimated from drawing) Nadolol 3.8 Internal standard 5.1
Ref.	Kinney CD (1984) J. Chromatogr. Biomed. Appl. 305:489

HPLC 5

Sample	Serum
Extraction	Mobile phase containing 20% perchloric acid
Apparatus	Spectra-Physics 8100 LC with autosampler and Perkin-Elmer M 650-10 fluorescence detector
Column	Reversed-phase; Spherisorb ODS 5 μm, 250 × 4.5 mm ID
Pre-column	Reversed-phase; ODS 5 μm, 40 × 4.5 mm
Mobile phase	Methanol/0.05 M ammonium hydrogen phosphate buffer, pH 4.6 (30:70)
Flow-rate	2.0 ml min^{-1}
Temperature	36 °C
Internal standard	Pindolol
Injection size	100 μl
Detection	Fluorescence, λ = 227/299 nm; UV, λ = 227 nm
Recording	Spectra-Physics SP 4200 integrator
Retention times, min	(estimated from drawing) Internal standard 5.5 Nadolol 6.5
Ref.	Moncrieff J (1985) J. Chromatogr. Biomed. Appl. 342:206

HPLC 6

Sample	Nadolol and Bendroflumethiazide tablet formulations
Extraction	Methanol
Apparatus	Beckman-Altex M 110A LC pump and Perkin-Elmer LC-75 or Schoeffel M 770 UV detector
Column	Phenyl phase (Waters); 5 or 10 μm, 250–300 × 4.6 mm ID
Pre-column	Saturator 37 μm, 50 × 2 mm ID
Mobile phase	Aqueous acetate buffer (0.029 M acetic acid, 0.01 M sodium acetate, 0.04 M sodium chloride)/methanol (6:4)
Flow-rate	1.5–2.5 ml min^{-1}; 200–3000 psi
Temperature	32 °C
Injection size	20 μl, 0–2000 μg ml^{-1}
Detection	UV, λ = 270 nm
Recording	Perkin-Elmer Sigma 10B integrator
Retention times	Disulfonamide, nadolol and bendroflumethiazide, in ascending order (from drawing)
Ref.	Perlman S, Szyper M and Kirschbaum JJ (1984) J. Pharm. Sci. 73:259

128 Nalidixic Acid

CAS-No	389-08-2			
Chemical name	1-Ethyl-1,4-dihydro-7-methyl-4-oxo-1,8-naphthyridine-3-carboxylic acid			
Empirical formula	$C_{12}H_{12}N_2O_3$			
Molecular mass	232.2			
Proprietary names	Naxuril; Negram; Uropan			
Clinical use	Antibacterial (urinary)			
Solubility	Water	alcohol	chloroform	aq NaOH
	insoluble	1 in 910	1 in 35	soluble
Light absorption	λ_{max} = 257 nm in aqueous acid			

HPLC 1

Sample	Fish
Extraction	0.1% Metaphosphoric acid/methanol (6:4) followed by Sep-Pak C_{18} clean-up procedure with methanol
Apparatus	Shimadzu LC-5A pump, SPD-2A detector and RF-530 fluorescence detector
Column	Reversed-phase; Kaseisorb LC ODS 300-5, 250 × 4.6 mm ID

Mobile phase	5 mM Sodium dihydrogen phosphate/acetonitrile (6:4)
Flow-rate	0.5 ml min^{-1}
Injection size	10 µl containing 2–20 ng
Detection	UV, $\lambda = 280$ nm Fluorescence, $\lambda = 325/365$ nm
Recording	Shimadzu Chromatopac C-R3A data system
Retention times, min	Oxolinic acid 8.0 (UV, fluorescence) Nalidixic acid 11.0 (UV, fluorescence) Piromidic acid 13.5 (UV)
Ref.	Horie M, Saito K et al (1987) J. Chromatogr. 402:301

HPLC 2

Sample	Human blood plasma and urine
Extraction	Chloroform, 0.2 N sodium hydroxide and again with chloroform under acidic condition
Apparatus	Hewlett-Packard 1081B HPLC with 1036A UV detector
Column	Reversed-phase; C_8, 10 µm 250 × 4.6 mm ID
Mobile phase	Methanol/phosphate buffer (containing 7.5 g l^{-1} KH_2PO_4 and 2.5 g l^{-1} Na_2HPO_4; pH 8.2 with 1 N NaOH) (55:45) with 2 g l^{-1} N,N,N-trimethyl-acetylammonium bromide
Flow-rate	0.6 ml min^{-1}; 435 psi
Temperature	Ambient
Internal standard	Oxolinic acid
Injection size	20 µl
Detection	UV, $\lambda = 254$ nm
Retention times, min	7-Hydroxynalidixic acid 7.0 Oxolinic acid 8.8 7-Carboxynalidixic acid 13.8 Nalidixic acid 15.4
Ref.	Cuisinaud G, Ferry N et al (1980) J. Chromatogr. Biomed. Appl. 181:399

HPLC 3

Sample	Blood plasma
Extraction	Toluene under acidic condtion
Apparatus	Waters M 6000A LC pump, U6K injector and 440 absorbance detector
Column	A. Reversed-phase; Lichrosorb C_{18}, 10 µm, 300 mm long B. Reversed-phase; µ-Bondapak C_{18}, 10 µm, 300 mm long
Mobile phase	For A. 70% (v/v) Methanol in water For B. 63% (v/v) Methanol in water

Flow-rate	1.5 ml min^{-1}
Temperature	Ambient
Internal standard	Nalidixic acid propylester
Injection size	20 µl, 0.66–99.2 µg ml^{-1}
Detection	UV, λ = 313 nm
Recording	Spectra-Physics SP 4000 data system
Retention times, min	4.0–8.0
Ref.	Sorel RHA and Roseboom H (1979) J. Chromatogr. Biomed. Appl. 162:461

HPLC 4

Sample	Chicken blood plasma
Extraction	Acetonitrile
Apparatus	Trirotar-V LC pump, Uvidec-100-V UV detector (Japan Spectroscopic) and Rheodyne M 7125 injector
Column	Reversed-phase; Nucleosil C_8, 5 µm, 250 × 4.6 mm ID
Mobile phase	Methanol/0.1 M citric acid/acetonitrile (6:7:1)
Flow-rate	1.0 ml min^{-1}
Temperature	36 °C
Injection size	20 µl
Detection	UV, λ = 260 nm; 0.04 AUFS
Recording	Hewlett-Packard 3390A integrator
Retention times, min	Oxolinic acid 7.6
	Nalidixic acid 9.1
Ref.	Hamamoto K (1986) J. Chromatogr. Biomed. Appl. 381:453

129 Naloxone

CAS-No	465-65-6; 357-08-4 (hydrochloride, anhydrous)			
Chemical name	17-Allyl-6-deoxy-7,8-dihydro-14-hydroxy-6-oxo-17-normorphine			
Empirical formula	Naloxone	Hydrochloride		
	$C_{19}H_{21}NO_4$	$C_{19}H_{21}NO_4$, HCl		
Molecular mass	327.4	363.8		
Proprietary names	Nalonee; Narcan; Narcanti			
Clinical use	Narcotic Antagonist			
Solubility,	Water	alcohol	chloroform	ether
Naloxone			soluble	insoluble
Hydrochloride	soluble	slightly sol	insoluble	insoluble
Light absorption	λ_{max} = 281 nm in aqueous acid			

HPLC 1

Sample	Injectable solutions
Apparatus	Micromeretics M 6000 HPLC system with Schoeffel SF-770 UV detector
Column	Reversed-phase; μ-Bondapak 10 μm, 300 × 6.3 mm ID
Mobile phase	Methanol containing 0.1% (w/v) aqueous ammonium carbonate (45:55)
Flow-rate	3.0 ml min^{-1}; 3000 psi
Temperature	Ambient
Internal standard	Papaverine, 0.1 mg ml^{-1}
Injection size	100 μl
Detection	UV, λ = 220 nm; 2 × 2 AUFS
Recording	Hewlett-Packard 3385A integrator
Retention times, min	Methylparaben 1.81; Propylparaben 4.60; Naloxone hydrochloride 6.65; Papaverine 9.62
Ref.	Hanna S, Insler M et al (1980) J. Chromatogr. 200:277

HPLC 2

Sample	Blood plasma
Extraction	Mobile phase on Bond-Elut CN extraction column
Apparatus	Waters M 6000A HPLC pump with pulse dampener, U6K injector and ESA Coulochem M 5100A electrochemical detector
Column	Phenyl phase; Radial-Pak 4 μm, 100 × 8 mm ID
Pre-column	ESA Graphite filter, 0.2 μm
Mobile phase	18% (v/v) Acetonitrile and 0.0045% (v/v) orthophosphoric acid-85% (w/v) in 5 mM aqueous sodium pentanesulfonic acid, monohydrate
Flow-rate	2 ml min^{-1}; 1653 psi
Temperature	Ambient
Injection size	400 μl containing 10 ng
Detection	EC, + 0.35 V and + 0.45 V
Recording	Hewlett-Packard 3392A integrator
Retention time, min	Naloxone 8.7
Ref.	Albeck H, Woodfield S and Kreek MJ (1989) J. Chromatogr. Biomed. Appl. 488:435

HPLC 3

Sample	Injectable preparations
Column	Reversed-phase; ODS 5–10 μm, 250 × 4.6 mm ID
Mobile phase	0.68 g Sodium 1 octanesulfonate, 1.0 g sodium chloride, 550 ml water, 450 ml methanol and 1.0 ml phosphoric acid

Flow-rate	1 ml min^{-1}
Internal standard	Acetaminophen
Injection size	Appr. 20 µl
Detection	UV, λ = 229 nm
Retention times, min	(relative) Acetaminophen 0.6 Naloxone 1.0
Ref.	Anonym (1985): The United States Pharmacopeia XXI and The National Formulary XVI, Supplement 4, 1986. United States Pharmacopeial Convention Inc., Rockville, p 2195

130 Nandrolone

CAS-No	434-22-0; 360-70-3 (decanoate)			
Chemical name	17β-Hydroxyestr-4-en-3-one			
Empirical formula	Nandrolone $C_{18}H_{26}O_2$	Decanoate $C_{28}H_{44}O_3$		
Molecular mass	274.4	428.7		
Proprietary names	Anabolin LA-100; Androlone-D; Deca-Durabol			
Clinical use	Anabolic Steroid			
Solubility,	Water	ethanol	chloroform	ether
Nandrolone		soluble	soluble	soluble
Decanoate	insoluble	1 in 1	freely soluble	freely soluble
Light absorption	λ_{max} = 240 nm in dehydrated alcohol (nandralone decanoate)			

HPLC 1

Sample	Oily injections
Extraction	85% Ethanol
Apparatus	Varian M 5020 LC with Valco injection valve and Varian UV-50 detector
Column	Reversed-phase; Lichrosorb C_8, 7 µm, 250 × 4.0 mm ID
Mobile phase	I. Methanol/tetrahydrofuran/water (82:3:15, v/v) II. Methanol/tetrahydrofuran/water (90:2:8, v/v)
Flow-rate	1 ml min^{-1}
Temperature	Ambient
Internal standard	Diphenyl
Injection size	10 µl, 40–80 µg ml^{-1}
Detection	UV, λ = 240 nm; 0.2–0.05 AUFS

Retention times, min (estimated from drawing)

Mobile phase	I.	II.
Nandrolone	3.8	3.8
Diphenyl (IS)	5.0	
Nandrolone phenylpropionate	6.6	5.0
Nandrolone decanoate		7.0
Nandrolone undecanoate		8.0

Ref. Cavrini V, Di Pietra AM et al (1987) J. Pharm. Biomed. Anal. 5:21

HPLC 2

Sample	Bovine urine
Extraction	Diethyl ether after enzymatic hydrolysis with glucuronidase/sulfatase
Apparatus	Waters M 6000 and LC-45 pumps, 660 gradient controller, WISP autosampler and 450 UV detector
Column	Reversed phase; Shandon ODS 5 µm, 150 × 4.6 mm ID
Pre-column	Corasil C_{18}, 75 × 2.1 mm ID
Mobile phase	Methanol/water (60:40)
Flow-rate	2.0 ml min^{-1}
Injection size	250 µl
Recording	Waters printer-plotter integrator (data module)
Retention times, min	(estimated from drawing)

17β-Trenbolone	4.6	trans-Stilbestrol	7.4
19-Nortestosterone	5.7	17α-Methyltestosterone	9.0
Zeranol	6.3	Medroxyprogesterone	13.6

Ref. Jansen EHJM, Both-Miedema R et al (1984) J. Chromatogr. 299:450

HPLC 3

Sample	Tissues from meat animals
Extraction	Diethyl ether after enzymatic deproteinization with subtilisin A in 0.1 M-Tris buffer, pH 9.5 for 2 hours at 60 °C
Column	Reversed-phase; Lichrosorb C_{18}, 5 µm, 250 × 4.6 mm ID
Pre-column	Lipidex 5000, 145 × 6 mm ID
Mobile phase	Water/methanol (45:65)
Flow-rate	1.5 ml min^{-1}
Injection size	12 pg
Detection	UV, λ = 240 nm
Ref.	Van Peteghem C (1986) J. Chromatogr. 369:253

HPLC 4

Sample	Steroid anabolics in injection sites	
Extraction	Tetrahydrofuran	
Apparatus	Waters M 501 and 510 HPLC pumps with gradient controller, Hewlett-Packard 1040A photodiode array and Rheodyne M 710 injector	
Column	Reversed-phase; Lichrosorb C_{18}, 5 µm, 150 × 3 mm ID	
Pre-column	Reversed-phase; Lichrosorb C_{18}, 5 µm, 100 mm	
Mobile phase	A. Acetonitrile B. Water Gradient: A 50% to 75% in 10 min; A 75% to 90% in 15 min; hold 7 min; A 90% to 50% in 10 min	
Flow-rate	0.5 ml min^{-1} up to 32 min; 1.0 ml min^{-1} after 32 min	
Injection size	20 µl	
Detection	UV	
Retention times, min	(estimated from drawing)	
	3,17-β-Estradiol	13.0
	Testosterone	15.0
	Testosterone propionate	36.8
	3,17-β-Estradiol-3-benzoate	37.8
	Testosterone cypionate	46.0
	19-Nortestosterone decanoate	55.6
Ref.	Siegert K and Holling F (1989) Archiv Lebensmittelhyg. 40:25, 83	

---

131 Neomycin

CAS-No	1404-04-2; 1405-10-3 (sulfate)		
Chemically	Aminoglycoside antibiotic containing neomycin B, neomycin C and neomycin A (neamine) produced by Streptomyces fradiae		
Empirical formula	Neamycin B	Neomycin B Sulfate	Neomycin A
	$C_{23}H_{46}N_6O_{13}$		$C_{12}H_{26}N_4O_6$
Molecular mass	614.6		322.4
Proprietary names	Kaomycin; Neobiotic; Tribiotic		
Clinical use	Antibiotic		

Solubility,	Water	ethanol	chloroform, ether
Neomycin	soluble		insoluble
Sulfate	1 in 1	very slightly sol	insoluble

HPLC 1

Sample	Dosage forms (ointments)
Extraction	Wash first with methanol, then chloroform at 55 °C and lastly extraction with methanol/water (20:80)
Column	Normal phase; Lichrosorb Si 100, 5 μm, 250 × 4.6 mm ID
Mobile phase	Tetrahydrofuran/chloroform/water/glacial acetic acid (392:598:8:2)
Flow-rate	1.0 ml min^{-1}
Derivatization	Pre-column; with 2,4-dinitrofluorobenzene
Injection size	40 μl
Detection	UV, λ = 254 nm
Recording	Electronic integrator (Shimadzu chromatopac-E1A)
Retention times, min	(estimated from drawing) Neomycin B 12.15 Neomycine C 4.10 Hydrocortisone acetate 18.23
Ref.	Binns RB and Tsuji K (1984) J. Pharm. Sci. 73:69

HPLC 2

Sample	Raw materials
Apparatus	Perkin-Elmer Series 4 and Waters M 6000A pumps or LDC minipump, Waters WISP 710B injector and LDC M 1203 UV III detector
Column	A. Normal phase; Perkin-Elmer P-E HS-5, 125 × 4.6 mm ID B. Normal phase; Lichrosorb Si 60, 250 × 4.6 mm ID
Mobile phase	For A: Chloroform/methanol/acetic acid (950:23:25, v/v) For B: 85% Chloroform, 12% acetonitrile and 3% methanol/acetic acid (10:1)
Flow-rate	1.7 ml min^{-1}
Temperature	Ambient
Derivatization	Pre-column; with 2-naphthalenesulfonyl chloride
Internal standard	Prednisolone
Detection	UV, λ = 254 nm; 0.1–0.064 AUFS
Recording	Hewlett-Packard 3390A integrator
Retention times, min	see below

Column	A	B
Neamine	2.5	
Internal standard	6.2	11.0
Neomycin B	8.3	8.7
Neomycin C	12.3	20.6

Ref.	Tsuji K and Jenkins KM (1986) J. Chromatogr. 369:105

HPLC 3

Sample	Raw materials
Apparatus	Perkin-Elmer Series 4 and Waters M 6000A pumps or LDC minipump, Waters WISP 710B injector and LDC UV III detector
Column	A. Normal phase; 5 μm Spherisorb-5 (Brownlee), 250 × 4.6 mm ID B. Normal phase; Zorbax SIL 5 μm
Mobile phase	For A: Chloroform/methanol/acetic acid (95:2.5:2.5, v/v) For B: Chloroform/methanol/acetic acid (96:2:2, v/v)
Flow-rate	1.7 ml min^{-1}
Temperature	Ambient
Derivatization	Pre-column; with 2-naphthalenesulfonyl chloride
Internal standard	Prednisolone
Detection	UV, $\lambda = 254$ nm; 0.1–0.064 AUFS
Recording	Hewlett-Packard 3390A integrator
Retention times, min	Column A. B. Internal standard 13.7 21.1 Neomycin B 16.2 26.0 Neomycin C 24.2 31.2
Ref.	Tsuji K and Jenkins KM (1986) J. Chromatogr. 369:105

HPLC 4

Sample	Pharmaceutical preparations
Apparatus	Orlita M 1515 HPLC pump with pulse dampener, Rheodyne M 7010 injector and Varian fluorimeter
Column	Reversed-phase; Lichrosorb C_2, 5 μm, 150 × 4.6 mm ID
Pre-column	Reversed-phase; Lichrosorb C_{18}, 10 μm, 20 × 4.6 mm ID; between pump and injection value
Mobile phase	75 mg l^{-1} Sodium dodecyl sulfate, 0.5 M Na_2SO_4 and 0.015 M sodium acetate buffer, pH 7.0
Flow-rate	1.8 ml min^{-1}
Derivatization	Post-column; with *o*-phthalaldehyde
Injection size	20 μl, 0.5–1.5 μg ml^{-1}
Detection	Fluorescence, $\lambda = 7–54$, 7–60/filter 4–76, 3–73
Retention times, min	Neomycin A 2.5 Neomycin C 9.0 Neomycin B 14.8 (estimated from drawing)
Ref.	Apffel JA, Van der Louw J et al (1985) J. Pharm. Biomed. Anal. 3:259

HPLC 5

Sample	Animal tissues
Extraction	Phosphate buffer, pH 8.0 and deproteinization by heat
Apparatus	Waters M 6000A LC pump with M 730 system controller, 710B WISP autosampler and Perkin-Elmer M 650S fluorescence detector
Column	Reversed-phase; Supelcosil LC-8-DB, 5 μm, 150 × 4.6 mm ID
Pre-column	Lichrosorb RP-18, 10 μm, 30 × 4.6 mm ID
Mobile phase	Aqueous solution containing 0.01 M 1-pentane-sulfonate, 0.056 M sodium sulfate, 0.007 M acetic acid and 1.5% methanol
Flow-rate	1.5 ml min^{-1}
Derivatization	Post-column; with *o*-phthalaldehyde
Internal standard	Paromomycin
Injection size	100 ng
Detection	Fluorescence, $\lambda = 340/455$ nm
Recording	Waters M 720 data module
Retention times, min	(estimated from drawing)
	Streptomycin 3.0, 6.0, 7.8
	Dihydrostreptomycin 9.0
	Paromomycin 18.29
	Neomycin 21.71
Ref.	Shaikh B, Allen EH and Gridley JC (1985) J. Assoc. Off. Anal. Chem. 68:29

132 Nicotinamide

CAS-No	98-92-0
Chemical name	Pyridine-3-carboxamide
Empirical formula	$C_6H_6N_2O$
Molecular mass	122.1
Proprietary names	Farmobion PP; Nicobion; Nicotilamida
Clinical use	Vitamin
Solubility	Water 1 in 1; alcohol 1 in 1.5; glycerol 1 in 10; chloroform, ether slightly soluble
Light absorption	$\lambda_{max} = 261$ nm in aqueous acid

HPLC 1

Sample	Rat urine
Extraction	Diethyl ether under alkaline condition

Apparatus	Shimadzu LC-4A pump and SPD-2AS UV detector
Column	Reversed-phase; 7 μm-ODS-L, 250 × 4.6 mm ID
Mobile phase	10 mM Potassium hydrogen phosphate/acetonitrile (96:4, v/v), pH 3.0 with concentrated H_3PO_4
Flow-rate	1 ml min^{-1}
Temperature	25 °C ± 1 °C
Internal standard	Isonicotinamide, 1.81 nmol
Injection size	5–10 μl containing 1.88 nmol
Detection	UV, λ = 260 nm; 0.01–0.001 AUFS
Recording	Shimadzu C-R3A integrator
Retention times, min	Isonicotinamide (IS) 5.78 Nicotinamide 6.73 N^1-Methyl-2-pyridone-5-carboxamide 8.94 N^1-Methyl-4-pyridone-3-carboxamide 11.10
Ref.	Shibata K, Kawada T and Iwai K (1988) J. Chromatogr. Biomed. Appl. 424:23

HPLC 2

Sample	Blood and pharmaceutical preparations
Extraction	Chloroform/methanol (1:1) after deproteinizing with acetone/methanol (1:1) (blood)
Apparatus	Knauer HPLC system with spectrophotometer detector and loop injection valve
Column	Reversed-phase; Lichrosorb C_{18}, 5 μm, 250 × 4 mm ID
Mobile phase	Methanol/water (80:20)
Flow-rate	2 ml min^{-1}; 2900 psi
Internal standard	Vitamin B_2
Injection size	60–100 ng
Detection	UV, λ = 254 nm; 1.28 AUFS
Retention times, min	Ascorbic acid 1.0–1.2 Nicotinamide 1.4–1.6 Vitamin B_2 1.7–1.9
Ref.	Amin M and Reusch J (1988) Pharm. Ind. 50:1307

HPLC 3

Sample	Rat tissue and blood
Extraction	Diethyl ether
Apparatus	Shimadzu M LC-4A pump, SPD-2AS ultraviolet detector and Rhoedyne M 7125 injector
Column	Reversed-phase; Chemco Scientific 5-ODS-H, 5 μm, 150 × 4.6 mm ID
Mobile phase	10 mM Potassium dihydrogen phosphate/ acetonitrile (96:4, v/v), pH 4.5
Flow-rate	1 ml min^{-1}

Temperature	25 °C
Injection size	5–20 µl containing 426 pmol
Detection	UV, λ = 260 nm
Recording	Shimadzu Chromatopac C-R3A
Retention times, min	Nicotinamide 5.8
	N-Methylnicotinamide 10.8
Ref.	Shibata K, Kawada T and Iwai K (1987) J. Chromatogr. Biomed. Appl. 422:257

HPLC 4

Sample	Premixes
Extraction	Water
Apparatus	HPLC with UV detector
Column	Reversed-phase; C_{18}, 7 µm, 250 × 4.6 mm ID
Mobile phase	0.05 M KH_2PO_4/methanol/acetic acid (90:9:1)
Flow-rate	1.0 ml min^{-1}
Temperature	Ambient
Injection size	10 µl
Detection	UV, λ = 264 nm
Retention times, min	Nicotinic acid 6.2
	Nicotinamide 12.3
Ref.	Degussa Prospect on Nicotinamide for Animal Nutrition (1989): Degussa, Frankfurt, p 24

133 Nicotine

CAS-No	54-11-5
Chemical name	3-(1-Methylpyrrolidin-2-yl)pyridine
Empirical formula	$C_{10}H_{14}N_2$
Molecular mass	162.2
Proprietary name	Nicorette
Clinical use	Insecticide/Ectoparasiticide
Solubility	Water soluble; alcohol soluble; chloroform soluble; ether soluble
Light absorption	λ_{max} = 259 nm in aqueous acid

HPLC 1

Sample	Blood plasma
Extraction	Chloroform
Apparatus	Rainin Instruments HP/HPX LC pump and Milton Roy UV Monitor III

Column	Normal phase; Microsorb-Si, 15 × 4.6 mm ID
Mobile phase	Dioxane/isopropanol/ammonium hydroxide (80:0.3:0.4)
Flow-rate	1 ml min^{-1}; 2000 psi
Temperature	Ambient
Internal standard	Cotinine
Injection size	20 µl, 0.05–10 µg ml^{-1}
Detection	UV, λ = 254 nm; 0.1 AUFS
Retention times, min	Internal standard 2.10 Nicotine 2.75 (estimated from drawing)
Ref.	Hefner JE, Williams AG et al (1988) J. Liq. Chromatogr. 11:2375

HPLC 2

Sample	Urine
Extraction	Chloroform under alkaline condition
Apparatus	Spectra-Physics 8770 HPLC, SP 8300 UV detector and Kontron MSI T660 autosampler
Column	Reversed-phase; Spherisorb 5, C_{18}, 220 mm
Mobile phase	650 ml Water, 290 ml methanol, 40 ml 1 M sodium acetate, pH 4.6 and 20 ml acetonitrile, pH 6.4 with triethylamine
Flow-rate	1.3 ml min^{-1}
Injection size	25 µl, 1250 ng ml^{-1}
Detection	UV, λ = 254 nm
Retention times, min	Nicotine 4.0 Caffeine 5.0 Cotinine 4.5 Pyridine 6.6
Ref.	Godin J and Hellier G (1989) J. Chromatogr. Biomed. Appl. 488:487

HPLC 3

Sample	Allergenic extracts of tobacco leaf
Apparatus	Spectra-Physics 3500B HPLC with Schoeffel M 770 UV detector
Column	Reversed-phase; Ultrasphere C_8, 5 µm, 150 × 4.6 mm ID
Mobile phase	A. Methanol/0.02 M (NH_4Cl–NH_4OH) buffer, pH 9.2 (20:80) B. Methanol Gradient: B 5% for 2 min; B 5% to 50% in 6 min
Flow-rate	1.2 ml min^{-1}
Internal standard	4-Chloroaniline
Injection size	10 µl
Detection	UV, λ = 254 nm

Retention times, min	(estimated from drawing) Internal standard 4.7 Nicotine 6.4
Ref.	Sudan BJL, Brouillard C et al (1984) J. Chromatogr. 288:415

134 Nifedipine

CAS-No	21829-25-4
Chemical name	1,4-Dihydro-2,6-dimethyl-4-(2-nitrophenyl) -3,5-pyridinedicarboxylic acid dimethyl ester
Empirical formula	$C_{17}H_{18}N_2O_6$
Molecular mass	346.3
Proprietary names	Adalat; Nifclat; Procardia
Clinical use	Antianginal/Vasodilator
Solubility	Water: insoluble; alcohol: slightly sol; acetone, chloroform: soluble
Light absorption	λ_{max} = 238 nm in aqueous acid

HPLC 1

Sample	Human blood serum
Extraction	Water-saturated chloroform under alkaline condition
Apparatus	Waters M 6000A HPLC with U6K injector and 481 UV detector
Column	Reversed-phase; μ-Bondapak C_{18}
Mobile phase	0.01 M Phosphate buffer, pH 6.1/methanol/ acetonitrile (20:35:45)
Flow-rate	1 ml min^{-1}
Internal standard	17-α-Ethinylestradiol
Injection size	10 μl, 5–100 ng ml^{-1}
Detection	UV, λ = 235 nm
Retention times, min	(estimated from drawing) Nifedipine 10.32 Internal standard 14.42
Ref.	Gurley BJ, Buice RG and Sidhu P (1985) Ther. Drug Monit. 7:321

HPLC 2

Sample	Blood plasma
Extraction	Deproteinization with acetonitrile/methanol (2:0.1, v/v), and then with acetone/chloroform (1:1, v/v)
Apparatus	Hitachi M 635A HPLC with 638-41 UV detector

Column	Reversed-phase; Zorbax ODS, 4–6 µm, 250 × 4.6 mm ID
Mobile phase	0.01 M Disodium hydrogen phosphate buffer, pH 6.1 with 50% phosphoric acid/methanol (45:55)
Flow-rate	0.8 ml min^{-1}; 1015 psi
Temperature	55 °C
Internal standard	Butamben
Injection size	20–30 µl, 40 ng ml^{-1}
Detection	UV, λ = 280 nm; 0.0025 AUFS
Retention times, min	(estimated from drawing)
	Nifedipine 14.53
	Internal standard 18.57
Ref.	Miyazaki K, Kohri N et al (1984) J. Chromatogr. Biomed. Appl. 310:219

HPLC 3

Sample	Blood plasma and pharmaceutical preparations
Extraction	Ethyl acetate (blood plasma)
Apparatus	Waters M 6000A HPLC pump with U6K injector and 440 UV detector
Column	Reversed-phase; µ-Bondapak C_{18}, 10 µm, 300 × 4.0 mm ID
Pre-column	Bondapak C_{18} Corasil
Mobile phase	0.01 M Disodium hydrogen phosphate buffer (pH 6.1 with 50% H_3PO_4)/methanol (45:55), pH 7.2
Flow-rate	2 ml min^{-1}; 3200 psi
Temperature	Ambient
Internal standard	4-Dimethylaminobenzaldehyde
Injection size	10 µl
Detection	UV, λ = 254 nm
Recording	Waters M 730 data module
Retention times, min	(estimated from drawing)
	Metabolite-1 1.37
	Metabolite-2 3.29
	Internal standard 4.57
	Degradation product-1 5.35
	Degradation product-2 6.37
	Nifedipine 7.75
Ref.	Pietta P, Rava A and Biondi P (1981) J. Chromatogr. 210:516

HPLC 4

Sample	Blood plasma
Extraction	On-line, pre-column-solid-phase system
Apparatus	LDC/Milton Roy Constametric III pumps with pulse dampener and Kratos M 773 UV detector

Column	Reversed-phase; Hypersil ODS 5 μm, 125 × 4 mm ID
Pre-column	Reversed-phase; Lichroprep C_2, 25–40 μm, 10 × 4 mm ID
Mobile phase	Pump A. 0.02 M Ammonium carbaminate Pump B. Acetonitrile/0.02 M acetate buffer, pH 4.0 (1:1, v/v)
Flow-rate	3 ml min^{-1} (A); 1 ml min^{-1} (B)
Temperature	Ambient
Injection size	650 μl
Detection	UV, λ = 355 nm; 0.001–0.005 AUFS
Recording	LDC/Milton Roy CI-10 integrator
Retention time, min	Nifedipine 2.4
Ref.	Nitsche V, Schuetz H and Eichinger A (1987) J. Chromatogr. Biomed. Appl. 420:207

HPLC 5

Sample	Human blood plasma
Column	Reversed-phase; Nucleosil C_{18}, 5 μm, 200 × 4 mm ID
Mobile phase	Water/methanol/acetonitrile (55:36:9)
Flow-rate	1 ml min^{-1}
Detection	UV, λ = 235 nm
Retention times, min	(estimated from drawing) Nitropyridine derivate 9.8 Nifedipine 14.8
Ref.	Dokladalova J and Tykal JA et al (1982) J. Chromatogr. 231:451

HPLC 6

Sample	Serum or blood plasma
Extraction	Acetonitrile on Bond-Elut C_{18} extraction column and then with toluene
Apparatus	Perkin-Elmer Series 2/2 LC pump, and LC-65T detector-oven
Column	Reversed-phase; Ultrasphere 5 μm, ODS, 150 × 4.6 mm ID
Mobile phase	A. Acetonitrile/methanol/10 mM sodium phosphate buffer, pH 6.1 (25:20:55) B. Methanol/tetrahydrofuran/10 mM sodium phosphate buffer, pH 6.1 (25:14:61) C. Methanol/10 mM sodium phosphate buffer, pH 6.1 (50:50)
Flow-rate	1.5 ml min^{-1}
Temperature	50 °C
Internal standard	11-Ketoprogesterone

Injection size	30 μl, 0–400 ng ml^{-1}			
Detection	UV, $\lambda = 235$ nm			
Recording	Shimadzu C-RIA data processor			
Retention times, min	Mobile phase	A.	B.	C.
	Metabolite	5.68	4.28	4.56
	Nifedipine	6.28	6.16	5.98
	Internal standard	8.29	4.93	8.71
Ref.	Bach PR and Clinical Investigation of Duchenne Dystrophy Group (1983). Clin. Chem. 29:1344			

---

135 Nitrofurantoin

CAS-No	67-20-9; 54-87-5 (sodium salt)
Chemical name	1-(5-Nitrofurfurylideneamino)hydantoin
Empirical formula	Nitrofurantoin $C_8H_6N_4O_5$; Sodium Salt $C_8H_5N_4NaO_5$
Molecular mass	238.2; 260.1
Proprietary names	Furadantine; Ivadantin; Urantoin
Clinical use	Antibacterial (urinary)
Solubility, Nitrofurantoin	Water 1 in 5000; alcohol 1 in 2000; acetone 1 in 200; dimethylformamide 1 in 16
Light absorption	$\lambda_{max} = 266$ nm and 267 nm in dimethylformamide diluted with water

---

HPLC 1

Sample	Human blood plasma and urine
Extraction	0.33 N Perchloric acid
Apparatus	Spectra-Physics 3500B HPLC with M 770 spectrophotometric detector
Column	Reversed-phase; Lichrosorb C_8, 5 μm, 150 × 4.6 mm ID
Mobile phase	Water/ethanol (95:5)
Flow-rate	1.6 ml min^{-1}; 2557 psi
Injection size	100 μl, 0.02 μg ml^{-1}
Detection	UV, $\lambda = 370$ nm; 0.004 AUFS
Retention times, min	Metronidazole 5.2; Hydromethylnitrofurantoin 7.7; Nitrofurantoin 7.7; Nitrofural 9.7; Nitrofurazolidone 10.3 (estimated from drawing)
Ref.	Vree TB, Hekster YA et al (1979) J. Chromatogr. Biomed. Appl. 162:110

HPLC 2

Sample	Blood plasma and urine
Extraction	Deproteinization of serum with acetonitrile and diluting the urine with water
Apparatus	Waters ALC/GPC M 244 HPLC, U6K injector and UV detector
Column	Reversed-phase; Lichrosorb C_{18}, 5 µm, 250 × 3.2 mm ID
Mobile phase	Methanol/0.02 M acetic acid (2.5:97.5, v/v)
Flow-rate	2 ml min^{-1}
Temperature	Ambient
Internal standard	5-Nitro-2-furoic acid
Injection size	30–120 µl
Detection	UV λ = 280 and 365 nm; 0.01 AUFS
Retention times, min	Internal standard 2.0 Cyano-metabolite 6.0 Amino-metabolite 8.0
Ref.	Hoener B-A and Wolff JL (1980) J. Chromatogr. 182:246

HPLC 3

Sample	Raw materials
Column	Reversed-phase; ODS 5–10 µm, 300 × 3.9 mm ID
Mobile phase	Phosphate buffer, pH 7.0 (containing 6.8 g potassium dihydrogen phosphate, and appr. 30 ml 1.0 N sodium hydroxide for pH 7.0 in 1 liter water)/tetrahydrofuran (9:1)
Flow-rate	1.6 ml min^{-1}
Injection size	60–100 µl
Detection	UV, λ = 375 nm; 0.1 FS
Retention time, min	Nitrofurantoin 10.5
Ref.	Anonym (1985): The United States Pharmacopeia XXI and The National Formulary XVI, Supplement 6, 1987. United States Pharmacopeial Convention Inc., Rockville, p 2591

136 Nitrofurazone

CAS-No	59-87-0
Chemical name	5-Nitro-2-furaldehyde semicarbazone
Empirical formula	$C_6H_6N_4O_4$
Molecular mass	198.1
Proprietary names	Acutol; Furacin; Furesol

Clinical use		Antibacterial/Antiprotozoal (vet)		
Solubility	Water	alcohol	dimethylformamide	chloroform
	1 in 4200	1 in 600	soluble	insoluble
Light absorption		λ_{max} = 261 nm in aqueous acid		

HPLC 1

Sample	Chicken and pork tissues
Extraction	First methanol/water (50:50), and then with 20% methanol in ethyl acetate on alumina column
Apparatus	Waters ALC/GPC with M 6000A HPLC pump, 440 UV detector and U6K injector or Beckmann HPLC system with UV detector
Column	Reversed-phase; Brownlee C_8, 10 μm, 250 × 4.6 mm ID
Pre-column	Reversed-phase; RP-GU C_8, 10 μm
Mobile phase	Acetonitrile/water (40:60) for muscle and (30:70) for liver and kidney, containing 0.01 M ammonium acetate, each
Flow-rate	1.5 ml min^{-1}
Injection size	200 μl containing 2–40 ppb
Detection	UV, λ = 365 nm

Retention times, min

Sample	muscle	liver and kidney
Nifursol	2.62	
Nitrofurantoin	2.63	
Nitrofurazone	2.75	3.91
Nihydrazone	3.04	
Furazolidone	3.73	5.71

Ref. Sugden EA, Macintosh AI and Vilim AB (1983) J. Assoc. Off. Anal. Chem. 66:874

HPLC 2

Sample	Milk
Extraction	Ethyl acetate under acidic condition
Apparatus	Waters ALC/GPC 204 LC with UV detector
Column	Reversed-phase; μ-Bondapak C_{18}, 300 × 3.9 mm ID
Pre-column	Reversed-phase; Vydac 30–44 μm, 50 × 3.9 mm ID
Mobile phase	Methanol/water (30:70)
Flow-rate	2.0 ml min^{-1}
Injection size	40 μl, 5–200 ng ml^{-1}
Detection	UV, λ = 365 nm; 0.005 AUFS
Retention time, min	Nitrofurazone 4.4 (estimated from drawing)
Ref.	Vilim AB and Macintosh AI (1979) J. Assoc. Off. Anal. Chem. 62:19

HPLC 3

Sample	Eggs, milk and meats
Extraction	Dichloromethane
Apparatus	Varian M 5000 LC, Valco 6-way-valve and Knauer 8700 UV detector
Column	Cyano phase; Zorbax CN, 250 × 4.6 mm ID
Pre-column	Reversed-phase; Vydac-201, HPLC-Sorb, 40 × 4.6 mm ID
Mobile phase	0.01 M Sodium acetate and 0.01 M acetic acid (2 + 1), pH 5.0/acetonitrile (7:3)
Flow-rate	1.5 ml min^{-1}; 1740 psi
Temperature	30 °C
Injection size	50 µl
Detection	UV, λ = 360 nm
Retention times, min	Nitrofurazone 3.4, Furazolidone 5.2, Nihydrazone 3.8, Furaltadone 6.0, Nitrofurantoin 4.2
Ref.	Petz M (1982) Dtsch. Lebensm.-Rundsch. 78:396

HPLC 4

Sample	Chicken tissues
Extraction	Homogenization with $CHCl_3$/ethyl acetate/dimethyl sulfoxide (50:50:0.8) and then on alumina column with 0.05 M phosphate buffer, pH 6.0 (with less EDTA)/methanol (1:1)
Apparatus	Altex M 100A LC pump with Bioanalytical Systems LC-4B amperometric detector
Column	Reversed-phase; Supelcosil LC-18, 5 µm, 250 × 4.6 mm ID
Mobile phase	0.05 M KH_2PO_4 containing 0.001 M EDTA, pH 6.0 with 1N NaOH/methanol (57.5:42.5)
Flow-rate	1 ml min^{-1}
Injection size	50 µl
Detection	EC, GC electrode, − 0.8 V vs Ag/AgCl
Retention times, min	Furazolidone 4.0, Zoalene 6.2, Nitrofurazone 4.1, Nitromide 6.7, Aklomide 5.3, Sulfanitran 3.2
Ref.	Parks OW (1989) J. Assoc. Off. Anal. Chem. 72:567

137 Nitroglycerin (Glyceryl Trinitrate)

CAS-No	55-63-0
Chemical name	Propane-1,2,3-triol trinitrate

Empirical formula	$C_3H_5N_3O_9$
Molecular mass	227.1
Proprietary names	Cardabid; Nitrol; Nitrolingual
Clinical use	Vasodilator
Solubility	Water 1 in 800; alcohol 1 in 4; almond oil 1 in 6; carbon disulfide 1 in 120

HPLC 1

Sample	Diluted nitroglycerin
Column	Reversed-phase; ODS 5–10 μm, 300 × 3.9 mm ID
Mobile phase	Methanol/water (50:50, v/v)
Flow-rate	1 ml min^{-1}
Internal standard	Pentaerythritol tetranitrate
Injection size	Appr. 20 μl
Detection	UV, λ = 220 nm
Retention times	(relative) Internal standard 0.7 Nitroglycerin 1.0
Ref.	Anonym (1985): The United States Pharmacopeia XXI and The National Formulary XVI, Supplement 1, 1984. United States Pharmacopeial Convention Inc., Rockville, p 1750

HPLC 2

Sample	Cardiotonics
Apparatus	HPLC low-pressure ternary gradient system
Column	Reversed-phase; Superspher 100 C_{18}, 125 × 4 mm ID
Mobile phase	Acetonitrile/triethylamine acetate, pH 4.0 (50:50, v/v)
Flow-rate	1.5 ml min; 2500 psi
Injection size	10 μl containing 10 μg
Detection	UV, λ = 220 nm; 0.16 AUFS
Retention times	Propranolol, isosorbide dinitrate and nitroglycerin, in ascending order (from drawing)
Ref.	Merck Application Manual on HPLC, TLC and HPTLC (1986). E. Merck, Darmstadt, p 77

HPLC 3

Sample	Sublingual, sustained-release and ointment dosage forms
Apparatus	Altex M 100A LC pump with 155–40 UV detector
Column	Reversed-phase; μ-Bondapak C_{18}, 300 × 3.9 mm ID
Pre-column	Reversed-phase; μ-Bondapak C_{18}/Corasil, 40 × 4.6 mm ID

Mobile phase	Methanol/water (40:60)
Flow-rate	1 ml min^{-1}; 1600 psi
Internal standard	Pentaerythritol tetranitrate
Injection size	20 µl
Detection	UV, λ = 220 nm
Retention times, min	Isosorbide dinitrate 12.0
	Nitroglycerin 18.0
	Internal standard 32.0
Ref.	Olsen CS and Scroggins HS (1983) J Pharm. Sci. 72:963

HPLC 4

Sample	Ointment
Extraction	Direct dissolution in warm dimethyl sulfoxide
Apparatus	Altex M 100A LC with Du Pont M 837 spectrophotometer
Column	Reversed-phase; Hypersil ODS 5 µm, 150 × 4 mm ID
Pre-column	Reversed-phase; µ-Bondapak C_{18}, 5 µm Corasil, 40 × 4.6 mm ID
Mobile phase	Methanol/water (40:60)
Flow-rate	0.8 ml min^{-1}; 800 psi (I) and 2.0 ml min^{-1}; 4400 psi (II)
Internal standard	Anisole
Injection size	10 µl
Detection	UV, λ = 216 or 254 nm
Retention times, min	(estimated from drawing) (flow-rate = 2 ml min^{-1})
	Nitroglycerin 6.5
	Internal standard 9.0
Ref.	Gelber L (1980) J. Pharm. Sci. 69:1084

HPLC 5

Sample	Raw materials
Apparatus	Perkin-Elmer 2/2 HPLC with LC-55B UV detector
Column	Alkyl-phenyl phase; µ-Bondapak 10 µm, 300 × 3.9 mm ID
Mobile phase	Acetonitrile/water (36:64, v/v)
Flow-rate	2 ml min^{-1}; < 2000 psi
Internal standard	Isosorbide dinitrate
Injection size	100 µl, 2.5–20 µg ml^{-1}
Detection	UV, λ = 218 nm; 0.005 AUFS
Recording	Hewlett-Packard 3354 integrator
Retention times, min	(estimated from drawing)
	Mononitroglycerin 2.0 Isosorbide dinitrate 7.5
	Dinitroglycerin 3.8 Trinitroglycerin 12.0

Ref. Baaske DM, Karnatz NN and Carter JE (1983) J. Pharm. Sci. 72: 194

138. Norepinephrine

CAS-No	51-41-2; 69815-49-2 (bitartrate, monohydrate)		
Chemical name	(*R*)-2-Amino-1-(3,4-dihydroxyphenyl)ethanol		
Empirical formula	Norepinephrine	Bitartrate	
	$C_8H_{11}NO_3$	$C_8H_{11}NO_3$, $C_4H_6O_6$, H_2O	
Molecular mass	169.2	337.3	
Proprietary names	Adrenor; Levophed; Noradrec		
Clinical use	Sympathomimetic		
Solubility,	Water	ethanol	ether
Norepinephrine	slightly soluble		slightly soluble
Bitartrate	1 in 2.5	1 in 300	insoluble
Light absorption	λ_{max} = 296 nm in aqueous alkali		

HPLC 1

Sample	Raw materials (enantiomers)
Apparatus	Shimadzu LC-3A pump with SGA-1A gradient programmer and Shimadzu RF-500 LC fluorimeter
Column	Reversed-phase; Develosil ODS-5, 100 × 6.0 mm ID
Mobile phase	Isocratic: 50 mM Phosphate buffer, pH 6.5/ methanol (55:45)
	Gradient: 50 mM Phosphate buffer, pH 6.5 (A) and methanol (B)
	B 40% at 1 min; B 40% to 50% at 2 min; B 50% to 60% at 3 min; B 60% to 35% at 9 min
Flow-rate	1 ml min^{-1}
Derivatization	Pre-column; with *o*-phthalaldehyde-N-acetyl-L-cysteine
Injection size	5 µl containing 10 ng
Detection	Fluorescence, λ = 340/450 nm
Retention times, min	(estimated from drawing)

System	isocratic	gradient
L-Norepinephrine		7.5
D-Noreprinephrine		8.5
L-Dopa	9.0	5.0
D-Dopa	11.0	6.0
Dopamine		11.0

Ref.	Nimura N, Iwaki K and Kinoshita T (1987) J. Chromatogr. 402:387

HPLC 2

Sample	Blood plasma and urine
Apparatus	Perkin-Elmer Series 10 pump and LC-4B electrochemical detector (Bioanalytical Systems)
Column	Reversed-phase; Microsorb C_{18}, 3 μm, 100 × 4.6 mm ID
Mobile phase	0.1 mol Formic acid, 0.35 mmol sodium octyl sulfate, 0.1 mmol citric acid, 0.1 mmol EDTA and 0.05 mmol diethylamine per liter
Flow-rate	1 ml min^{-1}
Internal standard	3,4-Dihydroxybenzylamine
Injection size	100 μl
Detection	EC, GC electrode, + 650 mV
Retention times, min	Vanillylmandelic acid 2.5; Norepinephrine 3.1; Epinephrine 4.3; Internal standard 5.6; Normetanephrine 6.7 (estimated from drawing)
Ref.	Foti A, Kimura S et al (1987) Clin. Chem. 33:2209

HPLC 3

Sample	Blood plasma
Extraction	0.1 M Perchloric acid using aluminia column for sample clean-up
Apparatus	Beckmann M 340 LC pump and Bioanlytical Systems LC-4B electrochemical detector
Column	Reversed-phase; Supelcosil 5 μm LC-18, 50 × 4.6 mm ID
Pre-column	Brownlee cartridge C_{18}
Mobile phase	0.067 M Monochloroacetic acid buffer, pH 4.0 with 3 M NaOH, containing 0.54 g/l sodium octyl sulfate and 9% acetonitrile
Flow-rate	0.55 ml min^{-1}
Temperature	Ambient
Internal standard	3,4-Dihydroxybenzylamine (DHBA)
Injection size	20 μl, 424 pg ml^{-1}
Detection	EC, + 0.4 V vs Ag/AgCl RE
Recording	Hewlett-Packard 3390A integrator
Retention times, min	(estimated from drawing) Norepinephrine 4.5; Epinephrine 5.8; Internal standard 7.3; Dopamine 9.0
Ref.	Koch DD and Polzin GL (1987) J. Chromatogr. 386:19

HPLC 4

Sample	Brain tissue from rats
Extraction	First 1 M formic acid/acetone (15:85) and then with diethyl ether as well as sample clean up procedure using boronic acid gel column
Apparatus	Bioanalytical Systems HPLC with column-switching valve, amperometric detector and Rheodyne injector
Column	Reversed-phase; Chemcosorb 7 ODS-L-300, 7 μm, 75 × 4.6 mm ID × 2
Mobile phase	A. 0.01 M Sodium acetate, pH 5.0 with citric acid containing 1.2 mM trimethylammonium chloride and 30 mg l^{-1} sodium 1-octanesulfonate, with B B. 0.2 M Potassium phosphate buffer, pH 8.5, through T
Flow-rate	1 ml min^{-1} (A) and 0.6 ml min^{-1} (B) together
Temperature	20–25 °C
Internal standards	Ethylhomocholine and 3,4-dihydroxybenzylamine
Injection size	10–20 μl containing 1–2500 pmol
Detection	EC, GC alectrode, + 0.65 V vs Ag/AgCl RE (A) and + 0.5 V vs Ag/AgCl RE (B)

Retention times, min		
	Choline	4.5
	L-Noradrenaline	4.5
	Ethylhomocholine	6.7
	Adrenaline	6.9
	Acetylcholine	7.8
	3,4-Dihydroxybenzylamine	7.9
	Dopamine	12.6
	5-Hydroxystryptamine	23.0

Ref. Kaneda N, Asano M and Nagatsu T (1986) J. Chromatogr. 360:211

139 Novobiocin

CAS-No	303-81-1; 1476-53-5 (sodium salt)			
Chemically	Antimicrobial substance produced by Streptomyces niveus or Streptomyces spheroides or by other means			
Empirical formula	Novobiocin	Sodium Salt		
	$C_{31}H_{36}N_2O_{11}$	$C_{31}H_{35}N_2NaO_{11}$		
Molecular mass	612.6	634.6		
Proprietary names	Albamycin; Cathomycine; Robiocina			
Clinical use	Antibacterial			
Solubility,	Water	ethanol	methanol	chloroform, ether
Novobiocin	insoluble	soluble	soluble	insoluble
Sodium salt	1 in 5	1 in 7	1 in 3	insoluble

HPLC 1

Sample	Milk, blood and tissues
Extraction	Methanol/water (1:1) on extraction column
Apparatus	Varian M LC 5000 and UV-50 detector
Column	Reversed-phase; Supelcosil LC-18-DB, 150 × 4.6 mm ID
Pre-column	Reversed-phase; LC-18-DB
Mobile phase	0.01 M H_3PO_4/acetonitrile/methanol (50:0:50) (0–1 min) to (20:80:0) in 20 min
Flow-rate	1 ml min^{-1} for run; reset, 2 ml min^{-1} for 5 min and 1 ml min^{-1} for 5 min (run)
Injection size	200 μl containing 2 μg
Detection	UV, λ = 340 nm; 1 AUFS
Retention time, min	Novobiocin 21 (estimated from drawing)
Ref.	Moats WA and Leskinen L (1988) J. Assoc. Off. Anal. Chem. 71:776

HPLC 2

Sample	Novobiocin powder and veterinary mastitis product in oil
Apparatus	LDC M 19-60066-022 HPLC minipump, LDC UV III monitor and Waters WISP M 710B autosampler or Rheodyne M 70-10 injector
Column	Normal phase; Lichrosorb SI-100, 5 μm, 250 × 4.6 mm ID
Mobile phase	Butyl chloride (50% water-saturated)/tetrahydrofuran/methanol/acetic acid (88:5:4:3)
Flow-rate	1 ml min^{-1}
Internal standard	Prednisone, 0.3 mg ml^{-1}
Injection size	20 μl, 0.2–0.6 mg ml^{-1}
Detection	UV, λ = 254 nm; 0.032 AUFS or 340 nm
Retention times, min	(estimated from drawing)

Novobiocic acid	7.6	Isonovobiocin	14–17.0
Descarbamylnovobiocin	8.2	Novobiocin	15–19.0
Internal standard	10.0	Desmethyldescarbamylnovobiocin	24.2

Ref.	Tsuji K, Rahn PD and Kane MP (1982) J. Chromatogr. 235:205

HPLC 3

Sample	Mastitis product in oil sterilized by ^{60}Co irradiation
Apparatus	Varian M 4200 HPLC; Altex M 150 UV monitor and Rheodyne M 701 sample-loop-injector
Column	Reversed-phase; Zorbax ODS, 250 × 4.6 mm ID
Pre-column	Whatman Co:Pell ODS, 50 × 2 mm ID

Mobile phase	0.005 M Sodium 1-heptanesulfonate in methanol/water (80:20)
Flow-rate	1 ml min^{-1}
Internal standard	2,6-Dichloro-4-nitroaniline, 1 μg ml^{-1}
Injection size	10 μl containing 25 μg
Detection	UV, λ = 254 nm; 0.2 AUFS
Retention times, min	(estimated from drawing)

Novenamine	2.8
Hydroxynovobiocin	4.9
Internal standard	5.2
Novobiocic acid	9.1
Desmethyldescarbamylnovobiocin	10.8
Dihydronovobiocin	11.0
Novobiocin	12.0
Isonovobiocin	12.0
Descarbamylnovibiocin	13.6

Ref. Hornish RE (1982) J. Chromatogr. 236:481

HPLC 4

Apparatus	LDC modular liquid chromatograph with Milton Roy minipump and LDC M 709 pulse dampener
Column	Zipax HCP, 1000 × 2.1 mm ID
Mobile phase	Methanol/0.02 M monobasic and dibasic sodium phosphate buffer, pH 7.0 (15:85)
Flow-rate	0.85 ml min^{-1}; 1000 psi
Temperature	Ambient
Retention times, min	

Novobiocic acid	3.5
Novenamine	4.0
Desmethyldescarbamyl-novobiocin	5.5
Descarbamylnovobiocin	12.5
Isonovobiocin	19.0
Novobiocin	23.5
Dihydronovobiocin	32.0

Ref. Tsuji K and Robertson JH (1974) J. Chromatogr. 94:245

HPLC 5

Sample	Test substance
Apparatus	Varian 8500 LC and Variscan UV detector
Column	Reversed-phase; Micropac CH, 10 μm, 250 × 2.1 mm ID
Mobile phase	Methanol/1% acetic acid in water (60:40)
Flow-rate	2.0 ml min^{-1}; 4703 psi
Temperature	Ambient
Injection size	2 μl

Detection	UV, λ = 307 nm; 0.5 AUFS
Retention time, min	Calcium novobiocin 2.1 (estimated from drawing)
Ref.	Varian Aerograph Application Notes No. 6 (1974) Varian, California

140 Nystatin

CAS-No	1400-61-9
Chemically	Mixture of polyenes produced by certain strains of Streptomyces noursei or by some other means
Empirical formula	$C_{47}H_{75}NO_{17}$
Molecular mass	926.1
Proprietary names	Moronal; Mycostatin; Nyspes
Clinical use	Antifungal
Solubility	Water very slightly soluble; alcohol sparingly soluble; ether insoluble; formamide freely soluble
Light absorption	λ_{max} = 305 nm in aqueous alkali

HPLC 1

Sample	Raw materials
Apparatus	Gilson M 302 LC pump, Kratos spectrophotometric detector and Rheodyne M 7413 sample valve
Column	A. Normal phase; Zorbax BP, Sil 7–8 µm, 500 × 1 mm ID B. Reversed-phase; Rosil C_{18} HL, 3 µm, 250 × 1 mm ID
Mobile phase	For A: (1) Methanol/water/acetic acid (95:4.5:0.5) For A: (2) Methanol/dimethylformamide/water/acetic acid (72:25:3:0.4) For B: Methanol/water/acetic acid (65:34:1)
Flow-rate	125 µl min^{-1} (A); 40 µl min^{-1} (B)
Injection size	200 µl containing 4 mg
Detection	UV, λ = 308 nm
Retention times, min	(estimated from drawing)

Mobile phase	A (1)	A (2)	B
Nystatin	14.5	13.5	22.0
Impurities			16.0, 31.0

Ref.	Milhaud J, Gareil P and Rosset R (1986) J. Chromatogr. 358:284

HPLC 2

Sample	Raw materials
Apparatus	LDC Constametric II G LC with gradient master M 1601 and Cecil CE 272 UV detector
Column	Reversed-phase; Hypersil MOS 3 μm, 100 × 4 mm ID
Pre-column	Reversed-phase; Hypersil MOS 5 μm, 35 × 4 mm ID
Mobile phase	Methanol/0.005 M sodium acetate buffer, pH 5.8 (55:45, v/v), pH 6.8 Gradient: Methanol 55% to 58.75% in 15 min; hold 10 min; methanol 58.75% to 67% in 10 min; continue 15 min
Flow-rate	1.0 ml min^{-1}
Injection size	5 μl, 3–5 mg ml^{-1}
Detection	UV, λ = 304 nm; 0.5 AUFS
Ref.	Thomas AH, Newland P and Sharma NR (1982) Analyst 107:849

HPLC 3

Sample	Test substance
Apparatus	HPLC system with UV detector
Column	Reversed-phase; Vydac 30–40 μm, 91.44 × 2.4 mm ID
Mobile phase	Water/methanol/tetrahydrofuran (420:90:50)
Flow-rate	0.9 ml min^{-1}; 1000 psi
Temperature	30 °C
Injection size	8 μl, 1 mg ml^{-1}
Detection	UV, λ = 350 nm; 0.002 A
Retention times, min	(estimated from drawing) Nystatin II 9.0 Nystatin I 5.7 Nystatin III 13.0
Ref.	Tsuji K (1978) in: Tusji K (ed) GLC and HPLC Determination of Therapeutic Agents, Part 2. Marcel Dekker Inc., New York and Basel, p 739

141 Oleandomycin

CAS-No	3922-90-5; 7060-74-4 (phosphate)	
Chemically	Macrolide antibiotic produced by Streptomyces antibioticus	
Empirical formula	Oleandomycin	Phosphate
	$C_{35}H_{61}NO_{12}$	$C_{35}H_{61}NO_{12}$, H_3PO_4
Molecular mass	687.9	785.9
Proprietary names	Matromycin; Mittamycin; Triolmicina	

Clinical use	Antibiotic		
Solubility,	Water	alcohol	acetone, methanol
Oleandomycin	moderately sol	freely sol	freely soluble
Phosphate	1 in 2.5	1 in 3	
Light absorption	λ_{max} = 242 nm in aqueous acid		

HPLC 1

Sample	Serum and urine
Extraction	Acetonitrile/0.05 M phosphate buffer (1:1) on C_{18} extraction column
Apparatus	Waters M 6000A pump, WISP 710B autosampler and Kratos M 769 UV detector
Column	Reversed-phase; Novapak C_{18}, 5 μm ODS, 150 × 3.9 mm ID;
Mobile phase	Acetonitrile/0.05 M phosphate buffer (containing 3.2 ml phosphoric acid in 1 liter water, pH 6.3) (300:700, v/v), pH 7.0 with sodium hydroxide
Flow-rate	1 ml min^{-1}; 1450 psi
Temperature	35 °C
Internal standard	Erythromycin
Injection size	1–5 μl
Detection	UV, λ = 200 nm; 0.01 AUFS
Retention times, min	(estimated from drawing) Oleandomycin 4.29 Internal standard 6.17
Ref.	Stubbs C, Haigh JM and Kanfer I (1986) J. Chromatogr. 353:33

HPLC 2

Sample	Human serum
Extraction	Methyl tert.-butyl ether under slight alkaline condition
Apparatus	Spectra-Physics 8700 XR LC pump and LC-4 amperometric detector (Bioanalytical Systems)
Column	PRP-1; 5 μm, 150 × 4.1 mm ID
Mobile phase	I. 10 mM Potassium phosphate buffer, pH 9.5/ acetonitrile (60:40, v/v) II. 25 mM Potassium phosphate buffer, pH 9.5/ acetonitrile (60:40, v/v) III. 50 mM Potassium phosphate buffer, pH 9.5/ acetonitrile (60:40, v/v) IV. 100 mM Potassium phosphate buffer, pH 9.5/ acetonitrile (60:40, v/v)
Flow-rate	0.70 ml min^{-1}; 2248 psi
Temperature	25 °C

Injection size	25 µl containing 25 mmol
Detection	EC, + 1.0–1.1 V vs Ag/AgCl RE
Recording	Spectra-Physics 4270 integrator

Capacity factor, *k'*

	Erythromycin	Oleandomycin
Mobile phase IV	6.1	3.7
Mobile phase III	6.8	4.0
Mobile phase II	7.3	4.3
Mobile phase I	8.7	4.9

Ref. Nilsson L-G, Walldorf B and Paulsen O (1987) J. Chromatogr. Biomed. Appl. 423:189

142 Oxazepam

CAS-No	604-75-1
Chemical name	7-Chloro-1,3-dihydro-3-hydroxy-5-phenyl-2*H*-1,4-benzodiazepin-2-one
Empirical formula	$C_{15}H_{11}ClN_2O_2$
Molecular mass	286.7
Proprietary names	Isodin; Oxanid; Sobril
Clinical use	Tranquillizer

Solubility

Water	alcohol	ether	dioxane
insoluble	1 in 220	1 in 2200	soluble

Light absorption λ_{max} = 230 nm in ethanol

HPLC 1

Sample	Benzodiazepines and their metabolites
Apparatus	Waters M 6000A LC pump and Perkin-Elmer LC-75 UV detector
Column	Normal phase; Spherisorb S5W, 5 µm, 250 × 5 mm ID
Mobile phase	Methanol/perchloric acid (999.9:0.1, v/v)
Flow-rate	2 ml min^{-1}
Injection size	20 µl
Detection	UV, λ = 240 nm

Capacity factor, *k'*

7-Aminonitrazepam	0	Triazolam	1.83
Desmethylclobazam	0.01	7-Acetamido-nitrazepam	1.93
Clobazam	0.03		
Demoxepam	0.03	Nordazepam	1.99
Ketazolam	0.04	Clorazepic acid	2.00
Lormetazepam	0.08	Prazepam	2.19
Lorazepam	0.14	Desmethylchlor-diazepoxide	2.39
Clonazepam	0.35		
Flunitrazepam	0.47	Diazepam	2.49

Temazepam	0.60	Alprazolam	2.79
Oxazepam	0.73	Chlordiazepoxide	2.87
N-1-Hydroxyethyl-flurazepam	1.43	Bromazepam	2.99
Nitrazepam	1.49	Medazepam	4.44
N-1-Desalkyl-flurazepam	1.52	Desoxychlor-diazepoxide	4.55
		Midazolam	5.90
		Flurazepam	6.50

Ref. Gill R, Law B and Gibbs JP (1986) J. Chromatogr. 356:37

HPLC 2

Sample: Serum

Extraction: Hexane/ethyl acetate (70:30, v/v)

Apparatus: Waters M 45 LC pump with 445 UV detector

Column: Reversed-phase; RCM 100 C_{18}, 10 μm, 100 × 5 mm ID

Pre-column: Reversed-phase; C_{18}, 10 μm, 30 mm

Mobile phase: Buffer (containing 4.5 ml triethylamine in 1000 ml water, pH 3.5–4.0 with appr. 2.2 ml H_3PO_4)/methanol/isopropanol (300:650:50, v/v)

Flow-rate: 1 ml min^{-1}

Internal standard: Prazepam

Injection size: 20 μl

Detection: UV, λ = 254 nm

Recording: Perkin-Elmer Sigma 10 data station

Retention times, min:

Heptabarbital	2.7	Methaqualone	4.2
Pentobarbital	2.7	Chlorazepate	4.2
Carbamazepine	3.0	Chlordiazepoxide	4.6
Secobarbital	3.0	Diazepam	5.5
Oxazepam	3.2	Prazepam (IS)	8.6
N-desmethyl-diazepam	4.2		

Ref. Kinberger B and Wahrgren P (1982) Anal. Lett. 15 (B6):549

HPLC 3

Sample: Urine and serum

Extraction: Methanol and then with diethyl ether on RP-18 Baker 10SPE extraction column after enzymatic hydrolysis with β-glucuronidase

Apparatus: Perkin-Elmer Series 3B-LC with LS-2-filter fluorimeter

Column: Reversed-phase; 5 μm Bondapak, C_8, 120 mm

Mobile phase: Acetonitrile/1 mM acetate buffer, pH 4.0 (1:1)

Derivatization: Pre-column; with 2 M NaOH and PbO_2 or dimethyl sufoxide and PbO_2

Internal standard	2-Aminobenzophenone
Detection	Fluorescence, λ = 260/430 nm
Retention times, min	Clonazepam 4.3 Oxazepam 6.7
	Flurazepam 5.9 Prazepam 12.9
Ref.	De Giovanni N and Chiarotti M (1988) J. Chromatogr. Biomed. Appl. 428:321

HPLC 4

Sample	Serum
Extraction	Toluene/heptane/isoamyl alcohol (76:20:4)
Apparatus	Waters M 6000A LC pump, U6K injector and 450 UV detector
Column	Reversed-phase; Supelcosil C_{18}, 5 µm, 250 × 4.5 mm ID
Pre-column	Perisorb RP 18
Mobile phase	10 mM Potassium phosphate buffer, pH 6.0 containing 2.6 g l^{-1} sodium pentane-sulfonate/methanol/acetonitrile (45:35:20, v/v)
Flow-rate	1 ml min
Internal standard	Nitrazepam
Injection size	5–10 µl
Detection	UV, λ = 254 nm; 0.04 AUFS
Retention times, min	(estimated from drawing)
	Demoxepam 7.31
	Nitrazepam 8.35
	Norchlordiazepoxide 9.29
	Oxazepam 10.23
	Chlordiazepoxide 12.13
	Nordiazepam 14.80
	Diazepam 18.90
Ref.	Haver VM, Porter WH et al (1986) Ther. Drug Monit. 8:352

HPLC 5

Sample	Serum, urine and saliva
Extraction	Diethyl ether under alkaline, acidic and again under alkaline condition
Apparatus	Orlita DMP 1515 HPLC with Waters M 440 UV detector and U6K injector
Column	Methyl-silica 60, 7-8 µm, 100 × 2.8 mm ID
Mobile phase	I. Methanol/0.05 M phosphate buffer, pH 6.0 (1:1) for test solution
	II. Methanol/0.05 M phosphate buffer, pH 7.0 (2:3) for serum, saliva and urine

Injection size	10–60 μl containing at least 0.2 ng				
Detection	UV, λ = 254 nm; 0.001 AUFS				
Recording	Kipp & Zonen BD8 potentiometric recorder and Spectra-Physics computing integrator				
Retention times, min	(estimated from drawing)				
	Mobile phase	I	II	II	II
	Flow-rate, ml min^{-1}	0.558	0.174	0.216	0.216
	Sample	test solution	serum	saliva	urine
	7-Aminonitrazepam	1.8			
	Bromazepam	3.2			
	Nitrazepam	3.7			
	Oxazepam	4.7	18.0	14.0	15.0
	Desmethylchlordiazepoxide	5.5			
	Hydroxydiazepam	6.0	22.5	17.0	18.5
	Chlordiazepoxide	7.0			
	Desmethyldiazepam	7.8	33.0	23.0	23.0
	Diazepam	10.5	37.5	35.0	
	Caffeine			3.5	
Ref.	Tjaden UR, Meeles MTHA et al (1980) J. Chromatogr. Biomed. Appl. 181:227				

---

143 Oxytetracycline Dihydrate

CAS-No	79-57-2 (anhydrous); 2058-46-*0* (hydrochloride)			
Chemical name	5β-Hydroxytetracycline dihydrate			
Empirical formula	Oxytetracycline Dihydrate	Hydrochloride		
	$C_{22}H_{24}N_2O_9$, $2H_2O$	$C_{22}H_{24}N_2O_9$, HCl		
Molecular mass	496.5	496.9		
Proprietary names	Macocyn; Terramycin; Unimycin			
Clinical use	Antibiotic			
Solubility,	Water	ethanol	chloroform	aq acids
Oxytetracycline	1 in 4150	sparingly soluble	insoluble	soluble
Hydrochloride	1 in 2	1 in 45	insoluble	
Light absorption	λ_{max} = 268 nm in aqueous acid			

HPLC 1

Sample	Fish tissue
Extraction	C_8 extraction column with water/acetone (5:95 and 10:90) after homogenization with hexane/dichloromethane (1:3, v/v)

Apparatus	Perkin-Elmer Series 400 LC pump, ISS 100 sampling system and LC 85 UV detector
Column	Reversed-phase; Supelcosil 5 μm, LC-18-DB, 150 × 4.6 mm ID
Pre-column	Reversed-phase; Supelcosil 5 μm, LC-18-DB, 20 × 4.6 mm ID
Mobile phase	0.005 M Phosphate buffer, pH 2.0/acetonitrile/tetrahydrofuran (81:10:9)
Flow-rate	0.9 ml min^{-1}
Internal standard	Demeclocycline
Injection size	25 μl
Detection	UV, $\lambda = 357$ nm
Recording	Laboratory computing integrator LC-I 100 (Perkin-Elmer)
Retention times, min	(estimated from drawing) Oxytetracycline 5.0–5.2 Internal standard 8.3
Ref.	Rogstad A, Hormazabal V and Yndestad M (1988) J. Liq. Chromatogr. 11:2337

HPLC 2

Sample	Premixes and veterinary products
Extraction	Methanol with 2% HCl
Apparatus	Instrument Specialistics M LC-825 pump and 226 UV detector
Column	Reversed-phase; Alltech Econosphere C_8, 5 μm, 150 × 4.6 mm ID
Pre-column	Brownlee C_8, 15 × 3.2 mm ID and silica gel, 250 × 4.5 mm ID
Mobile phase	Acetonitrile/0.1 M phosphate buffer (containing 13.8 g NaH_2PO_4, 2.28 g EDTA and 8 ml N,N-dimethyloctylamine per liter, pH 8.0 with NaOH) (25:75)
Flow-rate	1.5 ml min^{-1}
Injection size	20 μl, 2 μg ml^{-1}
Detection	UV, $\lambda = 280$ nm
Recording	Hewlett-Packard 3390A integrator
Retention times, min	epi-Oxytetracycline 2.3 Tetracycline 4.3 Oxytetracycline 3.5 Chlortetracycline 10.9
Ref.	Chappell GS, Houglum JE and Kelley WN (1986) J. Assoc. Off. Anal. Chem. 69:28

HPLC 3

Sample	Blood plasma
Extraction	Protein precipitation with 1.25 M perchloric acid or with ethyl acetate

Apparatus	Spectra Physics 8000B HPLC, Rheodyne M 7126 injector and Pye-Unicam LC-3 UV detector
Column	A. XAD-2, 6–7 μm, 100 × 3.0 mm ID B. PRP-1, 10 μm, 100 × 3.0 mm ID
Mobile phase	For A. Acetonitrile/dichloromethane/0.2 M acetate buffer, pH 3.6 plus 0.025 M EDTA (10:1:90, v/v) For B. Acetonitrile/dichloromethane/0.05 acetate buffer, pH 3.6 plus 0.025 M EDTA (12.5:1:87.5, v/v)
Flow-rate	0.8 ml min^{-1} (A); 0.6 ml min^{-1} (B)
Temperature	25 °C
Injection size	20 μl
Detection	UV, λ = 268 or 357 nm
Retention times, min	

	Column	
(estimated from drawing)	A.	B.
epi-Oxytetracycline	1.6	1.0
Oxytetracycline	1.6	1.4
α-apo-Oxytetracycline	1.8	1.6
β-apo Oxytetracycline	9.0	9.0

Ref.	Reeuwijk HJEM and Tjaden UR (1986) J. Chromatogr. 353:339

HPLC 4

Sample	Raw materials
Apparatus	Spectra-Physics 8100 HPLC system with autosampler and SP 8440 UV-VIS detector
Column	Reversed-phase; styrene-divinyl benzene polymer, PRP-1
Mobile phase	A: Citrate-phosphate buffer, pH 5.0/ propan-2-ol/tetrahydrofuran (82:11:7, v/v); B: As A, but (87:8:5, v/v) C: As B but containing 1% dichloromethane, namely (86:8:5:1, v/v)
Flow-rate	0.8 or 1 ml min^{-1}
Temperature	42 °C
Injection size	10 μl
Detection	UV, λ = 272 nm; 0.08 AUFS
Recording	Spectra-Physics 4200 computing integrator
Retention times, min	(estimated from drawing)

Mobile phase	A.	B.	C.
Oxytetracycline	2.7	4.0	3.0
Tetracycline	2.7	4.2	3.0
6-Dimethylchlortetracycline	3.7	6.0	4.0
Methacycline	4.8	9.5	5.5
Minocycline	6.2	14.8	9.0

Ref.	Yasin A and Jefferies TM (1988) J. Pharm. Biomed. Anal. 6:867

HPLC 5

Sample	Bulk materials and dosage forms (tablets)
Column	A. Reversed-phase; Spherisorb S5 C_8, 5 μm, 200 × 4.6 mm ID B. Cation exchange; 5 μm, 200 × 5 mm ID
Pre-column	For A: Reversed-phase; Lichrosorb C_{18}, 10 μm, 30 × 4.5 mm ID
Mobile phase	For A: 0.2 M Ammonium oxalate/0.1 M Na_2EDTA/dimethylformamide (55:20:25) For B: 350 ml Methanol and 650 ml water containing 3.4 g KH_2PO_4 and 9.3 g Na_2 EDTA, pH 5.3
Flow-rate	1.0 ml min^{-1} (A); 0.7 ml min^{-1} (B)
Internal standards	Paracetamol (A); sulfadimidine (B)
Injection size	10 μl, 0.25 mg ml^{-1} (A)
Detection	UV, λ = 280 nm (A); 263 nm (B)
Retention times, min	(estimated from drawing)

Column	A.	B.
Internal standards	3.75	3.00
4-epi-Oxytetracycline	4.29	
Oxytetracycline	4.87	15.49
α-apo-Oxytetracycline	8.52	12.3
β-apo-Oxytetracycline	12.66	9.3
andro-Oxytetracycline	23.13	24.00

(B = method of British Pharmacopoea; A = method for bulk materials)

Ref. Barnes WN, Ray A and Bates LJ (1985) J. Chromatogr. 347:173

HPLC 6

Sample	Blood plasma from rainbow trouts
Extraction	Deproteinization with trifluoroacetic acid and at 37 °C in water bath
Apparatus	Waters M 590 LC pump with 481 UV detector
Column	Cyano phase; Cyano Spherisorb-5 μm, 100 × 4.6 mm ID
Pre-column	Cyano Spherisorb-5 MPLC, 30 × 4.6 mm ID
Mobile phase	0.02 M Oxalic acid (pH 2.0 with 25%-NH_3)/methanol/N,N-dimethylformamide (950:50:50, v/v)
Flow-rate	1.0 ml min^{-1}
Temperature	22–23 °C
Injection size	20–100 μl, 0.2–3.0 μg ml^{-1}
Detection	UV, λ = 350 nm
Retention time, min	Oxytetracycline 4.3
Ref.	Iversen B, Aanesrud A et al (1989) J. Chromatogr. Biomed. Appl. 493:217

144 Oxytocin

CAS-No	50-56-6	
Chemical name	[3-L-Isoleucine,8-L-leucine]-vasopressin	
Empirical formula	Oxytoxin	Citrate
	$C_{43}H_{66}N_{12}O_{12}S_2$	$C_{43}H_{66}N_{12}O_{12}S_2, C_6H_8O_7$
Molecular mass	1007.2	1199.3
Proprietary names	Orasthin; Pitocin; Syntocinon	
Clinical use	Pituitary Hormone	
Solubility	Soluble in water and in 1-butanol	
Light absorption	λ_{max} = 293 nm in aqueous alkali	

HPLC 1

Sample	Pharmaceutical solutions
Apparatus	Waters M 45 LC pumps, Isco ISIS autosampler and Kratos FS 970 fluorescence detector and M 783 UV detector
Column	Reversed-phase; Whatman Partisphere C_{18}, 5 µm, 125 × 4.6 mm ID
Pre-column	Alltec C_{18} universal guard cartridge × 2
Mobile phase	Acetonitrile/0.1% phosphoric acid (21:79)
Flow-rate	1.3 ml min^{-1}
Temperature	Ambient
Derivatization	Post-column; with fluorescamine
Injection size	6.5 ml, 20 ng ml^{-1}
Detection	Fluorescence, λ = 250/418 nm; UV λ = 215 nm
Recording	Hewlett-Packard 3357 integrating computer
Retention times, min	(estimated from drawing)
	Oxytocin 12.6
	Degradation products 14.5, 15.8, 18.0, 20.8
Ref.	Brown DS and Jenke DR (1987) J. Chromatogr. 410:157

HPLC 2

Sample	Injectable dosage forms
Apparatus	Hewlett-Packard UFC-1000 LC with Orlita pump and Perkin-Elmer LC-55 UV detector
Column	Reversed-phase; Nucleosil C_{18}, 10 µm, 250 × 4 mm ID
Mobile phase	20% Acetonitrile in 4/15 M phosphate buffer, pH 7
Flow-rate	4.0 ml min^{-1}; 2938 psi
Temperature	Ambient
Injection size	50 µl and 750 µl, 10 IU ml^{-1}
Detection	UV, λ = 210 nm; 0.02–0.1 AUFS

Recording	Hewlett-Packard 3352B laboratory data system
Retention times, min	(estimated from drawing)
	Bye-Products 1.5, 2.0, 2.5, 5–7.5
	Oxytocin 4.4
	Trichlorobutanol 12.3
Ref.	Krummen K and Frei RW (1977) J. Chromatogr. 132:429

HPLC 3

Sample	Injectable dosage forms
Apparatus	Hewlett-Packard UFC-100 with Orlita pump and Perkin-Elmer LC-55 UV detector
Column	A. Reversed-phase; Nucleosil C_8, 5 μm, 150 × 4 mm ID B. Reversed-phase; C_8, 10 μm, 250 × 3 mm ID
Mobile phase	Acetonitrile/0.066 M phosphate buffer, pH 7.0 (20:80)
Flow-rate	1.8 ml min^{-1} (A); 3.0 ml min^{-1} (B)
Injection size	50 μl, 200 IU ml^{-1} (A); 120 μl, 10 IU ml^{-1}
Detection	UV, λ = 220 nm; 0.5 AUFS (A) and λ = 215 nm; 0.04 AUFS (B)
Recording	Hewlett-Packard 3352B laboratory data system
Retention times, min	(estimated from drawing)

Column	A.	B.
Pye-Products 1, 2, 3	2.5, 3.0, 5.2	1.25, 1.5, 1.75
Oxytocin	7.0	2.2
Pye-Products 4, 5	8.4, 13.0	2.88 (4)
Trichlorobutanol	18.0	5.75

Ref.	Krummen K and Frei RW (1977) J. Chromatogr. 132:429

HPLC 4

Sample	Raw materials from synthetic oxytocin
Apparatus	Hewlett-Packard 1010B LC, Rheodyne M 905-42 loop injector and Perkin-Elmer LC 55 UV and Schoeffel M 770 spectrophotometer detectors
Column	Reversed-phase; Lichrosorb C_8, 5 μm, 150 × 3.2 mm ID
Mobile phase	A. 0.015 M Phosphate buffer, pH 3.0/ acetronitrile (82:18) B. 0.015 M Phosphate buffer, pH 7.0/ acetonitrile (82:18)
Flow-rate	0.7 ml min^{-1}; 2030 psi
Temperature	Ambient (20–22 °C)
Injection size	20 μl

Detection	UV, λ = 215 nm
Recording	Hewlett-Packard 3353 laboratory data system
Retention times, min	(estimated from drawing)
	Mobile phase A. B.
	Oxytocin 7.5 11.7
	Nonapeptideamide 10.0 13.0
Ref.	Nachtmann F (1979) J. Chromatogr. 391:391

HPLC 5

Sample	Pharmaceutical preparations (injectable)
Apparatus	Seven different HPLC systems, samplers and detectors
Column	Reversed-phase; Hypersil C_{18}, 5 μm, 125 × 4.6 mm ID
Mobile phase	A. 50% Acetonitrile B. 0.1 M Sodium dihydrogen phosphate Gradient: 30% A to 60% A in 30 min
Flow-rate	1 ml min^{-1}
Temperature	Ambient
Injection size	1–200 μl containing 1–200 IU ml^{-1} and (2–400 μg ml^{-1})
Detection	UV, λ = 220 nm
Retention times, min	(estimated from drawing)
	Oxytocin 9.0
	[D-Gln^4]-oxytocin 9.8
	[D-Tyr^2]-oxytocin 11.0
	[D-Tyr^2-D-Gln^4]-oxytocin 12.4
	Oxytocin-α-dimer 15.5
	Oxytocin-β-dimer 18.5
	Chlorobutanol 20.0
Ref.	Maxl F and Siehr W (1989) J. Pharm. Biomed. Anal. 7:211

145 Pantothenic Acid

CAS-No	79-83-4; 137-08-6 (calcium pantothenate)
Chemical name	(+)-(*R*)-3-(2,4-Dihydroxy-3,3-dimethyl butyramido)propionic acid
Empirical formula	Pantothenic Acid: $C_9H_{17}NO_5$; Calcium Pantothenate: $(C_9H_{16}NO_5)_2Ca$
Molecular mass	219.2; 476.5
Proprietary names	Cal-Pan; Cantothen Injection; Pantholin

Clinical use	Vitamin			
Solubility,	Water	ethyl acetate	ether	chloroform
Pantothenic acid	freely soluble	freely soluble	soluble	insoluble
Ca-pantothenate	1 in 3		insoluble	insoluble

HPLC 1

Sample	Multivitamin preparations and raw materials
Apparatus	Beckman M 110A LC pump, Rheodyne M 7126 loop injector and Kratos M 773 UV detector
Column	Reversed-phase; Hypersil C_{18}, ODS, 5 µm, 150 × 4.6 mm ID
Mobile phase	0.25 M Sodium dihydrogen phosphate, pH 2.5 with phosphoric acid/acetonitrile (97:3, v/v)
Flow-rate	2.0 ml min^{-1}
Temperature	Ambient
Injection size	20 µl, 0.02–0.1 mg ml^{-1}
Detection	UV, $\lambda = 205$ nm; 0.1 AUFS
Recording	Hewlett-Packard 3357 digital computer
Retention times, min	(estimated from drawing) Pantothenate 5.0; Saccharin 7.0; 2-Sulfamoyl-benzoic acid 7.6
Ref.	Timmons JA, Meyer JC et al (1987) J. Assoc. Off. Anal. Chem. 70:510

HPLC 2

Sample	Multivitamin preparations
Apparatus	Waters M 6000A HPLC pump with R 400 differential refractometer
Column	Reversed-phase; Nucleosil 7, C_{18}, 200 mm
Mobile phase	Glacial acetic acid/water (50:950, v/v)
Flow-rate	2 ml min^{-1}
Temperature	35 °C (detector)
Injection size	100 µl containing 50–1000 µg ml^{-1}
Detection	RI; 1–128 FS
Recording	Hewlett-Packard 3390A integrator
Retention time, min	Calcium pantothenate 4.50
Ref.	Jonvel P, Andermann G and Barthelemy JF (1983) J. Chromatogr. 281:371

HPLC 3

Sample	Multivitamin preparations
Apparatus	HPLC-low-pressure ternary gradient system
Column	Reversed-phase; Lichrosorb C_{18}, 250 × 4 mm ID
Mobile phase	Aqueous phosphoric acid, pH 3.0/methanol (95:5, v/v) plus 2 ml triethylamine per liter

Flow-rate	2.0 ml min^{-1}; 2000 psi
Injection size	20 µl containing 20 µg
Detection	UV, λ = 270 nm; 0.08 AUFS
Retention times, min	(estimated from drawing)

Thiamine	1.68	Nicotinic acid	3.35
Ascorbic acid	2.15	Nicotinamide	5.39
Pyridoxine	2.63	Calcium-D(+)-pantothenate	9.09

Ref. Merck Application Manual on HPLC, TLC and HPTLC (1986). E. Merck, Darmstadt, p 12

HPLC 4

Sample	Multivitamin preparations and Vitamin B-complex tablets
Apparatus	Waters M 6000A LC pumps with 710B autosampler and Perkin-Elmers LC-75 UV detector
Column	Aminopropyl-phase; Lichrosorb NH_2 10 µm, 250 × 4.6 mm ID
Mobile phase	0.005 M Monobasic potassium phosphate, pH 4.5/acetonitrile (13:87, v/v)
Flow-rate	2 ml min^{-1}
Temperature	25 °C
Injection size	10 µl, 50–500 µg ml^{-1}
Detection	UV, λ = 200–225 nm
Recording	Waters M 730 data module
Retention times, min	(estimated from drawing)

Sample	Multivitamin-multimineral	B-complex tablets
Niacinamide	1.8	1.8
Vitamin B_6	2.0	2.0
Vitamin B_2	3.3	2.3
Vitamin B_1	4.8	3.7
Unknown	5.7	4.3
Unknown	9.0	
Pantothenic acid		6.2
Unknown		17.5

Ref. Hudson TJ and Allen RJ (1984) J. Pharm. Sci. 73.113

146 Paracetamol

CAS-No	103-90-2
Chemical name	4′-Hydroxyacetanilide

Empirical formula	$C_8H_9NO_2$
Molecular mass	151.2
Proprietary names	Acephen; Ben-u-ron; Valadol
Clinical use	Analgesic
Solubility	Water 1 in 70; alcohol 1 in 7–10; acetone 1 in 13; glycerol 1 in 40
Light absorption	λ_{max} = 245 nm in aqueous acid

HPLC 1

Sample	Synthetic mixture
Apparatus	Perkin-Elmer LC-pump, gradient controller and LC-55B UV detector,
Column	Reversed phase; ODS 10 μm, C_{18}, 250×4.6 mm ID
Mobile phase	Methanol/Water (30:70) Gradient: 30% Methanol to 95% methanol in 13 min
Flow-rate	1.8 ml min^{-1}
Temperature	Ambient
Internal standard	Phenacetin
Injection size	10 μl
Detection	UV, λ = 270 nm; 0.2–0.4 AUFS
Recording	Perkin-Elmer strip-chart recorder and integrator
Retention times, min	(estimated from drawing)

Paracetamol	4.0	Internal standard	9.6
Caffeine	6.4	Ethylparaben	11.4
Salicylamide	8.0	Propyphenazone	12.2

Ref. Mamolo MG, Vio L and Maurich V (1985) J. Pharm. Biomed. Anal. 3, 157

HPLC 2

Sample	Elixir
Apparatus	Du pont M 850 HPLC, autosampler, Waters M 440 UV detector and Hewlett-Packard 1040A photodiode array detector
Column	Reversed phase; μ-Bondapak TM C_{18}, 300×3.9 mm ID
Mobile phase	Buffer (containing 2.4 g 1-butanesulfonic acid sodium, 2.04 g KH_2PO_4 and 2 ml triethylamine per liter water for pH 4.8)/methanol (80:20, v/v)
Flow-rate	2 ml min^{-1}
Temperature	50 °C
Injection size	20 μl
Detection	UV, λ = 214 nm; 0.08 AUFS
Recording	Du Pont M 4100 integrator

Retention times, min	Paracetamol 2.6 Sodium benzoate 3.8 Codeine phosphate 4.7
Ref.	Sisco WR, Rittenhouse CT et al (1986) J. Chromatogr. 354, 355

HPLC 3

Sample	Biological fluids
Apparatus	Waters M 6000A LC and 45 pumps with gradient programmer and Hitachi M 220-S UV detector
Column	Reversed phase; Lichrosorb C_8, 5 μm, 250 × 4.6 mm ID or μ-Bondapak C_{18}, Radial-Pak
Pre-column	30 μm, C_8, 50 × 4.6 mm ID
Mobile phase	A: 10 mM KH_2PO_4, pH 3.5 B: 60% (v/v) Acetonitrile in 10 mM KH_2PO_4, pH 3.5 Gradient: 10% B to 80% B in 5 min; hold up to 10 min
Flow-rate	1.5 ml min^{-1}
Temperature	25 °C
Injection size	25 μl
Detection	UV, λ = 215 nm
Retention times, min	Isoniazid 2.6 Pyrazinamide 3.3 Paracetamol 5.3
Ref.	Chan K (1988) J. Liq. Chromatogr. 11, 1051

HPLC 4

Sample	Solid dosage forms
Apparatus	Waters M 6000A LC pump, 480 variable-wavelength UV detector and 710B autosampler
Column	Radialpak Cyanopropylsilane; 10 μm, 100 × 8 mm ID and cyano-phase; μ-Bondapak CN, 10 μm, 300 × 3.9 mm ID
Mobile phase	Water/acetonitrile/1 M dibutylamine phosphate (90:10:1)
Flow-rate	1.5 ml min^{-1}
Internal standard	Hydrocodone, 0.05 mg ml^{-1}
Injection size	5–50 μl, 2–8 mg ml^{-1}
Detection	UV, λ = 215 nm
Recording	Hewlett-Packard 3390A integrator and Fisher Recordall 5000 chart-recorder
Retention times, min	(estimated from drawing) Paracetamol 1.9 Internal standard 7.2
Ref.	Fatmi AA and Williams Sr GV (1987) J. Liq. Chromatogr. 10, 2461

HPLC 5

Sample	Raw materials and serum
Apparatus	Waters M 6000 LC pump, Pye-Unicam LC-UV detector and Rheodyne M 7125 injection valve
Column	Reversed-phase; Hypersil ODS 5 μm, 160 × 5 mm ID
Mobile phase	Isopropanol/0.1 M potassium dihydrogen phosphate/formic acid (17:1000:1, v/v)
Flow-rate	2.0 ml min^{-1}
Internal standard	N-Propionyl-*p*-aminophenol
Injection size	10 μl
Detection	UV, λ = 240 nm; 0.01 A
Retention times, min	(estimated from drawing)
	Paracetamol glucuronide 1.4
	Paracetamol sulfate 2.9
	Paracetamol cysteine 3.0
	Paracetamol 4.0
	Internal standard 8.5
	Paracetamol mercapturide 9.8
Ref.	Stevens HM and Gill R (1986) J. Chromatogr. 370:39

HPLC 6

Sample	Blood plasma
Extraction	Diethyl ether/hexane (1:1, v/v) under alkaline condition and then with ethyl acetate
Apparatus	Varian M 5000 HPLC with UV detector and Waters M 710B autosampler
Column	Reversed-phase; Spherisorb 5 ODS 5 μm, 250 × 4.6 mm ID
Mobile phase	190 ml Methanol and 810 ml 1/15 M potassium dihydrogen phosphate, pH 5.0
Flow-rate	2.0 ml min^{-1}
Temperature	30 °C
Internal standard	Salicylamide
Injection size	30 μl, 0.75–3.02 μg ml^{-1}
Detection	UV, λ = 254 nm; 0.002 AUFS
Retention times, min	(estimated from drawing)
	Paracetamol 2.2
	Internal standard 5.0
Ref.	Kinney CD and Kelly JG (1987) J. Chromatogr. Biomed. Appl. 419:433

HPLC 7

Sample	Post-mortem tissues and urine
Extraction	Diethyl ether and re-extraction with dichloromethane

Apparatus	Hewlett-Packard 1080 HPLC with UV detector
Column	Reversed-phase; Microbore C_{18}, 100 × 4.6 mm ID
Mobile phase	Acetonitrile/dilute phosphoric acid (7:93)
Flow-rate	0.3 ml min^{-1}
Temperature	40 °C
Internal standard	2-Acetamidophenol
Detection	UV, λ = 271 nm
Retention times, min	Paracetamol 2.43 Theophylline 3.19 Internal standard 4.87
Ref.	Mathis DF and Budd RD (1988) J. Chromatogr. 439:466

HPLC 8

Sample	Blood plasma
Extraction	Ethyl acetate
Apparatus	LKB M 2150 LC pump, ISS-100 autoinjector and LCA-15 electrochemical detector
Column	Reversed-phase; Hypersil 5 ODS 5 µm, 100 × 4.6 mm ID
Mobile phase	0.1 M Sodium acetate and 0.001 M ammonium acetate/methanol (77:23), pH 4.75 with glacial acetic acid
Flow-rate	2.0 ml min^{-1}
Internal standard	N-Propionyl-*p*-aminophenol
Injection size	10 and 25 µl
Detection	EC, + 0.80 V vs Ag/AgCl RE
Recording	Hewlett-Packard 3388A integrator
Retention times, min	Paracetamol 1.4 Internal standard 2.2
Ref.	Palmer JL (1986) J. Chromatogr. Biomed. Appl. 382:338

HPLC 9

Sample	Tablets
Apparatus	LC with UV detector
Column	Reversed-phase; ODS 10 µm, 300 × 4.6 mm ID
Mobile phase	Methanol/0.75% acetic acid (1:3)
Flow-rate	2 ml min^{-1}
Temperature	Ambient
Injection size	20 µl
Detection	UV, λ = 280 nm; 1.28 AUFS
Ref.	Krieger DJ (1987) J. Assoc. Off. Anal. Chem. 70:212

147 Penicillin V (Phenoxymethylpenicillin)

CAS-No	87-08-1; 132-98-9 (potassium salt)				
Chemical name	(6*R*)-6-(2-Phenoxyacetamido) penicillanic acid				
Empirical formula	Penicillin V	Potassium Salt			
	$C_{16}H_{18}N_2O_5S$	$C_{16}H_{17}KN_2O_5S$			
Molecular mass	350.4	388.5			
Proprietary names	Isocillin; Oracilline; Uticillin				
Clinical use	Antibiotic				
Solubility,	Water	ethanol	acetone	chloroform	ether
Penicillin V	1 in 1700	1 in 7	1 in 6	soluble	
Potassium salt	1 in 1.5	1 in 150	insoluble		insol
Light absorption	λ_{max} = 268 nm in water (potassium salt)				

HPLC 1

Sample: Raw materials

Apparatus: Konik M KNK-500A LC, Rheodyne M 7125 loop injector and Waters M 441 UV detector

Column: Reversed-phase; Spherisorb ODS 10 μm, 250 × 4.9 mm ID

Pre-column: Reversed-phase; Spherisorb ODS 10 μm, 30 × 4.6 mm ID

Mobile phase: Aqueous 25 mM KH_2PO_4 and 25 mM K_2HPO_4/isopropanol (80:20, v/v), pH 7.25

Flow-rate: 1.0 ml min^{-1}

Temperature: 15, 25, 35, 45 and 55 °C

Injection size: 20 μl

Detection: UV, λ = 229 nm

Recording: Varian 4290 computing integrator

Retention times, min

Temperature	15	25	35	45	55 °C
Amoxicillin	2.56	2.47	2.40	2.36	2.31
Ampicillin	3.51	3.24	3.07	2.94	2.84
Piperacillin	5.49	4.18	3.66	3.29	3.01
Penicillin G	6.12	5.01	4.37	3.89	3.58
Penicillin V	10.75	7.58	6.14	5.05	4.36
Cloxacillin	16.49	10.78	8.25	6.47	5.41

Ref.: Martin J, Mendez R and Negro A (1988) J. Liq. Chromatogr. 11:1707

HPLC 2

Sample: Raw materials

Apparatus: Haskel Engineering M 26980-4 LC pump, Altex M 153 ultraviolet detector and Rheodyne M 70-10 injector

Column	Reversed-phase; Lichrosorb 5 μm, C_8, 150 × 4.6 mm ID
Mobile phase	Methanol/0.01 M sodium dihydrogen phosphate (35:65)
Flow-rate	1.0 ml min^{-1}; 1523 psi
Temperature	Ambient
Injection size	20 μl containing 4.0 μg
Detection	UV, λ = 225 nm; 0.1 AUFS
Retention times, min	Ampcillin 3.8 Penicillin V 17.5 Methicillin 7.9 Oxacillin 21.8 Penicillin G 9.7 (estimated from drawing)
Ref.	White ER and Zarembo JE (1981) J. Antibiotics 34:836

HPLC 3

Sample	Tablets and powders for oral solution
Apparatus	Waters M 510 LC pump, U6K injector valve and 481 variable-wavelength detector
Column	Reversed-phase; μ-Bondapak C_{18}, 10 μm, 300 × 3.9 mm ID
Pre-column	Co-Pell ODS LC, 40 mm (Whatman)
Mobile phase	Acetonitrile/methanol/0.01 M KH_2PO_4 (21:4:75, v/v)
Flow-rate	1 ml min^{-1}
Internal standard	Sulfadimethoxine
Injection size	20 μl containing 25–225 μg ml^{-1}
Detection	UV, λ = 225 nm; 0.1 AUFS
Recording	Waters M 730 integrating recorder
Retention times, min	Degradation products (6) 2.0–13.0 Penicillin V 9.0–11.0 Internal standard 13.0–15.0 (estimated from drawing)
Ref.	Mopper B (1987) J. Assoc. Off. Anal. Chem. 70:39

HPLC 4

Sample	Penicillins
Apparatus	HPLC low-pressure ternary gradient system
Column	Reversed-phase; Lichrosorb C_{18}, 125 × 4 mm ID
Mobile phase	Acetonitrile/0.1 M sodium dihydrogen phosphate (30:70, v/v)
Flow-rate	1.5 ml min^{-1}; 2900 psi
Injection size	20 μl containing 3 μg
Detection	UV, λ = 220 nm; 0.04 AUFS
Retention	Amoxicillin, penicillin V, penicillin G, nafcillin, in ascending order (from drawing)

Ref.	Merck Application Manual on HPLC, TLC and HPTLC (1986). E. Merck, Darmstadt, p 71

HPLC 5

Apparatus	LDC modular liquid chromatograph with 1285 UV monitor and Milton Roy minipump with LDC 709 pulse dampener
Column	Anion exchange; Du Pont, 1000 × 2.1 mm ID
Mobile phase	0.02 M Sodium nitrate in 0.01 M sodium borate, pH 9.15
Flow-rate	0.45 ml min^{-1}; 500 psi
Injection size	2 μl
Detection	UV, λ = 254 nm
Retention times, min	(estimated from drawing) Penicillin G 9.5 Ampicillin 6.5 Penicillin V 15.0
Ref.	Tsuji K and Robertson JH (1975) J. Pharm. Sci. 64:1542

148 Pentobarbital

CAS-No	76-74-4; 57-33-0 (sodium salt)				
Chemical name	5-Ethyl-5-(1-methylbutyl)barbituric acid				
Empirical formula	Pentobarbital	Sodium Salt			
	$C_{11}H_{18}N_2O_3$	$C_{11}H_{17}N_2NaO_3$			
Molecular mass	226.3	248.3			
Proprietary names	Mintal; Nembutal; Repocal				
Clinical use	Hypnotic				
Solubility,	Water	alcohol	acetone	ether	methanol
Pentobarbital	insoluble	1 in 4.5	1 in 1	1 in 10	1 in 1
Sodium salt	1 in 1	1 in 1		insoluble	
Light absorption	λ_{max} = 239 nm in 0.05 M borax buffer pH 9.2				

HPLC 1

Sample	Serum
Extraction	Protein precipitation with acetonitrile
Apparatus	Beckmann M 110A LC-pump, Micrometrics M 785 UV detector and Rheodyne injection valve
Column	PRP-1; 5 μm, 150 × 4.6 mm
Mobile phase	A. Acetonitrile/15 mmol phosphate buffer, pH 11.8 (12:88) B. Acetonitrile/15 mmol phosphate buffer, pH 6.7 (40:60)

Injection size	2 μl, 5–50 μg ml^{-1}
Detection	UV, λ = 254 nm; 0.005 AUFS
Retention times, min	Pentobarbital 6.0 (A and B)
Ref.	Shihabi ZK, Dyer RD and Scaro J (1987) J. Liq. Chromatogr. 10:663

HPLC 2

Sample	Body fluids
Extraction	Blood: Deproteinization with 40% perchloric acid Urine: With diethyl ether under acidic condition
Apparatus	Waters M 510 LC pump, 490 UV/VIS detector and WISP 712 autosampler
Column	Reversed-phase; μ-Bondapak C_{18}, 10 μm, 300 × 3.9 mm ID
Mobile phase	Acetonitrile/water (35:65, v/v)
Flow-rate	1 ml min^{-1}
Temperature	Ambient
Internal standard	Aprobarbital and methaqualone
Injection size	20 μl, 2–20 μg ml^{-1}
Detection	UV, λ = 198 nm; 0.1–0.01 AUFS and 230 nm (thiopental and thialbarbital)
Recording	Waters M 840 data and chromatography control station

Retention times, min

Proxibarbital	3.80	Amobarbital	11.20
Barbital	4.50	Pentobarbital	11.50
Mephebarbital	6.30	Vinylbital	12.35
Phenobarbital	7.10	Secobarbital	13.50
Cyclobarbital	7.30	Aprobarbital	13.60
Butobarbital	7.48	Methaqualone	17.00
Butabarbital	8.00	Thialbarbital	20.50
Butalbital	8.00	Thiopental	21.00
Heptabarbitol	9.70	Methohexital	
Hexobarbital	10.14	sodium	27.0

Ref.	Rop PP, Spinazzola J et al (1988) J. Chromatogr. Biomed. Appl. 427:172

HPLC 3

Sample	Barbiturate injections
Apparatus	Waters M 6000A LC pump, LDC Spectromonitor III ultraviolet detector and Rheodyne M 70-10 injector
Column	Reversed-phase; Partisil ODS-3, C_{18}, 10 μm, 250–300 × 4.0 mm ID
Mobile phase	Acetate buffer (containing 4.1 g sodium acetate and 3 ml acetic acid in water to 1000 ml)/methanol/propylene glycol (45:55:4)
Flow-rate	2 ml min^{-1}
Injection size	20 μl, 0.91 mg ml^{-1}

Detection	UV λ = 230 nm; 0.4 AUFS
Retention times, min	(estimated from drawing)

2-Ethyl-2-(1-methylbutyl) propanediamide	1.8	Pentobarbital	3.0
N-(Aminocarbonyl)-2-carboxy-2-ethyl-3-methylhexanamide	2.0	N-(Aminocarbonyl) 2-ethyl-3-methyl-hexanamide	5.0

Ref.	Reif VD, Kaufmann KL et al (1986) J. Pharm. Sci. 75:714

HPLC 4

Sample	Blood plasma
Extraction	Chloroform under acidic condition
Apparatus	Waters M 501 LC pump, 490 ultraviolet detector and manual injector
Column	Reversed-phase; μ-Bondapak C_{18}, 5 μm, 150 × 3.9 mm ID
Pre-column	Reversed-phase; μ-Bondapak C_{18}, 10 μm
Mobile phase	Water/acetonitrile/1.75 M phosphoric acid (798:200:2, v/v) at pH 3.4
Flow-rate	0.8 ml min^{-1}; 1000–200 psi
Internal standard	Barbital
Injection size	25 μl, 1–20 μg ml^{-1}
Detection	UV, λ = 195 nm
Recording	Professional digital computer
Retention times, min	

Barbital	3.9	Butabarbital	9.8–10.5
Allobarbital	6.2–6.3	Butalbital	11.9–12.3
Phenobarbital	9.0–9.7	Pentobarbital	18.9–19.0

Ref.	Lopez-Rivadulla M, Fernandez P et al (1988) Anal. Lett. 21:2253

149 Phenacetin

CAS-No	62-44-2
Chemical name	4′-Ethoxyacetanilide
Empirical formula	$C_{10}H_{13}NO_2$
Molecular mass	179.2
Proprietary names	Coriforte; Fiorinal; Percodan
Clinical use	Analgesic

Solubility	Water	ethanol	chloroform	ether
	1 in 1300	1 in 20	1 in 20	slightly soluble

Light absorption	λ_{max} = 244 nm in aqueous acid

HPLC 1

Sample	Raw materials
Apparatus	Waters M 6000 LC pump and Cecil CE-212 variable-wavelength UV detector
Column	Reversed-phase; μ-Bondapak C_{18}, ODS 10 μm, 300 × 4 mm ID
Mobile phase	55% Methanol in 0.025 M phosphate buffer for pH 7.0
Flow-rate	2 ml min^{-1}
Temperature	Ambient
Injection size	2 μl containing 1–10 μg
Detection	UV, $\lambda = 220$ nm

Retention times, min

Acetylsalicylic acid	1.3	Phenytoin	2.6
Paracetamol	1.7	Glutethimide	2.8
Barbital	1.8	Quinalbarbital	3.3
Phenobarbital	2.0	Chlordiazepoxide	4.6
Phenacetin	2.3	(estimated from drawing)	

Ref. Twitchett PJ and Moffat AC (1975) J. Chromatogr. 111:149

HPLC 2

Sample	Human blood plasma and urine
Extraction	Serum: Deproteinization with 5% perchloric acid Urine: Diluting with water
Apparatus	Spectra-Physics 8100 LC system, SP 8320 UV detector and 8110 autosampler with Valco-Valve injector
Column	Reversed-phase; Lichrosorb C_{18}, 5 μm, 250 × 4 mm ID
Mobile phase	I. 1% Glacial acetic acid/0.1 M KH_2PO_4 (3:97) II. Methanol/1% acetic acid/0.1 M KH_2PO_4 (33:3:64)
Flow-rate	1.2 ml min^{-1}
Temperature	45 °C
Internal standard	4-Fluorophenol
Injection size	100 μl, 0.01–1.3 μg ml^{-1}
Detection	UV, $\lambda = 254$ nm
Recording	Spectra-Physics 4100 computing integrator

Retention times, min (estimated from drawing)

Mobile Phase	I.	II.
Phenacetin		9.0
Acetaminophen glucuronide	4.0	
Acetaminophen sulfate	9.0	
Acetaminophen cysteine	11.0	
Acetaminophen	13.3	
Acetaminophen mercapturic acid	26.6	
Internal standard	30.0	14.2

Ref.	Mineshita S, Eggers R et al (1986) J. Chromatogr. Biomed. Appl. 380:407

HPLC 3

Sample	Raw materials
Apparatus	Waters M 6000A LC U6K manual loop injector and Jasco Uvidec-100-III detector
Column	Normal phase; Lichrosorb Si 100 TM, 5 μm, 150 × 4.5 mm ID
Mobile phase	Hexane/ethyl acetate (50:50, v/v)
Flow-rate	0.9 ml min^{-1}
Internal standard	Benzophenone, 0.01 mg ml^{-1}
Injection size	10 μl containing 5–150 ng
Detection	UV, λ = 247 nm
Retention times, min	Benzophenone 0.40; *p*-Phenetidine 1.8; Acetanilide 1.30; Phenacetin 2.0; *p*-Chloracetanilide 1.50 (estimated from drawing)
Ref.	Quaglia MG, Mazzeo P and Lopez A (1985) J. Pharm. Biomed. Anal. 3:187

HPLC 4

Sample	Human blood plasma
Extraction	Diethyl ether at pH 6.5
Apparatus	Hitachi M 635 LC with UV detector and Rheodyne M 7125 injector
Column	Reversed-phase; Lichrosorb C_{18}, 5 μm, 250 × 4 mm ID
Mobile phase	6.5% Tetrahydrofuran and 33.5% methanol in water
Flow-rate	1.0 ml min^{-1}
Temperature	45 °C
Injection size	20 μl, 2.5–100 μg ml^{-1}
Detection	UV, λ = 210 nm; 0.08 AU
Recording	System Instruments 5000 E integrator
Retention times, min	(estimated from drawing) Chloramphenicol 7.8; Primidone 4.0; Carbamazepine 8.7; Phenacetin 6.0; Phenytoin 13.0; Phenobarbital 6.9
Ref.	Kushida K, Chiba K and Ishizaki T (1983) Ther. Drug Monit. 5:127

HPLC 5

Sample	Raw materials and dosage forms
Apparatus	Hewlett-Packard 1084 LC with UV detector
Column	Reversed-phase; Nucleosil C_{18}, 10 μm, 250 × 4 mm ID
Mobile phase	Acetonitrile/0.05 M ortho-phosphoric acid (25:75)

Flow-rate	2.0 ml min^{-1}		
Temperature	40 °C		
Internal standard	Caffeine		
Injection size	5–30 µl		
Detection	UV, λ = 249 nm		
Recording	Hewlett-Packard LC-Terminal 79850B integrator		
Retention times, min	Codeine phosphate 1.49	Acetyl salicylic acid 2.84	
	Caffeine 2.13	Phenacetin 4.13	
Ref.	Rehm KD and Steinigen M (1981) Pharm. Ztg. 126:99		

150 Phenazone (Antipyrine)

CAS-No	60-80-0			
Chemical name	1,5-Dimethyl-2-phenyl-4-pyrazolin-3-one			
Empirical formula	$C_{11}H_{12}N_2O$			
Molecular mass	188.2			
Proprietary names	Auralgicin; Auraltone; Sedonan			
Clinical use	Analgesic			
Solubility	Water	ethanol	chloroform	ether
	1 in 1	1 in 1	1 in 1	1 in 50
Light absorption	λ_{max} = 230 nm in aqueous acid			

HPLC 1

Sample	Blood plasma	
Extraction	Methanol	
Apparatus	Milton Roy LC pump and Waters U6K injector and 440 UV detector	
Column	Reversed-phase; µ-Bondapak C_{18}, 10 µm, 300 × 4 mm ID	
Mobile phase	Acetonitrile/acetic acid 1% in water (35:65)	
Flow-rate	1 ml min^{-1}; 986 psi	
Temperature	Ambient (21–24 °C)	
Internal standard	Benzoic acid	
Injection size	25 µl, 20 µg ml^{-1}	
Detection	UV, λ = 254 nm	
Retention times, min	Antipyrine	5.5
	Benzoic acid	7.7
Ref.	Campbell TM, Murdaugh EW and Killenberg PG (1979) J. Chromatogr. Biomed. Appl. 163:236	

HPLC 2

Sample	Blood plasma and saliva
Extraction	Chloroform under acidic condition
Apparatus	Waters M 6000A LC pump, WISP 710B autosampler and 480 UV detector
Column	Normal phase; Lichrosorb Si 60, 5 μm, 150 × 4.8 mm ID
Mobile phase	Ethanol/20% ammonia/chloroform (1.6:0.05:98.35)
Flow-rate	4 ml min^{-1}; 2175 psi
Internal standard	Dimethyldiazepam
Injection size	50 μl, 0–30 μg ml^{-1}
Detection	UV, λ = 280 nm; 0.02 AUFS
Retention times, min	Internal standard 1.2
	Antipyrine (Phenazone) 2.4
Ref.	Doare L, Colin C and Diquet B (1983) J. Chromatogr. Biomed. Appl. 274:403

HPLC 3

Sample	Biological fluids
Extraction	Dichloromethane/n-pentane (3:7, v/v), dichloromethane/isopropanol (9:1, v/v) and chloroform/ethanol (9:7, v/v)
Apparatus	Waters M 45 LC pump, WISP 710B autosampler and 440 UV detector
Column	Reversed-phase; MOS-Hypersil 5 μm, 100 × 3 mm ID
Mobile phase	A. 0.02 M Phosphate buffer, pH 7.2/acetonitrile (100:10) with sodium pyrosulfite, 2 g l^{-1} B. 0.02 M Phosphate buffer, pH 6.5/acetonitrile (100:8) for urine C. 0.0067 M Phosphate buffer, pH 7.2/acetonitrile (100:18) for blood plasma and saliva
Flow-rate	2.0 ml min^{-1}; 1958 psi
Internal standard	Phenacetin
Injection size	15–25 μl
Detection	UV, λ = 254 nm
Recording	Hewlett-Packard 3390 integrator

Retention times, min

Mobile phase	A.	B.	C.	D.
Sample	Human urine	standard solution	rat urine	plasma, saliva
3-Hydroxymethylantipyrine	0.74	0.89	1.36	
Norantipyrine	1.04	1.41	3.85	
Antipyrine (Phenazone)	1.71	2.14	4.40	1.06
4-Hydroxyantipyrine	2.86	3.35	6.17	
Internal standard	4.29	4.89	8.14	2.31
4,4-Dihydroxyantipyrine			2.43	

Ref.	Teunissen MWE, Meerburg-van der Torren JE et al (1983) J. Chromatogr. Biomed. Appl. 278:367

HPLC 4

Sample	Urine
Extraction	Ethyl acetate under acidic condition
Apparatus	Waters M 6000A LC pump, 712 WISP autosampler and Altex M 153 UV detector
Column	Reversed-phase; μ-Bondapak C_{18}, 10 μm, 150 × 3.9 mm ID
Pre-column	Reversed-phase; Supelguard C_{18}, 5 μm, 20 × 4.6 mm ID
Mobile phase	7.5% Acetonitrile, 92.5% 0.1 M sodium acetate and 0.5 mM diethylamine (v/v), pH 6.6 with glacial acetic acid
Flow-rate	2.0 ml min^{-1}
Internal standard	Phenacetin
Injection size	20 μl
Detection	UV, $\lambda = 254$ nm
Recording	Spectra-Physics 4290 plotting-integrator
Retention times, min	Hydroxymethyl-antipyrine 3.71; Norantipyrine 5.08; Antipyrine 8.49; 4-Hydroxyanti-pyridine 13.07; Internal standard 18.93
Ref.	St. Peter JV and Awani WM (1989) J. Chromatogr. Biomed. Appl. 494:424

151 Pheniramine

CAS-No	86-21-5; 132-20-7 (maleate)			
Chemical name	*NN*-Dimethyl-3-phenyl-3-(2-pyridyl)propylamine			
Empirical formula	Pheniramine	Maleate		
	$C_{16}H_{20}N_2$	$C_{16}H_{20}N_2, C_4H_4O_4$		
Molecular mass	240.3	356.4		
Proprietary names	Avil; Inhiston; Trimeton			
Clinical use	Antihistamine			
Solubility,	Water	ethanol	chloroform	ether
Pheniramine	insoluble	soluble	soluble	soluble
Maleate	1 in 0.3	1 in 2.5	1 in 1.5	very slightly sol
Light absorption	$\lambda_{max} = 266$ nm in aqueous acid			

HPLC 1

Sample	Dosage forms (tablets, capsules, solutions and sprays)
Apparatus	Kratos spectroflow M 400 LC pump, Rheodyne M 7125 manual injector and Milton Roy/LDC spectromonitor D
Column	β-Cyclodextrin; Cyclobond I, 100 × 4.6 mm ID
Mobile phase	0.05 M NaH_2PO_4, pH 7.0 with 0.1 M NaOH/ methanol (70:30)
Injection size	50 µl, 10-80 µg ml^{-1}
Detection	UV, λ = 254 nm
Recording	Hewlett-Packard 3392A integrator
Retention times, min	Phenylpropanolamine-HCl 2.0 Pheniramine maleate 10.0 Pyrilamine maleate 14.0
Ref.	El Gezawi S, Omar N et al (1988) J. Pharm. Biomed. Anal. 6:393

HPLC 2

Sample	Cough Syrup
Apparatus	Spectra-Physics 8000 LC with gradient controller and 8440 UV-VIS detector
Column	Hamilton PRP-1, 305 × 7 mm ID
Mobile phase	A. 0.1% Triethylamine in water B. 0.1% Triethylamine in acetonitrile Gradient: B 40% for 7 min; B 40% to 80% in 5 min; run 5 min; B 80% to 40 % in 6 min; run 2 min
Flow-rate	3.5 ml^{-1}
Temperature	40 °C
Injection size	500 µl
Detection	UV, λ = 254 nm; 0.16 AUFS
Recording	Spectra-Physics 4270 integrator
Retention times, min	Phenylephrine 2.0 Guaifenesin 2.8 Phenylpropanol-amine 3.2 Hydrocodone 6.9 Pheniramine 12.0 Etafedrine 12.4 Doxylamine 13.6 Pyrilamine 16.0 Diphenylpyraline 16.6
Ref.	Black DB, By AW and Lodge BA (1986) J. Chromatogr. 358:438

HPLC 3

Sample	Blood plasma
Extraction	Methylene chloride
Column	Normal phase; Lichrosorb Si 60, 5 µm, 250 × 2.16 mm ID

Mobile phase	0.1% (v/v) Acetic acid and 20 % (v/v) 0.1 M aqueous ammonium acetate in acetonitrile
Flow-rate	1 ml min^{-1}
Temperature	Ambient
Injection size	100 μl
Detection	UV, λ = 254 nm; 0.02 AUFS
Retention times, min	(estimated from drawing) Procainamide 4.9 Pheniramine 3.8 N-Acetyl-procainamide 5.8
Ref.	Butterfield AG, Cooper JK and Midha KK (1978) J. Pharm. Sci. 67:839

152 Phenobarbital

CAS-No	50-06-0; 57-30-7 (sodium salt)				
Chemical name	5-Ethyl-5-phenylbarbituric acid				
Empirical formula	Phenobarbital	Sodium Salt			
	$C_{12}H_{12}N_2O_3$	$C_{12}H_{11}N_2NaO_3$			
Molecular mass	232.2	254.2			
Proprietary names	Fenemal; Gardenal; Luminal				
Clinical use	Hypnotic/Sedative				
Solubility,	Water	alcohol	ether	chloroform	aq NaOH
Phenobarbital	1 in 1000	1 in 10	1 in 40	1 in 40	soluble
Sodium salt	1 in 3	1 in 25	insoluble	insoluble	
Light absorption	λ_{max} = 239 nm in 0.05 M borax buffer, pH 9.2				

HPLC 1

Sample	Serum
Extraction	Chloroform/methanol (4:1)
Apparatus	Waters M 6000A HPLC pumps with U6K injector and 440 variable-wavelength detector
Column	Reversed-phase; μ-Bondapak C_{18}, 250×4.6 mm ID
Mobile phase	420 ml Methanol, 280 ml acetonitrile and 1300 ml phosphate buffer, pH 4.4 (containing 0.60 ml 1 M KH_2PO_4 and 0.2 ml of 0.9 M H_3PO_4 in 4 liter water, pH 4.4 with 0.1 M NaOH or 0.9 M H_3PO_4)
Flow-rate	2 ml min^{-1}
Temperature	Ambient
Internal standard	Hexobarbital
Injection size	20 μl, 2–80 μg ml^{-1}
Detection	UV, λ = 195 and 254 nm

Retention times, min	Primidone	2.5	Diphenylhydantoin	10.0
	Phenobarbital	4.5	Carbamazepine	12.0
	Hexobarbital	8.2	(estimated from drawing)	
Ref.	Gerson B, Bell F et al (1984) Clin. Chem. 30:105			

HPLC 2

Sample	Barbiturate injections
Apparatus	Waters M 6000A LC pump, LDC spectromonitor III UV detector and Rheodyne M 70-10 injector
Column	Reversed-phase; μ-Bondapak C_{18}, 10 μm, 250–300 × 4.0 mm ID
Mobile phase	Acetate buffer (4.1 g sodium acetate + 3 ml acetic acid in water to 1000 ml)/methanol/propylene glycol (65:35:4)
Flow-rate	2 ml min^{-1}
Injection size	20 μl, 1.1 mg ml^{-1}
Detection	UV, λ = 230 nm; 0.4 AUFS
Retention times, min	Ethylphenylpropanediamide 2.2 Phenobarbital 5.2 α-Phenylbutyric acid 7.8 N-(Aminocarbonyl)-2-phenylbutanamide 12.0 (estimated from drawing)
Ref.	Reif, VD, Kaufmann KL et al (1986) J. Pharm. Sci. 75:714

HPLC 3

Sample	Serum
Extraction	Ethyl acetate
Apparatus	Hewlett-Packard HPLC 1090 gradient programmer, Rheodyne injection system and UV detector
Column	Reversed-phase; Hypersil 5 μm ODS, 200 × 2.1 mm ID
Mobile phase	A. Perchloric acid (diluted with water for pH 4.0)/acetonitrile (9:1, v/v) B. Perchloric acid pH 4.0 as in A/acetonitrile (4:6, v/v) Gradient: B 15% to 45% in 9 min; B 45% to 15% in 0.5 min; run 2 min
Flow-rate	0.3 ml min^{-1}
Temperature	65 °C
Internal standard	Tolybarb
Injection size	1.5 μl, 15.90 μg ml^{-1}
Detection	UV, λ = 207 nm

Retention times, min	Primidone	4.00
	Pentobarbital	6.55
	Internal standard	9.51
	Phenytoin	10.79
Ref.	Juergens U (1987) J. Liq. Chromatogr. 10:507	

HPLC 4

Sample	Human brain
Extraction	Dichloromethane/propanol-2 (97:3) on Extrelut (R)-III extraction column after treating with acidic buffer
Column	Reversed-phase; ODS 5 μm, 200 × 2.1 mm ID
Mobile phase	A. Acetonitrile/0.05 M ammonium dihydrogen phosphate, pH 4.4 (10:90) B. Acetonitrile/0.05 M $NH_4H_2PO_4$, pH 4.4 (60:40) Gradient: At 0.5 min 15% B; at 9.5 min 45% B; at 10.0 min 45% B; at 10.5 min 15% B
Flow-rate	0.3 ml min^{-1}
Temperature	65 °C
Internal standard	Tolybarb
Injection size	10 μl, 2.39 μg ml^{-1}
Detection	UV, λ = 207 nm
Retention times, min	Phenobarbital 6.20; Carbamazepine-10,11-epoxide 7.34; Internal standard 9.05; Phenytoin 10.16; Carbamazepine 10.61
Ref.	Juergens U and Rambeck B (1987) J. Liq. Chromatogr. 10:1847

HPLC 5

Sample	Raw materials
Apparatus	Jasco Twincle or Trirotar SR2 HPLC and Uvidec-100 IV variable-wavelength detector
Column	I. Reversed-phase; 1G-ODS, 150 × 4.6 mm ID II. Reversed-phase; 2G-ODS, 150 × 4.6 mm ID
Mobile phase	Acetonitrile/buffer (containing 2 parts 0.01 M citric acid and 1 part 0.01 M Na_2HPO_4) (3:7)
Flow-rate	1.0 ml min^{-1}
Detection	UV, λ = 205 nm

Retention times, min

Column	I	II
Phenobarbital	3.8	4.5
Diphenylhydantoin	7.3	5.9
Sodium valproate	15.8	11.8

(estimated from drawing)

Ref. Okamoto M, Jinno K et al (1987) Chromatographia 23:373

HPLC 6

Sample	Human serum
Extraction	Methanol
Apparatus	Jasco HPLC with UV detector (Uvidec 100 IV)
Column	1G-ODS-TMS and 2G-ODS-TMS, 150 × 4.6 mm ID
Mobile phase	Acetonitrile/water (15:85)
Flow-rate	2.0 ml min^{-1}
Injection size	10 or 20 µl
Detection	UV, $\lambda = 203$ nm
Retention times, min	(estimated from drawing) Trimethadione 3.4 Dimethadione 2.0 Phenobarbital 15.8
Ref.	Okamoto M, Nobuhara K et al (1987) Chromatographia 23:325

153 Phenol

CAS-No	108-95-2
Chemical name	Hydroxybenzene
Empirical formula	$C_6H_5{\cdot}OH$
Molecular mass	94.11
Proprietary names	Chloraseptic; Oily Phenol Injection; Paoscle
Clinical use	Preservative/Disinfectant
Solubility	Water 1 in 13; alcohol 6 in 1; chloroform 2 in 1; ether 5 in 1
Light absorption	$\lambda_{max} = 287$ nm in aqueous alkali

HPLC 1

Sample	Rain-Water
Extraction	Liquid-Liquid extraction with dichloromethane
Apparatus	Waters M 45 LC pump, Rheodyne M 7125 injection valve and Kontron SFM-22LC spectrofluorimeter
Column	Aminosilica; Hypersil APS 3 µm, 100 × 4 mm ID
Mobile phase	1.5% Isopropanol in hexane
Flow-rate	2.0 ml min^{-1}
Injection size	50–100 µl containing 56 ng
Detection	Fluorescence, $\lambda = 277/300$ nm
Retention times, min	4-tert.-Butylphenol 3.0 *m*- (and) *p*-Cresol 8.2 *o*-Cresol 5.8 Phenol 10.0 (estimated from drawing)
Ref.	Czuczwa J, Leuenberger C et al (1987) J. Chromatogr. 403:233

HPLC 2

Sample	Urine
Extraction	Steam-distillation
Apparatus	HPLC system with ESA 5100 A Coulochem detector and UV detector
Column	Reversed-phase; R-Sil C_{18}LL, 250 × 4.6 mm ID
Mobile phase	Methanol/20 mM monosodium phosphate buffer, pH 4.0 (48:52)
Flow-rate	0.7 ml min^{-1}
Internal standard	2,6-Dimethylphenol
Injection size	2 ng
Detection	EC, + 1.0 and + 0.75 V; UV, λ = 285 nm
Recording	Hewlett-Packard 3390A integrator
Retention times, min	Phenol 8.2 *p*-Cresol 12.5 (295 nm) Internal standard 18.0 (estimated from drawing)
Ref.	Schaltenbrand WE and Coburn SP (1985) Clin. Chem. 31:2042

HPLC 3

Sample	Pharmaceutical dosage forms
Apparatus	Waters ALC M 202 HPLC, U6K injector and UV detector
Column	Reversed-phase; μ-Bondapak, C_{18}, 300 × 4 mm ID
Mobile phase	0.02 M NH_4HCO_3 in 10% by volume of methanol in water, pH 7.67
Flow-rate	3.0 ml min^{-1}; 2200 psi
Temperature	Ambient
Injection size	10–20 μl
Detection	UV, λ = 254 nm; 0.16 AUFS
Recording	Omniscribe 5213-12 recorder equipped with an integrator
Retention times, min	(estimated from drawing) Resorcinol 7.0 Phenol 9.8
Ref.	Das Gupta V (1976) J. Pharm. Sci. 65:1706

HPLC 4

Sample	Mixture of phenolic derivatives
Apparatus	Merck M 655A-11 HPLC pump and 655A-22 UV detector
Column	Reversed-phase; Lichrosorb C_8, 250 × 4 mm ID
Mobile phase	Water/acetonitrile (63:37, v/v)
Flow-rate	2.5 ml min^{-1}; 2000 psi
Injection size	20 μl containing 20 μg

Detection	UV, $\lambda = 254$ nm			
Retention times, min	(estimated from drawing)		*m*-Nitrophenol	3.84
	p-Methoxyphenol	2.60	2,5-Dichlorophenol	6.32
	Phenol	3.06	2-Ethoxyphenol	7.80
Ref.	Merck Application Manual on HPLC, TLC and HPTLC (1986). E. Merck, Darmstadt, p 6			

HPLC 5

Sample	Human urine
Apparatus	Micrometrics M 7000 HPLC and Schoeffel SF 770 spectrophotometer detector
Column	Reversed-phase; Nucleosil C_{18}, 10 μm, 250 × 4.6 mm ID
Mobile phase	Methanol/water (50:50, v/v)
Flow-rate	1 ml min^{-1}; 1000 psi
Temperature	Ambient
Injection size	5 μl
Detection	UV, $\lambda = 268$ nm
Retention time, min	Phenol 5.1 (estimated from drawing)
Ref.	Schoofs T, Gossler K and Schaller KH (1978) GIT 22:608

154 Phenprocoumon

CAS-No	435-97-2			
Chemical name	4-Hydroxy-3(1-phenylpropyl)coumarin			
Empirical formula	$C_{18}H_{16}O_3$			
Molecular mass	280.3			
Proprietary names	Liquamar; Marcoumar; Marcumar			
Clinical use	Anticoagulant			
Solubility	Water	chloroform	methanol	aq NaOH
	insoluble	soluble	soluble	soluble
Light absorption	$\lambda_{max} = 310$ nm in aqueous alkali			

HPLC 1

Sample	Human blood plasma and urine
Extraction	Plasma: 1,2-Dichloromethane under acidic condition Urine: As plasma after enzymatic hydrolysis with β-glucuronidase
Apparatus	Waters M 6000A LC pump, U6K injector, 660 solvent programmer and 440 ultraviolet detector

Column	Reversed-phase; Lichrosorb C_{18}, 10 µm, 250 × 4 mm ID
Pre-column	Reversed-phase; Corasil C_{18}, 35–50 µm, 30 × 4 mm ID
Mobile phase	I (plasma): Acetonitrile/water/acetic acid (600:400:5) II (urine) A: As in I. II (urine) B: Acetonitrile/acetic acid (1000:5) Gradient: B 0% to B 100% in 30 min (urine)
Flow-rate	2.0 ml min^{-1}; 943–1450 psi
Internal standard	*p*-Chlorophenprocoumon
Injection size	20 µl, 0.1–5.0 µg ml^{-1}
Detection	UV, λ = 313 nm; 0.05 AUFS
Recording	Spectra-Physics Autolab System 1 computing integrator
Retention times, min	Mobile phase — I (plasma) — II (urine) Phenprocoumon — 3.40 — 11.02 Internal standard — 4.70 — 13.29
Ref.	De Vries JX, Harenberg J et al (1982) J. Chromatogr. Biomed. Appl. 231:83

HPLC 2

Sample	Human blood plasma
Column	Reversed-phase; Nucleosil C_{18}, 10 µm, 250 × 4.6 mm ID
Mobile phase	0.05 M Potassium phosphate buffer, pH 2.5/acetonitrile (43:57, v/v)
Flow-rate	1 ml min^{-1}
Internal standard	Chlorphenprocoumon
Injection size	100 µl
Detection	UV, λ = 284 nm
Retention times, min	Phenprocoumon 9.33 Chlorphenprocoumon 13.03
Ref.	Erdweg M and Hoffmann (1987): in High Performance Liquid Chromatography. Application No. 228, Macherey-Nagel, Dueren, p 202

HPLC 3

Sample	Pure racemates, blood plasma and urine
Extraction	Dichloromethane under acidic condition (plasma) and with methanol on a Sep-Pak C_{18} column after enzymatic hydrolysis with β-glucuronidase (urine)
Apparatus	System 1: HPLC with photodiode array detector System 2: HPLC with fluorescence detector
Column	Nucleosil Chiral-2, 5 µm, 250 × 4 mm ID

Mobile phase	System 1: n-heptane/dioxane/trifluoroacetic acid (100:25:0.1, v/v) System 2: n-heptane/dioxane/trifluoroacetic acid (100:30:0.1, v/v)
Flow-rate	1 ml min^{-1}; 1450 psi
Injection size	25–80 µl
Detection	Diode array, λ = 310 and 276 nm Fluorescence, λ = 310/360 nm

Retention times, min

Racemates/System	(S)	(R)	system
Phenprocoumon	18.77	19.77	1
p-Chlorophen-procoumon	18.82	19.82	1
Warfarin	31.22	33.02	2

Ref. De Vries JX and Voelker U (1989) J. Chromatogr. Biomed. Appl. 493:149

---

155 Phenylbutazone

CAS-No	50-33-9; 129-18-0 (sodium salt)
Chemical name	4-Butyl-1,2-diphenylpyrazolidine-3,5-dione
Empirical formula	Phenylbutazone $C_{19}H_{20}N_2O_2$; Phenylbutazone Sodium $C_{19}H_{19}N_2NaO_2$
Molecular mass	308.4; 330.4
Proprietary names	Butazolidin; Malgesic; Spondyril
Clinical use	Analgesic/Anti-inflammatory

Solubility,	Water	alcohol	chloroform	aq NaOH
Phenylbutazone	insoluble	1 in 28	1 in 1.25	freely soluble
Sodium salt	soluble			soluble

Light absorption: λ_{max} = 264 nm in aqueous alkali

HPLC 1

Sample	Blood plasma
Extraction	n-Hexane under acidic condition
Column	SIL-X (Perkin-Elmer), 100 × 1.8 mm ID
Mobile phase	0.002% Glacial acetic acid and 10% tetrahydrofuran in n-hexane
Flow-rate	1 ml min^{-1}; 475 psi
Temperature	35 °C
Injection size	10 µl, 30–200 ng ml^{-1}
Detection	UV, λ = 254 nm; 0.04 AUFS
Retention time, min	Phenylbutazone 6.3 (estimated from drawing)

Ref.	Pound NJ, McGilveray IJ and Sears RW (1974) J. Chromatogr. 89:23

HPLC 2

Sample	Blood plasma from horse
Apparatus	Beckman M LC-110A pump and UV detector
Column	ISRP (internal surface reversed phase); GFF-S5-80, 5 µm, 150 × 4.6 mm ID
Mobile phase	A. 0.1 M KH_2PO_4/isopropanol/tetrahydrofuran (84:10:6), pH 6.8 B. 0.1 M KH_2PO_4/isopropanol (90:10) C. 0.1 M KH_2PO_4/acetonitrile (90:10) D. 0.1 M KH_2PO_4/tetrahydrofuran (90:10)
Flow-rate	0.6 ml min^{-1}
Temperature	Ambient
Injection size	10 µl, 20 µg ml^{-1}
Detection	UV, $\lambda = 241$ nm; 0.05 AUFS
Retention times, min	(estimated from drawing)

	Phenylbutazone	Oxyphenbutazone
Eluent A	7.0	5.0
Eluent B	9.3	6.5
Eluent C		6.8
Eluent D	7.5	5.0

Ref.	Pinkerton TC, Perry JA and Rateike JD (1986) J. Chromatogr. 367:412

HPLC 3

Sample	Pharmaceutical dosage forms
Apparatus	HPLC and variable-wavelength UV detector
Column	Reversed-phase; 5 µm, C_{18}, 150 × 3.9 mm ID
Mobile phase	0.05 M Acetate buffer, pH 5.2/acetonitrile (40:60, v/v)
Flow-rate	1 ml min^{-1}
Temperature	Ambient
Internal standard	Hexylparaben 0.2 mg ml^{-1}
Injection size	20 µl, 0.02–0.12 mg ml^{-1}
Detection	UV, $\lambda = 254$ nm; 0.02 AUFS
Retention times, min	Phenylbutazone 3.2 Hexylparaben 4.4
Ref.	Chi SC, Jun HW and Foda NH (1988) J. Liq. Chromatogr. 11:1545

HPLC 4

Sample	Blood or plasma
Extraction	Methylene chloride under acidic condition

Apparatus	Altex M 334 HPLC with gradient system and Hitachi M 11-30 variable-wavelength detector
Column	Reversed-phase; Altex ODS 5 µm, 150 × 4.6 mm ID
Mobile phase	Acetonitrile/water/glacial acetic acid (450:550:3.2, v/v)
Flow-rate	2.0 ml min^{-1}
Internal standard	Phenolphthalein, 62.5 µg ml^{-1}
Injection size	20 µl
Detection	UV, λ = 240 nm; 0.05 AUFS
Recording	Hewlett-Packard 3385 integrator
Retention times, min	Salicylate 1.5 Oxyphenbutazone 3.3 Phenylbutazone 9.4
Ref.	Levine B and Caplan YH (1985) Clin. Chem. 31:346

HPLC 5

Sample	Biological fluids and horse tissue
Extraction	n-Hexane under acidic condition
Column	Reversed-phase; ODS 10 µm, 250 × 6.25 mm ID
Mobile phase	Water/acetonitrile/methanol/glacial acetic acid (5:4:2:0.03)
Flow-rate	1.5 ml min^{-1}
Internal standard	Hexestrol
Detection	UV, λ = 235 nm
Ref.	Lees P, Taylor JBO et al (1987) Cornel Vet. 77:192

156 Phenytoin

CAS-No	57-41-0; 630-93-3 (sodium salt)			
Chemical name	5,5-Diphenylimidazolidine-2,4-dione			
Empirical formula	Phenytoin $C_{15}H_{12}N_2O_2$	Sodium Salt $C_{15}H_{11}N_2NaO_2$		
Molecular mass	252.3	274.3		
Proprietary names	Dantoin; Epanutin; Phenhydan			
Clinical use	Anticonvulsant			
Solubility,	Water	ethanol	ether	aq NaOH
Phenytoin	insoluble	slightly soluble	1 in 600	soluble
Sodium salt	soluble	soluble	insoluble	soluble
Light absorption	λ_{max} = 258 nm in methanol			

HPLC 1

Sample	Liver perfusate	
Apparatus	High pressure liquid chromatography with Perkin-Elmer LC-75 spectrophotometric detector	
Column	Reversed-phase; Spherisorb ODS 5 μm, 250 × 3.6 mm ID	
Mobile phase	Acetonitrile/water (29:71)	
Flow-rate	1.4 ml min^{-1}	
Temperature	Ambient	
Detection	UV, λ = 214 nm	
Retention times, min	Parahydroxy phenytoin	3.2
	Phenacetin	4.7
	Phenytoin	7.5
Ref.	Maguire TA, Swanton JG and Temple DJ (1988) Eur. J. Drug. Metab. Pharmacokinet. 13:35	

HPLC 2

Sample	Serum			
Extraction	Ethyl acetate			
Apparatus	Hewlett-Packard HPLC 1090 with gradient controller, Rheodyne injection system and UV detector			
Column	Reversed-phase; Hypersil 5 μm ODS, 200 × 2.1 mm ID			
Mobile phase	A. Perchloric acid (diluted with water for pH 4.0)/acetonitrile (9:1, v/v) B. Perchloric acid, pH 4.0 as in A./acetronitrile (4:6, v/v) Gradient: B 15% to 45% in 9 min; B 45% to 15% in 0.5 min; run 2 min			
Flow-rate	0.3 ml min^{-1}			
Temperature	65 °C			
Internal standard	Tolybarb			
Injection size	1.5–5 μl			
Detection	UV, λ = 207 nm			
Retention times, min	Caffeine	2.40	Internal standard	9.57
	Ethosuximide	3.45	Phenytoin	10.79
	Primidone	4.00	Carbamazepine	11.34
	Phenobarbital	6.57		
Ref.	Juergens U (1987) J. Liq. Chromatogr. 10.507			

HPLC 3

Sample	Blood plasma
Extraction	Dichlormethane

Apparatus	Kontron M 600 HPLC system with Uvikon 720 LC UV-VIS detector
Column	Reversed-phase; Spherisorb 5 ODS, 250 × 4.6 mm ID
Pre-column	Co: Pell ODS, 50 × 4.6 mm ID (Whatman)
Mobile phase	Acetonitrile/methanol/phosphate buffer, pH 4.0 (21:24:55, v/v)
Flow-rate	3.0 ml min^{-1}
Temperature	Ambient
Injection size	20 μl, 2.5–30 μg ml^{-1}
Detection	UV, λ = 195 nm
Recording	Hewlett-Packard 3390A integrator
Retention times, min	(estimated from drawing) Phenytoin 4.2 Ethosuximide 1.5 Carbamazepine 5.2 Phenobarbital 2.5
Ref.	Soto-Otero R and Sierra-Marcuno G (1984) Clin. Chem. 30:817

HPLC 4

Sample	Serum and test substances	
Extraction	Chloroform/isopropanol (95:5) (serum)	
Column	Reversed-phase; Supelcosil LC-1, 5 μm, 250 × 4.6 mm ID	
Pre-column	Supelguard LC-1, 5 μm, 20 × 4.6 mm ID	
Mobile phase	10 mM Phosphate buffer (containing 2.72 g KH_2PO_4 in 2 liter water, pH 6.3 with 1 M NaOH)/ methanol/acetonitrile (65:17.5:17.5, v/v)	
Flow-rate	2 ml min^{-1}; 2400 psi	
Temperature	Ambient	
Internal standard	Tolybarb	
Injection size	6–10 μl	
Detection	UV, λ = 204 nm; 0.08 AUFS	
Retention times, min	Theophylline	1.95
	Acetaminophen	2.09
	Caffeine	2.18
	Ethosuximide	2.57
	Primidone	2.71
	3-Isobutyl-1-methyl-xanthine	3.15
	Phenobarbital	3.45
	Chloramphenicol	3.85
	Tolybarb (IS)	4.80
	Phenytoin	6.28
	Carbamazepine	6.68
Ref.	Meatherall R and Ford D (1988) Ther. Drug Monit. 10:101	

157 Physostigmine

CAS-No	57-47-6; 57-64-7 (salicylate)			
Chemically	Alkaloid obtained from the calabar bean, the seed of Physostigma venenosum			
Empirical formula	Physostigmine	Salicylate		
	$C_{15}H_{21}N_3O_2$	$C_{15}H_{21}N_3O_2$, $C_7H_6O_3$		
Molecular mass	275.3	413.5		
Proprietary names	Antilirium; Fisostin; Isopto-Eserine			
Clinical use	Parasympathomimetic			
Solubility,	Water	alcohol	chloroform	ether
Physostigmine	1 in 75	1 in 10	1 in 1	1 in 30
Salicylate	1 in 75–90	1 in 25	1 in 6	1 in 250
Light absorption	λ_{max} = 246 nm in aqueous acid			

HPLC 1

Sample	Raw materials
Apparatus	Spectra-Physics 3500B LC pump and 770 ultraviolet detector
Column	Reversed-phase; μ-Bondapak C_{18}, 10 μm, 300 × 3.9 mm ID
Mobile phase	Methanol/acetic acid/triethylamine/water (35:1.5:0.5:63)
Flow-rate	1.5 ml min^{-1}
Injection size	10 μl, 1.0 mg ml^{-1}
Detection	UV, λ = 261 nm; 0.32 AUFS
Recording	Hewlett-Packard 3380A integrator
Retention times, min	(estimated from drawing)
	Physostigmine 5.8
	Salicylic acid 6.0
Ref.	Roos RW and Lau-Cam CA (1986) J. Chromatogr. 370:403

HPLC 2

Sample	Eye-drops
Apparatus	Waters LC pump with variable-wavelength detector and LDC Spectromonitor III
Column	Reversed-phase; μ-Bondapak C_{18}, 10 μm, 300 × 3.9 mm ID
Mobile phase	Methanol/0.005 M heptanesulphonic acid (40:60), pH 3.6

Flow-rate	1 ml min^{-1}
Injection size	8 μg
Detection	UV, λ = 292 nm; 0.002 AUFS
Retention times, min	(estimated from drawing) Salicylate 9.5 Rubreserine 6.0 Physostigmine 12.8
Ref.	Kneczke M (1980) J. Chromatogr. 198:529

HPLC 3

Sample	Blood plasma and cerebrospinal fluid
Extraction	Diethyl ether (blood plasma) and chloroform (cerebrospinal fluid)
Apparatus	Bioanalytical Systems PM-30 pump, LC-4B amperometric detector and Rheodyne M 7125 injector
Column	Normal phase; Spherisorb silica 5 μm, 150 × 4.6 mm ID
Pre-column	Alltech pellicular silica
Mobile phase	10 mM Sodium acetate buffer, pH 4.6/methanol (1:10, v/v)
Flow-rate	2.0 ml min^{-1}
Injection size	190 μl containing 0.5–40 ng ml^{-1}
Detection	EC, GC electrode, + 0.25 and + 0.95 V, vs Ag/AgCl RE
Retention times, min	Eseroline (metabolite) 3.33 Physostigmine 4.33
Ref.	Unni LK, Hannant ME et al (1989) Clin. Chem. 35:292

HPLC 4

Sample	Cat brain tissue
Extraction	Mobile phase
Apparatus	Pye-Unicam LC3-XP reciprocating LC pump, LC-3 UV detector and Rheodyne M 7105 injector
Column	Reversed-phase; Lichrosorb C_8, 5 μm, 125 × 3 mm ID
Mobile phase	0.005 M NaH_2PO_4, 0.3 (w/v) sodium dodecyl sulfate and 0.1 M $NaClO_4$ in 500 ml methanol and 500 ml water, pH 3.00
Flow-rate	0.6 ml min^{-1}
Injection size	100 μl
Detection	UV, λ = 245 nm
Retention time, min	Physostigmine 8.5 (estimated from drawing)
Ref.	De Wildt DJ, Porsius AJ and Van Rooy HH (1981) J. Chromatogr. Biomed. Appl. 225:381

158 Pilocarpine

CAS-No	92-13-7; 54-71-7 (hydrochloride)			
Chemical name	3-Ethyldihydro-4-[(1-methyl-1*H*-imidazol-5-yl)methyl]-2(3*H*)-furanone			
Empirical formula	Pilocarpine		Hydrochloride	
	$C_{11}H_{16}N_2O_2$		$C_{11}H_{16}N_2O_2$, HCl	
Molecular mass	208.3		244.7	
Proprietary names	Isopto Carpine; Pilotonina; PV-Carpine			
Clinical use	Parasympathomimetic			
Solubility,	Water	alcohol	ether	chloroform
Pilocarpine	soluble	soluble	sparingly soluble	soluble
Hydrochloride	1 in >1	1 in 3	insoluble	1 in 360
Light absorption	$\lambda_{max} = 216$ nm in methanol			

HPLC 1

Sample	Dosage forms
Apparatus	Shimadzu HPLC system with ultraviolet detector
Column	A. Normal phase; μ-Porasil SI-OH 10 μm, 250 × 3.9 mm ID B. Amino phase; μ-Bondapak NH_2 10 μm, 250 × 3.9 mm ID
Mobile phase	Hexane/isopropanol with 2% ammonium hydroxide (75:25)
Flow-rate	1.5 ml min^{-1}
Injection size	20 μl
Detection	UV
Retention times, min	
Ref.	Durif C, Ribes M et al (1988) Pharm. Acta Helv. 63:294

Column	A.	B.
Pilocarpine	15.0	8.8
Isopilocarpine	17.0	10.0

HPLC 2

Sample	Dosage forms
Apparatus	Shimadzu HPLC system with ultraviolet detector
Column	A. Reversed-phase; Lichrosorb C_8, 10 μm, 250 × 4.6 mm ID B. Reversed-phase; Lichrosorb C_{18}, 7 μm, 250 × 4.6 mm ID
Mobile phase	Tampon Phosphate, pH 2.5/methanol (97:3)
Flow-rate	1.5 ml min^{-1}
Injection size	10 μl
Detection	UV

Retention times, min	Column	A.	B.
	Isopilocarpine	11.6	15.1
	Pilocarpine	12.5	16.4
	Pilocarpic acid	14.9	18.6
	Isopilocarpic acid	16.5	20.7
Ref.	Durif C, Ribes M et al (1988) Pharm. Acta Helv. 63:294		

HPLC 3

Sample	Eye-drops
Apparatus	Waters LC pump with LDC Spectromonitor III UV detector
Column	Reversed-phase; μ-Bondapak C_{18}, 10 μm, 300 × 3.9 mm ID
Mobile phase	Methanol/0.005 M heptanesulfonic acid, (40:60), pH 3.6
Flow-rate	1 ml min^{-1}; 2 ml min^{-1} after 15 min
Injection size	80 μg
Detection	UV, λ = 235 nm; 0.005–0.1 AUFS
Retention times, min	Pilocarpine 6.8 Physostigmine 12.0 Salicylate 8.0 Propylparaben 26.0 Methylparaben 10.8 (estimated from drawing)
Ref.	Kneczke M (1980) J. Chromatogr. 198:529

HPLC 4

Sample	Ophthalmic dispersions
Apparatus	Du Pont M 850 gradient pump, Rheodyne M 7010 injector and Waters M 481 UV detector
Column	Reversed-phase; Radial Z-Module cartridge, Resolve C_{18}, 10 μm, 100 × 8 mm ID
Mobile phase	Acetonitrile/diethylamine/water (5:1:94, v/v)
Flow-rate	3 ml min^{-1}
Temperature	Ambient
Injection size	20 μl
Detection	UV, λ = 235 nm
Recording	Waters data module
Retention times, min	Pilocarpine 10.0 (from polymer dispersion) Isopilocarpine 13.0 (from polymer dispersion containing 0.1% isopilocarpine nitrate) Pilocarpine 15.0 (estimated from drawing)
Ref.	Ibrahim H, Boye T et al (1987) J Pharm. Biomed. Anal. 5:379

HPLC 5

Sample	Aqueous humor
Extraction	Methylene chloride before derivatization

Apparatus	Waters M 6000A LC pump, 440 UV detector and U6K injector
Column	Reversed-phase; μ-Bondapak C_{18}, 10 μm, 300 × 3.9 mm ID
Mobile phase	Methanol/water (80:20, v/v) containing 0.001 M sodium octanesulfonate
Flow-rate	1.6 ml min^{-1}; 1000–1500 psi
Temperature	Ambient
Internal standard	Phenylephrine hydrochloride
Derivatization	Pre-column; with *p*-nitrobenzyl bromide
Injection size	Appr. 100 μl, 0.3–10.0 μg ml^{-1} (nitrate)
Detection	UV, λ = 254 nm; 0.005 AUFS
Retention times, min	(estimated from drawing) Internal standard 5.0 Pilocarpine (derivatised) 11.0 Isopilocarpine (derivatised) 12.0
Ref.	Mitra AK, Baustian CL and Mikkelson TJ (1980) J. Pharm. Sci. 69:257

HPLC 6

Sample	Raw materials
Apparatus	Altex M 310 HPLC with UV detector
Column	Reversed-phase; Lichrosorb C_{18}, 10 μm, 250 × 4.6 mm ID
Mobile phase	Water/methanol (97:3) containing 5% potassium dihydrogen phosphate, pH 2.5
Flow-rate	1.5 ml min^{-1}
Detection	UV, λ = 216 nm
Retention times, min	(estimated from drawing) Pilocapric acid 14.0 Isopilocarpine 9.0 Isopilocapric acid 17.0 Pilocarpine 10.7
Ref.	O'Donnell JJ, Sandman R and Drake MV (1980) J. Pharm. Sci. 69:1096

HPLC 7

Sample	Ophthalmic preparations
Apparatus	Waters M 6000A LC pump with U6K injector and Pye-Unicam LC-3 UV detector
Column	Reversed-phase; Nucleosil C_{18}, 150 × 4 mm ID
Mobile phase	Water/methanol (97:3), pH 2.5 with phosphoric acid
Flow-rate	1.5 ml min^{-1}
Detection	UV, λ = 215 nm
Ref.	Noordam A, Maat L and Beyerman HC (1981) J. Pharm. Sci. 70:96

159 Piperacillin

CAS-No	61477-96-1; 59703-84-3 (sodium salt)		
Chemically	Semisynthetic penicillin		
Empirical formula	Piperacillin	Sodium Salt	
	$C_{23}H_{27}N_5O_7S$	$C_{23}H_{26}N_5NaO_7S$	
Molecular mass	517.6	539.5	
Proprietary names	Pentcillin; Pipril		
Clinical use	Antibiotic		
Solubility,	Water	alcohol	methanol
Sodium salt	1 in 1.4	1 in 5	1.4

HPLC 1

Sample	Biological fluids
Apparatus	Constametric III-G LC pump and ultriviolet detector
Column	Reversed-phase; Nucleosil 5 μm, C_{18}, 200 × 4.0 mm ID
Mobile phase	Acetonitrile/57.4 mM K_2HPO_4 (23:77), pH 5.2 with H_3PO_4
Flow-rate	1 ml min^{-1}
Temperature	Ambient
Injection size	20 μl, 5–250 μg ml^{-1}
Detection	UV, λ = 220 nm
Recording	Computing integrator (LDC-301)
Retention times, min	Penicilloate 5.0
	Piperacillin 10.5 (estimated from drawing)
Ref.	Knoeller J, Koenig W et al (1988) J. Chromatogr. Biomed. Appl. 427:257

HPLC 2

Sample	Aqueous humor of the eye
Extraction	Chloroform/1-pentanol (3:1)
Apparatus	Beckman M 110A pump, 210 injection valve and 153 ultraviolet detector
Column	Reversed-phase; Ultrasphere ODS 5 μm, 250 × 4.5 mm ID
Mobile phase	120 ml Morpholinopropanesulfonic acid with 0.05 M triethylamine, pH 6.7, 480 ml water and 400 ml methanol, final pH 6.7 at 30 °C
Flow-rate	1.0 ml min^{-1}
Temperature	32 °C
Internal standard	Cephalothin
Injection size	20 μl, 0.13–25.2 μg ml^{-1}
Detection	UV, λ = 254 nm

Recording	Hewlett-Packard 3390 integrator recorder			
Retention times, min	Cefoxitin	3.0	Azlocillin	5.4
	Sulfamethoxazole	3.3	Cefamandole	
	Theophylline	3.8	nafate	5.5
	Ampicillin	4.0	Cefuroxime	5.5
	Caffeine	4.6	Mezlocillin	6.7
	Salicylate	4.7	Scopolamine	
	Acetaminophen	4.8	hydrobromide	19.0
	Ticarcillin	5.3	Timolol	23.7
Ref.	Riegel MA and Ellis PP (1988) J. Chromatogr. Biomed. Appl. 424:177			

HPLC 3

Sample	Blood plasma and urine
Extraction	Deproteinization with acetonitrile
Apparatus	Waters M 6000A LC pump and ultraviolet detector and Rheodyne M 7125 injector
Column	Reversed-phase; Partisil ODS-3, 250 × 4.6 mm ID
Mobile phase	Acetonitrile/water (32:68), pH 2.5 with 85% phosphoric acid
Flow-rate	2.0 ml min^{-1}
Internal standard	*p*-Nitrobenzene sulfonamide
Injection size	10–20 µl, 4.7–188.6 µg ml^{-1}
Detection	UV, $\lambda = 254$ nm; 0.05–0.01 AUFS
Recording	Hewlett-Packard 3390A computing integrator
Retention times, min	Internal standard 4.2 Piperacillin 6.0
Ref.	Jung D and Mahajan NK (1984) Clin. Chem. 30:122

HPLC 4

Sample	Biological materials
Extraction	Centrifugation
Apparatus	Perkin-Elmer Series 2-LC pump with LC-75 UV detector
Column	Reversed-phase; µ-Bondapak C_{18}, 10 µm, 300 × 3.9 mm ID
Pre-column	Supelcosil, C_{18}, 10 µm, pellicular
Mobile phase	22% Acetonitrile in 0.1 M sodium acetate buffer, pH 4.6 with glacial acetic acid
Flow-rate	2.0 ml min^{-1}
Temperature	50 °C
Internal standard	Tolybarb
Injection size	30 µl

Detection	UV, λ = 230 nm
Recording	Perkin-Elmer Sigma 10B data system
Retention times, min	Piperacillin 3.7
	Internal standard 6.5
Ref.	Aravind MK, Miceli JN and Kauffman RE (1982) J. Chromatogr. Biomed. Appl. 233:423

HPLC 5

Sample	Raw materials (sodium salt)
Column	Reversed-phase; ODS, 5 to 10 μm, 250 × 4.6 mm ID
Mobile phase	450 ml Methanol, 100 ml of 0.2 M NaH_2PO_4, 3 ml tetrabutylammonium hydroxide solution (1 in 10) and make 1000 ml with water, pH 5.5 with phosphoric acid
Flow-rate	1 ml min^{-1}
Internal standard	Ampicillin
Injection size	Appr. 10 μl, 0.4 mg ml^{-1}
Detection	UV, λ = 254 nm
Retention times, min	Ampicillin appr. 3.5
	Piperacillin appr. 12.0
Ref.	Anonym (1985): The USP XXI and The NF XVI. The United States Pharmacopeial Convention Inc., Rockville, p 841

160 Polymyxin B Sulfate

CAS-No	1404-26-8 (polymyxin B); 1405-20-5 (sulfate)		
Chemically	A mixture of the sulfates of some specific polypeptides produced by Bacillus polymyxa		
Empirical	Polymyxin B Sulfate	Polymyxin B_1	Polymyxin B_2
formula	$C_{43}H_{82}N_{16}O_{12}$ $7H_2SO_4$	$C_{56}H_{98}N_{16}O_{13}$	$C_{55}H_{96}N_{16}O_{13}$
Molecular mass	1702.0	1203.5	1189.4
Proprietary names	Aerosporin; Lidosporin; otobione		
Clinical use	Antibiotic		
Solubility	Water	alcohol	ether
	very soluble	slightly soluble	insoluble

HPLC 1

Sample	Raw materials
Apparatus	Du Pont M 830 LC with 837 spectrophotometeric detector

Column	Reversed-phase; Nucelosil 5 μm, C_{18}, 150 × 4.6 mm ID
Mobile phase	Acetonitrile/buffer (0.023 M H_3PO_4, 0.01 M CH_3COOH and 0.05 M Na_2SO_4), pH 2.5 with triethylamine (23:77)
Flow-rate	0.9 ml min^{-1}; 1204 psi
Injection size	20 μl containing 8–100 μg
Detection	UV, λ = 220 nm
Recording	Hewlett-Packard 3380A integrator
Retention times, min	Polymyxin B_2 6.5–7.8 Polymyxin B_1 14–15 (estimated from drawing)
Ref.	Elverdam I, Larsen P and Lund E (1981) J. Chromatogr. 218:653

HPLC 2

Sample	Raw materials
Apparatus	LDC modular LC, high-pressure Milton Roy minipump (max. 5000 psi) with a pulse dampener, UV detector and ultragrad gradient mixer
Column	Reversed-phase; μ-Bondapak C_{18}, 300 × 4.6 mm ID
Mobile phase	A. Acetonitrile/water/200 mM phosphate buffer, pH 2.0 (200:700:100) B. Acetonitrile/absolute methanol/water/ 200 mM phosphate buffer with Na_2HPO_4, pH 2.0 (500:200:290:10) Linear gradient: 20% Acetonitrile to 50% acetonitrile and 20% methanol, both pH 2.0, in 16 min
Flow-rate	1 ml min^{-1}; 1000 psi
Temperature	Ambient
Injection size	8 μl, 16 mg ml^{-1}
Detection	UV, λ = 254 nm; 0.04 AUFS
Retention times, min	Polymyxin B_1 14.0 Polymyxin B_2 10.5 (estimated from drawing)
Ref.	Tsuji K and Robertson JH (1975) J. Chromatogr. 112:663

HPLC 3

Sample	Raw materials
Apparatus	Altex M 110A LC pump, Rheodyne M 7010 sampling valve and Perkin-Elmer LC-55 variable-wavelength detector
Column	Reversed-phase; Lichrosorb C_{18}, ODS 5 μm, 250 × 4.6 mm ID
Mobile phase	0.05 M Triethanolammonium phosphate, pH 2.2 containing 0.025 M 1-butanesulfonic acid and 25% (v/v) acetonitrile

Flow-rate	1.0 ml min^{-1}; 2800 psi
Temperature	22 °C
Injection size	20 μl containing 60 μg
Detection	UV, λ = 220 nm; 0.08 AUFS
Retention times, min	Colistin B 10.0 Colistin A 18.5 Polymyxin B_2 13.0 Polymyxin B_1 24.0 (estimated from drawing)
Ref.	Kalász H and Horváth C (1981) J. Chromatogr. 215:295

HPLC 4

Sample	Papain-cleaved derivative of Polymyxin B
Apparatus	Varian M 5000 LC, UV-100 variable-wavelength detector and Rheodyne loop injector
Column	Vydac 218 TP C_{18}, 5 μm, 250 × 4.6 mm ID
Mobile phase	A. 0.1% (v/v) Trifluoroacetic acid in water B. 0.075% (v/v) Trifluoroacetic acid in acetonitrile Linear gradient: B 10% to 80% in 20 min
Flow-rate	1.2 ml min^{-1}
Injection size	10 μl containing 50–100 μg
Detection	UV, λ = 215 nm; 0.25 AUFS
Retention times, min	Polymyxin B nonapeptide 7.9–8.0 impurity 11.9–12.0
Ref.	Vaara M (1988) J. Chromatogr. 441:423

HPLC 5

Sample	Pharmaceutical formulations
Apparatus	LDC Constametric II LC pump and Spectroflow SF-770 UV detector (Schoeffel)
Column	Reversed-phase; Hypersil ODS 5 μm, 250 × 3 mm ID
Mobile phase	A. Acetonitrile/H_2SO_4 (v/v)/tetramethylammonium chloride (w/v)/water (22.5:0.2:0.5:76.3, v/v), pH 2.7 with 0.5 M K_2HPO_4 B. As in A but (25:0.2:0.5:74.3, v/v), pH 1.6
Flow-rate	1.09 ml min^{-1}
Injection size	Appr. 27.3 μg
Detection	UV, λ = 220 nm; 0.1 AUFS
Recording	Spectra-Physics IV integration system
Retention times, min	(estimated from drawing)

Column	A.	B.
Polymyxin B_2	9.5	8.5
Polymyxin B_1	21.0	15.0

Ref.	Fong GWK and Kho BT (1979) J. Liq. Chromatogr. 2:957

HPLC 6

Sample	Raw materials of polymyxin B_1
Apparatus	Waters M 6000A LC pump, 660 solvent programmer and 440 UV detector
Column	Reversed-phase; Vydac C_{18}, 300 × 3.6 mm ID
Mobile phase	(1) 0.01 M Pi (monosodium phosphate, (2) 0.05 M Pi and (3) 0.01 M HSA (1-heptanesulfonic acid sodium salt), each with 80% MeOH (methanol), 60% AcCN (acetonitrile) and 40% THF (tetrahydrofurn) and pH each 2.0, 3.4, and 6.6.
Flow-rate	1 ml min^{-1}
Temperature	Ambient
Injection size	1–10 μl
Detection	UV, λ = 254 nm; 0.2 A
Recording	Computing integrator
Retention times, min	Mobile phase

Buffer	Modifier	pH		
		2.0	3.4	6.6
0.01 M Pi	AcCN	3.05	3.64	3.21
	THF	3.25	3.83	3.54
	MeOH	4.39	5.33	8.94
0.05 M Pi	AcCN	3.34	3.50	3.10
0.01 M HSA	AcCN	3.36	4.48	4.39
	THF	4.78	4.97	3.95
	MeOH	9.34	12.57	9.42

Ref. Aszalos A and Aquilar A (1984) J. Chromatogr. 290:83

161 Polythiazide

CAS-No	346-18-9
Chemical name	6-Chloro-3,4-dihydro-2-methyl-3-[[(2,2,2-trifluoroethyl)thio]methyl]-2*H*-1,2,4-benzothiadiazine-7-sulfonamide 1,1-dioxide
Empirical formula	$C_{11}H_{13}ClF_3N_3O_4S_3$
Molecular mass	439.9
Proprietary names	Drenusil; Nephril; Renese
Clinical use	Diuretic
Solubility	Water insoluble; alcohol 1 in 40; acetone soluble; methanol soluble; chloroform insoluble
Light absorption	λ_{max} = 268 nm in methanol

HPLC 1

Sample	Dosage forms
Apparatus	LDC Constametric LC-III pump, Spectromonitor III ultraviolet detector and Rheodyne M 7125 injector
Column	A. Reversed-phase; μ-Bondapak C_{18}, 10 μm, 300 × 3.9 mm ID or B. Reversed-phase; ASI C_{18}, 10 μm, 300 × 3.9 mm ID
Mobile phase	Methanol/water/glacial acetic acid (55:44:1; v/v)
Flow-rate	1.5 ml min^{-1}
Temperature	Ambient
Injection size	10 μl, 80 μg ml^{-1}
Detection	UV, λ = 254 nm
Recording	Hewlett-Packard 3380A integrator
Retention times, min	Methylchlothiazide 2.51 Vanillin 2.52 Polythiazide 3.46 Benzthiazide 3.53 Polythiazide impurity 4.49 3-Isoreserpine 6.77 Reserpine 8.86 Rescinnamine 10.64 3,4-Dehydro-reserpine 12.06
Ref.	Bachman WJ (1986) J. Liq. Chromatogr. 9:1463

HPLC 2

Sample	Dosage forms (capsules)
Apparatus	LDC Constametric LC-I pump and Spectromonitor II ultraviolet detector
Column	A. Cyano phase; IBM CN 5 μ, 250 × 4.5 mm ID B. Cyano phase; Nucleosil CN, 10 μ, 250 × 4.6 mm ID C. Cyano phase; Bondapak CN, 10 μ, 300 × 3.9 mm ID D. Cyano phase; Spherisorb CN, 5 μ, 250 × 4.6 mm ID
Mobile phase	For A. Acetonitrile/0.05 M NaH_2PO_4 (35:65, v/v), pH 3.0 with H_3PO_4 For B. Acetonitrile/0.05 M NaH_2PO_4 (30:70, v/v), pH 4.8 For C. Acetonitrile/0.05 M NaH_2PO_4 (30:70, v/v), pH 4.8 For D. Acetonitrile/0.05 M NaH_2PO_4 (16:84, v/v), pH 5.5
Flow-rate	1.7 ml min^{-1}
Temperature	Ambient
Internal standard	Benzophenone, 30 μg ml^{-1}
Injection size	10 μl, 100 μg ml^{-1}
Detection	UV, λ = 268 nm; 0.01 AUFS

Recording	Hewlett-Packard 3380A integrator				
Retention times, min	Column	A.	B.	C.	D.
	Prazosin	4.3	4.0	3.7	9.6
	Benzophenone (IS)	5.8	8.9	6.8	8.6
	Polythiazide	6.6	10.2	8.3	6.8
Ref.	Bachmann WJ (1986) J. Liq. Chromatogr. 9:1033				

162 Prednisolone

CAS-No	50-24-8 (anhydrous); 52-21-1 (acetate)				
Chemical name	11β,17α,21-Trihydroxypregna-1,4-diene-3,20-dione				
Empirical formula	Prednisolone	Acetate			
	$C_{21}H_{28}O_5$	$C_{23}H_{30}O_6$			
Molecular mass	360.4	402.5			
Proprietary names	Hostacortin H; Marsolone; Ulacort				
Clinical use	Corticosteroid				
Solubility,	Water	alcohol	acetone	chloroform	methanol
Prednisolone	1 in 1300	1 in 30	1 in 50	1 in 180	soluble
Acetate	insoluble	1 in 120	slightly soluble	1 in 150	
Light absorption	λ_{max} = 240 nm in ethanol				

HPLC 1

Sample	Raw materials
Apparatus	Spectra-Physics 3500B LC pump and 770 ultraviolet detector
Column	Reversed-phase; μ-Bondapak C_{18}, 10 μm 300 × 3.9 mm ID
Mobile phase	Methanol/acetic acid/triethylamine/water (60:1.5:0.5:38)
Flow-rate	1.5 ml min^{-1}
Temperature	Ambient
Injection size	10 μl, 0.25 mg ml^{-1}
Detection	UV, λ = 240 nm; 0.32 AUFS
Retention times, min	(estimated from drawing)

Prednisone	4.5	Hydrocortisone acetate	8.0
Prednisolone	5.8	Norethindrone	11.6
Prednisolone succinate	7.0	Methyltestosterone	18.2
		Progesterone	28.2

Ref.	Roos RW and Lau-Cam CA (1986) J. Chromatogr. 370:403

HPLC 2

Sample	Prednisolone decomposition products
Apparatus	HPLC with ultraviolet detector
Column	Reversed-phase; μ-Bondapak
Mobile phase	Methanol/water (50:50, v/v) with 1% acetic acid
Flow-rate	1 ml min^{-1}
Temperature	25 °C
Injection size	10 μl
Detection	UV, λ = 313 and 254 nm; 0.32 and 0.64 AUFS
Retention times, min	

D-Homosteroid	7.6	17-Deoxy-prednisolone I.	14.0
Prednisolone	10.2	17-Deoxy-prednisolone II.	15.0
17-Ketosteroid	11.7	(estimated from drawing)	

Ref.	Dekker D (1980) Pharm. Weekbl. Sci. 2:59

HPLC 3

Sample	Raw materials
Apparatus	Waters HPLC pump with M 440 UV detector
Column	Reversed-phase; μ-Bondapak C_{18}, 300 × 3.9 mm ID
Mobile phase	Water containing 0.02 M PIC-A reagent (tetrabutylammonium phosphate in phosphate buffer)/methanol (1:1, v/v)
Flow-rate	1.5 ml min^{-1}
Temperature	24 °C
Injection size	5 μl with 4.1 μg
Detection	UV, λ = 254 nm; 1 AUFS
Retention times, min	(estimated from drawing)
	Adducts formed 3.2, 5.3
	Prednisolone sodium phosphate 9.0
Ref.	Dijkstra J and Dekker D (1982) J. Chromatogr. 238:247

HPLC 4

Sample	Tablets and bulk drugs
Apparatus	Waters M 6000A LC pump and 440 UV detector
Column	Normal phase; Du Pont 5–6 μm, 250 × 4.6 mm ID
Mobile phase	60 ml 95% Methanol with 1 ml acetic acid and dilute to 1 liter with water-washed ethylene dichloride
Flow-rate	1.5 ml min^{-1}
Temperature	Ambient
Internal standard	Fluoxymesterone, 1.0 mg ml^{-1}
Injection size	10–15 μl, 0.05 mg ml^{-1}

Detection	UV, λ = 254 nm; 0.2 AUFS
Recording	Waters M 730 data module
Retention times, min	Internal standard 6.0
	Prednisolone 9.0
Ref.	Brower JF (1984) J. Assoc. Off. Anal. Chem. 67:674

HPLC 5

Sample	Bulk drug substances and dosage forms
Apparatus	Isocratic HPLC system with ultraviolet detector
Column	Reversed-phase; ODS 10 μm
Mobile phase	I. CH_3CN/0.1% H_3PO_4 (35:65), pH 2.2
	II. CH_3CN/0.01 M phosphate buffer, pH 4.3 (35:65)
	III. CH_3CN/water (35:65), pH 5.8
Flow-rate	2.0 ml min^{-1}
Temperature	Ambient, 30 °C, 40 °C, 50 °C
Detection	UV, λ = 254 nm

Retention times, min	I	II	III	I	II	III
		Ambient		30 °C	40 °C	50 °C
Prednisolone alcohol	3.36	3.36	3.37	3.37	3.29	3.13
Dexamethasone alcohol	4.99	4.97	5.00	5.00	4.75	4.37
Prednisolone acetate	7.21	7.19	7.22	7.22	6.67	5.96
Dexamethasone acetate	12.23	12.21	12.27	12.27	10.90	9.37

Ref. Ng LL (1987) J. Assoc. Off. Anal. Chem. 70:829

163 Prednisone

CAS-No	53-03-2; 125-10-0 (acetate)			
Chemical name	17α,21-Dihydroxypregna-1,4-diene-3,11,20-trione			
Empirical formula	Prednisone	Acetate		
	$C_{21}H_{26}O_5$	$C_{23}H_{28}O_6$		
Molecular mass	358.4	400.5		
Proprietary names	Deltacortone; Sterapred; Ultracorten			
Clinical use	Corticosteroid			
Solubility,	Water	alcohol	chloroform	methanol
Prednisone	insoluble	1 in 190	1 in 200	slightly soluble
Acetate	insoluble	1 in 120	1 in 6	
Light absorption	λ_{max} = 240 nm in ethanol			

HPLC 1

Sample	Blood plasma
Apparatus	Waters M 6000A LC pump, U6K injector and 440 UV detector

Column	Normal phase; Lichrosorb SI-60, 5 μm, 250 × 3.2 mm, ID
Mobile phase	Glacial acetic acid/ethanol/methylene chloride/n-hexane (0.2:3.5:30:66.3)
Flow-rate	2.5 ml min^{-1}; 2000 psi
Internal standard	Dexamethasone
Detection	UV, λ = 254 nm
Retention times, min	Prednisone 7.2 Hydrocortisone 11.8 Dexamethasone 8.7 Prednisolone 17.2
Ref.	Loo JCK and Jordan N (1977) J. Chromatogr. Biomed. Appl. 143:314

HPLC 2

Sample	Swine blood plasma
Extraction	Solid-phase, with acetonitrile on Bond-Elut C_{18} column or liquid-liquid extraction with methylene chloride/diethyl ether (50:50)
Apparatus	Waters M 6000A HPLC pump with WISP 610B automatic injector and 420 fixed wavelength detector
Column	Normal phase; Lichrosorb Si-60, 5 μm, 250 × 4.6 mm ID
Mobile phase	Methylene chloride/water-saturated methylene chloride/tetrahydrofuran/methanol/glacial acetic acid (664.5:300:10:25:0.5)
Flow-rate	0.8 ml min^{-1}
Temperature	Ambient
Internal standard	Dexamethasone, 60 ng in 100 μl
Injection size	100 μl
Detection	UV, λ = 254 nm; 0.005 AUFS
Recording	Waters M 730 data module
Retention times, min	Prednisolone acetate 8.5 Dexamethasone (IS) 23.0 Cortisone 13.0 Hydrocortisone 27.5 Prednisone 17.0 Prednisolone 37.0
Ref.	Prasad VK, Ho B and Haneke C (1986) J. Chromatogr. Biomed. Appl. 378:305

HPLC 3

Sample	Swine blood plasma
Extraction	Ethyl acetate or with methyl tert.-butyl ether
Apparatus	Waters M 6000A HPLC pump with 710B automatic injector and Kratos M 773 variable-wavelength detector
Column	Reversed-phase; Zorbax C_{18}, 6 μm, 250 × 4.6 mm ID

Pre-column	Reversed-phase; Brownlee Spherical C_{18}, 5 µm, 30 × 4.6 mm ID
Mobile phase	Tetrahydrofuran/water (25:75, v/v)
Flow-rate	1 ml min^{-1}; 3002–3205 psi
Temperature	Ambient (25 °C)
Injection size	100 µl containing 10 ng
Detection	UV, λ = 240–242 nm; 0.004 AUFS
Recording	Waters 730 data module
Retention times, min	Prednisone 10.20 Cortisone 11.30 Prednisolone 12.30 Hydrocortisone 13.30 Methylprednisolone 19.70 Dexamethasone 26.10 Prednisolone acetate 38.10 Cortisone acetate 39.30 Hydrocortisone acetate 42.30
Ref.	Carlin AS, Prasad VK et al (1988) J. Chromatogr. Biomed. Appl. 425:162

HPLC 4

Sample	Bulk drugs and elixirs
Apparatus	HPLC system, ultraviolet detector and Valco Instruments injector valve
Column	Normal phase; Zorbax-Sil 5 µm, 250 × 4.6 mm ID
Pre-column	Normal phase; silica 10 µm, 30 × 4.6 mm ID
Mobile phase	Glacial acetic acid/methanol/methylene chloride (1:45:954, v/v)
Flow-rate	1.6 ml min^{-1}
Injection size	15–30 µl
Detection	UV, λ = 254 nm
Recording	Shimadzu C-R1A integrator with strip-chart recorder

Retention times, min

Cortisone acetate	2.91	Dexamethasone	7.35
Hydrocortisone acetate	3.35	Betamethasone	8.13
Cortisone	4.65	Prednisolone	10.04
Prednisone	5.25		

Ref. Bunch EA (1987) J. Assoc. Off. Anal. Chem. 70:967

HPLC 5

Sample	Mixture of synthetic corticosteroids
Apparatus	Varian M 4000 HPLC with UV detector
Column	Normal phase; Micropak Si-10, 250 × 2.1 mm ID
Mobile phase	5% Ethanol in methylene chloride
Flow-rate	1.55 ml min^{-1}; 1000 psi
Injection size	1–2 µl

Detection	UV, λ = 254 nm; 0.08 AUFS
Retention times, min	Fluocinolone acetonide 1.28
	Triamcinolone acetonide 1.28
	Prednisone 1.62
	Dexamethasone 2.19
	Hydrocortisone 2.67
	6-Methylprednisolone 3.43
	Triamcinolone 4.90
Ref.	Gaetani E and Laureri CF (1976) Varian Liquid Chromatography at work 29. Varian, California

HPLC 6

Sample	Urine
Extraction	Ethyl acetate on Chem-Elut cartridge
Column	Normal phase; silica 5 µm, 250 × 4.6 mm ID
Mobile phase	Methylene chloride/methanol/tetrahydrofuran/glacial acetic acid (96.9:2.0:1.0:10.1)
Flow-rate	1.3 ml min^{-1}
Internal standard	Dexamethasone
Injection size	50 µl
Detection	UV, λ = 254 nm
Retention times, min	Prednisone 6.4; Prednisolone 12.4
	Dexamethasone 8.0; 6β-Hydroxyprednisolone 21.0
Ref.	Teng R-L and Benet LZ (1989) J. Chromatogr. Biomed. Appl. 493:421

164 Procainamide

CAS-No	51-06-9; 614-39-1 (hydrochloride)			
Chemical name	4-Amino-*N*-(2-diethylaminoethyl)benzamide			
Empirical formula	Procainamide	Hydrochloride		
	$C_{13}H_{21}N_3O$	$C_{13}H_{21}N_3O$, HCl		
Molecular mass	235.3	271.8		
Proprietary names	Novocamid; Procapan; Pronestyl			
Clinical use	Antiarrhythmic			
Solubility,	Water	ethanol	chloroform	ether
Hydrochloride	1 in 0.25	1 in 2	1 in 140	insoluble
Light absorption	λ_{max} = 275 nm in aqueous alkali			

HPLC 1

Sample	Blood plasma
Extraction	Methylene chloride under alkaline condition

Apparatus	Altex M 110A HPLC pump with JY3 spectrofluorimeter (Jobin et Yvon)
Column	Reversed-phase; Nucleosil C_{18}, 5 μm, 150 × 4.8 mm ID
Mobile phase	Methanol/acetic acid/triethylamine/water (200:10:5:785, v/v)
Flow-rate	2 ml min^{-1}
Temperature	Ambient
Internal standard	N-Propionylprocainamide
Injection size	100 μl, 4.9 μg ml^{-1}
Detection	Fluorescence, λ = 290/335 nm
Retention times, min	Procainamide 1.8 N-Acetylprocainamide 3.0 Internal standard 4.5
Ref.	Raphanaud D, Borensztejn M et al (1986) Ther. Drug Monit. 8:365

HPLC 2

Sample	Serum
Extraction	Diethyl ether under alkaline condition
Apparatus	Varian 5020 LC pump with UV detector
Column	Reversed-phase; Micropak MCH-10, 300 × 4 mm ID
Mobile phase	0.1 M KH_2PO_4, 0.02 M tetrametyhlammonium chloride and 13% acetonitrile
Flow-rate	2 ml min^{-1}
Temperature	30 °C
Internal standard	N-Propionylprocainamide
Injection size	10 μl containing at least 50 ng
Detection	UV, λ = 280 nm; 0.02 AUFS
Retention times, min	(estimated from drawing) Procainamide 1.9 N-Acetyl procainamide 5.8 N-Propionyl-procainamide 10.0
Ref.	Larson L: Varian Liquid Chromatography at works LC 89. Varian, California

HPLC 3

Sample	Raw materials
Extraction	Dichloromethane
Apparatus	Waters M 45 LC pump, U6K injector and 440 UV detector
Column	Reversed-phase; Lichrosorb C_8, 10 μm, 250 × 4.6 mm ID
Mobile phase	Water/tetrahydrofuran/triethylamine (96:4:0.25), pH 2.9 with 2 M nitric acid
Flow-rate	2 ml min^{-1}

Internal standard	N-Propionylprocainamide
Injection size	30 µl
Detection	UV, λ = 254 nm; 0.020 AUFS
Retention times, min	(estimated from drawing) Caffeine 5.8 Procainamide 3.0 Internal standard 9.0 N-Acetylprocainamide 5.0
Ref.	Willems HJJ, Van der Horst A et al (1985) Pharm. Weekbl. Sci. 7:150

HPLC 4

Sample	Raw materials and capsules
Column	Reversed-phase; ODS 10 µm, 300 × 3.9 mm ID
Mobile phase	Acetonitrile (3 in 20) and triethylamine (1 in 1000) in 0.02 M aqueous sodium acetate previously adjusted pH 4.5 with glacial acetic acid
Flow-rate	1.2 ml min^{-1}
Internal standard	Procaine hydrochloride
Injection size	Appr. 10 µl
Detection	UV, λ = 254 nm
Retention times	(relative) Procainamide hydrochloride 0.5 Procaine hydrochloride 1.0
Ref.	Anonym (1985): The United States Pharmacopeia XXI and The National Formulary XVI. United States Pharmacopeial Convention Inc., Rockville, p 880

HPLC 5

Sample	Serum
Extraction	Methylene chloride under alkaline condition, and then with mobile phase
Apparatus	Waters M 6000 LC pump, U6K injector and Perkin-Elmer LC-55 UV detector
Column	Reversed-phase; µ-Bondapak C_{18}
Mobile phase	0.02 M Sodium acetate, pH 4.5 with acetic acid/acetonitrile (85:15)
Flow-rate	1.5 ml min^{-1}
Temperature	Ambient
Internal standard	Procaine
Injection size	25 or 50 µl
Detection	UV, λ = 275 or 280 nm; 0.002 AU
Retention times, min	(estimated from drawing) N′-Acetylprocainamide 4.8 Procainamide 4.2 Procaine 7.4
Ref.	Su SC and Au WYW (1978) J. Liq. Chromatogr. 1:783

HPLC 6

Sample	Serum and urine
Extraction	Methylene chloride/2-propyl alcohol (9:1)
Apparatus	Varian M 2010 LC pump, 2050 UV detector and Rheodyne M 7125 injector
Column	Phenyl phase; 5 µm, 150 × 4.5 mm ID
Mobile phase	0.075 M Acetate buffer, pH 4.3/acetonitrile (20:3)
Flow-rate	1.0 ml min^{-1}
Injection size	50 µl, 0.0781–7.50 µg ml^{-1}
Detection	UV, λ = 270 nm
Retention times, min	(estimated from drawing)

	Deethylprocainamide	4.0
	Deethyl-n-acetylprocainamide	5.7
	Caffeine	6.8
	Procainamide	9.5
	N-Acetylprocainamide	12.2

Ref. Coyle, JD, Mackichan JJ et al (1987) J. Pharm. Sci. 76:402

---

165 Procaine

CAS-No	59-46-1; 51-05-8 (hydrochloride)
Chemical name	2-Diethylaminoethyl 4-aminobenzoate
Empirical formula	Procaine $C_{13}H_{20}N_2O_2$; Hydrochloride $C_{13}H_{20}N_2O_2$, HCl
Molecular mass	236.3 ; 272.8
Proprietary names	Gero H3 Aslan; Novocain; Venocaina
Clinical use	Local anaesthetic/Analgesic/Geriatics

Solubility,	Water	alcohol	chloroform	ether
Procaine		soluble	soluble	soluble
Hydrochloride	1 in 1	1 in 25	slightly sol	insoluble

Light absorption λ_{max} = 296 nm in methanol

HPLC 1

Sample	Raw materials and dosage forms
Apparatus	Hewlett-Packard 1084 LC and UV detector
Column	Reversed-phase; Nucleosil C_8, 10 µm, 250 × 4 mm ID
Mobile phase	Acetonitrile/0.05 M ortho-phosphoric acid (50:50)
Flow-rate	3 ml min^{-1}
Temperature	40 °C
Internal standard	Hydroxycaine

Injection size	5–30 μl
Detection	UV, λ = 296 nm
Recording	Hewlett-Packard LC-Terminal 79850B integrator
Retention times, min	Procaine 1.90 Hydroxycaine 2.65
Ref.	Rehm KD and Steinigen M (1981) Pharm. Ztg. 126:99

HPLC 2

Sample	Antiarrhythmics
Apparatus	HPLC low-pressure ternary gradient system
Column	Reversed-phase; Superspher 100 C_{18}, 125 × 4 mm ID
Mobile phase	Acetonitrile/0.05% triethylamine acetate, pH 4 (20:80, v/v)
Flow-rate	1.5 ml min^{-1}; 3000 psi
Injection size	10 μl containing 10 μg
Detection	UV, λ = 220 nm; 0.16 AUFS
Retention times	Procaine, lidocaine, quinidine and benzocaine, in ascending order (from drawing)
Ref.	Merck Application Manual on HPLC, TLC and HPTLC (1986). E. Merck, Darmstadt, p 81

HPLC 3

Sample	Raw materials
Apparatus	Waters M 6000A LC pump, 440 UV detector and Valco sample loop
Column	Normal phase; Nucleosil 100, 10 μm, 200 × 4 mm ID
Mobile phase	Sodium phosphate buffer, ionic strength 0.1, pH 2.2 containing + 1% (v/v) 1-pentanol
Flow-rate	1.6 mm sec^{-1}; 7.7 MPa
Temperature	25 °C
Injection size	10 μl
Detection	UV, λ = 254 nm; 0.0005 A
Retention times, min	(estimated from drawing) Benzocaine 2.35 Procaine hydrochloride 5.20 Lidocaine hydrochloride 6.90 Amylocaine hydrochloride 8.20
Ref.	Crommen J (1979) J. Chromatogr. 186:705

HPLC 4

Sample	Mixtures of Cocaine, local anaesthetics and other amines
Apparatus	Waters M 6000A LC pump, U6K injector and 440 UV detector

Column	Reversed-phase; μ-Bondapak C_{18}, 300 × 3.9 mm ID
Pre-column	ODS; Co:Pell, 70 × 2.1 mm ID
Mobile phase	Phosphate buffer, pH 3.0 (containing 9.2 g NaH_2PO_4 in 1 liter water, pH 3.0 with 2 N H_3PO_4)/ methanol (2:1)
Flow-rate	2 ml min^{-1}
Temperature	Ambient
Injection size	10 μl
Detection	UV, $\lambda = 254$ and 280 nm
Retention times	(relative) cocaine = 1.00 (12.8 ml)

Acetaminophen	0.31
Procaine	0.32
Phenylpropanolamine	0.33
Ephedrine	0.34
Pseudoephedrine	0.34
Caffeine	0.58
Lidocaine	0.58
Prilocaine	0.58
Mepivacaine	0.59
Benzoylecgonine	0.78
Piperocaine	1.41
Benzocaine	1.59
cis-Cinnamoylcocaine	1.86
trans-Cinnamoylcocaine	2.81
Tetracaine	3.09
Dibucaine	11.56

Ref. Noggle, Jr. FT and Clark CR (1983) J. Assoc. Off. Anal. Chem. 66:151

166 Progesterone

CAS-No	57-83-0
Chemical name	Pregn-4-ene-3,20-dione
Empirical formula	$C_{21}H_{30}O_2$
Molecular mass	314.5
Proprietary names	Gestone; Lutogyl; Progestosol
Clinical use	Progestational Steroid

Solubility

Water	alcohol	ether	chloroform	arachis oil
insoluble	1 in 8	1 in 16	1 in < 1	1 in 60

Light absorption: $\lambda_{max} = 240$ nm in dehydrated alcohol

HPLC 1

Sample	Rat tissue
Apparatus	Perkin-Elmer Series 3B LC pump with gradient controller and ultraviolet detector
Column	Reversed-phase; Policonsult C_{18}, 5 µm, 250 × 4.6 mm ID
Pre-column	Supelguard LC-18, 5 µm, 20 mm
Mobile phase	Acetonitrile/water (binary solvent) Gradient (multistep): Acetonitrile 5%, 20%, 28%, 45% and 70% at 0, 10, 25, 50 and 60 min
Flow-rate	1.2 ml min^{-1}
Temperature	Ambient
Injection size	10–20 µl
Detection	UV, λ = 254 nm
Recording	Computing integrator
Retention times, min	Aldosterone 25.4 Testosterone 43.7 Progesterone 60.7
Ref.	Lagana A, D'Ascenzo G and Marino A (1987): Chromatographia 23, 796

HPLC 2

Sample	Rat blood plasma
Extraction	Solid-phase extraction
Apparatus	LKB M 2150 and Waters M 6000A pumping systems, Water 440 ultraviolet detector and Shimadzu SPD-6A variable-wavelength detector
Column	Reversed-phase; Hypersil ODS 5 µm, 100 × 1 mm ID
Mobile phase	I. Acetonitrile/water (35:65) II. Methanol/water (50:50) III. Acetonitrile/methanol/water (25:25:50)
Flow-rate	100 µl min^{-1}
Injection size	20 µl, 2 µg ml^{-1}
Detection	UV; 0.01 AUFS
Retention times, min	(estimated from drawing)

Mobile phase	I.	II.	III.
Testosterone	6.50	10.50	7.96
17α-Hydroxyprogesterone	8.94	13.00	9.12
Androstenedione	9.46	8.00	6.81
20α-Hydroxy-4-pregnen-3-one	15.00	28.73	15.23
Progesterone	27.93	28.73	20.20
Norethindrone	37.16	28.73	24.47

Ref.	Taylor RB, Kendle KE and Reid RG (1987): J. Chromatogr. 385, 383

HPLC 3

Sample	Blood plasma
Extraction	Centrifugation
Apparatus	Yanaco M L-100 and Yanaco L-4000W for pretreatment and analytical pumps respectively, ultraviolet detector and manual injector
Column	Reversed-phase; Yanapak ODS-A, 250 × 4.6 mm ID
Pre-column	Yanapak C_8, 30 × 4 mm ID for pretreatment
Mobile phase	For pretreatment: Water, methanol and 50 mM potassium phosphate buffer, pH 3.1/acetonitrile (8:2, v/v) For analysis: 50 mM Potassium phosphate buffer, pH 3.1/acetonitrile/methanol (10:7:3, v/v)
Flow-rate	1 ml min^{-1}
Temperature	25 °C
Internal standard	4′-Methoxyoctananilide
Injection size	98.2 ng
Detection	UV, λ = 245 nm; 0.001 AUFS
Recording	Data processor (System Instruments)
Retention times, min	Testosterone 20.5 Androstenedione 23.2 17-α-Hydroxyprogesterone 26.7 Internal standard 45.9 20-α-Hydroxyprogesterone 51.6 Progesterone 75.5
Ref.	Suzuki Y, Hayashi N and Sekiba K (1988) J. Chromatogr. Biomed. Appl. 426:33

HPLC 4

Sample	Serum from pregnant women
Extraction	Methylene chloride after deproteinization with acetonitrile
Apparatus	Waters M 6000A and 45 HPLC pumps, U6K injector, 680 gradient controller and 440 ultraviolet detector
Column	Reversed-phase; Vydac HS C_{18}, 10 μm, 250 × 4.6 mm ID
Pre-column	HS C_{18}
Mobile phase	A. 1 mM KH_2PO_4/methanol (48:52, v/v) B. Methanol/isopropanol (50:50, v/v) Linear gradient: 100% A for 8 min; B 0% to 34% in next 26 min
Flow-rate	1.5 ml min^{-1}
Internal standard	6β-Hydroxyprogesterone
Injection size	50 ng
Detection	UV, λ = 254 nm; 0.01 AUFS

Recording	LKB two-channel variable millivolt setting recorder	
Retention times, min	Aldosterone	8.3
	Cortisone	9.7
	Cortisol	11.4
	Prednisolone	11.4
	21-Desoxycortisol	16.2
	21-Desoxycortisone	17.0
	Betamethasone	17.2
	Dexamethasone	17.4
	Corticosterone	18.6
	11-Desoxycortisol	19.8
	16α-Hydroxyprogesterone	20.8
	11α-Hydroxyprogesterone	21.0
	6β-Hydroxyprogesterone	23.1
	Deoxycorticosterone	25.4
	11β-Hydroxyprogesterone	25.7
	Testosterone	25.8
	17α-Hydroxyprogesterone	27.0
	Deoxycorticosterone-hemisuccinate	29.8
	Depo-Provera	31.2
	Progesterone	34.8
	20β-Hydroxyprogesterone	36.0
Ref.	Kuhn RW and Deyman ME (1987) J. Chromatogr. Biomed. Appl. 421:123	

HPLC 5

Sample	Serum	
Extraction	Chloroform/methanol (90:10, v/v) on Supelcosil Carbopack B column	
Apparatus	Perkin-Elmer Series 3-LC, LC 95 UV detector and Rheodyne M 7125 injector	
Column	Reversed-phase; Policonsult C_{18}, 5 µm, 250 × 4.6 mm ID	
Pre-column	Policonsult C_{18}, 5 µm	
Mobile phase	Acetonitrile/water (46:54, v/v)	
Flow-rate	1.6 ml min^{-1}	
Injection size	20 µl containing 0.2 ng ml^{-1}	
Detection	UV, $\lambda = 242$ nm	
Retention times, min	18-Hydroxycorticosterone	2.1
	Aldosterone	2.4
	Prednisone	2.7
	Cortisol	2.7
	Cortisone	2.7
	7α-Hydroxytestosterone	2.7
	Prednisolone	2.7

	19-Hydroxyandrostenedione	3.4
	11-Dehydrocorticosterone	4.5
	21-Deoxycortisol	4.8
	Adrenosterone	5.1
	Corticosterone	5.4
	11-Deoxycortisol	6.0
	6α-Hydroxyprogesterone	7.2
	16α-Hydroxyprogesterone	7.4
	Testosterone	10.6
	Androstenedione	10.9
	11-Deoxycorticosterone	11.0
	17α-Hydroxyprogesterone	11.5
	20α-Dihydroprogesterone	12.6
	Progesterone	13.6
Ref.	Lagana A, D'Ascenzo G et al (1986) Clin. Chem. 32:508	

HPLC 6

Sample	Test solution of steroid hormones
Apparatus	Knauer 64.00 HPLC pump with 731.710 UV detector
Column	Reversed-phase; Nucleosil C_{18}, 3 μm, Microbore, 100 × 1 mm ID
Mobile phase	Methanol/tetrahydrofuran/water (15:25:60)
Flow-rate	35 μl min^{-1}
Injection size	2 μl, 0.01 mg ml^{-1}
Detection	UV, λ = 254 nm
Retention times, min	(estimated from drawing)

Aldosterone	2.5	Estrone	17.0
Estriol	4.0	17-β-Estradiol	19.5
Androstenedione	7.5	Progesterone	23.0
Testosterone	9.0		

Ref. Knauer Application Notes on Micro HPLC and Microbore HPLC Column. Knauer, Berlin

---

167 Promazine

CAS-No	58-40-2; 53-60-1 (hydrochloride)	
Chemical name	*NN*-Dimethyl-3-(phenothiazin-10-yl)propylamine	
Empirical formula	Promazine	Hydrochloride
	$C_{17}H_{20}N_2S$	$C_{17}H_{20}N_2S$, HCl
Molecular mass	284.4	320.9
Proprietary names	Prazine; Protactyl; Sparine	
Clinical use	Tranquillizer	

Solubility,	Water	alcohol	chloroform	ether
Hydrochloride	1 in 1	1 in 2	1 in 2	insoluble
Light absorption	λ_{max} = 251 nm in aqueous acid			

HPLC 1

Sample	Mixture of phenothiazine-derivatives
Column	Normal phase; silicagel 6.2 µm, 50 × 4 mm ID
Mobile phase	Diisopropylether/methanol (1:1) containing 2.6% water and 0.2% triethylamine
Flow-rate	0.8 ml min^{-1}; 218 psi
Detection	UV, λ = 254 nm
Retention times, min	(estimated from drawing)

3-Chlorphenothiazine	0.52
Chlorphenothiazine	0.85
Chlorpromazine	1.20
Promazine	1.85
5,5-Dioxy-chlorpromazine	2.28
Oxychlorpromazine	3.30
2-Chlor-10-(3-methyl-aminopropyl)-phenothiazine	3.89
N-Oxychlorpromazine	5.43

Ref. Meyer V (1984): Praxis der Hochleistungs-fluessigchromatographie, 3 edn. Diesterweg Salle and Sauerlaender, Frankfurt am Main and Salzburg, p 33

HPLC 2

Sample	Mixture of phenothiazine derivatives
Apparatus	Varian M 4000 HPLC
Column	Normal phase; Micropak Si 10 µm, 250 × 2.1 mm ID
Mobile phase	Methanol/methylene chloride (50:50)
Flow-rate	0.64 ml min^{-1}; 750 psi
Detection	UV, λ = 254 nm; 0.08 AUFS
Retention times, min	

Dixyrazine	1.98	Trifluperazine	3.15
Fluphenazine	1.98	Levomepromazine	4.20
Propericazine	2.34	Chlorpromazine	5.79
Butaperazine	3.15	Promazine	9.06

Ref. Gaetani E and Laureri CF (1976)
Varian Liquid Chromatography at Works 28.
Varian, California

HPLC 3

Sample	Test substance
Column	Normal phase; Econosphere Si 5 µm, 250 × 4.6 mm ID
Mobile phase	Ethanol/acetonitrile/t-butylamine (98:2:0.05)

Flow-rate	2.0 ml min^{-1}
Detection	UV; 0.2 AUFS
Retention times, min	(estimated from drawing) Promazine 3.8
	Trimipramine 2.2 Nortriptyline 4.8
	Amitriptyline 2.5 Desipramine 6.9
	Doxepin 2.7 Protriptyline 8.7
	Imipramine 3.0
Ref.	Alltech Drug Standards (1988) Alltech, Bulletin No. 153, State College, p 35

HPLC 4

Sample	Pharmaceutical formulations
Extraction	Methanol or dilute hydrochloric acid
Apparatus	Altex M 110 LC pump, Waters M 440 UV detector and Rheodyne injector
Column	Cyanopropyl phase; Zorbax CN 5 µm, 150 × 4.6 mm ID
Mobile phase	0.025 M Sodium acetate and acetic acid buffer, pH 4.5/acetonitrile/methanol (30:40:30, v/v)
Flow-rate	1 ml min^{-1}
Temperature	Ambient
Internal standard	Norephedrine
Injection size	10 µl, 0.01–0.04 mg ml^{-1}
Detection	UV, λ = 254 nm; 0.05 AUFS
Recording	Hewlett-Packard 3390A integrator
Retention times, min	Norephedrine 2.38 Desipramine 4.20
	Fluphenazine 4.15 Promazine 4.45
Ref.	Beaulieu N, Gagne C and Lovering EG (1986) J. Assoc. Off. Anal. Chem. 69:178

168 Promethazine

CAS-No	60-87-7; 58-33-3 (hydrochloride)				
Chemical name	1,*N*,*N*-Trimethyl-2-(phenothiazin-10-yl)-ethylamine				
Empirical formula	Promethazine		Hydrochloride		
	$C_{17}H_{20}N_2S$		$C_{17}H_{20}N_2S$, HCl		
Molecular mass	284.4		320.9		
Proprietary names	Atosil; Phenergan; Tixylin				
Clinical use	Antihistamine				
Solubility,	Water	alcohol	chloroform	acetone	ether
Promethazine	insoluble	1 in 1		1 in 1	1 in 2
Hydrochloride	1 in 0.6	1 in 9	1 in 2	insol	insoluble
Light absorption	λ_{max} = 249 nm in aqueous acid				

HPLC 1

Sample	Raw materials
Apparatus	Spectra-Physics 3500B LC pump and 770 ultraviolet detector
Column	Reversed-phase; μ-Bondapak C_{18}, 10 μm, 300 × 3.9 mm ID
Mobile phase	Methanol/acetic acid/triethylamine/water (60:1.5:0.5:38)
Flow-rate	1.5 ml min^{-1}
Temperature	Ambient
Injection size	10 μl, 0.25 mg ml^{-1}
Detection	UV, λ = 254 nm; 0.32 AUFS
Recording	Hewlett-Packard 3380A integrator
Retention times, min	(estimated from drawing)

8-Chlorotheophylline	2.3	Diphenylpyraline	8.7
Diphenhydramine	4.8	Promethazine	10.8

Ref. Roos RW and Lau-Cam CA (1986) J. Chromatogr. 370:403

HPLC 2

Sample	Antihistamines
Apparatus	HPLC low-pressure ternary gradient system
Column	Reversed-phase; Lichrosorb C_{18}, 125 × 4 mm ID
Mobile phase	Triethylamine acetate buffer, pH 4.0/0.1% acetonitrile (40:60, v/v)
Flow-rate	1.5 ml min^{-1}; 2700 psi
Injection size	15 μl containing 0.5 μg
Detection	UV, λ = 220 nm; 0.04 AUFS
Retention times	Phenylpropanolamine, diphenhydramine, pseudoephedrine and promethazine in ascending order (from drawing)
Ref.	Merck Application Manual on HPLC, TLC and HPTLC (1987). E. Merck, Darmstadt, p 73

HPLC 3

Sample	Serum
Extraction	Hexane under alkaline condition
Apparatus	Varian M 5020 HPLC with UV detector
Column	Reversed-phase; MCH-10, 300 × 4 mm ID
Mobile phase	Methanol/water (84:16, v/v)
Flow-rate	2.0 ml min^{-1}
Internal standard	Triflupromazine
Derivatization	Pre-column; with trichloroethyl chloroformate
Injection size	50 μl, 0–20 μg ml^{-1}
Detection	UV, λ = 254 nm

Retention times, min	(estimated from drawing)
	Promethazine 4.0
	Internal standard 6.0
Ref.	Wallace JE, Shimek EL et al (1981) Clin. Chem. 27:253

HPLC 4

Sample	Blood
Extraction	n-Heptane/dichloromethane/isoamylalcohol (88.5:10:1.5) under alkaline condition
Apparatus	Pye-Unicam LC-3 pump with LC-3 UV detector
Column	Reversed-phase; Hypersil 5-SAS, 100 × 4.8 mm ID
Mobile phase	Methanol containing 30% v/v of 0.05 M Sorensen's phosphate buffer, pH 7.4
Flow-rate	0.7 ml min^{-1}
Internal standard	Imipramine hydrochloride
Injection size	10–25 μl, 2–50 ng ml-1
Detection	UV, λ = 248 nm
Retention times, min	Promethazine sulfoxide 6.4
	Monodesmethylpromethazine 7.5
	Promethazine 9.3
	Internal standard 14.4
Ref.	Taylor G and Houston JB (1982) J. Chromatogr. Biomed. Appl. 230:194

HPLC 5

Sample	Whole blood and parotid saliva
Extraction	Heptane
Apparatus	Waters M 200 LC, loop injector and UV detector
Column	Cyano phase; μ-Bondapak CN, 300 × 3.9 mm ID
Mobile phase	1% Acetic acid in 0.005 M heptanesulfonic acid/methanol (50:50)
Flow-rate	2 ml min^{-1}
Temperature	Ambient
Internal standard	Promazine
Injection size	10 μl
Detection	UV, λ = 254 nm; 0.005 AUFS
Retention times, min	(estimated from drawing)
	Promethazine 12.5
	Promazine 13.7
Ref.	Digregorio GJ and Ruch E (1980) J. Pharm. Sci. 69:1457

169 Propranolol

CAS-No	525-66-6 and 13013-17-7 (±); 318-98-9 and 3506-09-0 (±) (hydrochloride)			
Chemical name	(±)1-Isopropylamino-3-(1-naphthyloxy)propan-2-ol			
Empirical formula	Propranolol	Hydrochloride		
	$C_{16}H_{21}NO_2$	$C_{16}H_{21}NO_2$, HCl		
Molecular mass	259.3	295.8		
Proprietary names	Dociton; Inderal; Tonum			
Clinical use	β-Adrenolytic			
Solubility,	Water	ethanol	chloroform	ether
Hydrochloride	1 in 20	1 in 20	slightly soluble	insoluble
Light absorption	λ_{max} = 290 nm in methanol			

HPLC 1

Sample	Blood plasma and breast milk
Extraction	Diethyl ether under alkaline and acidic conditions (plasma), and with chloroform/methanol (4:1) at pH 1.0 (breast milk)
Apparatus	Waters M 6000A HPLC pumps, Schoeffel FS 970 fluorescence detector and Rheodyne M 7120 injector
Column	Reversed-phase; μ-Bondapak C_{18}
Mobile phase	A. For plasma (alkaline): Methanol/butyl alcohol/acetonitrile/sodium sulfate/water (80:35:25:5 g to 500) B. For plasma (acidic): Methanol/CH_3COOH (glacial)/water (180:1:to 500) C. For milk (acidic): As in B but (170:2:to 500)
Flow-rate	2 ml min^{-1}
Internal standards	4-Methylpropranolol and 2-naphthoxyacetic acid
Injection size	20 μl
Detection	Fluorescence, λ = 242/370 nm
Retention times, min	(estimated from drawing)

Mobile phase/extract	A. Plasma (alkaline)	B. plasma (acidic)	C. milk (acidic)
4-Hydroxypropranolol	2.50		
N-Desisopropylpropranol	4.25		
Propranolol	6.75		12.0
Propranolol glycol	9.00		
4-Methylpropranolol (IS)	11.50		
α-Naphthol	14.00		
1-Naphthoxylactic acid		6.00	16.00
2-Naphthoxyacetic acid (IS)		7.20	22.00
1-Naphthoxyacetic acid		8.82	

Ref.	Smith MT, Livingstone IL et al (1983) Ther. Drug Monit. 5:87

HPLC 2

Sample	Blood plasma
Extraction	Ethyl acetate at pH 10.3
Apparatus	Waters M 6000 LC pump with WISP 710B autosampler and Schoeffel FS 970-LC fluorimeter
Column	Phenyl phase; μ-Bondapak 10 μm, 300 × 4 mm ID
Mobile phase	Acetonitrile/water/acetic acid/methnaol/ triethylamine (350:700:10:400:0.05, v/v), pH 3.4
Flow-rate	2 ml min^{-1}
Internal standard	4-Methylpropranolol hydrochloride
Injection size	50 μl containing 8 ng
Detection	Fluorescence, $\lambda = 238/360$ nm
Retention times, min	4-Hydroxy-propranolol 2.0 Internal standard 7.5 Propranolol 5.9 (estimated from drawing)
Ref.	Koshakji RP and Wood AJJ (1987) J. Chromatogr. Biomed. Appl. 422:294

HPLC 3

Sample	Dosage forms
Apparatus	Altex M LC-100A pumps and Hitachi M 155 UV detector
Column	Reversed-phase; Ultrasphere C_8, 5 μm, 250 × 4.6 mm ID
Mobile phase	0.5 g Dodecyl sodium sulfate in 18 ml 0.15 M H_3PO_4, 90 ml methanol, 90 ml acetonitrile and fill to 250 ml with water
Flow-rate	1.5 ml min^{-1}
Internal standard	Procainamide hydrochloride
Injection size	20 μl, 0.03–3.1 mg ml^{-1}
Detection	UV, $\lambda = 290$ nm
Recording	Hewlett-Packard computing integrator HP 3390A
Retention times, min	Internal standard 3.8 Propranolol 5.8
Ref.	Olsen CS and Scroggins HS (1988) J. Assoc. Off. Anal. Chem. 71:76

HPLC 4

Sample	Tablets
Apparatus	Waters M 501-LC pump, Perkin-Elmer LC-420B automatic injector and LC-95 variable-wavelength detector

Column	Cyanopropylsilane; Ultrasphere CN 5 μm, 250 × 4.5 mm ID
Mobile phase	Acetonitrile/0.05 M ammonium phosphate monobasic (15:85), pH 3.0 with H_3PO_4
Flow-rate	2 ml min^{-1}
Internal standard	*p*-Dimethyl-aminobenzoic acid
Injection size	20 μl, 0.016–0.160 mg ml^{-1}
Detector	UV, λ = 290 nm
Recording	Spectra Physics SP-4270 computing integrator
Retention times, min	Chlorothiazide (synthetic impurity) 4.0
	Hydrochlorothiazide 4.5
	Propranolol (thermally treated, polar) 9.0
	Propranolol hydrochloride 14.8
	Unidentified (thermally treated) 15.9
Ref.	Hitscherich ME, Rydberg EM et al (1987) J. Liq. Chromatogr. 10:1011

HPLC 5

Sample	Blood plasma and serum
Extraction	n-Heptane and isoamylalcohol (98.5:1.5)
Apparatus	Shimadzu LC-3A pump and RF-500 spectrofluorophotometer
Column	Hitachi Gel 3013; 5 μm (styrene-divinyl benzene), 250 × 4 mm ID
Mobile phase	Ethanol/0.02 M $HClO_4$-$NaClO_4$ (pH 2.0) (65:35)
Flow-rate	2 ml min^{-1}; 580 psi
Temperature	30 °C
Internal standard	Penbutolol, 0.6 mg ml^{-1}
Injection size	10 μl
Detection	Fluorescence, λ = 285/340 nm
Retention times, min	Propranolol 14.5
	Internal standard 19.5
Ref.	Yamamura Y, Uchino K et al (1986) J. Chromatogr. Biomed. Appl. 374:311

HPLC 6

Sample	Human and animal blood plasma
Extraction	1% 1-Butanol in heptane
Apparatus	Waters M 6000 LC pump, WISP 710B autoinjector and Schoeffel FS-970 LC-fluorimeter
Column	Phenyl phase; μ-Bondapak 10 μm, 300 × 4 mm ID
Mobile phase	A. Acetonitrile/water/acetic acid/triethylamine (450:600:10, v/v), pH 3.1
	B. Acetonitrile/water/acetic acid/triethylamine (450:600:10:0.15, v/v), pH 3.5
Flow-rate	2 ml min^{-1}

Temperature	Ambient
Internal standard	N-Ethyl-propranolol
Injection size	50 µl
Detection	Fluorescence, λ = ex. 228 nm
Recording	Waters data module
Retention times, min	Mobile phase A. B.
	Propranolol 8.73 4.38
	Internal standard 11.80 5.75
Ref.	Koshakji RP, Wood AJJ (1986) J. Pharm. Sci. 75:87

HPLC 7

Sample	Paracetamol bound to human serum protein
Extraction	Mobile phase
Apparatus	Waters M 6000A LC pump, U6K injector and 450 UV detector
Column	Lichrosorb diol 10 µm, 150 × 4.7 mm ID
Mobile phase	Phosphate buffer, pH 7.4
Flow-rate	1 ml min^{-1}
Temperature	37 °C
Injection size	25 µl, $1–8 \cdot 10^{-4}$ M
Detection	UV, λ = 254 nm
Recording	Tektronix 4051 computer
Retention times, min	Propranolol 1.66
Ref.	Bree F and Tillement J-P (1986) J. Chromatogr. Biomed. Appl. 375:416

170 Propylparaben

CAS-No	94-13-3; 35285-69-9 (sodium salt)			
Chemical name	Propyl *p*-hydroxybenzoate			
Empirical formula	Propylparaben		Sodium Salt	
	$C_{10}H_{12}O_3$		$C_{10}H_{11}NaO_3$	
Molecular mass	180.2		202.2	
Proprietary names	Nipasol M; Nipasol M sodium			
Clinical use	Preservative			
Solubility,	Water	alcohol	acetone	chloroform
Propylparaben	1 in 2500	1 in 3.5	1 in 3	1 in 4
Sodium salt	1 in 1	1 in 50		
Light absorption				

HPLC 1

Sample	Raw materials and dosage forms
Apparatus	Hewlett-Packard 1084 LC with UV detector

Column	Reversed-phase; Nucleosil C_8, 10 μm, 250 × 4 mm ID
Mobile phase	Acetonitrile/0.05 M ortho-phosphoric acid (45:55)
Flow-rate	1.8 ml min^{-1}
Temperature	40 °C
Internal standard	Ethylparaben
Injection size	5–30 μl
Detection	UV, λ = 258 nm
Recording	Hewlett-Packard LC-Terminal 79850B integrator
Retention times, min	Methylparaben 2.80 Ethylparaben 3.52 Propylparaben 4.74
Ref.	Rehm KD and Steinigen M (1981) Pharm. Ztg. 126:99

HPLC 2

Sample	Mixture of phenolic preservatives in cover girl make up
Apparatus	Bioanalytical Systems LC-154 and LC-4 amperometric detector, Schoeffel SF-770 UV detector and Bioanalytical Systems CV-5 cyclic voltammograph
Column	Reversed-phase; μ-Bondapak C_{18}, 10 μm, 150 × 4.2 mm ID
Mobile phase	0.1 M Ammonium acetate, pH 6.0/methanol (50:50)
Flow-rate	1 ml min^{-1}
Internal standard	Ethylparaben
Injection size	20 μl containing 10–200 ng
Detection	EC, GC electrode, TLC-5 + 1.1 V vs Ag/AgCl RE
Retention times, min	(estimated from drawing) Ethylparaben 5.2 Methylparaben 3.8 Propylparaben 8.8
Ref.	King WP, Joseph KT and Kissinger PT (1980) J. Assoc. Off. Anal. Chem. 63:137

HPLC 4

Sample	Sterile liquid formulations
Apparatus	Waters M 6000A LC pump with Milton Roy minipump, and Waters 440 and LDC M 1203 UV detector and autosampler
Column	Reversed-phase; μ-Bondapak C_{18}
Mobile phase	Water/acetonitrile (70:30) containing 0.01 M potassium acetate, pH 5.4 with glacial acetic acid
Flow-rate	2 ml min^{-1}
Internal standard	Ethylparaben
Injection size	10 μl, 0.15 mg ml^{-1}
Detection	UV

Retention times, min	(estimated from drawing)		Ethylparaben	4.2
	Sorbic acid	2.4	Propylparaben	6.0
	Methylparaben	3.2		
Ref.	Radus TP and Gyr G (1983) J. Pharm. Sci. 72:221			

HPLC 5

Sample	Cosmetic emulsion
Extraction	95% Ethanol
Column	Reversed-phase; Vydac-ODS, 300 × 2 mm ID
Mobile phase	Methanol/water (20:80)
Flow-rate	0.33 ml min^{-1}; 500 psi
Temperature	Ambient
Internal standard	Ethylparaben
Injection size	3 µl containing at least 30 ng
Detection	UV, λ = 254 nm

Retention times, min	(estimated from drawing)		Propylparaben	10.0
	Methylparaben	3.0	Butylparaben	26.0
	Ethylparaben	5.0		
Ref.	Nelson JJ (1978) in: Tsuji K (ed) GLC and HPLC Determination of Therapeutic Agents, Part 2. Marcel Dekker Inc., New York and Basel, p 807			

---

171 Prostaglandin $F_{2\alpha}$

CAS-No	551-11-1 (dinoprost); 38562-01-5 (dinoprost trometamol)	
Chemical name	9,11,15-Trihydroxyprosta-5,13-dien-1-oic acid	
Empirical formula	Prostaglandin $F_{2\alpha}$	Trometamol
	$C_{20}H_{34}O_5$	$C_{20}H_{34}O_5,C_4H_{11}NO_3$
Molecular mass	354.5	475.6
Proprietary names	Amoglandin; Minprostin 2α; Prostin F2 alpha	
Clinical use	Prostaglandin	

Solubility	Water	methanol	ethanol	chloroform
Prostaglandin $F_{2\alpha}$	slightly sol	freely soluble		freely sol
Trometamol	1 in 5	soluble		slightly sol

HPLC 1

Sample	Human seminal plasma
Extraction	Ethyl acetate at pH 3.0–3.5 with HCl
Apparatus	Hitachi 655A HPLC with sample injector and F-1000 fluorescence spectromonitor
Column	Reversed-phase; YMC Pack C_8, 10 µm, 150 × 6.0 mm ID

Mobile phase	A. Acetonitrile/methanol/water (35:10:55) B. Acetonitrile/methanol/water (35:30:35) Gradient: 100% A for 26 min; 100% B for the next 10 min
Flow-rate	2.0 ml min^{-1}; 1450 psi
Temperature	20–27 °C
Derivatization	Pre-column; with 3-bromomethyl-6,7-dimethoxy-1-methyl-2(1H)-quinoxalinone
Injection size	10 µl
Detection	Fluorescence, λ = 370/455 nm
Retention times, min	

6-Keto-$PGF_{1\alpha}$	6.8	PGE_1	20.4
$PGF_{2\alpha}$	14.0	16-Methyl-$PGF_{1\alpha}$	23.2
$PGF_{1\alpha}$	15.2	PGA_2 and PGB_2	30.8
PGD_2	16.4	PGA_1 and PGB_1	32.8
PGE_2	17.6	(PG = Prostaglandin)	

Ref. Yamaguchi M, Fukuda K et al (1986) J. Chromatogr. Biomed. Appl. 380:257

HPLC 2

Sample	Nasal secretion
Extraction	Methyl formate on C_{18} Sep-Pak extraction column
Apparatus	Kontron M 414 LC pumps with 200 gradient controller and Rheodyne M 7125 injector
Column	Reversed-phase; µ-Bondapak C_{18}, 300 × 3.9 mm ID
Mobile phase	Acetonitrile/water (35:65, v/v), pH 3.4
Flow-rate	1.5 ml min^{-1}
Injection size	100 µl
Detection	Liquid scintillation counter
Retention times, min	

6-oxo-Prostaglandin $F_{1\alpha}$	4.5	Prostaglandin E_2	10.7
Prostaglandin $F_{2\alpha}$	8.5	Prostaglandin D_2	12.9

Ref. Ramis I, Serra J et al (1988) J. Pharm. Biomed. Anal. 6:1009

HPLC 3

Sample	Mixture of prostaglandins
Apparatus	Jasco FLC-350 HPLC with Uvidec-100 UV detector and VL-611 loop injector
Column	Reversed-phase; Jasco SC-02, 250 × 4.6 mm ID
Pre-column	5 cm long
Mobile phase	0.02 M Potassium dihydrogen ortho-phosphate/acetonitrile (3:2)
Flow-rate	1.4 ml min^{-1}
Temperature	Ambient
Internal standard	Prednisolone
Injection size	2–3 µl

Detection	UV, λ = 215 nm; 0.04 AUFS
Retention times, min	Internal standard 13.7; Prostaglandin $F_{2\alpha}$ 18.0; Prostaglandin $F_{1\alpha}$ 19.2; Prostaglandin E_2 20.6; Prostaglandin E_1 22.6
Ref.	Nagayo K and Mizuno N (1979) J. Chromatogr. 178:347

HPLC 4

Sample	Prostaglandins
Column	Reversed-phase; Nucleosil 5 µm, C_{18}, 250 × 4 mm ID
Mobile phase	Water/acetonitrile/tetrahydrofuran (70:30:2)
Flow-rate	1 ml min^{-1}; 1160 psi
Detection	UV, λ = 208 nm; 0.05 AUFS
Retention times, min	(Prostaglandin = PG)

6K-PG$F_{1\alpha}$	2.00	PGA_2	7.20
PG$F_{2\alpha}$	3.30	PGB_2	7.20
PGE_2	4.00	PGA_1	9.30
PGE_1	4.40	PGB_1	9.30

Ref. Inayama S, Hori H et al (1980) J. Chromatogr. 194:85

HPLC 5

Sample	Prostaglandin $F_{2\alpha}$ tromethamine
Column	Normal phase; silica 5–10 µm, 300 × 4.6 mm ID
Mobile phase	Methylene chloride/1,3-butanediol/water (496:3.5:0.25)
Flow-rate	1.5 ml min^{-1}
Derivatization	Pre-column; with α-bromo-2′-acetonaphthone
Internal standard	Guaifenesin
Injection size	Appr. 20 µl
Detection	UV, λ = 254 nm
Retention times	(relative)

Internal standard	0.4	Prostaglandin $F_{2\alpha}$ tromethamine	1.0
15-R epimer	0.5	5,6-trans isomer	1.2

Ref. Anonym (1985): The United States Pharmacopeia XXI and The National Formulary XVI, Supplement 2. United States Pharmacopeial Convention Inc., Rockville, p 1835

HPLC 6

Sample	Biological materials and test substances
Extraction	Chloroform (urine)
Apparatus	Waters M 6000A LC pumps with 660 programmer and liquid scintillation spectrophotometer

Column	Reversed-phase; μ-Bondapak 10 μm, for fatty acid analysis (μ-Porasil)
Mobile phase	A. Gradient: Chloroform to 6% methanol and 0.6% acetic acid in chloroform in 60 min B. Isocratic: Acetonitrile/water/benzene/acetic acid (23:76.7:0.2:0.1)
Flow-rate	1 ml min^{-1} (A); 2 ml min^{-1} (B)
Injection size	1–2 μg or 0.05–0.1 μCi
Detection	Radioactivity, and with gas chromatography and mass spectrometry

Retention times, min

System (PG = Prostaglandin)	gradient	isocratic
$15KH_2E_2$	22.0	48.0
Prostaglandin PGD_2	28.0	30.0
PGE_1	36.0	40.0
PGE_2	36.0	30.0
$6K\text{-}PGF_{1\alpha}$	36.0	10.0
Thromboxane B_2	36.0	17.0
$PGF_{2\alpha}$	46.0	21.0

Ref. Whorton AR, Carr K et al (1979) J. Chromatogr. Biomed. Appl. 163:64

172 Pseudoephedrine

CAS-No	90-82-4; 7460-12-0 (sulfate)		
Chemical name	(1*S*,2*S*)-2-Methylamino-1-phenylpropan-1-ol		
Empirical formula	Pseudoephedrine	Sulfate	
	$C_{10}H_{15}NO$	$(C_{10}H_{15}NO)_2,H_2SO_4$	
Molecular mass	165.2	428.5	
Proprietary names	Actifed; Paragesic; Robidrine		
Clinical use	Sympathomimetic		
Solubility,	Water	ethanol	ether
Pseudoephedrine	sparingly soluble	soluble	soluble
Sulfate	soluble		
Light absorption	λ_{max} = 251 nm in aqueous acid		

HPLC 1

Sample	Blood plasma and urine
Extraction	Mixture of n-hexane and diethyl ether

Apparatus	Merck-Hitachi M 655A LC pump, Kratos SF 757 variable-wavelength detector and Waters M 710B autosampler
Column	Cyano phase; Lichrosorb 100 CN 5 µm, 250 × 4.0 mm ID (serum) and 125 × 4.0 mm ID (urine)
Mobile phase	20 g Methanol, 800 g water, 160 g acetonitrile, 10 g KH_2PO_4,1 g sodium pentanesulfonate and 1 g sodium heptanesulfonate
Flow-rate	1.5 ml min^{-1}
Temperature	Ambient
Internal standard	Acebutolol
Injection size	100–150 µl
Detection	UV, λ = 205; 0.02 AUFS
Retention times, min	

	serum	urine
Pseudoephedrine	4.1	2.3
Internal standard	5.9	3.0

Ref. Nieder M and Jaeger H (1988) J. Chromatogr. Biomed. Appl. 424:73

HPLC 2

Sample	Tablets
Apparatus	Perkin-Elmer Series 3B LC pump, LC-75 UV detector and M 420 autosampler
Column	Cyano phase; Zorbax CN 10 µm, 250 × 4.2 mm ID
Pre-column	Normal phase; silica 20 µm
Mobile phase	18.0 g Brij 35 + 3.46 g sodium lauryl sulfate + 45 ml n-propanol in 1 liter water, pH 2.25 with H_3PO_4
Flow-rate	1.5 ml min^{-1}; 2000 psi
Temperature	65 °C
Injection size	20 µl, 0.27–1.36 mg ml^{-1}
Detection	UV, λ = 220 nm, and 310 nm (paracetamol)
Recording	Sigma 15 data station (Perkin-Elmer)
Retention times, min	(estimated from drawing)

Paracetamol	5.0
Pseudoephedrine	11.8
Chlorpheniramine	14.2

Ref. Biemer TA (1987) J. Chromatogr. 410:206

HPLC 3

Sample	Raw materials and tablets
Column	Normal phase; 5 to 10 µm, 250 × 4.2 mm ID
Mobile phase	Alcohol/0.4% ammonium acetate solution (17:3)
Flow-rate	1.5 ml min^{-1}
Injection size	Appr. 10 µl
Detection	UV, λ = 254 nm

Ref.	Anonym (1985): The United States Pharmacopeia XXI and The National Formulary XVI. United States Pharmacopeial Convention Inc., Rockville, p 913

HPLC 4

Sample	Animal dosing formulations
Apparatus	HPLC with autosampler and variable-wavelength UV detector
Column	Cation-exchange; Partisil SCX 10 μm, 250 × 4.6 mm ID
Mobile phase	Acetonitrile/water (50:50, v/v) and make 0.024 M sodium phosphate, pH appr. 2.3
Flow-rate	1.5 ml min^{-1}
Temperature	Ambient
Injection size	20 μl, 0.6 μg ml^{-1}
Detection	UV, λ = 210 nm; 0.32 AUFS
Recording	Laboratory data controll system
Retention times, min	(estimated from drawing)
	Ibuprofen 2.0
	Terfenadine 3.3
	Pseudoephedrine 7.5
Ref.	George RC and Contario JJ (1988) J. Liq. Chromatogr. 11:475

HPLC 5

Sample	Cough and cold products
Apparatus	IBM Instruments LC/9533 ternary gradient LC with UV detector
Column	Adamantyl phase; modified Si 100, 10 μm, 150 × 4.6 mm ID
Mobile phase	A. 90% 0.01 M Phosphate buffer and 10% acetonitrile B. 80% 0.01 M Phosphate buffer and 20% acetonitrile
Flow-rate	3.0 ml min (A); 2.5 ml min'(B)
Temperature	Ambient
Injection size	20 μl
Detection	UV, λ = 216 nm
Recording	IBM Instruments M 9000 computer

Retention times, min

Column	A.	B.
Pseudoephedrine-HCl	2.0	1.2
Caffeine	5.0	2.0
Triprolidine-HCl	9.0	
Dextromethorphan-HBr		11.0

(estimated from drawing)

Ref.	Yang SS and Gilpin RK (1988) J. Chromatogr. Sci. 26:416

173 Pyrantel

CAS-No	15686-83-6; 22204-24-6 (embonate)	
Chemical name	1,4,5,6-Tetrahydro-1-methyl-2-[(*E*)-2-2-(2-thienyl)-vinyl]pyrimidine	
Empirical formula	Pyrantel	Embonate
	$C_{11}H_{14}N_2S$	$C_{11}H_{14}N_2S,C_{23}H_{16}O_6$
Molecular mass	206.3	594.7
Proprietary names	Antiminth; Combantrin; Strongid-P (vet)	
Clinical use	Anthelmintic	
Solubility,	Water, ethanol and methanol	dimethyl sulfoxide
Embonate	insoluble	soluble
Light absorption	λ_{max} = 315 nm in methanolic acid	

HPLC 1

Sample	Raw materials (1 = hydroxyzine pamoate, 2 = pyrantel pamoate and 3 = pyrvinium pamoate)
Apparatus	Spectra-Physics 3500B LC pump and 770 ultraviolet detector
Column	Reversed-phase; μ-Bondapak 10 μm, C_{18}, 300 × 3.9 mm ID
Mobile phase	A. Methanol/acetic acid/triethylamine/water (58:1.5:0.5:40) B. As in A but (50:1.5:0.5:48) C. As in A but (67:1.5:0.5:31)
Flow-rate	1.5 ml min^{-1}
Temperature	Ambient
Injection size	10 μl, 0.25 mg ml^{-1}
Detection	UV, λ = 254 and 230 nm
Recording	Hewlett-Packard 3380A integrator

Retention times, min

Mobile phase	A	B	C
Sample	1	2	3
λ =	230 nm	254 nm	230 nm
Pamoic acid	8.0	16.8	3.8
Hydroxyzine	10.5		
Pyrantel		4.0	
Pyrvinium			10.8

Ref.	Roos RW and Lau-Cam CA (1986) J. Chromatogr. 370:403

HPLC 2

Sample	Raw materials and tablets
Apparatus	Waters M 45 LC pump, U6K injector and 450 variable-wavelength UV detector
Column	Reversed-phase; Spherisorb octyl 5 μm, 150 × 4.6 mm ID
Mobile phase	Acetonitrile/water (42:58, v/v) containing 0.1 M n-butylamine, pH 3.0 with perchloric acid
Flow-rate	1.0 ml min^{-1}
Temperature	22 °C
Detection	UV, λ = 313 nm; 0.4 AUFS
Recording	Waters M 730 data module
Retention times, min	Oxantel 5.95
	Pyrantel pamoate 7.80
	Carbaryl 11.4
Ref.	Allender WJ (1988) J. Chromatogr. Sci. 26:470

HPLC 3

Sample	Swine feeds
Extraction	First methanolic NaCl, and then chloroform under alkaline condition
Apparatus	Waters M 98750 HPLC pump with 440 UV detector and Valco CV-6-UHPa-N60 injection valve
Column	Normal phase; Zorbax Sil, 250 × 4.6 mm ID
Pre-column	Corasil II
Mobile phase	Acetonitrile/water/acetic acid/diethylamine (94:2.5:2.5:1)
Flow-rate	1.6 ml min^{-1}
Injection size	25 μl
Detection	UV, λ = 313 nm; 0.2 AUFS
Retention times, min	(estimated from drawing)
	Tylosin 3.5
	Pyrantel tartrate 7.3
	cis-Pyrantel tartrate 9.4
Ref.	Goras JT (1981) J. Assoc. Off. Anal. Chem. 64:1291

174 Pyridostigmine Bromide

CAS-No	155-97-5 (pyridostigmine); 101-26-8 (bromide)
Chemical name	3-Dimethylcarbamoyloxy-1-methylpyridinium bromide
Empirical formula	$C_9H_{13}BrN_2O_2$
Molecular mass	261.1

Proprietary names	Mestinon; Regonol			
Clinical use	Anticholinesterase			
Solubility,	Water	ethanol	chloroform	ether
Bromide	1 in < 1	1 in < 1	1 in 1	insoluble
Light absorption	λ_{max} = 270 nm in aqueous acid			

HPLC 1

Sample	Biological fluids
Extraction	Water-saturated dichloromethane at pH 7.0
Apparatus	Varian M 5020 LC pumps with LDC M 1203 UV-III ultraviolet detector
Column	A. Reversed-phase; Lichrosorb C_{18}, 10 µm, 150 × 3.2 mm ID B. Reversed-phase; Ultrasphere octyl 5 µm, 150 × 4.6 mm ID
Pre-column	Reversed-phase; Perisorb C_2, 30–40 µm, 50 × 3.2 mm ID
Mobile phase	For A: 0.01 M Heptanesulfonic acid sodium and 0.01 M NaH_2PO_4 in acetonitrile/water (15:85, v/v), pH 3.0 For B: 0.01 M Heptanesulfonic acid sodium salt, 0.01 M NaH_2PO_4 and 0.0025 M tetramethyl-ammonium chloride in acetonitrile/water (17:83, v/v), pH 3.0 with concentrated H_2SO_4
Temperature	Ambient
Internal standard	Edrophonium
Injection size	50 µl, 0–400 ng ml^{-1}
Detection	UV, λ = 214 nm; 0.004 AUFS
Recording	Varian M 9176 potentiometric recorder

Retention times, min

Column	A	B
Flow-rate	1 ml min^{-1}	2 ml min^{-1}
Edrophonium		4.8
3-Hydroxy metabolite-1	2.0	
Pyridostigmine	3.3	3.4
3-Hydroxy metabolite-2	4.0	
Neostigmine	10.0	7.4

(estimated from drawing)

Ref. De Ruyter MGM, Cronnelly R and Castagnoli Jr N (1980) J. Chromatogr. Biomed. Appl. 183:193

HPLC 2

Sample	Blood plasma
Extraction	Methylene chloride

Apparatus	Tracor M 995 LC pump with 970 UV detector and Rheodyne M 7120 injector
Column	Reversed-phase; μ-Bondapak C_{18}, 10 μm
Mobile phase	37.5% Acetonitrile in water containing 0.001 M sodium lauryl sulfate and 1% acetic acid, pH appr. 4.0 with concentrated NH_4OH
Flow-rate	1.0 ml min^{-1}
Temperature	Ambient
Injection size	100 μl
Detection	UV, $\lambda = 269$ nm; 0.02 AUFS
Retention times, min	(estimated from drawing) Pyridostigmine 6.0 Neostigmine 7.7
Ref.	Yakatan GJ and Tien J-Y (1979) J. Chromatogr. Biomed. Appl. 164:399

HPLC 3

Sample	Blood plasma
Extraction	Deproteinization with 10% trichloroacetic acid, and then on Bond Elut column with 1 M HCl in methanol
Apparatus	Waters M 6000A and M 45 LC pumps with 680 gradient controller, 480 UV detector and U6K injector
Column	Cyano phase; Radial-Pak CN 10 μm, 100 × 5 mm ID
Mobile phase	0.1% Triethylamine, pH 3.2 with acetic acid in water/acetonitrile (50:50, v/v)
Flow-rate	1.0 ml min^{-1}
Temperature	Ambient
Internal standard	Pralidoxime iodide
Injection size	80 μl
Detection	UV, $\lambda = 270$ nm; 0.005 AUFS
Recording	Waters data module M 730
Retention times, min	Internal standard 6.2 Pyridostigmine 7.9
Ref.	Matsunaga H, Suehiro T et al (1987) J. Chromatogr. Biomed. Appl. 422:353

175 Pyridoxine (Vitamin B_6)

CAS-No	65-23-6; 58-56-0 (hydrochloride)
Chemical name	3-Hydroxy-4,5-bis(hydroxymethyl)-2-methyl-pyridine

Empirical formula	Pyridoxine		Hydrochloride	
	$C_8H_{11}NO_3$		$C_8H_{11}NO_3,HCl$	
Molecular mass	169.2		205.6	
Proprietary names	Benadon; Hexobion; Pydox			
Clinical use	Vitamin			
Solubility,	Water	alcohol	acetone	chloroform
Hydrochloride	1 in 5	1 in 100–115	insoluble	insoluble
Light absorption	λ_{max} = 290 nm in aqueous acid			

HPLC 1

Sample	Raw materials
Apparatus	LDC/Milton Roy 5000 psi minipump with pulse dampener, Valco injection system and UV detector
Column	Reversed-phase; Spherisorb ODS 10 µm, 250 × 4.6 mm ID
Pre-column	Whatman Co:Pell ODS
Mobile phase	0.033 M KH_2PO_4 in water, pH 2.2 with H_3PO_4/acetonitrile (99:1, v/v)
Flow-rate	1 ml min^{-1}; 1100 psi
Temperature	Ambient
Internal standard	4′-Deoxypyridoxine
Injection size	8 µl, 2 ng ml^{-1}
Detection	UV, λ = 254 nm
Retention times, min	(estimated from drawing)
	Pyridoxamine 2.5 Pyridoxol 5.5
	Pyridoxal 4.4 4′-Deoxypyridoxine 8.8
Ref.	Lim KL, Young RW and Driskell JA (1980) J. Chromatogr. 188:285

HPLC 2

Sample	Standard solutions
Apparatus	Waters M 45 LC pumps, 680 gradient controller, U6K injector and Hitachi M 650–60 fluorimeter
Column	Alltech Vydac 401 TP, 300 × 4.5 mm ID
Mobile phase	A. 0.02 N HCl
	B. 0.1 M NaH_2PO_4, pH 3.3 with 0.1 N H_3PO_4
	C. 0.5 M NaH_2PO_4, pH 5.9 with 0.5 M NaOH
	Gradient: A 100% for 13 min; B 0% to 100% from 13.1 min to 17 min; linear gradient, from B 100% to 88% and C 0% to 12% from 17 min to 25 min; B 88% to 0% and C 12% to 100% from 25 min to 30 min; hold up to 40 min
Flow-rate	1.0 ml min^{-1} for first 15 min, then 1.5 ml min^{-1}
Internal standard	4′-Deoxypyridoxine 5′-phosphate
Derivatization	Post-column, with sodium metabisulphite

Injection size	500 μl containing 2 ng
Detection	Fluorescence, λ = 330/400 nm
Recording	Spectra-Physics 4270 computing integrator
Retention times, min	(estimated from drawing)
	Pyridoxal 5′-phosphate 5.6
	4′-Deoxypyridoxine 5′-phosphate 8.5
	Pyridoxic acid 10.8
	Pyridoxamine 5-phosphate 21.0
	Pyridoxal 27.2
	Pyridoxine 29.9
	Pyridoxamine 33.6
Ref.	Shephard GS and Labadarios D (1986) Clinica Chimica Acta 160:307

HPLC 3

Sample	Rat blood plasma
Extraction	Cold methylene chloride after deproteinization with 100% trichloroacetic acid
Apparatus	Waters HPLC system with fluorescence detector
Column	Reversed-phase; μ-Bondapak ODS 10 μm, C_{18}, 300 × 3.9 mm ID
Mobile phase	A. Methanol/water (850:150, v/v) B. 0.005 M Heptanesulfonic acid in 1% acetic acid Concave gradient: A 0% to 40% in 12 min; hold 4 min; B 60% to 100% in 4 min; hold 10 min
Flow-rate	1.5 ml min^{-1}
Temperature	Ambient
Internal standard	Deoxypyridoxine
Injection size	250 μl containing 50 ng
Detection	Fluorescence, λ = 300/375 nm
Retention times, min	Pyridoxal 12.41 Internal standard 16.15 Pyridoxine 14.18 Pyridoxamine 20.05 (estimated from drawing)
Ref.	Hefferan TE, Charisley BM and Driskell JA (1986) J. Chromatogr. Biomed. Appl. 374:155

HPLC 4

Sample	Mixture of six vitamers of vitamin B_6
Apparatus	Waters M 45 LC pump, Rheodyne M 7125 injector and Kratos M 980 fluorescence detector
Column	Ion-exchange; Partisil-10 SCX, 250 × 4.6 mm ID
Pre-column	Partisil-10 SCX, 70 × 2.1 mm ID
Mobile phase	0.1 M $NH_4H_2PO_4$ in water, pH 4.0 with 85% H_3PO_4
Flow-rate	1.0 ml min^{-1}; 406 psi
Temperature	Ambient

Internal standard	4′-Deoxypyridoxine
Injection size	20 µl containing 25–100 pmol
Detection	Fluorescence, λ = 290/389 nm
Recording	Hewlett-Packard 3390A reporting integrator
Retention times, min	(estimated from drawing)

Pyridoxal 5′-phosphate	3.7	Pyridoxal	7.2
Pyridoxine 5′-phosphate	3.9	Pyridoxine	10.2
Pyridoxamine 5′-phosphate	4.7	4′-Deoxypyridoxine	14.3
Iso-pyridoxal	6.3	Pyridoxamine	18.3

Ref. Argoudelis CJ (1988) J. Chromatogr. Biomed. Appl. 424:315

HPLC 5

Sample	Blood plasma
Extraction	Deproteinization with 4 mM perchloric acid
Apparatus	LKB M 2150 HPLC pump with 2154–002 injector and Perkin-Elmer M L5-20 fluorimeter
Column	Reversed-phase; ODS 5 µm, 250 × 4.6 mm ID
Pre-column	Lichrosorb RP C_{18}, 30 mm
Mobile phase	75 mM NaH_2PO_4 buffer containing 75 mM $NaClO_4$, 8.5 ml acetonitrile and 0.5 ml triethanolamine per liter, pH 3.38 with concentrated $HClO_4$
Flow-rate	1.2 ml min^{-1}; 12000 kPa
Derivatization	Post-column; with sodium bisulfite
Injection size	2.6 ng
Detection	Fluorescence, λ = 325/400 nm
Retention times, min	(estimated from drawing)

		Pyridoxal phosphate	8.2
Pyridoxamine phosphate	3.2	Pyridoxal	11.5
Pyridoxamine	5.0	Pyridoxine	18.0
Pyridoxine phosphate	7.0	Pyridoxic acid	22.5

Ref. Edwards P, Liu PKS and Rose GA (1989) Clin. Chem. 35:241

176 Pyrimethamine

CAS-No	58-14-0
Chemical name	5-(4-Chlorophenyl)-6-ethylpyrimidine-2,4- diamine
Empirical formula	$C_{12}H_{13}ClN_4$
Molecular mass	248.7
Proprietary names	Daraprim; Maloprim; Supacox (vet)

Clinical use	Antiprotozoal			
Solubility	Water	alcohol	chloroform	acetone
	insoluble	1 in 200	1 in 125	slightly soluble
Light absorption	λ_{max} = 286 nm in aqueous alkali			

HPLC 1

Sample	Raw materials
Apparatus	Spectra-Physics 8700 XR HPLC pump, Waters WISP 710B autoinjector and Kratos M 757 UV detector
Column	Reversed-phase; Spherisorb ODS-1, 5 µm, 175 × 4.6 mm ID
Mobile phase	0.1 M Phosphate buffer/acetonitrile (45:55, v/v), pH 2.2
Flow-rate	1.0 ml min^{-1}
Injection size	25–100 µl
Detection	UV, λ = 285 nm
Recording	Spectra-Physics 4100 computing integrator

Retention times, min

Desethylamodiaquine	< 3.0	Chloroquine	3.58
Desethylchloroquine	< 3.0	Sulfadoxine	3.73
Halofantrine	< 3.0	Chlorcycloguanil	4.25
Primaquine	< 3.0	Pyrimethamine	4.34
Quinine	< 3.0	Metoprine	4.47
Cycloguanil	3.04	Proguanil	4.89
Amodiaquine	3.15	Chloroproguanil	6.93
		Mefloquine	8.01

Ref. Bergqvist Y, Eckerbom S et al (1988) J. Chromatogr. Biomed. Appl. 427:295

HPLC 2

Sample	Sulfadoxine and Pyrimethamine Tablets
Extraction	Acetonitrile and mobile phase
Column	Reversed-phase; ODS 5–10 µm, 300 × 3.9 mm ID
Mobile phase	Dilute glacial acetic acid (1 in 100)/ acetonitrile (4:1)
Flow-rate	2.0 ml min^{-1}
Internal standard	Phenacetin
Injection size	Appr. 10 µl
Detection	UV, λ = 254 nm

Retention times (relative)

		Phenacetin	1.0
Sulfadoxine	0.7	Pyrimethanine	1.3

Ref. Anonym (1985): The United States Pharmacopeia XXI and The National Formulary XVI. United States Pharmacopeial Convention Inc., Rockville, p 990

HPLC 3

Sample	Human blood plasma
Extraction	Ethylene dichloride
Apparatus	Pye-Unicam LC-XPD pump, Waters M 440 UV detector and WISP 710B autosampler
Column	Reversed-phase; μ-Bondapak C_{18}, 10 μm, 300 × 3.9 mm ID
Mobile phase	Methanol/acetonitrile/water (25:15:60) containing 0.005 M 1-pentanesulfonic acid, pH 3.4
Flow-rate	1.5 ml min^{-1}; 1668 psi
Temperature	Ambient
Internal standard	Quinine dihydrochloride
Injection size	40 μl
Detection	UV, λ = 254 nm; 0.005 AUFS
Retention times, min	Sulfadoxine 3.8; Quinine 7.4; N^4-Acetylsulfadoxine 4.7; Pyrimethamine 9.7
Ref.	Edstein M (1984) J. Chromatogr. Biomed. Appl. 305:502

HPLC 4

Sample	Human blood plasma after Fansidar or Fansidar-mefloquine administration
Extraction	n-Butyl chloride/dichloromethane (96:4, v/v)
Apparatus	Altex M 110A LC pump, Rheodyne M 7125 injector and Kontron SFM 22 fluorimeter
Column	Normal phase; Lichrosorb Si 60, 5 μm, 250 × 3.2 mm ID
Mobile phase	Methanol/acetonitrile/25% aqueous ammonia/diiso-propyl ether (6:25:0.1:71, v/v)
Flow-rate	0.7 ml min^{-1}; 1015 psi
Injection size	50 μl containing 50 ng
Detection	Fluorescence, λ = 290/345 nm
Retention time, min	Pyrimethamine appr. 4.6
Ref.	Timm U and Weidekamm E (1982) J. Chromatogr. Biomed. Appl. 230:107

177 Quinidine

CAS-No	56-54-2 (anhydrous); 63717-04-4 (dihydrate); 6591-63-5 (sulfate, dihydrate)
Chemical name	(8*R*,9*S*)-6′-Methoxycinchonan-9-ol dihydrate
Empirical formula	Quinidine $C_{20}H_{24}N_2O_2,2H_2O$; Quinidine Sulfate $(C_{20}H_{24}N_2O_2)_2H_2SO_4,2H_2O$

Molecular mass	360.5	782.9		
Proprietary names	Cardioquin; Kinidin Durules; Quinidex			
Clinical use	Antiarrhythmic			
Solubility,	Water	ether	chloroform	alcohol
Quinidine	1 in 2000	1 in 70		
Sulfate	1 in 80	insoluble	1 in 15	1 in 10
Light absorption	λ_{max} = 250 nm in aqueous acid			

HPLC 1

Sample	Cinchona alkaloids
Apparatus	Altex M 100 LC pumps with Pye-Unicam LC-3 detector and Rheodyne M 7125 valve injector
Column	Reversed-phase; Novapak C_{18}, 4 µm, 150 × 3.9 mm ID
Mobile phase	Acetonitrile/0.05 M hexylamine (7:93), pH 3.0 with ortho-phosphoric acid
Flow-rate	1 ml min^{-1}
Temperature	Ambient
Injection size	5 µl
Detection	UV, λ = 220 nm
Recording	Trivector M 2000 microcomputer data station
Retention times, min	(estimated from drawing)

Cinchonine	3.0	Quinidine	5.0
Cinchonidine	3.5	Quinine	6.4
Hydrocinchonine	4.0	Hydroquinidine	7.4
Hydrocinchonidine	4.6	Hydroquinine	8.6

Ref. McCalley DV (1986) J. Chromatogr. 357:221

HPLC 2

Sample	Biological fluids
Extraction	Methylene chloride/isopropanol (8:2, v/v) under alkaline condition
Apparatus	Beckman HPLC pump with Waters WISP autoinjector and Schoeffel SF 970 fluorescence detector or SF 773 spectrophotometer
Column	Reversed-phase; µ-Bondapak C_{18}, 10 µm, 300 × 4.7 mm ID
Pre-column	µ-Bondapak C_{18}, 10 µm, 40 × 0.2 mm ID
Mobile phase	Acetonitrile/acetic acid/water (7:4:89, v/v)
Flow-rate	2 ml min^{-1}
Temperature	25 °C
Internal standard	Cupreine 25 µg ml^{-1} or quinine sulfate 10 µg ml^{-1}
Injection size	50 µl, 0.1–3 µg ml^{-1}

Detection	Fluorescence, λ = 250/418 nm; UV, λ = 250 mm; 0.002 AUFS	
Recording	Icap 10 integrator (Delsi Instrument)	
Retention times, min	(Dihydroquinidine = DHQ)	
	11-Hydroxy DHQ, Metabolite 1	3.73
	(3S)-3-Hydroxy DHQ, Metabolite 2	4.50
	Cupreine (IS)	6.33
	DHQ N-oxide, Metabolite 3	11.07
	Quinine (IS)	13.42
	Dihydroquinidine (DHQ)	16.17
	(3S)-10-Hydroxy DHQ, minor metabolite	17.80
Ref.	Resplandy G, Roux A et al (1988) J. Liq. Chromatogr. 11:1495	

HPLC 3

Sample	Blood plasma from pregnant sheep	
Extraction	Hexane/ethyl acetate (9:1)	
Apparatus	Waters M 6000A HPLC, 481 Lambdamax UV detector and U6K injector	
Column	Reversed-phase; Rad-Pak Z-module, μ-Bondapak C_{18}, 10 μm, 100 × 8 mm ID	
Pre-column	Reversed-phase; μ-Bondapak C_{18}, 10 μm	
Mobile phase	Water/acetonitrile (91:9, v/v) containing 1% triethylamine, pH 2.5 with *o*-H_3PO_4	
Flow-rate	3.5 ml min^{-1}; 190 kPa	
Internal standard	Monoethylglycinexylidide	
Injection size	40 μl	
Detection	UV, λ = 254 nm; 0.005–0.01 AUFS	
Retention times, min	Internal standard	4.0
	Quinidine	5.0
	Quinine	6.0
Ref.	Mihaly GW, Hyman KM et al (1987) J. Chromatogr. Biomed. Appl. 415:177	

HPLC 4

Sample	Human blood plasma
Extraction	Ether/dichloromethane/isopropanol (6:4:1)
Apparatus	Varian 8500 HPLC with UV detector and different injectors
Column	Normal phase; Varian Micropak Si 10 LSC, 250 × 2.1 mm ID
Mobile phase	I Hexane/ethanol/ethanolamine (91.5:8.47:0.03) II. Hexane/ethanol/ethanolamine (92.97:7.0:0.03)
Flow-rate	1.0 ml min^{-1}; appr. 750 psi
Internal standard	Primaquine diphosphate
Injection size	10–20 μl

Detection	UV, λ = 235 nm
Recording	Hewlett-Packard 3380A integrator
Retention times, min	(estimated from drawing)

Mobile phase	I.	II.
Quinidine	4.0	6.2
Dihydroquinidine	5.0	7.2
2′-Quinidinone	6.1	9.0
Primaquine	7.5	11.0
Quinidine-oxide-R	9.5	14.2
3-OH-Quinidine	12.0	17.0

Ref.	Guentert TW, Coates PE et al (1979) J. Chromatogr. Biomed. Appl. 162:59

HPLC 5

Sample	Urine
Extraction	Centrifugation
Apparatus	Altex M 110A LC pump, LDC Spectromonitor III UV detector and Perkin-Elmer LC-1000 fluorescence detector and Rheodyne M 70–10 injector
Column	Reversed-phase; Lichrosorb C_8, 5 or 10 μm, 150 × 4.5 mm ID
Mobile phase	25% Acetonitrile in 0.095 M $NaClO_4$ and 0.005 $HClO_4$
Flow-rate	1.4 ml min^{-1}
Temperature	Ambient
Injection size	20 μl, 0.3 μmol l^{-1}
Detection	UV, λ = 254 nm; 0.005 AUFS Fluorescence, λ = 363/440 nm
Retention time, min	Qunidine 5.0 (estimated from drawing)
Ref.	Lagerstroem P-O (1981) J. Chromatogr. Biomed. Appl. 225:476

178 Quinine

CAS-No	130-95-0 (anhydrous); 6119-70-6 (sulfate, dihydrate)	
Chemical name	(8*S*,9*R*)-6′-Methoxycinchonan-9-ol-trihydrate	
Empirical formula	Quinine	Sulfate
	$C_{20}H_{24}N_2O_2$, $3H_2O$	$C_{20}H_{24}N_2O_2)_2H_2SO_4$, $2H_2O$
Molecular mass	378.5	782.5
Proprietary names	Kinine; Quine; Quinsan	
Clinical use	Antimalarial	

Solubility,	Water	alcohol	chloroform
Quinine	very slightly sol	1 in 1	1 in 3
Sulfate	1 in 810	1 in 95	slightly sol
Light absorption	λ_{max} = 250 nm in aqueous acid		

HPLC 1

Sample	Raw materials
Apparatus	Spectra-Physics 3500B LC pump and 770 ultraviolet detector
Column	Reversed-phase; μ-Bondapak C_{18}, 10 μm, 300 × 3.9 mm ID
Mobile phase	Methanol/acetic acid/triethylamine/water (40:1.5:0.5:58)
Flow-rate	1.5 ml min^{-1}
Temperature	Ambient
Injection size	10 μl, 0.25 ng ml^{-1}
Detection	UV, λ – 280 nm; 0.64 AUFS
Recording	Hewlett-Packard 3380A integrator
Retention times, min	(estimated from drawing)

Cinchonidine	9.8	Quinidine	14.5
Cinchonine	10.5	Quinine	18.2
Dihydrocinchonine	13.0	Dihydroquinidine	20.0
Dihydrocinchonidine	13.0	Dihydroquinine	24.2

Ref.	Roos RW and Lau-Cam CA (1986) J. Chromatogr. 370:403

HPLC 2

Sample	Adulterated heroin
Apparatus	Perkin-Elmer LC pump with gradient controller, photodiode array and electrochemical detector
Column	Reversed-phase; HS-5, C_{18}, 125 × 4.6 mm ID
Mobile phase	Start: Methanol/phosphate buffer, pH 2.2 with 0.023 M hexylamine (5:95) Gradient: Methanol 5% to 30% in 20 min; run 6 min; methanol 30% to 60% in 10 min; run 9 min
Flow-rate	1.5 ml min^{-1}
Injection size	50 μl
Detection	UV, λ = 218–240 nm; EC, GC electrode, + 1.0 V and 1.1 V vs Ag/AgCl
Recording	Perkin-Elmer computing integrator
Retention time, min	Quinine 9.7 (estimated from drawing)
Ref.	Lurie IS and McGuinness K (1987) J. Liq. Chromatogr. 10:2189

HPLC 3

Sample	Cinchona alkaloids
Apparatus	Altex M 100 LC pumps with Pye-Unicam LC-3 detector and Rheodyne M 7125 valve injector
Column	Reversed-phase; Lichrosorb 7 µm, C_8, 250 × 4.0 mm ID
Mobile phase	Acetonitrile/0.1 M potassium dihydrogen phosphate, pH 3.0 (15:85)
Flow-rate	1 ml min^{-1}
Temperature	Ambient
Injection size	5 µl
Detection	UV, λ = 220 nm
Recording	Trivector M 2000 microcomputer data station
Retention times, min	(estimated from drawing)

Cinchonine	6.0	Quinidine	9.1
Cinchonidine	6.5	Quinine	10.3
Hydrocinchonine	7.8	Hydroquinidine	11.1
Hydrocinchonidine	8.3	Hydroquinine	14.0

Ref. McCalley DV (1986) J. Chromatogr. 357:221

HPLC 4

Sample	Cinchona alkaloids
Apparatus	Perkin Elmer Series 3B LC pump; LC-75 UV detector and Rheodyne M 7125 valve injector
Column	Cyano phase; Spherisorb CN, 5 µm, 250 × 4.6 mm ID
Mobile phase	Buffer (0.0068 M H_3PO_4, pH 7.4 with 1 M NaOH) and organic modifiers, composition (%, v/v)

Buffer	Acetonitrile	Methanol	Tetrahydrofuran
a. 40	60		
b. 57			43
c. 48		52	
d. 52.5		26	21.5
e. 48.5	30		21.5
f. 44	30	26	
g. 48.4	20	17.3	14.3
h. 50	17	28.7	3.3

Flow-rate	1.5 ml min^{-1}
Temperature	50 °C
Injection size	20 µl
Detection	UV, λ = 231 nm
Recording	Perkin-Elmer Sigma 15 data control
Retention times, min	(mobile phase h.)

Quinine	17.26 (16.60–17.89)
Quinidine	19.29 (18.60–19.93)
Cinchonidine	21.63 (21.20–21.85)
Cinchonine	23.76 (23.25–24.02)

Ref.	Hobson-Frohock A and Edwards WTE (1982) J. Chromatogr. 249:369

179 Ranitidine

CAS-No	66357-35-7; 71130-06-8 (hydrochloride)	
Chemical name	*NN*-Dimethyl-5-[2-(1-methylamino-2-nitrovinyl-amino)-ethylthiomethyl]furfurylamine	
Empirical formula	Ranitidine	Hydrochloride
	$C_{13}H_{22}N_4O_3S$	$C_{13}H_{22}N_4O_3S,HCl$
Molecular mass	314.4	350.9
Proprietary names	Zantac	
Clinical use	Histamine H_2-Receptor Antagonist	
Solubility	Freely soluble in water (hydrochloride)	
Light absorption	λ_{max} = 313 nm in water	

HPLC 1

Sample	Blood plasma
Extraction	Centrifugation with 200 µl acetonitrile and 50 mg $ZnSO_4$ per milliliter plasma
Apparatus	Perkin-Elmer Series 410 LC pump, Rheodyne M 7275 injector and Kratos M 783 ultraviolet detector
Column	Reversed-phase; 5 µm PLRP-S, 50 × 4.1 mm ID
Pre-column	PLRP-S, C-130B (Upchurch Scientific)
Mobile phase	Acetonitrile/5.0 mM K_2HPO_4 and 0.50 mM tetrabutylammonium hydroxide in water (20:80, v/v), pH appr. 11.0
Flow-rate	1.0 ml min^{-1}
Injection size	200 µl, 0.010–40.0 µg ml^{-1}
Detection	UV, λ = 315 nm; 0.2 to 0.002 AUFS
Retention times, min	(estimated from drawing)
	Ranitidine N-oxide 3.4 · Desmethylranitidine 5.2
	Ranitidine S oxide 3.9 · Ranitidine 6.1
Ref.	Rustum AM (1988) J. Liq. Chromatogr. 11:2315

HPLC 2

Sample	Human blood plasma and urine
Extraction	Acetonitrile/ethyl acetate (1:1, v/v) under alkaline condition (plasma)
Apparatus	Hitachi M 655 A-11 HPLC, 655-A UV detector and Rheodyne M 7125 manual injector valve

Column	Cation-exchange; Partisil SCX 10 μm, 250 × 4.6 mm ID
Mobile phase	Acetonitrile/tetrahydrofuran/0.1 M sodium acetate buffer, pH 5 with glacial acetic acid (36:7.5:56.5, v/v)
Flow-rate	1.9 ml min^{-1}
Temperature	Ambient
Internal standard	Phenylpyramidol hydrochloride
Injection size	10–20 μl
Detection	UV, λ = 320 nm; 0.004–0.016 AUFS
Recording	Hitachi M 655–71 data processor
Retention times, min	Internal standard 2.8; Ranitidine 5.3 Ranitidine N-oxide 3.6; Ranitidine S-oxide 7.5 Desmethylranitidine 4.5
Ref.	Prueksaritanont T, Sittichai N et al (1989) J. Chromatogr. Biomed. Appl. 490:175

HPLC 3

Sample	Human urine and serum
Extraction	Urine: Diluting with water Serum: After clot formation
Apparatus	Spectra-Physics 740B LC pumps, Waters WISP autosampler and Pye LC3 UV detector
Column	Reversed-phase; Spherisorb ODS 5 μm, 100 × 5 mm ID
Mobile phase	600 ml Methanol and 400 ml water containing 0.0005 M sodium dihydrogen phosphate and 0.005 M sodium lauryl sulfate
Flow-rate	1 ml min^{-1}
Temperature	45 °C
Injection size	15 μl, 0–400 μg ml^{-1}
Detection	UV, λ = 320 nm; 0.01 AUFS
Recording	Spectra-Physics 4050 printer plotter and 4000 chromatography data system with 4100 computing integrator
Retention times, min	Metabolite I 1.9; Ranitidine 5.8 Metabolite III 3.9; Metabolite II 7.9
Ref.	Carey PF, Martin LE and Owen PE (1981) J. Chromatogr. Biomed. Appl. 225:161

HPLC 4

Sample	Blood plasma
Extraction	Octan-l-ol under alkaline codition, and re-extraction with 0.1 M phosphate buffer, pH 6.0
Apparatus	Spectra-Physics 740B LC pump, Waters WISP sample processor and Pye-Unicam LC3 UV detector

Column	Reversed-phase; Spherisorb ODS 5 μm, 100 × 5 mm ID
Mobile phase	Methanol/0.05 M phosphate buffer, pH 7.0 (60:40)
Flow-rate	0.9 ml min^{-1}
Temperature	40 °C
Internal standard	AH 20480 (Glaxo, United Kingdom)
Injection size	100 μl
Detection	UV, λ = 320 nm; 0.01 AUFS
Recording	Spectra-Physics 4100 computing integrator with SP 4000 data system
Retention times, min	Ranitidine 3.75 Internal standard 5.83
Ref.	Carey PF and Martin LE (1979) J. Liq. Chromatogr. 2:1291

HPLC 5

Sample	Raw materials and tablets
Extraction	30% Aqueous acetonitrile
Column	Cyano phase; Spherisorb-CN 3 μm, 150 × 4.6 mm ID
Mobile phase	25 mM Ammonium phosphate buffer, pH 5.0 /acetonitrile (3:7)
Flow-rate	1.0 ml min^{-1}
Injection size	5 μl, 0.4–140 μg ml^{-1}
Detection	UV, λ = 228 nm
Ref.	Beaulieu N, Lacroix PM et al (1988) J. Pharm. Sci. 77:889

HPLC 6

Sample	Whole blood or blood plasma
Extraction	Methanol in presence of zinc sulfate
Apparatus	Waters M 6000 LC pump, U6K injector and Kratos M 773 UV/VIS detector
Column	Reversed-phase; PRP-1, 10 μm, 60 × 4.6 mm ID
Pre-column	PRP-1, 20–30 μm, 40 × 2.0 mm ID
Mobile phase	Acetonitrile/5 mM K_2HPO_4 and 5 mM sodium 1-pentanesulfonate (18:82, v/v), pH 10.0 with 0.1 M NaOH
Flow-rate	1.0 ml min^{-1}
Temperature	Ambient (25 °C)
Injection size	100 μl, 25–300 ng ml^{-1}
Detection	UV, λ = 314 nm; 0.2–0.002 AUFS
Retention times, min	Ranitidine 5.0–6.0
Ref.	Rustum AM, Rahman A and Hoffman NE (1987) J. Chromatogr. Biomed. Appl. 421:418

180 Reserpine

CAS-No	50-55-5
Chemical name	Methyl 18-*O*-(3,4,5-trimethoxybenzoyl)-reserpate
Empirical formula	$C_{33}H_{40}N_2O_9$
Molecular mass	608.7
Proprietary names	Rausan; Sandril; Serpasil
Clinical use	Antihypertensive

Solubility	Water	alcohol	ether	chloroform	methanol
	insoluble	1 in 2000	insol	1 in 6	1 in 2000

Light absorption: $\lambda_{max} = 267$ nm in ethanol

HPLC 1

Sample	Rauwolfia serpentine powder and tablets
Extraction	First chloroform, then methanol using Celite- 0.1 N NaOH extraction column
Apparatus	Waters M 45 LC pump, Rheodyne M 7125 injector and Perkin-Elmer LS5 fluorescence spectrophotometer
Column	Normal phase; μ-Prasil, 300 × 3.9 mm ID
Mobile phase	Methanol
Flow-rate	1.5 ml min^{-1}
Injection size	100 μl containing 0.05–0.5 mg
Detection	Fluorescence: I. $\lambda = 280/360$ nm II. $\lambda = 330/435$ nm III. $\lambda = 280/370$ nm IV. $\lambda = 330/370$ nm
Retention times, min	Reserpine appr. 5.0 (I, II, III, IV) Rescinnamine appr. 5.0 (I, II, III, IV)
Ref.	Cieri UR (1987) J. Assoc. Off. Anal. Chem. 70:540

HPLC 2

Sample	Gastric fluid (artificial) containing anti-hypertensives
Column	Reversed-phase; Spherisorb C_{18}, 5 μm, 100 × 2.1 mm ID
Mobile phase	Acetonitrile/0.4 M phosphoric acid (85:15)
Flow-rate	3 ml min^{-1}
Injection size	10 μl
Detection	UV, $\lambda = 260$ nm
Retention times, sec	(estimated from drawing) Dihydroergocristine 13.0 Clopamide 10.0 Reserpine 18.0
Ref.	Meyer V (1984): Praxis der Hochleistungsfluessigchromatographie, 3rd edn. Diesterweg Salle and Sauerlaender, Frankfurt am Main and Salzburg, p 216

HPLC 3

Sample	Pharmaceutical dosage forms (tablets)
Apparatus	LDC HPLC with UV detector and Schoeffel FS 970 fluorescence detector
Column	Reversed-phase; Lichrosorb C_8, 10 μm, 250 × 4.6 mm ID
Mobile phase	Methanol/0.05 M sodium dihydrogen phosphate (6.9 g NaH_2PO_4 l^{-1} water), pH 4.5 (50:50, v/v)
Flow-rate	2.0 ml min^{-1}; 2000 psi
Internal standard	Propiophenone
Injection size	20 μl containing 0.25–40 μg
Detection	UV, λ = 254 nm Fluorescence, λ = 330/470 nm
Retention times, min	(estimated from drawing) Reserpine 14.5 Internal standard 6.8 3,4-Dehydroreserpine 21.0
Ref.	Vincent A and Awang DVC (1981) J. Liq. Chromatogr. 4:1651

HPLC 4

Sample	Commercial tablets of reserpine and hydro-chlorothiazide
Extraction	Methanol
Apparatus	Waters HPLC system with M 45 pump, Perkin-Elmer LS5 fluorescence detector and Kratos SF 769Z UV detector
Column	Normal phase; μ-Porasil, 300 × 3.9 mm ID
Mobile phase	Methanol
Flow-rate	1.5 ml min^{-1}
Injection size	100 μl
Detection	UV, λ = 345 nm Fluorescence, λ = 280/360 nm
Retention times, min	Hydrochlorothiazide appr. 2.0 (fluorimetric) Reserpine appr. 4.0 (UV, fluorimetric)
Ref.	Cieri UR (1988) J. Assoc. Off. Anal. Chem. 71:515

181 Ribavirin

CAS-No	36791-04-5
Chemical name	1-β-D-Ribofuranosyl-1*H*-1,2,4-triazole-3-carboxamide
Empirical formula	$C_8H_{12}N_4O_5$
Molecular mass	244.2
Proprietary names	Vilona; Viramid; Virazole

Clinical use	Antiviral
Solubility	Soluble in water

HPLC 1

Sample	Serum and Urine
Extraction	Dichloromethane after molecular filtration and centrifugation, then re-extraction with water using extraction column
Apparatus	Gilson M 302 pump with 811 gradient system, 440 variable-wavelength UV detector and Rheodyne M 7125 injector valve
Column	Reversed-phase; Lichrosorb C_{18}, 7 µm, 250 × 5 mm ID
Pre-column	Reversed-phase; Guard-Pak C_{18}
Mobile phase	Water
Flow-rate	1 ml min^{-1}
Internal standard	Uridine
Injection size	100 µl, 20–1000 ng ml^{-1}
Detection	UV, $\lambda = 207$ nm
Retention times, min	1,2,4-triazole-3-carboxamide 4.0
	Ribavirin 6.9
	Uridine (IS) 13.0
Ref.	Paroni R. Sirtori CR et al (1987) J. Chromatogr. Biomed. Appl. 420:189

HPLC 2

Sample	Biological fluids and tissues
Extraction	Deproteinization by ultrafiltration, and then with 0.1 M formic acid on phenyl boronate affinity gel column at pH 8.8 (sample containing 2.5 M ammonium acetate buffer)
Apparatus	Waters M 840 HPLC system with 510 LC pumps, 710B WISP autosampler and 490 UV detector
Column	Reversed-phase; Microsorb TM C_{18}, 5 µm, 250 × 4.6 mm ID
Pre-column	Reversed-phase; Guard-Pak TM C_{18}
Mobile phase	0.02 M Ammonium phosphate buffer; pH 5.10 (adjusted with 10% NH_4OH or 8% H_3PO_4) containing 1% methanol
Flow-rate	1 ml min^{-1}
Temperature	Ambient
Injection size	1–500 ng
Detection	UV, $\lambda = 207$ nm
Retention time, min	Ribavirin 5.5 (estimated from drawing)

Ref.	Smith RHA and Gilbert BE (1987) J. Chromatogr. Biomed. Appl. 414:202

182 Riboflavine (Vitamin B_2)

CAS-No	83-88-5; 130-40-5 (sodium phosphate, anhydrous)			
Chemical name	7,8-Dimethyl-10-(1-D-ribityl) isoalloxazine			
Empirical formula	Riboflavine $C_{17}H_{20}N_4O_6$	Sodium Phosphate $C_{17}H_{20}N_4NaO_9P$		
Molecular mass	376.4	478.4		
Proprietary names	Beflavin; Berivine; Vitamin B_2			
Clinical use	Vitamin			
Solubility,	Water	ethanol	chloroform, ether	aq NaOH
Riboflavine	insoluble	insoluble	insoluble	very sol
Sodium phosphate	1 in 20	very slightly soluble	insoluble	
Light absorption	λ_{max} = 267 nm in aqueous acid			

HPLC 1

Sample	Blood plasma and urine
Extraction	Deproteinization with acetonitrile and then with chloroform (plasma)
Apparatus	Beckman M 110A LC pump and Perkin-Elmer M 650-10 fluorescence detector
Column	Reversed-phase; PRP-1, 10 μm, 250 × 4.6 mm ID
Pre-column	Reversed-phase; PRP-1, 10 μm, 70 × 2.1 mm ID
Mobile phase	Acetonitrile/water/10% (v/v) trifluoroacetic acid/phosphoric acid (14:84:1.5:0.09), pH 1.8 with 10% NaOH
Flow-rate	1 ml min^{-1}
Temperature	Ambient (22 °C)
Injection size	50 μl, 10 ng ml^{-1}
Detection	Fluorescence, λ = 470/525 nm
Recording	Hewlett-Packard 3390 integrator
Retention times, min	Riboflavine-5-phosphate 5.70 Riboflavine 7.47
Ref.	Lopez-Anaya A and Mayersohn M (1987) J. Chromatogr. Biomed. Appl. 423:105

HPLC 2

Sample	Urine
Extraction	Centrifugation

Apparatus	Spectra-Physics 3500 HPLC system, Valco sample injector and LDC M 1209 spectrofluorimeter
Column	Reversed-phase; μ-Bondapak C_{18}, 10 μm, 300 × 4 mm ID
Pre-column	Reversed-phase; Whatman Co:Pell ODS C_{18}, 70 × 2.1 mm ID
Mobile phase	0.01 M KH_2PO_4, pH 5.0/methanol (65:35)
Flow-rate	2.0 ml min^{-1}
Injection size	50 μl, 10 ng ml^{-1}
Detection	Fluorescence, $\lambda = 320–400/400–700$ nm
Recording	Hewlett-Packard 7120A chart recorder and Spectra-Physics Autolab minigrator
Retention times, min	Riboflavine-5-phosphate 2.5 Riboflavine 4.5 (estimated from drawing)
Ref.	Smith MD (1980) J. Chromatogr. Biomed. Appl. 182:285

HPLC 3

Sample	Milk from goats
Extraction	Methanol on Sep-Pak C_{18} extraction column
Apparatus	Waters M 510 LC pump, Rheodyne M 7125 injector and Hewlett-Packard 1040A photodiode array detector
Column	Reversed-phase; Novapak 5 μm, C_{18}, 8 mm ID with Waters radial RCM-100
Pre-column	Reversed-phase; μ-Bondapak C_{18}
Mobile phase	0.0005 M Hexanesulfonate (containing Waters one low-UV PIC B_6 reagent vial in 1000 ml water)-methanol (70:30)
Flow-rate	2.0 ml min^{-1}; 2000 psi
Injection size	20 μl
Detection	UV, $\lambda = 214$ nm
Recording	Hewlett-Packard 9000 Series 300 technical computer
Retention times, min	Thiamine 4.22 Riboflavine 5.65
Ref.	Lavigne C, Zee JA et al (1987) J. Chromatogr. 410:201

HPLC 4

Sample	Foods
Extraction	0.1 M HCl
Apparatus	Beckman M 110A LC pump, Waters WISP 710B autosampler and Varian SF-330 spectrofluorimeter
Column	Reversed-phase; Lichrosorb 10 μm, C_8, 250 × 4.0 mm ID

Mobile phase	0.005 M 1-Hexanesulfonic acid in water/ methanol (60:40)
Flow-rate	1.0 ml min^{-1}
Temperature	Ambient
Injection size	50 μl containing 2.5–15 ng
Detection	Fluorescence, λ = 440/565 nm
Retention time, min	Riboflavine appr. 6.0
Ref.	Reyes ESP, Norris KM et al (1988) J. Assoc. Off. Anal. Chem. 71:16

HPLC 5

Sample	Pharmaceutical formulations
Extraction	Acetate buffer, pH 4.0
Column	Reversed-phase; μ-Bondapak C_{18}
Mobile phase	1.2 g Camphorsulfonic acid in 300 ml water, pH 7.0 with 0.5 N NaOH plus 275 ml methanol
Flow-rate	1.5 ml min^{-1}
Injection size	15 μl, 2–7 μg ml^{-1}
Detection	UV, λ = 280 nm
Retention times, min	(estimated from drawing) Riboflavine 3.0 Thiamine 5.6
Ref.	Wittmer DP and Haney Jr. WG (1979) in: Tsuji K (ed) GLC and HPLC determination of therapeutic agents, Part 3. Marcel Dekker Inc., New York and Basel, p. 1235

---

183 Rifampicin

CAS-No	13292-46-1		
Chemical name	3-[[(4-Methyl-1-piperazinyl)imino]methyl] rifamycin		
Empirical formula	$C_{43}H_{58}N_4O_{12}$		
Molecular mass	823.0		
Proprietary names	Rifa; Rimactane; Tibirim		
Clinical use	Antitubercular		
Solubility	Water and ethanol slightly soluble	chloroform freely soluble	methanol soluble
Light absorption	λ_{max} = 231 nm in aqueous acid		

HPLC 1

Sample	Biological fluids
Extraction	Dichloromethane/diethyl ether (2:3) pH 4.2

Apparatus	Waters M 6000A and 45 LC pumps with gradient programmer and Hitachi M 220S UV detector
Column	I. Reversed-phase; Lichrosorb 5 μm, C_8, 250 × 4.6 mm ID II. Reversed-phase; μ-Bondapak C_{18} Radial-Pak
Pre-column	Waters 30 μm, C_8, 50 × 4.6 mm ID
Mobile phase	A. 10 mM KH_2PO_4 at pH 3.5 B. 60% (v/v) Acetonitrile in 10 mM KH_2PO_4, pH 3.5 Gradient: 10% B to 80% B in 5 min and run up to 10 min
Flow-rate	1.5 ml min^{-1}
Temperature	25 °C
Internal standards	Paracetamol (1), butyrylaminophenol (2), and p-acetamidobenzoic acid (3)
Injection size	25–75 μl
Detection	UV, λ = 215 nm; 0.05 AU

Retention times, min

(estimated from drawing)	column I	column II
Pyrazinamide	3.2	3.0
Internal standard (3)	7.2	
Rifampicin	9.5	12.7
Internal standard (1)		5.2
Internal standard (2)		8.0
25-Desacetyl rifampicin		10.6

Ref. Chan K (1988) J. Liq. Chromatogr. 11:1051

HPLC 2

Sample	Serum
Extraction	Dichloromethane/diethyl ether (2:3) with 2% ascorbic acid, pH 4.2
Apparatus	Waters M 6000A and 45 pump with 660 gradient programmer, U6K injector and Hitachi M 220S UV detector
Column	Reversed-phase; Lichrosorb C_8, 5 μm, 250 × 4.6 mm ID
Pre-column	Reversed-phase; C_8, 30 μm, 50 × 4.6 mm ID
Mobile phase	A. 10 mM Phosphate buffer, pH 3.5 B. 10 mM Phosphate buffer, pH 3.5/acetonitrile (40:60) Gradient: B 10% to 80% in 5 min, hold 10 min
Flow-rate	1.5 ml min^{-1}
Temperature	25 °C
Internal standard	*p*-Acetamidobenzoic acid
Injection size	75 μl, 15.5 μg ml^{-1}
Detection	UV, λ = 215 nm; 0.1 AUFS
Recording	Hitachi chart recorder

Retention times, min	Isoniazid 2.6; Pyrazinamide 3.3; *p*-Aminosalicylic acid 5.0; Internal standard 7.4; Rifampicin 9.3
Ref.	Woo J, Wong CL et al (1987) J. Chromatogr. Biomed. Appl. 420:73

HPLC 3

Sample	Biological fluids
Extraction	Diluting (1:6) with water
Apparatus	ACS M 351 LC pump, Rheodyne M 7125 injector, Pye-Unicam 4025 UV detector and LCA 15 electrochemical detector (EDT Research)
Column	Reversed-phase; Hypersil C_{18}, 3 μm, 150 × 4.6 mm ID
Pre-column	Reversed-phase; Lichroprep C_{18}, 25–40 μm, 25 × 4.6 mm ID
Mobile phase	70% Aqueous methanol with 50 mmol KNO_3
Flow-rate	0.7 ml min^{-1}
Temperature	Ambient
Injection size	20 μl containing 2.6–14.4 μg ml^{-1}
Detection	UV, λ = 334 nm; 0.01 AUFS and EC, GC electrode, 0.45 V
Retention times, min	25-Desacetylrifampicin 5.2; Rifampicin 7.8 (estimated from drawing)
Ref.	Oldfield S, Berg JD et al (1986) J. Chromatogr. Biomed. Appl. 377:423

HPLC 4

Sample	Rifampicin and its quinone
Apparatus	Varian 4100 HPLC and Variscan UV/VIS detector
Column	Alkylamine; Micropak-NH_2 10 μm, 250 × 2 mm ID
Mobile phase	Chloroform/methanol (97:3), pH 7.38
Flow-rate	0.7 ml min^{-1}; 3.4 MPa
Injection size	2–10 μl, 200–400 μg ml^{-1}
Detection	UV/VIS, λ = 235, 255, 333, 474 nm (rifampicin); 259, 332, 539 (quinone)
Retention times, min	Quinone 1.6; Rifampicin 4.7
Ref.	Vlasáková V, Benes J and Zivny K (1978) J. Chromatogr. 151:199

HPLC 5

Sample	Serum, polymorphonuclear cells and alveolar macrophages
Extraction	First acetone and then re-extraction with butanol/chloroform (30:70)

Apparatus	Waters M 6000A HPLC pump, U6K injector and 440 UV detector
Column	Reversed-phase; μ-Bondapak C_{18}, 10 μm, 300 × 3.9 mm ID
Mobile phase	Methanol/0.05 M ammonium formate (65:35, v/v), pH 7.3
Flow-rate	2.0 ml min^{-1}
Injection size	20 μl, 17 ng ml^{-1}
Detection	UV, λ = 254 nm; 0.02 AUFS
Retention times, min	(estimated from drawing) Desacetylrifampicin 4.1–4.4 Rifampicin 6.8–7.0
Ref.	Guillaumont M, Leclercq M et al (1982) J. Chromatogr. Biomed. Appl. 232:369

HPLC 6

Sample	Human blood plasma
Extraction	Ethyl acetate under acidic condition, and re-extraction with acetonitrile using n-heptane
Apparatus	Hewlett-Packard and Waters HPLC systems with UV detectors and autosamplers
Column	Reversed-phase; C_8, 10 μm, 250 × 4.6 mm ID
Mobile phase	A. Acetonitrile/0.1 M KH_2PO_4, pH 3.5 with 0.2 M H_3PO_4 (38:62) (Hewlett-Packard system) B. Acetonitrile containing 10% water/0.1 M KH_2PO_4, pH 3.5 with 0.2 M H_3PO_4 (40:60) (Waters system) C. Acetonitrile/isopropanol (1:1, v/v) (Waters system)
Flow-rate	2 ml min^{-1}
Temperature	30 °C (A); ambient (B and C)
Internal standard	Butylparahydroxybenzoate
Injection size	10 μl containing 50–250 ng
Detection	UV, λ = 254 nm
Recording	Hewlett-Packard 3380A computing reporting integrator system

Retention times, min

System	A.	C.
25-Desacetylrifampicin	2.25	4.11
3-Formyl-25-desacetyl-rifampicin SV		5.75
N-Desmethylrifampicin		8.48
Rifampicin	9.32	9.17
3-Formylrifampicin SV		24.04
Internal standard	27.27	27.35

Ref. Ratti B, Rosina Parenti R et al (1981) J. Chromatogr. Biomed. Appl. 225:526

HPLC 7

Sample	Capsules
Apparatus	Waters HPLC system with M 6000A pump and U6K injector and Altex M 155-40 UV detector
Column	Reversed-phase; C_2, 10 μm
Mobile phase	480 ml Methanol to 470 ml 0.05 M ammonium formate and 50 ml tetrahydrofuran, pH 7.3 with appr. 3 drops of 2 N NH_4OH
Flow-rate	2 ml min^{-1}; 1200 psi
Internal standard	Diphenyl phthalate
Injection size	5 μl containing 32 μg
Detection	UV, λ = 254 nm; 0.05 AUFS
Recording	Spectra-Physics 4000 chromatography data system
Retention times, min	25-Desacetylrifampicin 4.4
	3-Formylrifampicin SV 5.8
	Rifampicin 8.4
	Rifampicin quinonc 13.3
	Internal standard 17.2
Ref.	Graham KC, Lebelle MJ and Wilson WL (1979) J. Liq. Chromatogr. 2:365

184 Salicylamide

CAS-No	65-45-2
Chemical name	2-Hydroxybenzamide
Empirical formula	$C_7H_7NO_2$
Molecular mass	137.1
Proprietary names	Amid-Sal; Delimon; Salizell
Clinical use	Analgesic
Solubility	Water 1 in 500; ethanol 1 in 15; chloroform 1 in 100; ether 1 in 35; aq NaOH soluble
Light absorption	λ_{max} = 235 nm in aqueous acid

HPLC 1

Sample	Biological fluids
Extraction	Deproteinization with 1 M perchloric acid, and or with ethyl acetate (conjugates)
Apparatus	Waters HPLC with M 6000A pump, WISP 710B autosampler, 680 gradient controller and 441 spectrophotometric detector

Column	Reversed-phase; μ-Bondapak C_{18}, 10 μm, 250 × 4.6 mm ID
Pre-column	Reversed-phase; μ-Bondapak 40 μm, C_{18}
Mobile phase	System I (plasma, bile, urine): 0.085 M Potassium dihydrogen phosphate, pH 3.35 with glacial acetic acid Gradient: 0.8 ml min^{-1} for 34 min; 0.2–1.8 ml min^{-1} from 34–39 min and hold upto 50 min System II (unconjugated forms): 0.5% Acetic acid/methanol (60:40) with constant flow-rate, 0.6 ml min^{-1}
Injection size	30–100 μl containing 1.09 nmol
Detection	UV, λ = 280 nm (system I) 313 nm (system II)
Recording	Hewlett-Packard 5880A chart recorder

Retention times, min

System	I
Gentisamide (GAM)	38
Salicylamide (SAM)	48
Metabolites (assigned)	
Peak I (GAM-2-sulfate)	18
Peak II (GAM-5-glucuronide	21
Peak III (SAM-2-sulfate)	26
Peak IV (SAM-2-glucuronide)	29
Peak V (GAM-5-sulfate) and SAM-sulfate)	34

Ref. Xin Xu and Pang KS (1987) J. Chromatogr. Biomed. Appl. 420:313

HPLC 2

Sample	Biological fluids
Extraction	Ether (serum and saliva) at pH 5.0 and dilution with water (urine)
Apparatus	Waters M 6000A LC pump, 440 UV detector and WISP 710B autosampler
Column	Reversed-phase; μ-Bondapak C_{18}
Mobile phase	A. (urine): 3 mM Tetrabutylammonium hydroxide in a mixture of methanol/7% acetic acid (8:92, v/v) B. (serum): 1% Acetic acid in 15% (v/v) methanol/water
Flow-rate	1.6–2.0 ml min^{-1}
Temperature	Ambient
Internal standard	*o*-Methoxybenzoic acid
Injection size	25 μl
Detection	UV, λ = 254 nm (A) and 313 nm (B)
Recording	Waters data module

Retention times, min (estimated from drawing)

Mobile phase	A. (serum)	B. (urine)
Gentisamide glucuronide	3.0	3.5
Salicylamide glucuronide	4–5.0	5.0
Gentisamide	4.2	
Salicylamide	6.7	7.0
Internal standard	9.5–12.0	10–13.0
Salicylamide sulfate	15–19.0	14–16.0
Gentisamide sulfate		6.2

Ref. Morris ME and Levy G (1983) J. Pharm. Sci. 72:612

HPLC 3

Sample	Tablets containing different analgesics
Extraction	Chloroform/methylene chloride (42:50, v/v)
Column	Waters Corasil II, 600 × 2.1 mm ID
Mobile phase	Acetic acid/chloroform/dichloromethane (8:42:50, v/v)
Flow-rate	1.0 ml min^{-1}
Temperature	Ambient
Injection size	20 µl
Detection	UV, λ = 280 nm; 0.08 AUFS
Retention times, min	(estimated from drawing)

Aspirin	1.58	Phenacetin	3.68
Salicylamide	2.24	Caffeine	10.13
		Acetaminophen	16.45

Ref. Cornish DW, Jacobs AL et al (1978) in: Tsuji K (ed) GLC and HPLC Determination of Therapeutic Agents, Part 2. Marcel Dekker Inc., New York and Basel, p 519

HPLC 4

Sample	Raw materials and serum
Apparatus	Waters M 6000 LC pump, Pye-Unicam LC-UV detector and Rheodyne M 7125 injection valve
Column	Reversed-phase; Hypersil-ODS 5 µm, 160 × 5 mm ID
Mobile phase	Isopropanol/0.1 M potassium dihydrogen phosphate/formic acid (176:1000:1, v/v)
Flow-rate	1.5 ml min^{-1}
Injection size	10 µl
Detection	UV, λ = 240 nm

Capacity factor, k'

Amidopyrine	* 0.32	Phenacetin	4.4
Paracetamol	* 0.32	Etenzamide	4.6
Dipyrone	* 0.45	Salicylic acid	4.6
Nifenazone	* 0.45	Choline salicylate	4.8

Phenazone	* 0.95	Glycol salicylate	7.3
Morazone	* 2.05	Piroxicam	7.7
Acetanilide	2.3	Propyphenazone	11.0
Salicylamide	2.5	Benorylate	22.4
Acetylsalicylic acid	2.7		

*Tailing peaks

Ref. Stevens HM and Gill R (1986) J. Chromatogr. 370:39

---

185 Salicylic Acid

CAS-No	69-72-7; 54-21-7 (sodium salicylate)
Chemical name	2-Hydroxybenzoic acid
Empirical formula	Salicylic Acid $C_7H_6O_3$; Sodium Salt $C_7H_5NaO_3$
Molecular mass	138.1; 160.1
Proprietary names	Fomac; Myoflex; Xseb
Clinical use	Analgesic

Solubility,	Water	ethanol	chloroform	ether	olive oil
Salicylic acid	1 in 550	1 in 4	1 in 45	1 in 3	1 in 80
Sodium salt	1 in 1	1 in 11	insoluble	insoluble	

Light absorption: λ_{max} = 236 nm in aqueous acid

HPLC 1

Sample	Blood plasma
Apparatus	Milton Roy/LDC 711-26/711-46 solvent delivery system with pulse dampener, Rheodyne M 7120 injector and Schoeffel FS 970 fluorimeter and LDC Spectromonitor III UV detector
Column	Reversed-phase; Lichrosorb 5 μm, C_{18} coated with tributyl phosphate, 100 × 4.6 mm ID
Pre-column	Reversed-phase; Lichrosorb 5 μm, C_{18}, 50 × 3.2 mm ID
Mobile phase	Methanol/citrate buffer, pH 5.4 (1:9)
Flow-rate	0.8 ml min^{-1}
Temperature	25 °C
Injection size	10–50 μl
Detection	Fluorescene, λ = 295/ > 370 nm
Retention time, min	Salicylic acid 4.7 (estimated from drawing)
Ref.	Wahlund K-G (1981) J. Chromatogr. 218:671

HPLC 2

Sample	Blood plasma
Apparatus	Milton Roy/LDC 711-46 or 711-46 solvent delivery system with pulse dampener, Spectromonitor III, Schoeffel FS 970 fluorescence detector and M 7120 Rheodyne injector
Column	Reversed-phase; Lichrosorb 5 μm, C_{18}, 100 × 4.6 mm ID
Pre-column	Reversed-phase; Lichrosorb 5 μm, C_{18}, 50 × 3.2 mm ID
Mobile phase	Phosphate buffer, pH 5.9/1.02 · 10^{-4} M sodium octyl sulfate
Flow-rate	1 ml min^{-1}
Temperature	25 °C
Injection size	10–50 μl
Detection	Fluorescence, λ = 295/370 nm UV, λ = 296 nm; 0.005 A
Retention time, min	(estimated from drawing) Salicylic acid 6.4
Ref.	Wahlund K-G (1981) J. Chromatogr. 218:671

HPLC 3

Sample	Raw materials and serum
Apparatus	Waters M 6000 LC pump, Pye-Unicam LC-UV detector and Rheodyne M 7125 injection valve
Column	Reversed-phase; Hypersil-ODS 5 μm, 160 × 5 mm ID
Mobile phase	Isopropanol/0.1 M potassium dihydrogen phosphate/formic acid (176:1000:1, v/v)
Flow-rate	1.5 ml min^{-1}
Injection size	10 μl
Detection	UV, λ = 240 nm
Retention times, min	(estimated from drawing)

Paracetamol	1.3	Salicylic acid	6.4
Phenazone	2.1	Piroxicam	10.0
Salicylamide	3.9	Propyphenazone	14.0

Ref. Stevens HM and Gill R (1986) J. Chromatogr. 370:39

HPLC 4

Sample	Blood plasma
Extraction	Diethyl ether on glass extraction column
Apparatus	Waters M 6000A LC pump connected with Altech T-piece system, U6K injector and Schoeffel FS 970 fluorescence detector
Column	Reversed-phase; Spherisorb ODS-2, 5 μm, 50 × 4.6 mm ID

Mobile phase	Methanol/0.085% phosphoric acid (40:60)
Flow-rate	1 ml min^{-1}
Derivatization	Post-column; with 0.5 M NaOH;
Injection size	50 µl, 50–2000 ng ml^{-1}
Detection	Fluorescence, λ = 310/389 nm
Retention times, min	(estimated from drawing) Salicyluric acid 5.2 Acetylsalicylic acid 7.0 Salicylic acid 11.5
Ref.	Siebert DM and Bochner F (1987) J. Chromatogr. Biomed. Appl. 420:425

HPLC 5

Sample	Blood and blood plasma
Extraction	Ethyl acetate under acidic condition
Apparatus	Waters HPLC system with M 600 and 45 pumps, U6K injector, 680 gradient controller and 480 ultraviolet detector
Column	Reversed phase; Hypersil ODS 5 µm, 200 × 4.6 mm ID
Pre-column	Guard-Pak
Mobile phase	10^{-3} M Phosphate buffer, pH 3.2/acetonitrile (80:20) Linear gradient: Acetonitrile 20% to 40% in 10 min; hold 6 min; acetonitrile 40% to 20% in 4 min
Flow-rate	3.0 ml min^{-1}
Temperature	60 °C
Internal standard	Allobarbital, 76 µg ml^{-1}
Injection size	20 µl
Detection	UV, λ = 202 nm
Recording	Hewlett-Packard 3350A data system
Retention times, min	Caffeine 1.4 Allobarbital 2.9 Acetylsalicylic acid 3.1 Salicylic acid 3.6 (3.46–3.70) Butalbital 5.0 (4.93–5.08)
Ref.	Drost ML and Walter L (1988) J. Anal. Toxicol. 12:322

HPLC 6

Sample	Human blood plasma
Extraction	Chloroform under acidic condition
Apparatus	Spectra-Physics 8000 HPLC with 770 UV detector
Column	Reversed-phase; C_{18}, 10 µm, 250 × 4.6 mm ID
Mobile phase	Methanol/0.072% ortho-phosphoric acid (55:45, v/v)
Flow-rate	1.5 ml min^{-1}; 1363 psi

Temperature	40 °C	
Internal standard	3,4-Dimethylbenzoic acid	
Injection size	10 µl, 36.3 µg ml^{-1}	
Detection	UV, λ = 234 nm	
Retention times, min	Acetylsalicylic acid	2.75
	Salicylic acid	4.10
	Internal standard	7.22
Ref.	Lo LY and Bye A (1980) J. Chromatogr. Biomed. Appl. 181:473	

---

186 Secobarbital (Quinalbarbitone)

CAS-No	76-73-3; 309-43-3 (sodium salt)		
Chemical name	5-Allyl-5-(1-methylbutyl)barbituric acid		
Empirical formula	Secobarbital	Sodium Salt	
	$C_{12}H_{18}N_2O_3$	$C_{12}H_{17}N_2NaO_3$	
Molecular mass	238.2	260.3	
Proprietary names	Seconal; Quinbar; Tuinal		
Clinical use	Hypnotic/Sedative		
Solubility,	Water	alcohol	ether and aq NaOH
Secobarbital	very slightly sol	freely sol	freely soluble
Sodium salt	1 in 3	1 in 5	insoluble
Light absorption	λ_{max} = 239 nm in 0.05 M borax-buffer, pH 9.2		

HPLC 1

Sample	Pharmaceuticals (dosage forms)		
Apparatus	Du Pont M 820 HPLC with 410 UV detector		
Column	Strong anion-exchange; SAX, Du Pont, 1000 × 2.1 mm ID		
Mobile phase	I. 0.01 M Sodium borate plus 0.03 M sodium nitrate II. 0.01 M Citric acid		
Flow-rate	0.48 ml min^{-1}; 500 psi (I) and 0.60 ml min^{-1}; 1000 psi (II)		
Temperature	25 °C		
Injection size	5 µl containing 1.0 µg (I) and 10.0 µg (II)		
Detection	UV, λ = 254 nm; 0.08 AUFS		
Retention times, min	(estimated from drawing)		
	Mobile phase	I.	II.
	Barbital	4.00	
	Vinbarbital	5.18	4.93
	Aprobarbital	5.65	

	Butabarbital	6.59	
	Heptabarbital	7.53	
	Phenobarbital	9.18	7.71
	Pentobarbital	10.82	
	Meprobarbital	13.88	10.07
	Secobarbital	16.24	13.07
	Allobarbital		4.07
	Hexobarbital		6.21
Ref.	Roos RW (1972) J. Pharm. Sci. 61:1979		

HPLC 2

Sample	Barbiturate injections
Apparatus	Waters M 6000A LC pump, LDC Spectromonitor III ultraviolet detector and Rheodyne M 70-10 injector
Column	A. Reversed-phase; Chromegabond C_{18}, 5 µm, 75–150 × 4.6 mm ID B. Reversed-phase; Zorbax ODS, 7 µm, 75–150 × 4.6 mm ID
Mobile phase	Acetate buffer (containing 4.1 g sodium acetate + 3 ml acetic acid in water to 1000 ml)/ methanol/polyethyleneglycol 300 (400:600:4)
Flow-rate	1 ml min^{-1}
Injection size	20 µl, 0.91 mg ml^{-1}
Detection	UV, λ = 230 nm; 0.4 AUFS
Retention times, min	(estimated from drawing)

	Degradation product I	3.0
	Degradation product II	3.2
	Secobarbital	5.4
	Degradation product III	10.4
Ref.	Reiff VD, Kaufmann KL et al (1986) J. Pharm. Sci. 75:714	

HPLC 3

Sample	Mixture of anticonvulsive agents
Column	Reversed-phase; Microspher C_{18}, 50 × 4.6 mm ID
Mobile phase	Acetonitrile/0.05 M sodium dihydrogen phosphate (70:30, v/v)
Flow-rate	7 ml min^{-1}
Injection size	10 µl
Detection	UV, λ = 254 nm
Retention times, sec	(estimated from drawing)

			Carbamazepine	25.97
	Phenobarbital	13.60	Phenytoin	29.68
	Butobarbital	16.58	Secobarbital	37.11

Ref.	Chrompack Application Notes on High performance Column (1987). Chrompack, Middelburg

HPLC 4

Sample	Test substance
Column	Reversed-phase; Adsorbosphere HS C_{18}, 7 μm, 150 × 4.6 mm ID
Mobile phase	Methanol/water (51:49)
Flow-rate	2.0 ml min^{-1}
Detection	UV, λ = 220 nm; 0.2 AUFS
Retention times, min	(estimated from drawing)
	Barbital 1.0
	Allobarbital 1.5
	Phenobarbital 1.8
	Aprobarbital 2.3
	Butabarbital 2.8
	Mephobarbital 4.0
	Amobarbital 5.5
	Secobarbital 7.5
Ref.	Alltech Drug Standards (1988) Alltech Bulletin No. 153, State College, p 32

HPLC 5

Sample	Test substance
Apparatus	Waters ALC/GPC M 201 LC with 6000 pumps, U6K injector, 600 programmer and Varian M 635 UV/VIS detector
Column	A. Reversed-phase; Corasil C_{18}, 37–70 μm, 600 × 2 mm ID
	B. As A but with 1% squalene loading on Corasil II
	C. As A but with 1% octalon loading on Corasil II
Mobile phase	Acetate buffer, pH 5.00 with 10% acetonitrile
Flow-rate	1–3 ml min^{-1} (A); 0.4 ml min^{-1} (B and C)
Injection size	Appr. 1.0 μl, 0.1–1 mg ml^{-1}
Detection	UV, λ = 245 nm
Retention times	Barbital, amobarbital, pentobarbital, secobarbital and phenobarbital (in A, B, C) in ascending order
Ref.	Henry D, Block JH et al (1976) J. Med. Chem. 19:619

187 Serotonin

CAS-No	50-67-9
Chemical name	3-(2-Aminoethyl)-1*H*-indol-5-ol
Empirical formula	$C_{10}H_{12}N_2O$
Molecular mass	176.2
Proprietary names	5-HT; Thrombocytin; Thrombotonin
Solubility	Soluble in water
Light absorption	λ_{max} = 276 nm in aqueous acid

HPLC 1

Sample	Raw materials
Apparatus	Waters M 6000A LC pump and Perkin-Elmer M 204A spectrophotofluorimeter
Column	Reversed-phase; Ultrasphere ODS 5 μm, 150 × 4.6 mm ID
Mobile phase	425 ml 0.067 M phosphate buffer, pH 7.0 with absolute methanol to make 1000 ml
Flow-rate	1.3 ml min^{-1}
Derivatization	Pre-column; with fluorescamine
Injection size	30–50 μl
Detection	Fluorescence, λ = 395/485 nm
Recording	Hewlett-Packard 3380A electronic integrator
Retention times, min	5-Hydroxytryptamine 2.1 Phentermine 5.7 Thyroxine 8.8
Ref.	Sedman AJ and Gal J (1982) J. Chromatogr. Biomed. Appl. 232:315

HPLC 2

Sample	Rat and human brain and biological fluids
Apparatus	Waters Micromeritics M 750 pump, WISP 710B autosampler and Bioanalytical Systems LC-4 amperometric detector
Column	Reversed-phase; Spherisorb ODS 5 μm
Mobile phase	0.1 M K_2HPO_4/citric acid buffer (1:0.6, v/v; pH 4.73) containing 5% methanol and 0.1 mM EDTA
Flow-rate	1 ml min^{-1}
Temperature	30 °C
Injection size	50–100 μl containing 15 pg
Detection	EC, + 0.6 V vs Ag/AgCl RE
Recording	Spectra-Physics 4100 computing integrator
Retention times, min	(estimated from drawing)

Norepinephrine	1.9	Dopamine	3.0
Epinephrine	2.5	Serotonin	7.8
Dopa	2.5	Homovanillic acid	10.0

Ref.	Semerdjian-Rouquier L, Bossi L and Scatton B (1981) J. Chromatogr. 218:663

HPLC 3

Sample	Brain tissue from rats
Extraction	0.1 M Perchloric acid
Apparatus	Beckman M 110A LC pump and Bioanalytical Systems LC-4B amperometric detector
Column	Reversed-phase; Ultrasphere ODS C_{18}, 5 μm, 250 × 4.6 mm ID

Pre-column	Waters Guard-Pak, C_{18}
Mobile phase	Methanol/buffer (containing 50 mM citric acid, 100 mM disodium hydrogen ortho-phosphate, 0.38 mM sodium octyl sulfate and 0.5 mM EDTA) (15:85, v/v), pH 4.2
Flow-rate	1.2 ml min; 2500 psi^{-1}
Temperature	Ambient
Injection size	20 µl containing 0.8 ng
Detection	EC, GC electrode, + 0.8 V vs Ag/AgCl RE
Recording	Hewlett Packard 3390A recording integrator
Retention times, min	Norepinephrine 3.91; Epinephrine 4.79; Dopamine 9.83; 5-Hydroxyindol-acetic acid 10.78; Morphine 20.76; 5-Hydroxystrep-tamine 23.28
Ref.	Kim C, Speisky MB and Kalant H (1986) J. Chromatogr. 370:303

HPLC 4

Sample	Brain tissue from rats
Apparatus	Beckman M A-112 LC pump, Perkin-Elmer ISS-100 automatic injector and Metrohm ED 666 electro-chemical detector
Column	Reversed-phase; Ultrasphere ODS 5 µm, 150 × 4.5 mm ID
Mobile phase	I. 0.1 M KH_2PO_4 containing 0.1 M EDTA and 5 mM sodium heptanesulfonate/methanol (92.5:7.5, v/v), pH 3.75 II. 90 ml 0.2 M KH_2PO_4 containing 0.1 mM EDTA and 10 ml methanol, pH 5.4 III. 97 ml of a solution containing 0.1 M CH_3COONa, 0.1 M KH_2PO_4 and 0.1 mM EDTA and 3 ml acetonitrile, pH 5.4
Flow-rate	1 ml min^{-1}
Injection size	0.4 ng
Detection	EC, GC electrode + 0.8 V vs Ag/AgCl RE,
Retention times, min	(estimated from drawing) (Mobile phase I)

4-Hydroxy,3-methoxy-phenylglycol	3.0	Dopamine	11.0
Norepinephrine	4.5	5-Hydroxyindol-acetic acid	15.0
Epinephrine	5.0	Homovanillic acid	21.0
5-Hydroxytryptophan	6.5	L-Tryptophan	24.3
Normetanephrine	8.0	3-Methoxytyramine	25.3
3,4-Dihydroxyphenyl-acetic acid	9.0	5-Hydroxytryp-tamine (serotonin)	30.0
Metanephrine	10.0		

Ref.	Trouvin JH, Gardier A et al (1987) J. Liq. Chromatogr. 10:121

HPLC 5

Sample	Mixture of serotonin and its precursors in biological materials
Extraction	Deproteinization with 0.2 M–0.5 M perchloric acid
Apparatus	LDC HPLC with Kontron SFM-23 L-fluorimeter and Pye-Unicam 4022 electrochemical detector
Column	Reversed-phase; Polygosil C_{18}, 5 µm, 100 × 3 mm ID
Mobile phase	Phosphate buffer, pH 1.95/methanol (8:2)
Flow-rate	1.1 ml min^{-1}
Injection size	100 µl containing 2–3 nmol
Detection	Fluorescence, λ = 296/333 nm and EC, + 0.8 V vs Ag/AgCl RE
Retention times, min	(estimated from drawing) Tryptophan 5.0
	5-Hydroxytryptophan 2.2 Tryptamine 7.4
	5-Hydroxytryptamine 3.1 5-Hydroxyindol-acetic acid 10.8
Ref.	De Jong J, Tjaden UR et al (1983) J. Chromatogr. 282:443

188 Spectinomycin

CAS-No	1695-77-8; 23312-56-4 (sulfate, anhydrous)		
Chemically	Aminoglycoside anitbiotic produced by Streptomyces spectabilis		
Empirical formula	Spectinomycin		Sulfate
	$C_{14}H_{24}N_2O_7$		$C_{14}H_{24}N_2O_7,H_2SO_4,4H_2O$
Molecular mass	332.4		502.5
Proprietary names	Kempi; Togamycin; Trobicine		
Clinical use	Antibiotic		
Solubility,	Water	alcohol	acetone, chloroform
Spectinomycin	soluble	soluble	insoluble
Sulfate	freely soluble	very slightly soluble	very slightly soluble

HPLC 1

Sample	Raw materials
Apparatus	LDC Constametric III pump, ESA 5100A Coulochem detector and Rheodyne M 7125 injection valve

Column	Reversed-phase; Nucleosil C_{18}, 10 µm, 250 × 4.6 mm ID
Mobile phase	Aqueous buffer (containing 0.02 M sodium citrate and 0.0015 octyl sodium sulfate, pH 6.1 with perchloric acid)/acetonitrile (100:4)
Flow-rate	1.5 ml min^{-1}; 1800–2100 psi
Injection size	50 µl containing 5 ng
Detection	EC, GC electrode, + 0.85 V
Recording	Shimadzu C-R3A integrator
Retention times, min	Actinamine 4.28 Spectinomycin 8.56 (estimated from drawing)
Ref.	Elrod Jr L, Bauer JF and Messner SL (1988) Pharm. Res. 5:664

HPLC 2

Sample	Mixture of spectinomycin, actinospectinoic acid and actinamine
Apparatus	Spectra-Physics 740B LC pump, Perkin-Elmer M 204A fluorimeter and Valco CV-6UPa-N60 injection valve
Column	Reversed-phase; Lichrosorb C_8, 250 × 4.6 mm ID
Pre-column	Reversed-phase; Perisorb C_8, 43 × 4.2 mm ID
Mobile phase	0.02 M Sodium heptanesulfonate, 0.2 M sodium sulfate and 0.1% acetic acid in water
Flow-rate	2 ml min^{-1}
Temperature	Ambient
Derivatization	Post-column; with *o*-phthalaldehyde
Injection size	15 µl, 0.05–0.9 mg ml^{-1}
Detection	Fluorescence; λ = 350/450 nm
Recording	Hewlett-Packard 3354A data system and 7153A recorder
Retention times, min	(estimated from drawing) Actinamine 2.9 Actinospectinoic acid 3.8 Spectinomycin 7.5
Ref.	Myers HN and Rindler JV (1979) J. Chromatogr. 176:103

---

189 Spironolactone

CAS-No	52-01-7
Chemical name	7α-Acetylthio-3-oxo-17α-pregn-4-ene-21,17β-carbolactone

Empirical formula	$C_{24}H_{32}O_4S$
Molecular mass	416.6
Proprietary names	Aldactone; Laractone; Spirolone
Clinical use	Diuretic
Solubility	Water insoluble; alcohol 1 in 80; chloroform 1 in 3; ethyl acetate soluble
Light absorption	λ_{max} = 242 nm in aqueous acid

HPLC 1

Sample	Guinea pig blood plasma
Extraction	Ethyl acetate
Apparatus	Waters M 6000A HPLC pumps, 710B autosampler, 720 system controller and 440 absorbance detector
Column	Reversed-phase; Radial Pak 5 μm, C_{18}
Mobile phase	30 Min concave gradient of 65–100% methanol/water
Flow-rate	1.0 ml min^{-1}
Temperature	Ambient
Internal standard	Progesterone
Injection size	100 ng
Detection	UV, λ = 254 nm; 0.05 AUFS
Recording	Waters M 730 data module
Retention times, min	(estimated from drawing)

Spironolactone	11.0	Canrenone	17.2
7α-Thiospironolactone	12.5	Progesterone (IS)	33.0
7α-Thiomethylspironolactone	14.5		

Ref. Sherry AH, O'Donnell JP and Colby HD (1986) J. Chromatogr. Biomed. Appl. 374:183

HPLC 2

Sample	Urine
Extraction	Ethyl acetate at pH 5.0–5.5 or at pH 9.0–9.5
Apparatus	Hewlett-Packard 1090 HPLC with photodiode array UV detector
Column	I. Reversed-phase; Hypersil ODS C_{18}, 5 μm, 200 × 4.6 mm ID II. Reversed-phase; Lichrosorb C_{18}, 5 μm, 200 × 4.6 mm ID
Mobile phase	A. 0.05 M Phosphate buffer, pH 3.0 with 0.016 M propylamine hydrochloride (containing 6.9 g NaH_2PO_4, H_2O and 1.59 propylamine hydrochloride per liter water) B. Acetonitrile Gradient: B 15% at 2 min; B 15% to 80% at 20 min

Flow-rate	1 ml min^{-1}		
Temperature	Ambient (22 °C)		
Injection size	5 µl containing 0.5 µg		
Detection	Diode array UV, λ = 230 nm		
Recording	Integrator		
Retention times, min	Column	I.	II.
	β-Hydroxyethyltheophylline	3.45	4.40
	Amiloride	3.71	3.95
	Acetazolamide	4.30	5.17
	Chlorothiazide	5.37	6.43
	Hydrochlorothiazide	6.03	7.07
	Quinethazone	6.83	7.77
	Triamterene	7.68	7.55
	Flumethiazide	7.68	8.73
	Hydroflumethiazide	8.30	9.36
	Chlorthalidone	9.02	9.79
	Dichlorphenamide	9.37	10.30
	Trichloromethiazide	11.10	12.03
	Methyclothiazide	11.68	12.60
	Furosemide	12.16	12.92
	Metolazone	12.16	12.92
	Benzthiazide	13.52	14.28
	Cyclothiazide	13.85	14.76
	Polythiazide	14.52	15.24
	Bendroflumethiazide	14.63	15.33
	Ethacrynic acid	15.10	15.31
	Bumetanide	15.25	15.78
	Probenecid	15.78	16.09
	Spironolactone	16.25	17.60
	Canrenone	16.36	18.38
Ref.	Cooper SF, Massé R and Dugal R (1989) J. Chromatogr. Biomed. Appl. 489:65		

HPLC 3

Sample	Body fluids containing canrenone (metabolite)
Extraction	Chloroform (canrenone metabolite)
Apparatus	Pye-Unicam LC-20 pump with LC-3 UV detector and Perkin-Elmer M 1000 fluorimeter
Column	Normal phase; Lichrosorb Si-100, 10 µm, 250 × 2.5 mm ID
Mobile phase	I. (serum): Chloroform/n-hexane (50:50) II. (urine): As in I. but (55:45)
Flow-rate	2.0 ml min^{-1}; 7 MPa
Injection size	50–100 µl containing 5 ng ml^{-1}
Detection	UV, λ = 283 nm Fluorescence, λ = 450/457 nm

Retention times, min	Canrenone (metabolite of spironolactone) 4.5 (serum) 4.0 (urine)
Ref.	Neurath GB and Ambrosius D (1979) J. Chromatogr. Biomed. Appl. 163:230

HPLC 4

Sample	Raw materials
Column	Reversed-phase; ODS 5–10 μm, 300 × 4 mm ID
Mobile phase	Acetonitrile/0.02 M dibasic ammonium phosphate (55:45)
Flow-rate	1 ml min^{-1}
Injection size	Appr. 20 μl, 250 μg ml^{-1}
Detection	UV, λ = 254 nm
Retention times, min	Spironolactone appr. 5.0
Ref.	Anonym (1985): The United States Pharmacopeia XXI and The National Formulary XVI. United States Pharmacopeial Convention Inc., Rockville, p 981

190 Stilbestrol (Diethylstilbestrol)

CAS-No	56-53-1; 130-80-3 (dipropionate)			
Chemical name	(*E*)-*αβ*-Diethylstilbene-4,4′-diol			
Empirical formula	Stilbestrol	Dipropionate		
	$C_{18}H_{20}O_2$	$C_{24}H_{28}O_4$		
Molecular mass	268.4	380.5		
Proprietary names	Cyren-A; Dicorvin; Stilbilium			
Clinical use	Sex Hormone			
Solubility,	Water	alcohol	ether	arachinis oil
Stilbestrol	insoluble	1 in 5	1 in 3	1 in 40
Dipropionate	slightly soluble		1 in 6	soluble
Light absorption	λ_{max} = 241 nm in ethanol			

HPLC 1

Sample	Bovine urine
Extraction	Methanol on C_{18} and alumina B extraction columns after enzymatic hydrolysis with β-glucuronidase
Apparatus	Varian M LC-5560, UV-200 detector and Rheodyne automatic injector
Column	Micropak, MCH-N-Cap-5 μm, 300 × 4 mm ID
Mobile phase	Water/acetonitrile/tetrahydrofuran (50:45:5, v/v)
Flow-rate	1 ml min^{-1}

Temperature	35 °C
Injection size	100 µl containing 0.90 ng
Detection	UV, λ = 341 nm up to 10 min, then 244 nm
Recording	Varian Vista CDS 402 integrator
Retention times, min	Trenbolone 7.22
	Diethylstilbestrol 13.15
Ref.	Simonella A, Torreti L et al (1987) Intn. Analyst 1:26

HPLC 2

Sample	Biological materials	
Extraction	Deproteinization with methanol/ethanol (1:1)	
Apparatus	Waters M 6000A LC pump with 660 gradient controller and 440 UV detector	
Column	Reversed-phase; Lichrosorb C_{18}, 5 µm, 250 × 10 mm ID	
Pre-column	Reversed-phase; Corasil C_{18}, 37–50 µm, 35 × 3.2 mm ID	
Mobile phase	Methanol/0.01 M ammonium acetate (10:90), pH 6.9 Convex gradient: Methanol 10% to 100% in 50 min	
Flow-rate	2 ml min^{-1}; 1600 psi	
Temperature	25 °C ± 2 °C	
Injection size	10 µg	
Detection	UV, λ = 254 nm	
Retention times, min	(estimated from drawing)	
	Diethylstibestrol diglucuronide	13.0
	trans-Diethylstibestrol monoglucuronide	22.0
	omega-1-Hydroxy-diethylstilbestrol	29.0
	trans-Diethylstilbestrol	35.5
	cis,cis-Dienestrol	38.0
	cis-Diethylstilbestrol	41.5
	Tetramethoxy-diethylstilbestrol	46.5
Ref.	Newport GD, Headley SK et al (1980) J. Liq. Chromatogr. 3:1053	

HPLC 3

Sample	Calf urine
Column	Reversed-phase; Ultrasphere ODS C_{18}, 5 µm, 250 × 4.6 mm ID
Mobile phase	A. Methanol/20 mM Tris-acetatepuffer, pH 7.2 (65:35, v/v)
	B. Acetonitrile/20 mM Tris-acetate buffer, pH 7.2 (42:58, v/v)
Flow-rate	1 ml min^{-1}
Temperature	25 °C
Injection size	0.2–2 µg

Detection: UV, λ = 230 nm and 340 nm (trenbolones and trendions)

Retention times, min

Mobile phase	A.	B.
Cortisol	5.34	5.18
Prednisolone	5.89	4.66
Flumethasone	6.47	6.53
Betamethasone	7.36	6.53
Dexamethasone	7.50	6.53
Triamcinolone acetonide	7.61	7.75
Trendion	9.47	14.26
Trenbolone-17β	10.03	9.26
19-Norandrostenedione	10.15	15.03
Trenbolone-17α	11.13	10.19
19-Norethinodrel	11.60	14.90
19-Norethisterone	11.60	14.91
Zeranole	11.63	11.98
Diethylstibestrol (trans)	11.87	25.73
17α-Methyltrienolone	12.10	11.03
19-Nortestosterone-17β	12.15	11.43
17α-Ethinylestradiol	12.31	16.87
Androstenedione	12.41	19.40
Estrone	12.46	19.61
15β,16β-Methylen-19-Nortestosterone-17β	12.47	11.30
Dienestrol	12.86	26.03
Estradiol-17β	12.96	12.83
Estradiol-17α	13.58	15.03
Hexestrol	13.96	27.65
Testosterone	14.72	14.23
19-Normethisterone	15.72	14.62
19-Nortestosterone-17α	15.99	15.03
19-Norgestrel	16.90	25.25
17α,18α-Dimethyl-15-en-19-Nortestosterone-17β	17.01	
18-Methyl-19-NT-17β	17.63	
Mibolerone	17.92	18.03
17α-Methyltestosterone	18.67	18.37
18-Methyl-15-methylen-19-Nortestosterone-17β	18.94	
Epitestosterone	20.95	19.61
17α-Allyltrienolone	21.73	23.81
Medroxyprogesteronacetate	24.90	> 30.60
17α-Ethyl-19-NT-17β	26.61	
Melengestrolacetate	26.74	30.00
Diethylstilbestrol (cis)	27.84	
Progesterone	30.37	

Ref.	Rapp M, Funcke P and Meyer HHD (1988) Archiv Lebensmittelhyg. 39:114

HPLC 4

Sample	Bovine urine
Apparatus	HPLC with off-line chemiluminescent immunochemical detection system
Column	Reversed-phase; Hypersil ODS 5 μm, 150 × 4.6 mm ID
Mobile phase	Methanol/water (6:4) for 8 min followed by pure methanol for 2 min
Flow-rate	2.0 ml min^{-1}
Detection	Chemiluminescent immunochemical
Retention times, min	Diethylstilbestrol 6.3
	Hexetrol 7.6
Ref.	Van den Berg RH, Jansen EHJM et al (1986) J. Chromatogr. 360:449

HPLC 5

Sample	Avian muscle tissue extract
Extraction	95% Acetone and clean-up on alumina and then on AGMP-1 anion-exchange resin columns
Column	Normal phase; silica 5 μm, 250 × 4.6 mm ID
Mobile phase	Hexane, methanol and propan-2-ol Gradient: 100% Hexane to 40% hexane/methanol/propan-2-ol (8:9:3) in 15 min
Flow-rate	2.0 ml min^{-1}
Injection size	5 ng
Detection	UV, λ = 280 nm
Ref.	Medina MB and Sherman JT (1986) Food Addit. Contam. 3:263

191 Streptomycin and Dihydrostreptomycin

CAS-No	57-92-1 (streptomycin) and 3810-74-0 (sulfate); 128-46-1 (dihydrostretomycin) and 5490-27-7 (sulfate)
Chemically	Aminoglycoside antibiotics produced by Streptomyces griseus

	Streptomycin	Sulfate	Dihydrostreptomycin Sulfate
Empirical formula	$C_{21}H_{39}N_7O_{12}$	$(C_{21}H_{39}N_7O_{12})_2$, $3H_2SO_4$	$(C_{21}H_{41}N_7O_{12})_2$, $3H_2SO_4$
Molecular mass	581.6	1457.0	1461.4

Proprietary names	Abiocine; Enterastrep; Novostrep			
Clinical use	Antibiotic			
Solubility,	Water	alcohol	acetone	chloroform
Streptomycin sulfate	soluble	insoluble	insoluble	insoluble
Dihydrostreptomycin sulfate	soluble	insoluble	insoluble	insoluble

HPLC 1

Sample	Raw materials
Apparatus	Waters M 6000A LC pump, 450 UV detector and Valco loop injection valve
Column	Reversed-phase; μ-Bondapak C_{18}, 10 μm, 300 × 3.9 mm ID
Pre-column	Bondapak C_{18}/Corasil, 250 × 4.6 mm ID
Mobile phase	0.02 M Sodium hexanesulfonate, 0.025 M Na_3PO_4 in acetonitrile/water (8:92, v/v), pH 6.00 with H_3PO_4
Flow-rate	1.3 ml min^{-1}
Temperature	25 °C
Injection size	5 μg
Detection	UV, λ = 195 nm; 0.2 AUFS

Retention times, min

Streptidine	6.00	Streptomycin A	20.4
Streptomycin B	17.00–22.5	Dihydro-streptomycin A	25.2

Ref. Whall TJ (1981) J. Chromatogr. 219:89

HPLC 2

Sample	Raw materials
Apparatus	Waters M 6000A LC pump, 450 UV detector and Valco loop injection valve
Column	Reversed-phase; Ultrasphere 5 μm, 250 × 4.6 mm ID
Pre-column	Bondapak C_{18}/Corasil, 250 × 4.6 mm ID
Mobile phase	0.02 M Sodium hexanesulfonate, 0.025 M Na_3PO_4 in acetonitrile/water (10:90, v/v), pH 6.0 with H_3PO_4
Flow-rate	1.3 ml min^{-1}
Temperature	25 °C
Injection size	5 μg
Detection	UV, λ = 195 nm; 0.2 AUFS

Retention times, min

Streptidine	6.0	Streptomycin A	17.0
Streptomycin B	14.5–16.0	Dihydro-streptomycin A	19.7

Ref. Whall TJ (1981) J. Chromatogr. 219:89

HPLC 3

Sample: Aminoglycoside antibiotics
Apparatus: Waters M 6000A pump, U6K injector and Jobin Yvon refractive index detector
Column: Reversed-phase; Ultrasphere C_{18}, 5 µm, 250 × 4.6 mm ID
Pre-column: Reversed-phase; Brownlee, Spherisorb 5 µm, C_{18}
Mobile phase:
A. Methanol/0.05 M pentafluoropropionic acid, pH 2.1 (42:58)
B. Methanol/0.05 M heptafluorobutyric acid, pH 2.1 (60:40)
C. Methanol/0.05 M camphorsulfonic acid, pH 2.05 (52:48)
D. 0.2 M Trifluoroacetic acid, pH 1.4
Flow-rate: 1 ml min^{-1}
Temperature: Ambient
Injection size: 35–40 µg
Detection: RI
Retention times, min (estimated from drawing)

Mobile phase	A	B	C	D
Streptomycin	5.0	4.0	5.0	
Butirosin	6.4		7.8	
Kanamycin	7.0	6.7	9.0	6.0
Lividomycin	11.5	8.4	16.8	6.8
Fortimicins A and B		11.0		15.8
Sisomicin				11.0
Gentamicin C_{1a}				12.0
Verdamicin				20.0
Gentamicin C_2				22.0
Gentamicin C_1				41.6

Ref. Inchauspé G and Samain D (1984) J. Chromatogr. 303:277

---

192 Strychnine

CAS-No	57-24-9; 60491-10-3 (sulfate, pentahydrate)	
Chemical name	Strychnidin-10-one	
Empirical formula	Strychnine	Sulfate
	$C_{21}H_{22}N_2O_2$	$(C_{21}H_{22}N_2O_2)_2,H_2SO_4,5H_2O$
Molecular mass	334.4	857.0
Proprietary name	Movellan	

Solubility,	Water	alcohol	chloroform	ether
Strychnine	1 in 7000	1 in 250	1 in 6	insoluble
Sulfate	1 in 50	1 in 135	slightly soluble	insoluble
Light absorption	λ_{max} = 254 nm in aqueous acid			

HPLC 1

Sample	Raw materials
Apparatus	Beckman M 110A pump, 153 UV detector, 421 microprocessor and Altex M 210 injector
Column	Normal phase; 10 μm silica gel, 250 × 2.6 mm
Mobile phase	Concentrated ammonium hydroxide (28.0–30.0% NH_3)/methanol (0.75:99.25, v/v)
Flow-rate	1.1 ml min^{-1}
Internal standard	Quinine, 10 μg ml^{-1}
Injection size	20 μl, 0.625–20 μg ml^{-1}
Detection	UV, λ = 254 nm; 0.02 AUFS
Retention times, min	Internal standard 2.2 Strychnine 4.8 (estimated from drawing)
Ref.	Alliot L, Bryant G and Guth PS (1982) J. Chromatogr. Biomed. Appl. 232:440

HPLC 2

Sample	Test solution containing procaine hydrochloride, promazine hydrochloride, ethoheptazine citrate, protriptyline and strychnine
Apparatus	HPLC equipments with ultraviolet detector
Column	Normal phase; Spherisorb S5W, 250 × 4.9 mm ID
Mobile phase	2700 ml Methanol and 300 ml aqueous ammonium nitrate buffer (containing 27 g NH_4NO_3, 90 ml concentrated NH_3, specific gravity 0.880 in 900 ml water), pH 9.3–9.5
Flow-rate	2.0 ml min^{-1}
Temperature	Ambient (19–33 °C)
Internal standard	Protriptyline
Injection size	5 μl
Detection	UV, λ = 254 nm
Recording	Electronic integrator
Retention times, min	Procaine 1.8 Protriptyline 4.7 Promazine 2.7 Strychnine 5.8 Ethoheptazine 3.4
Ref.	Gill R, Osselton MD et al (1987) J. Chromatogr. 386:65

HPLC 3

Sample	Raw materials
Apparatus	Pye-Unicam LC-XPS HPLC pump, Altex M 153-UV detector and Rheodyne M 7125 injector
Column	Normal phase; Spherisorb S5W, 5 μm, 250 × 5 mm ID
Mobile phase	90 ml Methanol and 10 ml ethylenediamine-ammonium nitrate buffer, pH 10.2 (containing 10.5 g ammonium nitrate, 15 ml ethylenediamine and 200 ml water, and dilute 10-fold) or pH 10.56 (using 15 ml ethylenediamine, 5.02 g NH_4NO_3 in 200 ml water and dilute 10-fold)
Flow-rate	2.0 ml min^{-1}
Temperature	30 °C
Internal standard	Protriptyline hydrochloride, 0.15 mg ml^{-1}
Injection size	5 μl, 0.07 mg ml^{-1}
Detection	UV, λ = 254 nm
Retention times, min	(estimated from drawing)

Procaine	2.3	Protriptyline	6.8
Promazine	3.6	Strychnine	8.5
Ethoheptazine	5.0		

Ref.	Smith RM and Rabuor JO (1989) J. Chromatogr. 464:117

HPLC 4

Sample	Illicit heroin
Apparatus	Spectra-Physics 8000 LC with UV detector
Column	Normal phase; Lichrosorb Si 60, 5 μm, 250 × 4.6 mm ID
Mobile phase	Hexane/dichloromethane/methanol (75:20:5); methanol contains 0.75% (v/v) diethylamine
Flow-rate	3.0 ml min^{-1}
Injection size	10 μl, 12.5–50 μg ml^{-1}
Detection	UV, λ = 254 nm
Ref.	Huizer H (1983) Pharm. Weekbl. Sci. 5:254

HPLC 5

Sample	Illicit dimorphine seizures
Column	Cation-exchange; Zipax SCX, 120 × 2.1 mm ID
Mobile phase	A. 0.2 M Aqueous H_3BO_3, pH 9.3 with 40% NaOH B. 0.2 M Aqueous H_3BO_3/acetontrile/n-propanol (86:12:2), pH 9.8 with 40% NaOH Linear gradient: B 0% to 100% in 6 min
Flow-rate	2 ml min^{-1}
Internal standard	Diphenylamine

Injection size	2 µl			
Detection	UV, $\lambda = 235$, 254 and 270 nm; 0.2 AUFS			
Retention times, min				
	Barbital 1.10		Lignocaine	7.15
	Caffeine 1.10		Diamorphine	7.40
	Morphine 1.80		Procaine	7.55
	Monoacetylmorphine 4.70		Ephedrine	8.30
	Strychnine 6.30		Quinine	9.30
	Dihydrocodeine 6.80		Cocaine	11.50
Ref.	Twitchett PJ (1975) J. Chromatogr. 104:205			

193 Sucrose

CAS-No	57-50-1			
Chemical name	*β*-D-Fructofuranosyl-*α*-D-glucopyranoside			
Empirical formula	$C_{12}H_{22}O_{11}$			
Molecular mass	342.3			
Proprietary names	Simple Linctus; Sucrose Eye-drops; Syrup			
Clinical use	Sweetening Agent			
Solubility	Water	alcohol	alcohol (70%)	chloroform
	2 in 1	1 in 370	freely soluble	insoluble

HPLC 1

Sample	Raw materials		
Apparatus	Shimadzu LC-6A HPLC system and Hitachi U-2000 spectrophotometer		
Column	β-Cyclodextrin; 5 µm, 250 × 4.6 mm ID		
Mobile phase	I. Acetonitrile/water (85:15, v/v)		
	II. Acetone/water (90:10, v/v)		
Flow-rate	1.0 ml min^{-1}		
Temperature	22 °C		
Injection size	31 µg		
Detection	UV, $\lambda = 195$ nm		
Retention times, min	Mobile phase	I.	II.
	Erythrose	3.12	4.23
	2-Deoxy-D-ribose	3.45	
	Glycerine	3.47	
	Glyceraldehyde	3.57	4.56
	Rhamnose	3.63	
	Ribose	3.74	4.60

Mobile phase	I.	II.
6-Deoxy-D-glucose	3.83	
Lyxose	4.04	4.84
2-Deoxy-D-glucose	4.05	
Erythritol	4.07	
Threitol	4.10	
Xylose	4.12	4.97
6-Deoxy-D-galactose	4.13	
2-Deoxy-D-galactose	4.17	
Talose	4.23	4.82
Arabinose	4.30	5.15
Tagatose	4.54	5.18
Fructose	4.73	5.43
Ribitol	4.73	
Arabitol	4.80	
Sorbose	4.81	5.44
Xylitol	4.88	
Allose	4.91	
Mannose	4.91	
Glucose	5.36	5.87
Galactose	5.57	6.17
Sorbitol	5.77	
Mannitol	5.80	
Galactitol	5.87	
D-gluco-D-gulo-Heptose	6.17	6.13
Sucrose	7.79	6.99
Turanose	8.51	7.38
myo-Inositol	8.62	
Cellobiose	9.02	8.23
Maltose	9.09	7.87
Maltitol	10.00	
Lactulose		8.66
Lactose	10.16	8.85
Gentiobiose	10.75	9.50
Melibiose	11.56	9.67
Melezitose	14.18	10.13
Maltotriose	16.44	11.61
Raffinose	17.20	12.29
Stachyose	39.0	26.0

Ref. Armstrong DW and Jin HL (1989) J. Chromatogr. 462:219

HPLC 2

Sample Sugars

Apparatus Merck HPLC low-pressure ternary gradient system

Column	Amino phase; Lichrospher NH_2, 250 × 4 mm ID
Mobile phase	Acetonitrile/water (76:24, v/v)
Flow-rate	2 ml min^{-1}
Injection size	20 μl containing 25.5 μg
Detection	UV, λ = 192 nm
Retention times, min	(estimated from drawing) Sucrose 6.50 Fructose 4.11 Maltose 8.57 Glucose 4.75 Lactose 9.61
Ref.	Merck Application Manual on HPLC, TLC and HPTLC (1986). E. Merck, Darmstadt, p 23

HPLC 3

Sample	Typical cough syrup
Apparatus	Waters ALC M 201 with 6000 pump, R-401 refractometer and U6K injector
Column	μ-Bondapak-carbohydrate, 300 × 4 mm ID
Mobile phase	Water/acetonitrile (20:80)
Flow-rate	2.0 ml min^{-1}
Temperature	25 °C
Injection size	5–40 μl
Detection	RI
Retention times, min	(estimated from drawing) Glucose 5.0 Glycerol 3.0 Sucrose 8.0 Fructose 4.5 Maltose 10.0
Ref.	Palmer JK (1979) in: Tsuji K (ed) GLC and HPLC Determination of Therapeutic Agents, Part 3. Marcel Dekker Inc., New York and Basel, p 1317

HPLC 4

Sample	Orange juice, apple juice, beverage and strawbery yogurt
Apparatus	LDC Constametric I or CM 4000 HPLC with ICM conductivity detector or differential refractometer
Column	A. Anion-exchange; RCX-10, 150 × 4.1 mm ID B. Ligand-exchange; RCX-20, 250 × 4.1 mm ID C. Lead form ligand-exchange; CHO 682, 300 × 7.8 mm ID D. Lead form ligand-exchange; Bio-Rad HPX-87P, 300 × 7.8 mm ID
Mobile phase	For A: 5 × 10^{-3} M Sodium acetate For B, C and D: Water
Flow-rate	2.0 ml (A); 0.5 ml (B); 0.4 ml (C) and 1.0 ml (D) min^{-1}
Temperature	80 °C (B); 85 °C (C) and 80 °C (D)
Detection	Indirect conductivity (A); refractive index (C)

Retention times, min	(estimated from drawing)				
	Column	A	B	C	D
	Glucose	2.3	3.3	16.0	12.0
	Fructose	2.6	3.5	22.0	16.0
	Sucrose	3.0		14.0	9.0
	Maltotriose		2.7		
	Maltose		3.1		
	Glycerol		4.0		
	Ethanol		6.7		
	Mannitol			33.0	
	Sorbitol			45.0	
	Lactose				10.0
	Galactose				13.0
Ref.	Lee DP and Bunker MT (1989) J. Chromatogr. Sci. 27:496				

194 Sulfadiazine

CAS-No	68-35-9; 547-32-0 (sodium salt)			
Chemical name	N^1-(Pyrimidin-2-yl) sulfanilamide			
Empirical formula	Sulfadiazine	Sodium Salt		
	$C_{10}H_{10}N_4O_2S$	$C_{10}H_9N_4NaO_2S$		
Molecular mass	250.3	272.3		
Proprietary names	Adiazine; S-Diazine; Theradia			
Clinical use	Chemotherapeutic			
Solubility,	Water	acetone	alcohol	chloroform
Sulfadiazine	insoluble	1 in 300	very slightly sol	insoluble
Sodium salt	1 in 2		slightly sol	insoluble
Light absorption	λ_{max} = 254 nm in aqueous alkali			

HPLC 1

Sample	Dosage forms and biological fluids
Extraction	Serum: Deproteinizing with acetonitrile Urine: Diluting with water
Apparatus	Pye-Unicam LC-3 HPLC with variable-wavelength detector and Rheodyne injection valve
Column	Reversed-phase; Lichrosorb C_{18}, 10 µm
Mobile phase	Methanol/water/glacial acetic acid (15:84:1, v/v)
Flow-rate	1 ml min^{-1}; 725 psi
Injection size	20 µl, 0.1-200 µg ml^{-1}
Detection	UV, λ = 254 nm
Retention time, min	Sulfadiazine 10.8

Ref.	Springolo V and Coppi G (1989) J. Pharm. Biomed. Anal. 7:57

HPLC 2

Sample	Raw materials
Column	Reversed-phase; Chromspher C_{18}, 5 μm, 100 × 3 mm ID
Mobile phase	0.02 M Sodium dihydrogen phosphate, pH 3.0/ acetonitrile (90:10)
Flow-rate	0.8 ml min^{-1}; 1000 psi
Injection size	0.1 μl
Detection	UV, λ = 254 nm
Retention times, min	(estimated from drawing) Sulfamethizole 4.69 Sulfisomidine 2.06 Sulfadimidine 5.44 Sulfadiazine 2.44 Sulfafurazole 14.81
Ref.	Chrompack Application Notes (1987) Antibiotics, Application 48-HPLC, Middelburg, p 5

HPLC 3

Sample	Dosage forms
Apparatus	Du Pont 8800 pump module with UV detector
Column	β-Cyclodextrin; Cyclobond I, 100 × 4.6 mm ID
Mobile phase	0.05 M Sodium ortho-phosphate (NaH_2PO_4), pH 7.0/methanol (85:15)
Flow-rate	0.5 ml min^{-1}
Temperature	35 °C
Injection size	50 μl, 10–30 μg ml^{-1}
Detection	UV, λ = 257 nm
Recording	Spectra-Physics computing integrator SP 4100
Retention times, min	(estimated from drawing) Sulfaphenazole 3.13 Sulfacetamide 1.56 Sulfaguanidine 3.72 Sulfadiazine 2.15 Sulfadimidine 4.70 Sulfamerazine 2.54 Sulfathiazole 9.38
Ref.	Ahmed A-E HN and El-Gizawy SM (1989) Analyst 114:571

HPLC 4

Sample	Body fluids
Extraction	Deproteinization with 0.33 M perchloric acid
Apparatus	Spectra-Physics 3500B HPLC with 770 spectrophotometric detector
Column	Reversed-phase; Lichrosorb C_8, 5 μm, 150 × 4.6 mm ID
Mobile phase	390 ml 0.067 M KH_2PO_4 with 10 ml 0.067 M Na_2HPO_4 pH 6.7 and 80 ml methanol
Flow-rate	1.6 ml min^{-1}; 1690 psi

Injection size	100 μl, 0.5 μg ml^{-1}			
Detection	UV, λ = 260 nm; 0.05 A			
Retention times, min	(estimated from drawing)		Sulfadoxine	8.8
	Sulfacetamide	1.7	Sulfapyridine	9.3
	Sulfanilamide	2.2	N^4-Acetylsulfamethoxazole	11.5
	Sulfadiazine	3.3		
	Sulfamethoxazole	5.3	Sulfamethoxypyridine	14.6
	Sulfathiazole	7.0		
			Sulfadimidine	16.4
Ref.	Vree TB, Hekster YA et al (1978) J. Chromatogr. Biomed. Appl. 146:103			

HPLC 5

Sample	Raw materials and tablets
Column	Reversed-phase; ODS 5–10 μm, 300 × 4 mm ID
Mobile phase	Water/acetonitrile/glacial acetic acid (87:12:1)
Flow-rate	2 ml min^{-1}
Injection size	Appr. 10 μl
Detection	UV, λ = 254 nm
Ref.	Anonym (1985): The United States Pharmacopeia XXI and The National Formulary XVI. United States Pharmacopeial Convention Inc., Rockville, p 989

195 Sulfaguanidine

CAS-No	57-67-0; 6190-55-2 (monohydrate)			
Chemical name	N^1-Amidinosulfanilamide monohydrate			
Empirical formula	$C_7H_{10}N_4O_2S,H_2O$			
Molecular mass	232.3			
Proprietary names	Aterian; Ganidan; Guabeta			
Clinical use	Chemotherapeutic			
Solubility	Water	alcohol	acetone	aq HCl
	1 in 1000	1 in 250	slightly sol	readily soluble
Light absorption	λ_{max} = 259 nm in aqueous alkali			

HPLC 1

Sample	Animal tissues
Extraction	Acetonitrile at pH 8.2 and then re-extraction with dichloromethane
Apparatus	Varian LC-5000, Knauer M 8700 UV detector and Shimadzu RF-530 fluorimeter

Column	Ion-exchange; 5 μm Nucleosil 5 SA, 200 × 4 mm ID
Pre-column	S-X3, 200–400 mesh, 30 × 4 mm ID (Bio-Beads)
Mobile phase	0.05 M NaH_2PO_4, pH 2 with 85% ortho-H_3PO_4/methanol (85:15, v/v)
Flow-rate	1 ml min^{-1}
Injection size	20 μl, 25 μg ml^{-1}
Detection	UV, λ = 275 nm Fluorescence, λ = 275/340 nm
Retention times, min	Sulfanilamide 7.5 Sulfaguanidine 12.5
Ref.	Petz M (1987) J. Chromatogr. Biomed. Appl. 423:217

HPLC 2

Sample	Sulfonamides
Apparatus	HPLC low-pressure ternary gradient system
Column	Reversed-phase; Lichrosorb C_{18}, 250 × 4 mm ID
Mobile phase	Methanol/water (40:60, v/v)
Flow-rate	0.75 ml min^{-1}; 3000 psi
Injection size	10 μl containing 20 μg
Detection	UV, λ = 230 nm; 0.04 AUFS
Retention times	Sulfaguanidine, sulfathiazole, sulfapyridine and sulfamethazine, in ascending order (from drawing)
Ref.	Merck Application Manual on HPLC, TLC and HPTLC (1986). E. Merck, Darmstadt, p 76

HPLC 3

Sample	Test substance
Apparatus	Waters ALC/GPC M 201 LC with M 6000 pumps, U6K injector, 600 programmer and Varian M 635 UV/VIS detector
Column	Reversed-phase; Corasil C_{18}, 37–70 μm with 1% squalene loading, 600 × 2 mm ID
Mobile phase	A. Acetate buffer, pH 4.0 B. Sorensen buffer, pH 6.5
Flow-rate	0.4 ml min^{-1}
Injection size	1.0 μl, 0.1–1 mg ml^{-1}
Detection	UV, λ = 255 nm
Retention times	(in ascending order)

pH 4.0 (A)	pH 6.5 (B)
Sulfaguanidine	Sulfacetamide
Sulfanilamide	Sulfanilamide
Sulfacetamide	Sulfisoxazole
Sulfadiazine	Sulfaguanidine
Sulfathiazole	Sulfamethoxazole
Sulfamerazine	Sulfadiazine

	Sulfamethoxazole	Sulfaethidole
	Sulfisoxazole	Sulfathiazole
	Sulfamethazine	Sulfamerazine
	Sulfamethoxy-pyridazine	Sulfamethazine
	Sulfaethidole	Sulfamethoxy-pyridazine
Ref.	Henry D, Block JH et al (1976) J. Med. Chem. 19:619	

196 Sulfamerazine

CAS-No	127-79-7; 127-58-2 (sodium salt)			
Chemical name	N^1-(4-Methylpyrimidin-2-yl)sulfanilamide			
Empirical formula	Sulfamerazine	Sodium Salt		
	$C_{11}H_{12}N_4O_2S$	$C_{11}H_{11}N_4NaO_2S$		
Molecular mass	264.3	286.3		
Proprietary names	Dosulfin; Romezin; Trinamide (vet)			
Clinical use	Chemotherapeutic			
Solubility,	Water	alcohol	acetone	aq HCl, aq NaOH
Sulfamerazine	1 in 6250	1 in 550	1 in 60	soluble
Sodium salt	1 in 3.5	slightly soluble		
Light absorption	$\lambda = 255$ nm in aqueous alkali			

HPLC 1

Sample	Pork tissue		
Extraction	Multistep, with ethanol, 1 N HCl, and then with methylene chloride		
Apparatus	Waters ALC/GPC M 204 LC with 6000A pump, U6K injector and 440 UV detector		
Column	Reversed-phase; Brownlee C_8, 10 μm		
Mobile phase	25% Methanol in 0.01 M ammonium acetate, pH 6.7 or 3.5		
Flow-rate	2.0 ml min^{-1}		
Temperature	Ambient		
Injection size	50 μl, 12.5–125 ng ml^{-1}		
Detection	UV, $\lambda = 254$ nm		
Recording	Hewlett-Packard 3385 reporting integrator		
Retention times	(relative retention volume)		
	pH, mobile phase	6.7	3.5
	Sulfamethazine	1.00	1.00
	Phthalylsulfacetamide	0.22	1.36
	Sulfaguanidine	0.25	0.26

Sulfanilamide	0.27	0.27
Succinylsulfathiazole	0.32	0.89
Sulfadiazine	0.38	0.50
Sulfachlorpyridazine	0.48	1.27
Sulfathiazole	0.52	0.54
Sulfapyridine	0.60	0.60
Sulfamerazine	0.63	0.71
Sulfadoxine	0.73	1.80
Sulfaquinoxaline	1.89	7.08
Sulfadimethoxine	2.07	5.02
Sulfaethoxypyridazine	2.47	2.63

(absolute retention volume of sulfamethazine is 14.24 ml at pH 6.7 and 14.04 ml at pH 3.5)

Ref. Vilim AB, Larocque L and MacIntosh AI (1980) J. Liq. Chromatogr. 3:1725

HPLC 2

Sample	Milk
Extraction	Chloroform
Apparatus	Knauer 52.00 HPLC, Metrohm E611 electrochemical detector and Rheodyne M 7120 injector
Column	Reversed-phase; C_2, 10 µm, 250 × 4.6 mm ID
Mobile phase	Methanol/water (25:75) and methanol/water (40:60, v/v) with 0.01 M $LiClO_4$
Flow-rate	2 ml min^{-1}; 1885–2030 psi
Injection size	10 µl containing 10 ng ml^{-1}
Detection	EC, GC electrode, + 1.1 V vs Ag/AgCl RE
Retention times, min	Sulfadiazine 3.8 Sulfadimidine 5.7 Sulfamerazine 4.8
Ref.	Alawi MA and Ruessel HA (1981) Fresenius Z. Anal. Chem. 307:382

HPLC 3

Sample	Chicken tissue
Extraction	Methanol, chloroform and then 85% acetonitrile on alumina column
Column	Reversed-phase; Nucleosil C_{18}, 5 µm, 250 × 4.6 mm ID
Pre-column	Reversed-phase; Nucleosil C_{18}, 10 µm, 50 × 4.6 mm ID
Mobile phase	Acetonitrile/acetic acid/water (14:0.5:86)
Flow-rate	1 ml min^{-1}
Temperature	45 °C
Injection size	10 µl, 2 µg ml^{-1}
Detection	UV, $\lambda = 270$; 0.01 AUFS and 350 nm (furazolidone); 0.01 AUFS

Retention times, min (estimated from drawing)

Clopidol	8.0	Dinitolmide	24.0
Trimethoprim	10.2	Sulfamethoxazole	27.0
Sulfamerazine	12.0	Sulfisoxazole	34.5
Ormethoprim	13.0	Pyrimethamine	46.0
Sulfamethoxy-pyridazine	15.8	Sulfadimithoxine	64.5
Furazolidone	17.5	Ethopamate	68.5
Sulfamonomethoxine	21.5	Sulfaquinoxaline	74.0

Ref. Nagata T and Saeki M (1988) J. Fd. Hyg. Soc. Jap. 29:13

HPLC 4

Sample: Test materials
Column: Reversed-phase; PRP-1, 10 μm, 150 × 4.1 mm ID
Mobile phase: 0.1 N Sodium phosphate/acetonitrile (28.6:71.4), pH 12.0
Flow-rate: 2.0 ml min^{-1}
Detection: UV
Retention times, min (estimated from drawing)

Sulfanilic acid	1.5	Sulfamerazine	2.0
Sulfadiazine	1.8	Sulfanilamide	3.0
		Sulfamethazine	3.5

Ref. Latek Labortechnik (1984) Katalog No 2, p 281

HPLC 5

Sample: Test substances
Apparatus: Spectra-Physics 3500B LC pump with 770 ultraviolet detector
Column:
A. Reversed-phase; μ-Bondapak 10 μm, C_{18}, 300 × 3.9 mm ID
B. Reversed-phase; Zorbax ODS 5 μm, 250 × 4.6 mm ID
C. Reversed-phase; Ultrasphere ODS 5 μm, 250 × 4.6 mm ID

Mobile phase: Methanol/acetic acid/triethylamine (TEA)/water [22.5 : 1.5 : (0 or 1 ml) : (76 or 75)]
Flow rate: 1.5 ml min^{-1}
Temperature: Ambient
Injection size: 10 μl, 0.05 μg ml^{-1}
Detection: UV, λ = 254 nm
Capacity factor, *k′*

TEA %	0	0	0	1	1	1
Column	A	B	C	A	B	C
Ephedrine	0.60	22.79	20.10	0.97	1.73	2.26
Sulfamerazine	1.75	3.26	2.43	1.73	3.20	2.46
Phenol	2.65	5.23	6.25	2.44	5.90	6.48

Ref.	Roos RW and Lau-Cam CA (1986) J. Chromatogr. 370:403

197 Sulfamethazine

CAS-No	57-68-1; 1981-58-4 (sodium salt)			
Chemical name	N^1-(4,6-Dimethylpyrimidin-2-yl)sulfanilamide			
Empirical formula	Sulfamethazine	Sodium Salt		
	$C_{12}H_{14}N_4O_2S$	$C_{12}H_{13}N_4NaO_2S$		
Molecular mass	278.3	300.3		
Proprietary names	Deladine; Diazil; S-Dimidine			
Clinical use	Chemotherapeutic			
Solubility,	Water	alcohol	acetone	aq NaOH
Sulfamethazine	very slightly soluble	1 in 120	1 in 30	soluble
Sodium salt	1 in 2.5	1 in 60		
Light absorption	λ_{max} = 258 nm in aqueous alkali			

HPLC 1

Sample	Cattle urine		
Extraction	Methanol		
Apparatus	Perkin-Elmer 1200 HPLC with digital controlled pump and LC-55 ultraviolet detector		
Column	Amino-bonded; Amino-Sil-X-I, 500 × 2.6 mm ID		
Mobile phase	Methanol		
Flow-rate	1.5 ml min^{-1}; 700 psi		
Temperature	Ambient		
Injection size	5 µl containing 0.2 µg		
Detection	UV, λ = 254 nm		
Retention times, min	Sample	test solution	urine
	Sulfamethazine	2.3	2.6
	Sulfamerazine	3.6	3.8
	Sulfamethazine acetyl derivate	5.7	5.8
	Sulfathiazole	8.1	9.0
	Sulfamerazine acetyl derivative	11.0	11.0
	Sulfathiazole acetyl derivative	13.8	15.6
Ref.	Sharma JP, Perkins EG and Bevill RF (1976) J. Pharm. Sci. 65:1606		

HPLC 2

Sample	Raw materials and dosage forms
Apparatus	Hewlett-Packard 1084 LC with UV detector
Column	Reversed-phase; Nucleosil C_{18}, 10 μm, 250 × 4 mm ID
Mobile phase	Acetonitrile/0.05 M ortho-phosphoric acid (30:70)
Flow-rate	2 ml min^{-1}
Temperature	40 °C
Injection size	5–30 μl
Detection	UV, λ = 270 nm
Recording	Hewlett-Packard 79850B LC-Terminal integrator
Retention times, min	Trimethoprim 1.54
	Sulfamethazine 2.00
	Sulfamethoxazole 2.94
Ref.	Rehm KD and Steinigen M (1981) Pharm. Ztg. 126:99

HPLC 3

Sample	Feed Premixes
Apparatus	LDC HPLC with Constametric III pump, Valco M 7000 PSI injector and Spectromonitor III UV detector
Column	Reversed-phase; Spherisorb ODS 5 μm, 200 × 4.7 mm ID
Mobile phase	A. Acetonitrile/0.05 N H_2SO_4 (21:79, v/v)
	B. Acetonitrile/0.05 N H_2SO_4 (24:76, v/v)
	C. Acetonitrile/0.05 N H_2SO_4 (27.5:72.5, v/v)
	D. Acetonitrile/0.05 N H_2SO_4/dimethylformamide (20.5:76.5:3, v/v)
	E. Acetonitrile/0.05 N H_2SO_4/dimethylformamide (19.5:74:6.5, v/v)
Flow-rate	1 ml min^{-1}
Temperature	Ambient
Injection size	25 μl, 200 μg ml^{-1} (sodium salt)
Detection	UV, λ = 254 nm; 0.5 AUFS
Recording	LDC/Milton Roy CI/10 integrator

Retention times, min

Mobile phase	A	B	C	D	E
Sulfamethazine sodium	5.5	3.5	3.0	5.0	4.16
Sulfamethoxypyridazine sodium	7.0	4.75	3.66	6.5	5.66
Oxytetracycline hydrochloride	9.25	6.25	4.5	7.75	6.92

Ref. Torel J, Cillard J et al (1985) J. Chromatogr. 330:425

HPLC 4

Sample	Milk
Extraction	Chloroform and then re-extraction with buffer solution using hexane

Apparatus	HPLC with Perkin-Elmer LC-95 UV/VIS detector
Column	Reversed-phase; Supelcosil DB C_{18}, 250 × 4.6 mm ID
Pre-column	Supelcosil 20 mm
Mobile phase	600 ml Methanol in 0.1 M KH_2PO_4 phosphate buffer (containing 27.2 g KH_2PO_4 in 2 liter water) up to 2 liter
Flow-rate	1.5 ml min^{-1}
Temperature	35 °C
Injection size	100 µl, 1 ng ml^{-1}
Detection	UV, $\lambda = 265$ nm
Retention time, min	Sulfamethazine 5.0
Ref.	Weber JD and Smedley MD (1989) J. Assoc. Off. Anal. Chem. 72:445

HPLC 5

Sample	Blood plasma and milk of dairy cows
Extraction	Deproteinization with 0.66 M perchloric acid and then neutralized with 40% KOH; in low concentration further with ethyl acetate
Apparatus	Kipp Analytica M 9208 HPLC pump with Kratos M 773 UV detector
Column	Reversed-phase; Lichrosorb C_8, 5 µm, 150 × 4.6 mm ID
Mobile phase	0.02 M Sodium acetate/0.2 M acetic acid/methanol/acetonitrile/water (50:3.6:15:4:27.4), pH 4.9
Flow-rate	2.0 ml min^{-1}
Temperature	20 °C
Injection size	100 µl, 0.01–0.05 µg ml^{-1}
Detection	UV, $\lambda = 272$ nm
Recording	Shimadzu C-R1B integrator
Capacity factor, *k'*	6-Hydroxymethylsulfadimidine 3.6
	5-Hydroxysulfadimidine 6.5
	Sulfadimidine 8.1
	N^4-Acetylsulfadimidine 14.2
Ref.	Nouws JFM, Vree TB et al (1985) Vet. Q. 7:177

HPLC 6

Sample	Biological materials from horse
Extraction	Diethyl ether after diluting (plasma) or homogenization with 0.9% sodium chloride solution (tissues)
Apparatus	Varian M 5000 HPLC with Vista 401 working station and UV 100 detector
Column	Reversed-phase; Hypersil 5 µm, 125 × 4.0 mm ID
Mobile phase	70 mM Sodium phosphate buffer, pH 6.8/methanol (85:15)

Flow-rate	1.0 ml min^{-1}; 1498 psi
Injection size	100 µl
Detection	UV, λ = 260 nm
Ref.	Dettwiler M, Straub R et al (1987) Schweiz. Arch. Tierheilk. 129:473

---

198 Sulfamethoxazole

CAS-No	723-46-6
Chemical name	*N*^{1}-(5-Methyl-3-isoxazolyl)-sulfanilamide
Empirical formula	$C_{10}H_{11}N_3O_3S$
Molecular mass	253.3
Proprietary names	Bactrim; Gantanol; Sinomin
Clinical use	Chemotherapeutic

Solubility

Water	alcohol	acetone	ether	aq NaOH
Very slightly soluble	1 in 50	1 in 3	insoluble	soluble

Light absorption: λ_{max} = 256 nm in aqueous alkali

HPLC 1

Sample	Raw materials (sulfonamides)
Apparatus	Cecil CE 210 LC and CE 212 variable-wavelength spectrophotomer
Column	Normal phase; Spherisorb S5W 5 µm, 250 × 4 mm ID
Mobile phase	Cyclohexane/anhydrous ethanol/glacial acetic acid (85.7:11.4:2.9)
Flow-rate	2 ml min^{-1}
Temperature	Ambient
Injection size	1 µl, 0.005 mg ml^{-1}
Detection	UV, λ = 260 nm

Retention times, min

Sulfachlorpyridazine	6.18
Sulfapyridine	7.05
Sulfadoxine (Sulphormethoxine)	7.90
Sulfaquinoxaline	8.40
Sulfamethoxazole	8.48
N^4-Acetylsulfamethoxazole	8.58
Sulfisoxazole (Sulfafurazole)	10.12
Sulfamethazine (Sulfadimidine)	11.78
Sulfamethoxypyridazine	12.33
Sulfacetamide	12.65
Sulamerazine (Sulfamethyldiazine)	13.25
Sulfameter (Sulfamethoxydiazine)	13.45

	Sulfadiazine	14.08
	Sulfanilamide	14.42
	N^4-Acetylsulfanilamide	15.43
	Sulfamoxole	19.82
	Sulfathiazole	20.90
	Phthalylsulfathiazole	21.88
	Succinylsulfathiazole	25.88
Ref.	Cobb PH and Hill GT (1976) J. Chromatogr. 123:444	

HPLC 2

Sample	Raw materials and blood plasma
Apparatus	Waters M 6000A LC pump and Perkin-Elmer M 204A spectrophotofluorimeter
Column	Reversed-phase; Ultrasphere ODS 5 µm, 150 × 4.6 mm ID
Mobile phase	425 ml 0.067 M Phosphate buffer, pH 7.0 and absolute methanol to make 1000 ml
Flow-rate	1.3 ml min^{-1}
Derivatization	Pre-column; with fluorescamine
Injection size	30–50 µl
Detection	Fluorescence, λ = 395/485 nm
Recording	Hewlett-Packard 3380A electronic integrator

Retention times, min				
	Sulfamethoxazole	1.5	Thyroxine	8.8
	Procainamide	3.1	Amphetamine	11.8
	Tocainide	6.0	Amantadine	18.9

Ref. Sedman AJ and Gal J (1982) J. Chromatogr. Biomed. Appl. 232:315

HPLC 3

Sample	Biological fluids
Extraction	Chloroform/ethyl acetate (75:25) at pH 6.2
Apparatus	Varian 5000 HPLC with UV detector
Column	Normal phase; Lichrosorb Si-60, 5 µm, 250 × 4 mm ID
Pre-column	Normal phase; Lichrosorb Si-60, 25–40 µm, 40 × 4 mm ID
Mobile phase	A. Chloroform/methanol/water/ammonia 28–30% (94.5:5:0.25:0.19, v/v) B. Chloroform/methanol/water/ammonia 28–30% (79:20:1:0.15, v/v) Gradient: A 100% for 3 min; B 50% at 3.5 min; B 100% in 4–12 min; B 50% at 12.5 min; A 100% in 13–16 min
Flow-rate	2 ml min^{-1}; 2610 psi
Temperature	Ambient

Internal standard	Sulfamoxole
Injection size	50 µl, 2.0–150 µg ml^{-1}
Detection	UV, λ = 280 nm; 0.02–0.5 AUFS
Retention times, min	Trimethoprim 5.8; Sulfamethoxazole 9.6; Internal standard 7.0; N^4-Sulfamethoxazole 12.6
Ref.	Spreux-Varoquaux O, Chapalain JP et al (1983) J. Chromatogr. Biomed. Appl. 274:187

HPLC 4

Sample	Body fluids
Extraction	Deproteinization with 0.33 M perchloric acid
Apparatus	Spectra-Physics 3500B HPLC with 770 spectrophotometric detector
Column	Reversed-phase; Lichrosorb C_8, 5 µm, 150 × 4.6 mm ID
Mobile phase	390 ml 0.067 M KH_2PO_4 with 10 ml 0.067 M Na_2HPO_4, pH 6.7 and 80 ml ethanol
Flow-rate	1.2 ml min^{-1}; 1249 psi
Injection size	100 µl, 0.5 µg ml^{-1}
Detection	UV, λ = 225 nm; 0.05 A
Retention times, min	(estimated from drawing) Sulfamethoxazole 6.3; Trimethoprim 5.4; N^4-Acetylsulfamethoxazole 7.8
Ref.	Vree TB, Hekster YA et al (1978) J. Chromatogr. Biomed. Appl. 146:103

HPLC 5

Sample	Sulfamethoxazole and trimethoprim tablets
Column	Reversed-phase; ODS 5–10 µm, 300 × 4 mm ID
Mobile phase	Diluted acetic acid (1 in 100)/acetonitrile (84:16)
Flow-rate	3.0 ml min^{-1}
Injection size	10–20 µl
Detection	UV, λ = 254 nm
Retention times, min	Trimethoprim 7.0; Sulfamethoxazole 13.0
Ref.	Anonym (1985): The United States Pharmacopeia XXI and The National Formulary XVI. United States Pharmacopeial Convention Inc., Rockville, p 995

---

199 Sulfanilamide

CAS-No	63-74-1
Chemical name	4-Aminobenzenesulfonamide

Empirical formula	$C_6H_8N_2O_2S$					
Molecular mass	172.2					
Proprietary names	Exoseptoplix; Rhinamid; Streptosil					
Clinical use	Chemotherapeutic					
Solubility	Water	alcohol	acetone	aq HCl, aq NaOH	ether	
	1 in 170	1 in 30	1 in 4	soluble	insol	
Light absorption	λ_{max} = 250 nm in aqueous alkali					

HPLC 1

Sample	Swine kidney and ham muscle
Extraction	Chloroform/acetone (1:1) on ion-exchange extraction column
Apparatus	LKB M 2150 pump, Rheodyne M 7125 injector and M 2128 Uvicord S UV detector
Column	Reversed-phase; CP TM Spher C_8, 8 μm, 250×4 mm ID
Pre-column	Reversed-phase; 75×2.1 mm ID
Mobile phase	Acetonitrile/10 mM ammonium acetate buffer, pH 4.6 (3:7, v/v)
Flow-rate	1.5 ml min^{-1}
Temperature	Ambient
Injection size	20 μl
Detection	UV, λ = 254 nm; 0.05 AUFS
Recording	Spectra-Physics 4270 printer/plotter integrator

Retention times, min				
(estimated from drawing)		Sulfamethazine	4.3
Sulfanilamide	2.6	Sulfadoxine	6.0
Sulfadiazine	3.3	Sulfaquinoxaline	9.2

Ref.	Haagsma N and Van De Water C (1985) J. Chromatogr. 333:256

HPLC 2

Sample	Raw materials
Apparatus	Spectra-Physics 3500B LC pump and UV 770 detector
Column	Reversed-phase; 10 μm Bondapak, C_{18}, 300×3.9 mm ID
Mobile phase	Methanol/acetic acid/triethylamine/water (20:1.5:0.5:78)
Flow-rate	1.5 ml min^{-1}
Temperature	Ambient
Injection size	10 μl, 0.25 mg ml^{-1}
Detection	UV, λ = 254 nm; 0.32 AUFS
Recording	Hewlett-Packard 3380A integrator

Retention times, min	(estimated from drawing)			
	Sulfanilic acid	2.0	Sulfamethizole	10.6
	Sulfanilamide	2.8	Sulfamethazine	12.0
	Sulfadiazine	5.5	Sulfamethoxazole	14.0
	Sulfapyridine	7.0	Sulfisoxazole	15.5
	Sulfamerazine	8.7	Sulfachlorpyridazine	18.5
Ref.	Roos RW and Lau-Cam CA (1986) J. Chromatogr. 370:403			

HPLC 3

Sample	Sulfonamides and blood plasma		
Extraction	Direct injection or deproteinization with methanol (1:1)		
Apparatus	Waters ALC/GPC M 204 LC with 440 UV detector and U6K universal injector		
Column	Reversed-phase; μ-Bondapak, C_{18}		
Pre-column	Corasil C_{18}		
Mobile phase	Water/methanol (60:40), pH 4.00 with glacial acetic acid and pH 7.45 without buffer		
Flow-rate	0.8 ml min^{-1} and 10–100 ng ml^{-1}		
Temperature	Ambient (25 °C)		
Injection size	20 μl, 10–100 μg ml^{-1}		
Detection	UV, $\lambda = 254$ nm		
Retention times, min	pH, mobile phase	4.00	7.45
	Sulfaguanidine	3.60	3.90
	Sulfanilamide	3.75	3.05
	Sulfathiazole	3.75	3.30
	Sulfapyridine	4.00	5.00
	Sulfamerazine	4.10	5.40
	Sulfisoxazole	4.20	4.20
	Sulfamethazine	4.25	6.10
	Acetylsulfisoxazole(N^4)	5.00	4.40
Ref.	Suber RL and Edds GT (1980) J. Liq. Chromatogr. 3:257		

HPLC 4

Sample	Test substances
Apparatus	Waters ALC/GPC M 201 HPLC with M 6000 pumps and Varian M 635 UV detector
Column	A. Reversed-phase; Corasil C_{18}, 37–70 μm, 600 × 2 mm ID B. As in A but with 1% octanol loading
Mobile phase	I. Acetate buffer, pH 4.0 II. Sorensen phosphate buffer, pH 6.5
Flow-rate	1–3 ml min^{-1} (A); 0.4 ml min^{-1} (B)

Injection size	1.0 µl, 0.1–1.0 mg ml^{-1}
Detection	UV, λ = 255 nm
Retention times	(retention positions)

Column	A.		B.	
pH, mobile phase	4.0	6.5	4.0	6.5
Sulfaguanidine	1	4	1	2
Sulfanilamide	2	5	2	8
Sulfacetamide	3	1	3	1
Sulfadiazine	4	6	4	4
Sulfathiazole	5	8		5
Sulfamerazine	6	9	5	6
Sulfamethazine	7	11	8	7
Sulfamethoxypyridazine	8	10	9	
Sulfamethoxazole	9	3	6	
Sulfisoxazole	10	2	7	3
Sulfaethidole	11	7	10	

Ref. Henry D, Block JH et al (1976) J. Med. Chem. 19:619

HPLC 5

Sample	Test substances
Apparatus	Waters M 6000A LC pump, 440 UV detector and Valco sample loop
Column	Normal phase; Nucleosil 100, 10 µm, 200 × 4 mm ID
Mobile phase	0.02 M Camphor-10-sulfonic acid, pH 1.8
Flow-rate	1.7 mm sec^{-1}; 8.4 MPa
Temperature	25 °C
Injection size	10 µl
Detection	UV, λ = 254 nm; 0.005 A
Retention times, min	(estimated from drawing)

Sulfanilamide	2.4	Sulfamerazine	3.9
Sulfadiazine	3.2	Sulfamethazine	5.7

Ref. Crommen J (1979) J. Chromatogr. 186:705

---

200 Sulfapyridine

CAS-No	144-83-2; 127-57-1 (sodium salt, anhydrous)	
Chemical name	N^1-(2-Pyridyl) sulfanilamide	
Empirical formula	Sulfapyridine	Sodium Salt
	$C_{11}H_{11}N_3O_2S$	$C_{11}H_{10}N_3NaO_2S$
Molecular mass	249.3	271.3

Proprietary names	Dagenan; M & B 693				
Clinical use	Chemotherapeutic				
Solubility,	Water	ethanol	acetone	chloroform	aq NaOH
Sulfapyridine	1 in 3500	1 in 440	1 in 65	insoluble	soluble
Sodium salt	1 in 3	1 in 10			
Light absorption	λ_{max} = 247 nm in aqueous alkali				

HPLC 1

Sample	Serum and urine	
Extraction	Deproteinization with methanol; the metabolites with isobutyl-methyl ketone at pH 5 and re-extraction with dichloromethane at pH 12.8 after enzymatic hydrolysis with β-glucuronidase	
Apparatus	Beckman M 112 LC pump, 210 injector and Waters M 440 UV detector or with Waters HPLC system	
Column	Reversed-phase; Nucleosil 10 μm, C_{18}, 250 × 4 mm ID	
Pre-column	Merck Perisorb C_{18}, 40 μm, 30 × 4 mm ID	
Mobile phase	Methanol/0.02 M phosphate buffer, pH 7.0 (20:80, v/v)	
Flow-rate	2.0 ml min^{-1}	
Internal standard	6-Methyl sulfapyridine	
Injection size	20 μl containing 21.8 μg	
Detection	UV, λ = 254 nm; 0.001-0.005 AUFS	
Recording	Spectra-Physics 4270 integrator	
Retention times, min	(estimated from drawing)	
	5-Hydroxysulfapyridine	6.0
	Sulfapyridine	7.0
	N-Acetylsulfapyridine-O-glucuronide	10.0
	6-Methyl-sulfapyridine	12.5
	N-Acetylsulfapyridine	14.5
Ref.	Ahnfelt N-O, Modin R et al (1986) J. Pharm. Biomed. Anal. 4:261	

HPLC 2

Sample	Blood plasma
Extraction	Dichloromethane under acidic condition
Apparatus	LDC Constametric III dual-piston pump, Spectromonitor III UV detector and Rheodyne injection system
Column	Reversed-phase; Spherisorb ODS 10 μm, 250 × 3.2 mm ID
Pre-column	Whatman Co:Pell ODS, 75 × 4.6 mm ID
Mobile phase	Methanol/0.05 M phosphate buffer, pH 7.4 (25:75)
Flow rate	1.2 ml min^{-1}; 20.7 MPa

Temperature	Ambient
Injection size	20 µl
Detection	UV, λ = 270 nm
Recording	Hewlett-Packard 3390A integrator
Retention times, min	Sulfamethoxazole 2.5–2.7 Acetylsulfamethoxazole 3.7 Sulfapyridine 4.2–5.2 Acetylsulfapyridine 5.5
Ref.	Astbury C and Dixon JS (1987) J. Chromatogr. Biomed. Appl. 414:223

HPLC 3

Sample	Blood plasma
Extraction	Chloroform from acetate buffer extract or with ether-acetone or chloroform
Apparatus	Spectra-Physics HPLC with UV detector
Column	Reversed-phase; Nucleosil 10 µm, C_{18}, 250 × 4.6 mm ID
Mobile phase	Methanol/1% acetic acid (20:80, v/v)
Flow-rate	1.4 ml min^{-1}
Temperature	Ambient
Internal standard	Sulfadimidine
Injection size	100 µl, 5–150 µg ml^{-1}
Detection	UV, λ = 254 nm
Retention times, min	(estimated from drawing) Sulfapyridine 6.5 5-Hydroxysulfapyridine 7.0 Sulfadimidine 8.5 N^4-Acetylsulfapyridine 11.0
Ref.	Fischer C and Klotz U (1978) J. Chromatogr. Biomed. Appl. 146:157

HPLC 4

Sample	Blood plasma
Extraction	Methylene chloride
Column	Reversed-phase; Nucleosil C_{18}, 10 µm, 250 × 3 mm ID
Mobile phase	50% Methanol containing 1 g l^{-1} N,N,N,-trimethylcetylammonium bromide
Flow-rate	1.3 ml min^{-1}
Internal standard	*p*-Aminosalicylic acid (PAS)
Injection size	100 µl
Detection	Fluorescence, λ = 310/430 nm
Retention times, min	(estimated from drawing) Aminosalicylate (metabolite) 20.0 Internal standard 25.0 N-Acetyl-aminosalicylate (metabolite) 40.0

Ref.	Fischer C and Klotz U (1979) J. Chromatogr. Biomed. Appl. 162:237

HPLC 5

Sample	Sulfonamide mixture	
Column	Normal phase; Lichrosorb SI 100, 10 µm, loaded with 0.3 M TBA HSO_4 (tetrabutylammonium hydrogen sulfate) in 0.1 M phosphate buffer, pH 6.8; 250 × 3.2 mm ID	
Mobile phase	Heptane/n-butanol (75:25)	
Flow-rate	0.4 ml min^{-1}	
Temperature	27 °C ± 0.5 °C	
Detection	UV, λ = 254 nm	
Retention times, min	(estimated from drawing)	
	Phthalylsulfathiazole	4.5
	Sulfabenzamide	7.0
	Sulfisoxazole	8.5
	Sulfachlorpyridazine	9.2
	Sulfaquinoxaline	10.0
	Sulfadimethoxine	15.3
	Sulfathiazole	21.2
	Sulfamethoxypyridazine	27.0
	Sulfamethazine	29.6
	Sulfamerazine	37.5
	Sulfacetamide	44.0
	Sulfadiazine	57.0
	Sulfapyridine	85.4
Ref.	Gilbert MT (1987): High Performance Liquid Chromatography. Wright, Bristol, p 230	

201 Sulfathiazole

CAS-No	72-14-0; 6791-71-5 (sodium salt, pentahydrate)				
Chemical name	N^1-(Thiazol-2-yl)sulfanilamide				
Empirical formula	Sulfathiazole	Sodium Salt			
	$C_9H_9N_3O_2S_2$	$C_9H_8N_3NaO_2S_2.5H_2O$			
Molecular mass	255.3	367.4			
Proprietary names	Cibazol; Sulzol; Thiazamide				
Clinical use	Chemotherapeutic				
Solubility,	Water	alcohol	acetone	chloroform	aq NaOH
Sulfathiazole	1 in 2500	1 in 120	1 in 50	insoluble	soluble
Sodium salt	1 in 3	1 in 20			
Light absorption	λ_{max} = 256 nm in aqueous alkali				

HPLC 1

Sample	Swine feed
Extraction	0.3 N HCl + 1.5% diethylamine in 25% methanol in water
Apparatus	Varian M 5020 LC pump, 8050 autosampler and Kratos SF 773 UV/VIS detector
Column	Reversed-phase; Lichrosorb C_{18}, 7 μm, 250 × 4.6 mm ID
Mobile phase	A. Acetonitrile/2% acetic acid (17:83) B. Acetonitrile/methanol (1 + 4) plus 2% acetic acid (20 + 80)
Flow-rate	1–1.3 ml min^{-1}
Derivatization	Post-column; with dimethylaminobenzaldehyde;
Internal standard	Sulfamerazine
Injection size	200 μl containing 80 ng
Detection	UV/VIS, λ = 450 nm; 0.02 AUFS
Retention times, min	

Mobile phase	A.	B.
Sulfathiazole	6.0	6.0
Sulfamerazine (IS)	7.0	8.0
Sulfamethazine	8.5	11.0
	(estimated from drawing)	

Ref. Smallidge RL, Kentzer EJ et al (1988) J. Assoc. Off. Anal. Chem. 71:710

HPLC 2

Sample	Raw materials
Column	Reversed-phase; Chromspher C_{18}, 5 μm, 200 × 3 mm ID
Mobile phase	Methanol/water (25:75)
Flow-rate	0.5 ml min^{-1}
Injection size	20 μl, 100 μg ml^{-1}
Detection	UV, λ = 254 nm; 0.05 AUFS
Retention times, min	(estimated from drawing) Sulfamerazine 5.1 Sulfathiazole 3.9 Sulfadimidine 7.5
Ref.	Chrompack Application Notes (1987) Antibiotics, Application 49-HPLC, Middelburg, p 5

HPLC 3

Sample	Sulfathiazole, sulfacetamide, and sulfabenzamide vaginal cream
Extraction	Acetone after warming under steam bath
Column	Reversed-phase; ODS, 5–10 μm, 300 × 3.9 mm ID
Mobile phase	Water/acetonitrile/1 M tetrabutylammonium hydroxide (78:22:10), pH 7.7, with diluted phosphoric acid (1:10 or 1:50) or 1 M tetrabutylammonium hydroxide

Flow-rate	1 ml min^{-1}			
Internal standard	Sulfapyridine			
Injection size	Appr. 10 µl			
Detection	UV, λ = 280 nm			
Retention times	(relative)		Sulfathiazole	1.8
	Sulfacetamide	0.8	Sulfabenzamide	2.5
	Sulfapyridine	1.0		
Ref.	Anonym (1985): The United States Pharmacopeia XXI and The National Formulary XVI. The United States Pharmacopeial Convention, Inc., Rockville, p 997			

HPLC 4

Sample	Human blood plasma and urine
Extraction	Serum: With acetonitrile by deproteinization Urine: With water by dilution
Apparatus	Hewlett-Packard 1084A HPLC with UV detector
Column	A. Serum: Reversed-phase; C_8, 5 µm, 120 × 7.5 mm ID B. Urine: reversed-phase; C_8, 10 µm, 250 × 4.7 mm ID
Pre-column	Serum: Whatman, Co:Pell ODS, 100 × 4.7 mm ID
Mobile phase	A. Serum: Titrisol buffer, pH 5.0/acetonitrile (80:16, v/v), 40 °C B. Urine: Titrisol buffer, pH 4.0/acetonitrile (92:8, v/v), 40 °C
Flow-rate	2.0 ml min^{-1}; 1160 psi (A) and 2.5 ml min^{-1}; 1450 psi (B)
Temperature	Ambient
Injection size	20 µl containing 20 µg ml^{-1}
Detection	UV, λ = 254 nm
Retention times, min	(estimated from drawing)

Sample	plasma	urine	aq. solution
Sulfathiazole	4.8	7.2	8.3
Acetyl-sulfathiazole			17.7

Ref. Sioufi A, Godbillon J and Caudal F (1980) J. Chromatogr. Biomed. Appl. 221:419

HPLC 5

Sample	Sulfonamides and blood plasma
Extraction	Direct injection or deproteinization with methanol (1:1)
Apparatus	Waters ALC/GPC M 204 LC with 440 UV detector and U6K universal injector
Column	Reversed-phase; µ-Bondapak, C_{18}
Pre-column	Corasil C_{18}

Mobile phase	Water/methanol (50:50), pH 4.00 with glacial acetic acid and pH 7.45 without buffer
Flow-rate	0.8 ml min^{-1}
Temperature	Ambient (25 °C)
Injection size	20 µl, 10–100 µg ml^{-1} and 10–100 ng ml^{-1}
Detection	UV, λ = 254 nm

Retention times, min	pH, mobile phase	4.00	7.45
	Sulfaguanidine	3.70	3.70
	Sulfanilamide	3.80	3.80
	Sulfathiazole	4.00	4.00
	Sulfapyridine	4.75	4.45
	Sulfamerazine	4.90	4.50
	Sulfamethazine	5.00	5.45
	Sulfisoxazole	5.20	3.20
	Acetylsulfisoxazole (N^4)	7.00	2.90

Ref. Suber RL and Edds GT (1980) J. Liq. Chromatogr. 3:257

---

202 Temazepam

CAS-No	846-50-4
Chemical name	7-Chloro-1,3-dihydro-3-hydroxy-1-methyl-5-phenyl-2*H*-1,4-benzodiazepin-2-one
Empirical formula	$C_{16}H_{13}ClN_2O_2$
Molecular mass	300.7
Proprietary names	Euhypnos; Levanxol; Restoril
Clinical use	Hypnotic

Solubility	Water	alcohol	chloroform
	insoluble	1 in 10	1 in 10

Light absorption λ_{max} = 230 nm in methanol

HPLC 1

Sample	Blood plasma and urine
Extraction	Absolute methanol on extraction column
Apparatus	Waters M 6000 or 6000A or Beckman M 110A LC pump, Waters WISP 710B or Perkin-Elmer ISS-100 autosampler and LDC Spectromonitor III ultraviolet detector
Column	Reversed-phase; C_{18}, 300 × 3.9 mm ID
Mobile phase	Methanol/tetrahydrofuran/water (62:6:32, v/v)
Flow-rate	1.5 ml min^{-1}
Temperature	Ambient

Internal standard	Diazepam
Injection size	10–100 μl
Detection	UV; 0.1–0.5 AUFS
Recording	Spectra-Physics 4100 integrator
Retention times, min	Temazepam 4.5 Diazepam 5.5
Ref.	Patterson SE (1986) J. Pharm. Biomed. Anal. 4:271

HPLC 2

Sample	Soft gelatine capsules
Apparatus	Applied Chromatographic Systems M 750-03 pump, Rheodyne M 7125 injector and Pye-Unicam M 4025 ultraviolet and Hewlett-Packard 1040A photodiode array detector
Column	Reversed-phase; Lichrosorb 10 μm, C_{18}, 250 × 4.6 mm ID
Mobile phase	Methanol/water (60:40)
Flow-rate	3.5 ml min^{-1}
Internal standard	Diazepam
Injection size	20 μl containing 2 μg
Detection	UV, λ = 258 nm; 0.08 AUFS
Recording	Philips PM 8251A recorder
Retention times, min	Temazepam 4.6 Diazepam 7.8
Ref.	Gordon SM, Freeston LK and Collins AJ (1986) J. Chromatogr. 368:180

HPLC 3

Sample	Raw materials
Column	Reversed-phase; Nucleosil C_{18}, 10 μm, 250 × 4.6 mm ID
Mobile phase	0.05 M Potassium phosphate buffer, pH 2.5/acetonitrile (60:40, v/v)
Flow-rate	1 ml min^{-1}
Detection	UV, λ = 207 nm
Retention times, min	Medazepam 7.62, Clobazam 15.89 Oxazepam 9.54, Clotiazepam 17.16 Di-K-chorazepat 11.78, Diazepam 20.14 Temazepam 14.21
Ref.	Erdweg M and Hoffmann M (1987) in: High Performance Liquid Chromatography. Application No. 253. Macherey-Nagel, Dueren, p 210

HPLC 4

Sample	Raw materials
Extraction	Dichloromethane 40% in cyclohexane

Apparatus	Waters M 45 LC pump, U6K injector and 440 UV detector
Column	Reversed-phase; Lichrosorb C_8, 10 µm, 250 × 4.6 mm ID
Mobile phase	Water/methanol/triethylamine (43:57:0.25), pH 6.0 with 1 M acetic acid
Flow-rate	2 ml min^{-1}
Internal standard	Methylnitrazepam
Injection size	20–75 µl, 200 ng ml^{-1}
Detection	UV, λ = 254 nm; 0.01–0.020 AUFS
Retention times, min	(estimated from drawing) Temazepam 7.4; Internal standard 5.2; Desmethyldiazepam 8.9; Oxazepam 6.2; Diazepam 10.8
Ref.	Willems HJJ, Van der Horst A et al (1985) Pharm. Weekbl. Sci. 7:150

HPLC 5

Sample	Serum
Extraction	First diethyl ether and 1 M borate buffer and then mobile phase using hexane
Apparatus	Waters M 510 dual-piston pump and Beckman M 163 UV detector
Column	Reversed-phase; Ultrasphere C_{18}, 5 µm, 150 × 2 mm ID
Pre-column	Water 2 µm
Mobile phase	Methanol/acetonitrile/0.056 M sodium acetate buffer, pH 4.0 with glacial acetic acid (50:5:45, v/v)
Flow-rate	0.3 ml min^{-1}; 2000 psi
Internal standard	Demoxepam
Injection size	20 µl
Detection	UV, λ = 228 nm; 0.002 AUFS
Recording	Perkin-Elmer LCI-100 injector
Retention times, min	(estimated from drawing) Temazepam 9.0; Demoxepam 3.8; Desmethyldiazepam 11.8; Oxazepam 6.8; Diazepam 13.0
Ref.	Lau CE, Dolan S and Tang M (1987) J. Chromatogr. Biomed. Appl. 416:212

203 Terbutaline

CAS-No	23031-25-6; 23031-32-5 (sulfate)
Chemical name	2-*tert.*-Butylamino-1-(3,5-dihydroxyphenyl)ethanol

Empirical formula	Terbutaline	Sulfate
	$C_{12}H_{19}NO_3$	$C_{12}H_{19}NO_3,1/2H_2SO_4$
Molecular mass	225.3	274.3
Proprietary names	Brethine; Bricanyl; Terbasmin	
Clinical use	Sympathomimetic	

Solubility,	Water	alcohol	choroform	ether
Sulfate	1 in 4	slightly soluble	insoluble	insoluble

Light absorption λ_{max} = 297 nm in aqueous alkali

HPLC 1

Sample	Pharmaceutical preparations
Apparatus	Waters M 45 LC pump, LC-4B amperometric detector (Bioanalytical Systems) and Rheodyne M 7125 injector
Column	Cyanopropyl phase; μ-Bondapak CN, 300 × 3.9 mm ID
Mobile phase	Methanol/0.1 M KH_2PO_4 containing 0.1 M sodium heptanesulfonate and 1 mM disodium-EDTA (15:85)
Flow-rate	1.0 ml min^{-1}
Temperature	Ambient
Injection size	20 μl, 0–10 μg ml^{-1} (sulfate)
Detection	EC, GC electrode, + 0.9 V vs Ag/AgCl RE
Recording	Spectra-Physics 4270 recording integrator
Retention time, min	Terbutaline 4.0
Ref.	Childress WL (1987) J. Assoc. Off. Anal. Chem. 70:974

HPLC 2

Sample	Dosage forms (injections and tablets)
Apparatus	HPLC system with multiple-wavelength detector
Column	Microphenyl phase; 300 × 3.9 mm ID
Mobile phase	Methanol/0.02 M KH_2PO_4 in water (8:92, v/v), pH 3.6 with 1.7% aqueous phosphoric acid
Flow-rate	2 ml min^{-1}
Temperature	Ambient
Internal standard	Salicylic acid, 50 μg ml^{-1}
Detection	UV, λ = 278 nm; 0.01 AUFS
Retention times, min	(estimated from drawing) Terbutaline 5.0 Internal standard 7.8
Ref.	Das Gupta V (1986) J. Liq. Chromatogr. 9:1065

HPLC 3

Sample	Human blood plasma
Extraction	On-line, over pre-column
Column	Reversed-phase; Nucleosil C_{18}, 10 μm, 200 × 5 mm ID

Pre-column	Sep-Pak C_{18}, 23 × 3.9 mm ID
Mobile phase	Methanol/50 mM sodium perchlorate buffer, pH 6.0 (12:88, v/v)
Flow-rate	2 ml min^{-1}
Detection	EC, + 0.9 V vs Ag/AgCl RE
Retention time, min	Terbutaline 7.3 (estimated from drawing)
Ref.	Bergquist S and Edholm L-E (1983) J. Liq. Chromatogr. 6:559

HPLC 4

Sample	Human blood plasma
Apparatus	HPLC system with ESA Coulochem electrochemical detector
Column	Reversed-phase; C_{18}
Mobile phase	Methanol/25 mM dibasic sodium phosphate, pH 7.4/isopropanol/5 mM octanesulfonic acid (19:77:4:5)
Flow-rate	1.3 ml min^{-1}
Detection	EC, + 450 mv and + 700 mv; guard cell + 700 mv
Retention time, min	Terbutaline 6.0 (estimated from drawing)
Ref.	ESA Coulochem Application Notes (1987) Terbutaline, No. CL6-2/86. ESA, Bedford

204 Testosterone

CAS-No	58-22-0; 57-85-2 (propionate)				
Chemical name	17β-Hydroxyandrost-4-en-3-one				
Empirical formula	Testosterone		Propionate		
	$C_{19}H_{28}O_2$		$C_{22}H_{32}O_3$		
Molecular mass	288.4		344.5		
Proprietary names	Andronaq; Testoral; Virormone				
Clinical use	Androgen				
Solubility,	Water	alcohol	chloroform	ether	fixed oils
Testosterone	insol		1 in 2	1 in 100	soluble
Propionate	insol	1 in 6	very soluble	freely soluble	soluble
Light absorption	λ_{max} = 240 nm in dehydrated alcohol				

HPLC 1

Sample	Rat tissue
Extraction	Multistep, methanol/chloroform (10:90) after using chloroform/methanol and water/methanol on Sep-Pak C_{18} cartridge and Carbopak B columns

Apparatus	Perkin-Elmer 3B LC with gradient system, LC 95 UV/VIS wavelength detector and Rheodyne M 7125 injector
Column	Reversed-phase; Policonsult C_{18}, 5 μm, 250 × 4.6 mm ID
Pre-column	Supelguard LC-18, 5 μm, 20 mm
Mobile phase	Acetonitrile/water (binary solvent gradient) (5:95) Multistep: Acetonitrile 5, 20, 28, 45 and 70% at 0, 10, 25, 50 and 60 min
Flow-rate	1.2 ml min^{-1}
Temperature	Ambient
Injection size	20 μl
Detection	UV, λ = 254 nm
Recording	Perkin-Elmer computing integrator

Retention times, min

Aldosterone	25.4	11α-Hydroxyprogesterone	40.5
Cortisol	28.3	Testosterone	43.7
Cortisone	29.3	11-Deoxycorticosterone	46.7
20β-Dehydoepiandrosterone	32.0	Androstenedione	48.1
20β-Dehydrocorticosterone	32.4	17α-Hydroxyprogesterone	49.4
11-Dehydrocorticosterone	34.4	20α-Dihydroprogesterone	54.6
21-Deoxycortisol	35.2	20β-Dihydroprogesterone	60.2
Corticosterone	35.9	Progesterone	60.7
Adrenosterone	36.8		
11-Deoxycortisol	37.6		
16α-Hydroxyprogesterone	38.1		

Ref. Lagana A, d'Ascenzo G and Marino A (1987) Chromatographia 23:796

HPLC 2

Sample	Gonadal steroids (testes)
Extraction	Dichlormethane on Extrelut (E. Merck) column, pH 7.5
Apparatus	Waters M 45 LC pump and 440 ultraviolet detector
Column	A. Reversed-phase; μ-Bondapak C_{18}, 5 μm, 300 × 3.9 mm ID B. Normal phase; μ-Porasil 10 μm, 300 × 3.9 mm ID
Mobile phase	For A. Methanol/water (4:1, v/v) For B. Dichloromethane/acetonitrile (4:1, v/v)
Flow-rate	1–2 ml min^{-1} for 20 MPa
Temperature	Ambient

Injection size	10 µl containing 5 ng		
Detection	UV, λ = 254 nm		
Capcity-factor, k′	Column	A.	B.
	(DHT = Dihydroxytestosterone)		
	17α-Hydroxyprogesterone	0.914	0.958
	Androstenedione	0.932	0.425
	17α, 20β-Dihydroxypregn-4-en-3-one	0.969	5.041
	17α, 20α-Dihydroxypregn-4-en-3-one	1.032	8.375
	Testosterone	1.051	2.270
	Epitestosterone	1.193	2.875
	20α-Hydroxypregn-4-en-3-one	2.165	1.125
	Progesterone	2.317	0.330
	5α-DHT	2.404	1.071
	20β-Hydroxypregn-4-en-3-one	2.705	1.205
	5α-epiDHT	3.652	1.175
	5β-epiDHT	3.787	2.125
Ref.	Watson TG, Windmill DJ and Bourne AR (1987) Chromatographia 23:31		

HPLC 3

Sample	Blood plasma from rats
Extraction	Solid-phase extraction
Apparatus	LKB M 2150 and Waters M 6000A HPLC pumps, Waters 440 ultraviolet detector and Shimadzu SPD-6A UV detector
Column	Reversed-phase; Hypersil ODS 5 µm, 100 × 1 mm ID
Mobile phase	Acetonitrile/methanol/water (25:25:50)
Flow-rate	100 µl min^{-1}
Injection size	20 µl, 400 ng ml^{-1}
Detection	UV, λ = 0.01 AUFS

Retention times, min				
	Andostenedione	5.5	20α-Hydroxy-4-pregnen-3-one	13.5
	Testosterone	6.0	Progesterone	16.5
	17α-Hydroxy-progesterone	8.0	Norethindrone	21.0
			(estimated from drawing)	

Ref.	Taylor RB, Kendle KE and Reid RG (1987) J. Chromatogr. 385:383

HPLC 4

Sample	Hydrolytic products of testosterone esters
Apparatus	Perkin-Elmer LC-65T LC with 2/2 pump module, UV detector and Rheodyne M 7105 injector
Column	Reversed-phase; Lichrosorb C_{18}, 10 µm, 250 × 4.6 mm ID

Mobile phase	Methanol/water (85:15)
Flow-rate	0.8 ml min^{-1}
Temperature	40 °C
Internal standard	Testosterone acetate
Injection size	10 µl, 1–8 µg ml^{-1}
Detection	UV, λ = 243 nm; 0.01 AUFS
Recording	Perkin-Elmer 3600 data station
Retention times, min	(estimated from drawing)
	Testosterone 4.8
	Testosterone acetate (IS) 7.2
	Testosterone phenylacetate 9.8
	Testosterone benzoate 12.5
	Testosterone phenylbutyrate 14.0
	Testosterone phenylvalerate 16.3
Ref.	Karunanithy R, Becket G et al (1985) J. Pharm. Biomed. Anal. 3:375

HPLC 5

Sample	Raw materials (steroid mixture)
Apparatus	Waters M 6000A LC pump with U6K injector and 441 fixed wavelength UV detector
Column	Reversed-phase; RCSS Radial-Pak, µ-Bondapak C_{18}, RCM-100
Mobile phase	A. 50% (v/v) Acetonitrile and 50% water B. 70% (v/v) Methanol and 30% water
Flow-rate	2.0 ml min^{-1}
Temperature	Ambient
Injection size	100 µl containing 100 ng
Detection	UV, λ = 254 nm; 0.005–0.01 AUFS
Retention times, min	

Mobile phase	A.	B.
Estriol	2.5	3.0
Testosterone	5.5	6.8
Estradiol-17β	5.6	6.3
Estrone	6.9	5.7
Progesterone	16.3	10.9

Ref. Erkoc FU, Oezsar S et al (1989) J. Chromatogr. Sci. 27:86

HPLC 6

Sample	Test substances
Apparatus	Knauer 64.00 HPLC pumps, 731.710 UV detector and 730.6641 or 730.6642 injector
Column	A. Reversed-phase, Nucleosil C_{18}, 10 µm, 250 × 4 mm ID B. Reversed-phase; Hypersil ODS 5 µm, 250 × 4 mm ID

	C. Reversed-phase; Nucleosil C_{18}, 5 μm, 120 × 4 mm ID
	D. Reversed-phase; Hypersil ODS 3 μm, 120 × 4 mm ID
Mobile phase	Methanol/water/tetrahydrofuran (15:60:25)
Flow-rate	1.5 ml min^{-1}
Injection size	20 μl, 0.008 mg ml^{-1}
Detection	UV, λ = 254 nm; 0.08 AUFS
Retention times, min	(estimated from drawing)

Column	A.	B.	C.	D.
Solvent	1.2	1.1	0.7	0.4
Aldosterone	1.9	1.6	0.8	0.8
Estriol	3.0	2.2	1.7	1.1
Androstenedione	5.1	3.5	2.8	1.7
Testosterone	5.8	4.6	3.2	1.8
Estrone	9.9	6.3	5.1	2.8
17-β-Estradiol	11.0	7.0	5.7	3.0
Progesterone	14.2	8.6	7.0	3.6

Ref. Knauer Application Notes on Micro-HPLC and Microbore-HPLC columns. Knauer, Berlin

205 Tetracycline

CAS-No	60-54-8; 64-75-5 (hydrochloride)			
Chemical name	4-(Dimethylamino)-1,4,4*a*,5,5*a*,6,11,12*a*-octahydro-3,6,10,12,12*a*-pentahydroxy-6-methyl-1,11-dioxo-2-naphthacenecarboxamide			
Empirical formula	Tetracycline $C_{22}H_{24}N_2O_8$	Hydrochloride $C_{22}H_{24}N_2O_8$, HCl		
Molecular mass	444.4	480.9		
Proprietary names	Achromycin; Panmycin; Tetracyn			
Clinical use	Antibiotic			
Solubility,	Water	alcohol	chloroform	aq acids
Tetracycline	1 in 2500	1 in 50	insoluble	freely soluble
Hydrochloride	1 in 10	1 in 100	insoluble	
Light absorption	λ_{max} = 270 nm in aqueous acid			

HPLC 1

Sample	Meat and fish
Extraction	First 1 N HCl and then with methanol on a Amberlite XAD-2 column

Apparatus	Shimadzu LC-2 pump with SPD-1 UV detector
Column	A. Wako Gel, DMS-10H, 10 μm, 250 × 4 mm ID B. Shimadzu Gel, PSG-100, 10 μm, 250 × 4 mm ID
Pre-column	Bondapak C_{18} Corasil, 50 × 4 mm ID
Mobile phase	For A. 0.05 M Phosphate buffer, pH 2.5/acetonitrile (90:10, v/v) For B. 0.04 M KH_2PO_4 Buffer/0.08 M EDTA/methanol (10:10:80, v/v)
Flow-rate	1.5 ml min^{-1}
Temperature	Ambient
Injection size	20 μl containing 10 ppm
Detection	UV, λ = 370 nm; 0.04–0.08 AUFS
Retention times, min	Oxytetracycline 4.5 Chlortetracycline 10.5 Tetracycline 6.0 (estimated from drawing)
Ref.	Onji Y, Uno M and Tanigawa K (1984) J. Assoc. Off. Anal. Chem. 67:1135

HPLC 2

Sample	Urine
Apparatus	Waters M 6000A LC pump with 440 absorbance detector
Column	Reversed-phase; Brownlee RP-10A C_8, 10 μm, 250 × 4.6 mm ID
Pre-column	Reversed-phase; Brownlee A C_8, 10 μm, 30 × 4.6 mm ID
Mobile phase	Isopropanol/1 M diethanolamine in water, pH 7.3 with H_3PO_4/tetraammonium EDTA/water (11:5:1:83)
Flow-rate	2 ml min^{-1}
Temperature	40 °C
Injection size	20 μl, 1–50 μg ml^{-1}
Detection	UV, λ = 365 nm; 0.005 AUFS
Retention time, min	Tetracycline 4.5
Ref.	Charles UG, Cole JJ and Ravenscroft PJ (1981) J. Chromatogr. Biomed. Appl. 222:152

HPLC 3

Sample	Blood plasma
Extraction	Deproteinization with 1.25 M perchloric acid or with ethyl acetate
Apparatus	Spectra-Physics 8000B LC, Rheodyne M 7126 injector and Pye-Unicam LC-UV 3 detector
Column	A. XAD-2, 6–7 μm, 100 × 3.0 mm ID B. PRP-1, 10 μm, 100 × 3.0 mm ID

Pre-column	XAD-2 or PRP-1, 250 × 4.6 mm ID
Mobile phase	For A. Acetonitrile/dichloromethane/0.2 M acetate buffer, pH 3.6 plus 0.025 M EDTA (10:1:90, v/v) For B. Acetonitrile/dichloromethane/0.05 M acetate buffer, pH 3.6 plus 0.025 M EDTA (12.5:1:87.5, v/v)
Flow-rate	0.8 ml min^{-1} (A); 0.6 ml min^{-1} (B)
Temperature	25 °C
Injection size	20 µl
Detection	UV, λ = 268 or 357 nm
Retention times, min	(estimated from drawing)

Column	A.	B.
epi-Tetracycline	1.6	1.0
Tetracycline	1.8	1.8
epi-anhydro-Tetracycline	7.5	2.4
anhydro-Tetracycline	14.5	14.2

Ref.	Reeuwijk HJEM and Tjaden UR (1986) J. Chromatogr. 353:339

HPLC 4

Sample	Tissue and serum from cattle and swine
Extraction	Water under acidic condition after deproteinization with acetonitrile
Apparatus	Varian M 5000 LC, UV-50 detector and Valco automatic loop injector
Column	PRP-1, 10 µm, 250 × 4.1 mm ID
Mobile phase	A. 0.01 M Phosphoric acid/methanol (80:20) B. 0.01 M Phosphoric acid/acetonitrile/ methanol (30:50:20) Gradient: 2 Min after the injection, 100% A to 100% B in 2 min; hold 23 min
Flow-rate	1 ml min^{-1}
Injection size	200 µl containing 2 ng
Detection	UV, λ = 355 nm; 0.1 AUFS
Retention times, min	Oxytetracycline 14.8 Tetracycline 15.8 Chlortetracycline 19.0
Ref.	Moats WA (1986) J. Chromatogr. 358:253

HPLC 5

Sample	Serum
Extraction	Trichloroacetic acid/methanol
Apparatus	Waters ALC/GPC 200 LC with M 6000 pump, U6K injector and Schoeffel SF 770 UV/VIS spectrophotometer detector

Column	Reversed-phase; μ-Bondapak C_{18}, 300×4 mm ID
Mobile phase	Methanol/0.005 M EDTA (3:7, v/v)
Flow-rate	2.8 ml min^{-1}; 4000 psi
Injection size	200 μl
Detection	UV, λ = 355 nm; 0.001 AU
Retention time, min	Tetracycline 3.75
Ref.	Nilsson-Ehle I, Yoshikawa TT et al (1976) Antimicrob. Ag. Chemother. 9:754

HPLC 6

Sample	Meats
Extraction	0.5% Metaphosphoric acid/methanol (8:2), and then with methanol on Sep-Pak C_{18} column
Column	Cyano phase; Nucleosil CN, 250×4.6 mm ID
Mobile phase	0.01 M Monobasic sodium phosphate/ acetonitrile (75:25), pH 2.5
Flow-rate	0.5 ml min^{-1}
Injection size	50 μl, 0.5–5 μg ml^{-1}
Detection	UV, λ = 268 nm; 0.08 AUFS
Retention times, min	(estimated from drawing)

Oxytetracycline	8.5	Chlortetracycline	12.5
Tetracycline	9.5	Tylosin	15.2

Ref.	Hoshino Y, Horie M et al (1984) J. Fd. Hyg. Soc. Jap. 25:430

HPLC 7

Sample	Tetracyclines
Apparatus	LDC modular LC, Milton Roy Minipump with pulse dampener and LDC M 1285 UV detector
Column	Polymer; Zipax-hydrocarbon polymer (HCP), 1000×2.1 mm ID
Mobile phase	A. 13% Methanol and 0.001 M EDTA in 0.02 M H_3PO_4 and 0.02 M Na_2HPO_4, pH 2.5 B. 3% Acetonitrile and 0.001 M EDTA in 0.05 M Na_2HPO_4, pH 4.5
Flow-rate	0.45 ml min^{-1}; 500 psi (A) and 0.80 ml min^{-1}; 1000 psi (B)
Temperature	25 °C
Internal standard	Prednisolone or 4-(N-butylamino)benzoic acid
Injection size	1.5 μl
Detection	UV, λ = 280 nm
Retention times, min	(estimated from drawing)

Mobile phase	A.	B.
4-epi-Tetracycline	3.9	4.0
Tetracycline	4.0	4.0

	Chlortetracycline	6.0	5.5
	Doxycycline	8.8	
	Prednisolone	13.8	
	4-epi-anhydro-Tetracycline	19.8	12.0
	anhdro-Tetracycline	22.0	14.5
	4-N-Butylaminobenzoic acid		7.0
Ref.	Tsuji K, Robertson JH and Beyer WF (1974) Anal. Chem. 46:539		

206 Tetroxoprim

CAS-No	53808-87-0		
Chemical name	5-[3,5-Dimethoxy-4-(2-methoxyethoxy)benzyl] pyrimidine-2,4-diyldiamine		
Empirical formula	$C_{16}H_{22}N_4O_4$		
Molecular mass	334.4		
Proprietary name	Sterinor		
Clinical use	Chemotherapeutic		
Solubility	Water	chloroform	n-octan
	0.26 in 100	6.9 in 100	0.16 in 100

HPLC 1

Sample	Dosage forms and biological fluids
Extraction	Deproteinizing the serum with acetonitrile and diluting the urine with water
Apparatus	Pye-Unicam LC-3 HPLC with variable-wavelength detector, and Rheodyne injection valve
Column	Reversed-phase; Lichrosorb C_{18}, 10 μm
Mobile phase	Methanol/water (50:50) with 1 mg ml^{-1} sodium pentanesulfonate
Flow-rate	1 ml min^{-1}; 870 psi
Injection size	20 μl, 0.1–200 μg ml^{-1}
Detection	UV, $\lambda = 235$ nm
Retention time, min	Tetroxoprim 5.32
Ref.	Springolo V and Coppi G (1989) J. Pharm. Biomed. Anal. 7:57

HPLC 2

Sample	Raw materials and dosage forms
Apparatus	Hewlett-Packard 1084 LC with UV detector
Column	Reversed-phase; Nucleosil C_{18}, 10 μm, 250 × 4 mm ID

Mobile phase	Acetonitrile/0.05 M ortho-phosphoric acid (15:85)
Flow-rate	3 ml min^{-1}
Temperature	40 °C
Internal standard	Sulfamethazine
Injection size	5–30 µl
Detection	UV, λ = 230 nm
Recording	Hewlett-Packard 79850B LC-Terminal integrator
Ref.	Rehm KD and Steinigen M (1981) Pharm. Ztg. 126:99

207 Thalidomide

CAS-No	50-35-1	
Chemical name	2-Phthalimidoglutarimide	
Empirical formula	$C_{13}H_{10}N_2O_4$	
Molecular mass	258.2	
Proprietary names	Contergan; Neurosedyn; Talimol	
Clinical use	Hypnotic/Immunosuppressant/Antileprosis	
Solubility	Water and ethanol slightly soluble	chloroform and ether insoluble
Light absorption	λ_{max} = 240 nm in aqueous acid	

HPLC 1

Sample	Human serum
Extraction	Various solvents on RP-18 extraction cartouche
Apparatus	Kontron M 414 pump, MS 660T autosampler and Uvikon 740 LC-absorbance detector
Column	Reversed-phase; C_8, 5 µm Lichrosorb, 125 × 4.6 mm ID
Pre-column	Reversed-phase, C_{18}, 10 µm Lichrosorb, 10 × 4.6 mm ID
Mobile phase	Water/methanol/n-propanol (180:30:20, v/v) plus 150 mg sodium EDTA and 230 mg tetramethylammonium bromide
Flow-rate	0.9 ml min^{-1}
Temperature	30 °C
Injection size	20 µl
Detection	UV, λ = 254 nm; 0.02 AUFS
Recording	Shimadzu C-R1B integrator

Retention times, min

Metabolite IV	3.0	Metabolite VIII	8.6
Metabolite III	4.4	Metabolite VII	9.6
Metabolite II	5.8	Metabolite IX	11.0
Metabolite V, VI	7.0	Thalidomide	18.7

Ref.	Czejka MJ and Koch HP (1987) J. Chromatogr. Biomed. Appl. 413:181

HPLC 2

Sample	Human serum
Extraction	Various solvents on RP-18 (Baker) extraction cartouche
Apparatus	Kontron M 414 pump, MS 660T autosampler and Uvikon 740 LC-absorbance detector
Column	Reversed-phase; Spherisorb S5 C_{18}, 126×4.6 mm ID
Mobile phase	Water/acetonitrile (80:20, v/v) plus 120 mg n-heptanesulfonic acid per ml and 150 μl of 30% hydrochloric acid per 100 ml
Flow-rate	1.5 ml min^{-1}
Temperature	30 and 50 °C
Injection size	20 μl
Detection	UV, λ = 290 nm; 0.02 AUFS
Recording	Shimadzu C-R1B integrator

Retention times, min	Temperature	30 °C	50 °C
	Metabolite IV	3.2	
	Metabolite VII	4.2	6.3
	Metabolite V, VI	4.6	
	Metabolite II, III	6.3	
	Metabolite VIII, IX	6.7	6.0 (VIII)
	Thalidomide	8.9	15.0

Ref.	Czejka MJ and Koch HP (1987) J. Chromatogr. Biomed. Appl. 413:181

HPLC 3

Sample	Enantiomers of thalidomide analoge EM 12 [2-(2,6-dioxopiperidine-3-yl)-phthalimidine]
Column	ODS Hypersil, 5 μm coupled with Nucleosil phenyl, 7 μm
Mobile phase	Phosphate buffer (containing 2.6 g H_3PO_4 per liter water, pH 2.0)/methanol (77.5:22.5, v/v)
Internal standard	Supidimide([3-O-sulfobenzoylimido)]-2-piperidone
Detection	UV, λ = 229 nm

Retention times, min		
	EM 356 (hydrolytic product)	8.0
	EM 27 (hydrolytic product)	10.0
	EM 12	12.0
	EM 138 (hydrolytic product)	14.1
	Internal standard	16.1

Ref.	Schmahl H-J, Heger W and Nau H (1989) Toxicol. Lett. 45:23

208 Theobromine

CAS-No	83-67-0
Chemical name	3,7-Dihydro-3,7-dimethylpurine-2,6(1*H*)-dione
Empirical formula	Theobromine $C_7H_8N_4O_2$
Molecular mass	180.2
Proprietary names	Seominal; Theosalvose
Clinical use	CNS-Stimulant
Solubility	Water 1 in 2000; alcohol 1 in 2500; ether insoluble; aq HCl, aq NaOH freely soluble
Light absorption	λ_{max} = 274 nm in aqueous alkali

HPLC 1

Sample	Raw materials
Apparatus	Spectra-Physics 3500B LC pump with 770 ultraviolet detector
Column	Reversed-phase; 10 μm Bondapak, C_{18}, 300 × 3.9 mm ID
Mobile phase	Methanol/acetic acid/triethylamine/water (25:1.5:0.5:73)
Flow-rate	1.5 ml min^{-1}
Temperature	Ambient
Injection size	10 μl, 0.25 mg ml^{-1}
Detection	UV, λ = 254 nm; 0.16 AUFS
Recording	Hewlett-Packard 3380A integrator
Retention times, min	(estimated from drawing) Theobromine 5.5; Dyphylline 6.0; Theophylline 7.0; Caffeine 10.7; 8-Chlorotheophylline 14.5
Ref.	Roos RW and Lau-Cam CA (1986) J. Chromatogr. 370:403

HPLC 2

Sample	Urine
Extraction	Ethyl acetate/chloroform/isopropanol (45:45:10) under alkaline condition
Apparatus	Gradient HPLC system with ultraviolet detector
Column	Reversed-phase; Hypersil 5 μm ODS, 250 × 4.5 mm ID
Mobile phase	A. 1% Tetrahydrofuran in 10 mM acetate buffer, pH 4.8 B. 15% Acetonitrile and 1% tetrahydrofuran in 10 mM acetate buffer, pH 4.8

	Gradient: 100% A for 5 min; B 5%, 10% and 2% per min; every step 5 min
Flow-rate	1.5 ml min^{-1}
Internal standard	Proxyphylline, 10 mg ml^{-1}
Injection size	50 µl, 10 mg l^{-1}
Detection	UV, $\lambda = 280$ nm
Retention times, min	Theobromine 9.7
	Internal standard 18.0
Ref.	Scott NR, Chakraborty J and Marks V (1986) J. Chromatogr. Biomed. Appl. 375:321

HPLC 3

Sample	Theophyllines and xanthines
Apparatus	Merck HPLC low-pressure ternary gradient system
Column	A. Reversed-phase; Superspher 100 C_{18}, 125 × 4 mm ID
	B. Reversed-phase; Superspher 100 C_{18}, 250 × 4 mm ID
Mobile phase	I. For A: Acetonitrile/0.4% sodium acetate (10:90, v/v)
	II. For B: Acetonitrile/0.4% sodium acetate (15:85, v/v)
Flow-rate	1.5 ml min^{-1} (A) and 0.7 ml min^{-1} (B); 2850 psi (A) and (B)
Injection size	10 µl containing 20 µg
Detection	UV, $\lambda = 254$ nm; 0.02 AUFS
Retention times, min	(estimated from drawing)

Column	A.	B.
Theobromine	1.34	4.40
Theophylline	1.89	5.41
β-Hydroxyethyltheophylline		5.90
7-β-Hydroxypropyltheophylline		8.20
Caffeine	4.00	9.30

Ref.	Merck Application Manual on HPLC, TLC and HPTLC (1986). E. Merck, Darmstadt, p 36–37

HPLC 4

Sample	Xanthine stimulants
Apparatus	Kipp M 771 LC with UV detector
Column	Reversed-phase; Nucleosil C_8, 5 µm, 150 × 4.6 mm ID
Mobile phase	Methanol/0.01 M phosphate buffer, pH 7.0 (20:80, v/v)
Flow-rate	1.2 ml min^{-1}; 1885 psi
Temperature	25 °C
Injection size	20 µl

Detection	UV, λ = 280 nm	
Retention times, min	(estimated from drawing)	Theophylline 3.2
	Theobromine 2.3	Caffeine 5.3
Ref.	Kipp Analytica Delft (1987) in: High Performance Liquid Chromatography. Application No. 424. Macherey-Nagel, Dueren, p 268	

HPLC 5

Sample	Serum and test materials	
Extraction	Chloroform/isopropanol (95:5) (serum)	
Column	Reversed-phase; Supelcosil LC-1, 5 μm, 250 × 4.6 mm ID	
Pre-column	Supelguard LC-1, 5 μm, 20 × 4.6 mm ID	
Mobile phase	10 mM Phosphate buffer (containing 2.72 g KH_2PO_4 in 2 liter water, pH 6.8 with 1 M NaOH)/methanol/acetonitrile (65:17.5:17.5, v/v)	
Flow-rate	2 ml min^{-1}; 2400 psi	
Temperature	Ambient	
Internal standards	Tolybarb and N-desmethylmethsuximide	
Injection size	6–10 μl	
Detection	UV, λ = 204, 273 and 285 nm; 0.08 AUFS	
Retention times, min	1-Methyluric acid	1.40
	3-Methyluric acid	1.40
	1,3-Dimethyluric acid	1.44
	p-Aminosalicylic acid	1.44 tails
	S-Aminosalicylic acid	1.44 tails
	Gentisic acid	1.48
	8-Chlorotheophylline	1.49
	Acetylsalicylic acid	1.50
	Salicyluric acid	1.54
	Sodium salicylate	1.61 tails
	1,3,7-Trimethyluric acid	1.62
	Ampicillin	1.67
	Sulfamethoxazole	1.73
	1-Methylxanthine	1.76
	3-Methylxanthine	1.76
	7-Methylxanthine	1.76
	N-Acetylcysteine	1.78
	Chlorpropamide	1.83
	Vancomycin	1.83
	Dyphylline	1.86
	Theobromine	1.86
	Chloramphenicol succinate	1.92 2.00
	Theophylline	1.95
	β-Hydroxyethyltheophylline	1.96
	Paraxanthine	1.97

Acetaminophen	2.09
Oxyphenbutazone	2.09
β-Hydroxypropyltheophylline	2.13
Caffeine	2.18
Naproxen	2.28
Tolmetin	2.28
Phenylethylmalonamide	2.35
Ketoprofen	2.40
p-Hydroxyphenobarbital	2.44
Barbital	2.50
Ethosuximide	2.57
Cimetidine	2.59
Sulindac	2.69
Primidone	2.71
Phenylbutazone	2.73
Salicylamide	3.02
Acetanilide	3.31
Ibuprofen	3.31
Methyprylon	3.31
Diflunisal	3.40 tails
Diazoxide	3.42
Phenobarbital	3.45
Normephenytoin (nirvanol)	3.52
Indomethacin	3.71
Acetophenetidin (phenacetin)	3.82
Chloramphenicol	3.85
N-desmethylmethsuximide	3.90
Trimethoprim	4.00
Butabarbital	4.06
Phensuximide	4.22
Carbamazepine-10,11-epoxide	4.24
Butalbital	4.55
Mefenamic acid	4.56
Mephenytoin	4.65
Morphine	4.80 tails
Tolybarb	4.80 IS
Heptabarbital	5.15
Hexobarbital	5.22
Methsuximide	5.30 IS
Procainamide	5.46 tails
Mephobarbital	5.60
Ethchlorvynol	5.63
Pentobarbital	5.86
Amobarbital	6.05
N-Acetylprocainamide	6.08 tails
Phenytoin	6.28

	Glutethimide	6.51
	Carbamazepine	6.68
	Methylsalicylate	7.17 tails
	Codeine	7.26 tails
	Secobarbital	7.31
	Methaqualone	7.57
	Thiopental	7.87
	Cyheptamide	9.05
	Diisopyramide	18.08 tails
Ref.	Meatherall R and Ford D (1988) Ther. Drug Monit. 10:101	

209 Theophylline

CAS-No	58-55-9; 317-34-0 (aminophylline, anhydrous)			
Chemical name	3,7-Dihydro-1,3-dimethylpurine-2,6(1*H*)-dione			
Empirical formula	Theophylline $C_7H_8N_4O_2$	Aminophylline $(C_7H_8N_4O_2)_2$, $C_2H_4(NH_2)_2$, $2H_2O$		
Molecular mass	180.2	456.5		
Proprietary names	Bronkodyl; Euphyllin; Theograd			
Clinical use	CNS-Stimulant/Bronchodilator			
Solubility,	Water	alcohol	chloroform	aq HCl, aq NaOH
Theophylline	1 in 120	1 in 80	1 in 200	freely soluble
Aminophylline	1 in 5	insoluble		
Light absorption	λ_{max} = 275 nm in aqueous alkali			

HPLC 1

Sample	Blood plasma and urine
Extraction	Propanol-2-ol/dichloromethane/diethyl ether (10:65:25, v/v)
Apparatus	Varian M 5000 HPLC, UV-100 ultraviolet detector and Waters WISP M 710B autosampler
Column	Reversed-phase; Sepralyte C_{18}, 3 µm, 50 × 4.6 mm ID
Mobile phase	0.05 M Ammonium dihydrogen phosphate and 0.01 M H_3PO_4 with N,N-dimethylformamide (1%, v/v) and methanol (4%, v/v)
Flow-rate	1.5 ml min^{-1}
Temperature	50 °C
Internal standard	β-Hydroxyethyltheophylline
Injection size	10 µl, 1.0 µg ml^{-1}

Detection	UV, λ = 276 nm; 0.005 AUFS
Recording	Spectra-Physics 4270 computing integrator
Retention times, min	Theophylline 3.7
	Internal standard 4.3
Ref.	Chiou R, Stubbs RJ and Bayne WF (1987) J. Chromatogr. Biomed. Appl. 422:281

HPLC 2

Sample	Urine
Extraction	Ethyl acetate at pH 5.0–5.5 or pH 9.0–9.5
Apparatus	Hewlett-Packard 1090 HPLC with photodiode array UV detector
Column	Reversed-phase; Hypersil ODS C_{18}, 5 μm, 200 × 4.6 mm ID
Mobile phase	A. 0.05 M Phosphate buffer, pH 3.0 containing 0.016 M propylamine hydrochloride (with 6.9 g NaH_2PO_4, H_2O and 1.59 g propylamine hydrochloride per liter water)
	B. Acetonitrile
	Gradient: B 15% at 2 min; B 15% to 80% at 20 min
Flow-rate	1 ml min^{-1}
Temperature	Ambient (22 °C)
Injection size	5 μl
Detection	Diode array UV, λ = 275 nm
Retention times, min	Theophylline 3.22
	Amiloride 3.67
	Caffeine 4.49
Ref.	Cooper SF, Massé R and Dugal R (1989) J. Chromatogr. Biomed. Appl. 489:65

HPLC 3

Sample	Blood plasma
Extraction	Protein precipitation with 20% trichloroacetic acid
Apparatus	Perkin-Elmer Series 2 pump with LC-75 ultraviolet detector
Column	Reversed-phase; 5 μm, C_8, 125 × 4.6 mm ID
Pre-column	μ-Bondapak C_{18}, Corasil
Mobile phase	4% Methanol in 10 mM sodium acetate and 0.005 M tetrabutylammonium phosphate, pH 4.5 with CH_3COOH
Flow-rate	2.0 ml min^{-1}; 10 MPa
Temperature	45 °C
Internal standard	β-Hydroxyethyltheophylline (BHET),

Injection size	100 µl			
Detection	UV, λ = 273 nm; 0.04 AU			
Retention times, min	3-Methylxanthine 2.73		Theophylline	6.28
	1-Methyluric acid 3.22		BHET (IS)	7.42
	1,3-Dimethyluric acid 3.90		Caffeine	11.19
Ref.	Kester MB, Saccar CL et al (1986) J. Chromatogr. Biomed. Appl. 380:99			

HPLC 4

Sample	Xanthines
Apparatus	Merck M 655A-11 HPLC pump with 655A-22 UV detector
Column	Reversed-phase; Lichrosorb C_{18}, 250 × 4 mm ID
Mobile phase	Aqueous phosphate buffer, pH 3.3/ acetonitrile (80:20, v/v)
Flow-rate	1.0 ml min^{-1}; 1500 psi
Injection size	20 µl containing 2 µg
Detection	UV, λ = 254 nm; 0.08 AUFS
Retention times, min	(estimated from drawing) Theophylline 2.85
	Theobromine 2.35 Caffeine 3.80
Ref.	Merck Application Manual on HPLC, TLC and HPTLC (1986). E. Merck, Darmstadt, p 30

HPLC 5

Sample	Blood plasma			
Extraction	Centrifugation and dilution with buffer			
Apparatus	LDC M 711-47 LC pump, Hewlett-Packard 1040A UV detector and Rheodyne M 7120 injector			
Column	Reversed-phase; Lichrosorb C_8, 5 µm, 100 × 4.6 mm ID			
Pre-column	Reversed-phase; Lichrosorb C_8, 10 µm, 10 × 3.2 mm ID			
Mobile phase	Methanol/phosphate buffer, pH 6.0 (7:93, v/v) containing 7.8 mM N,N dimethyl-n-octylamine			
Flow-rate	0.68 ml min^{-1}			
Injection size	10 µl			
Detection	UV, λ = 254 nm			
Recording	Hewlett-Packard 85-B personal computer			
Retention times, min	(estimated from drawing)			
	Hypoxanthine 3.4		Paraxanthine	8.8
	3-Methylxanthine 5.0		Theophylline	10.2
	Theobromine 6.0		1,3-Dimethyl uric acid	11.4
	1-Methyluric acid 7.9		Caffeine	14.4

Ref.	Daoud N, Arvidsson T and Wahlund K-G (1986) J. Pharm. Biomed. Anal. 4:253

210 Thiabendazole

CAS-No	148-79-8
Chemical name	2-(Thiazol-4-yl)-1*H*-benzimidazole
Empirical formula	$C_{10}H_7N_3S$
Molecular mass	201.3
Proprietary names	Equizole (vet); Mintezol; Thiprazole (vet)
Clinical use	Anthelmintic

Solubility	Water	alcohol	chloroform	acetone	aq HCl
	insoluble	1 in 150	1 in 300	slightly soluble	soluble

Light absorption	λ_{max} = 302 nm in aqueous acid

HPLC 1

Sample	Body fluids
Extraction	Diethyl ether
Column	Reversed-phase; ODS-Hypersil 10 μm, 100 × 8 mm ID
Mobile phase	Methanol/0.05 M ammonium carbonate (65:35)
Flow-rate	1.5 ml min^{-1}
Injection size	5 μl
Detection	UV, λ = 292 nm
Retention times, min	(estimated from drawing)

Oxfendazole	3.1	Oxibendazole	5.2
Thiabendazole	3.7	Albendazole	7.5
Cambendazole	4.3	Fenbendazole	10.1
Mebendazole	4.8	Parbendazole	10.9

Ref.	Borgan JA and Marriner S (1980) J Pharm. Sci. 69:422

HPLC 2

Sample	Test substance
Apparatus	HPLC low-pressure ternary gradient system
Column	Reversed-phase; Lichrosorb C_8, 250 × 4 mm ID
Mobile phase	Methanol/0.25% aqueous ammonia (60:40, v/v)
Flow-rate	1 ml min^{-1}; 3190 psi
Injection size	50 μl
Detection	UV, λ = 305 nm
Retention time, min	Thiabendazole 5.48 (estimated from drawing)

Ref.	Merck Application Manual on HPLC, TLC and HPTLC (1986). E. Merck, Darmstadt, p 28

HPLC 3

Sample	Marmalades
Extraction	Ethyl acetate
Apparatus	Pye-Unicam PCXPD LC pump, UV detector and Rheodyne M 7120 injector
Column	Normal phase; Polygosil 60/7.5/18
Mobile phase	A. Tetrahydrofuran/chloroform (1:9) (from other publication) B. 0.8% (w/v) Ammonium nitrate containing 0.7% (v/v) ammonia (buffered solution)/ methanol (3:2)
Flow-rate	1 ml min^{-1}
Temperature	22–26 °C
Injection size	20 µl
Detection	UV, λ = 305 nm
Retention times, min	Thiabendazole 7.3 (A); 8.3 (B)
Ref.	Collinge A and Noirfalise A (1983) J. Chromatogr. 257:416

HPLC 4

Sample	Fruits
Extraction	Multistep, first ethyl acetate and then with chloroform and again with ethyl acetate
Apparatus	Perkin-Elmer Series 3 LC with LC-55B UV detector
Column	A. Reversed-phase; HC-ODS Sil-X, 250 × 2.6 mm ID B. Normal phase; Silica-A/10, 250 × 2.6 mm ID
Mobile phase	For A. Methanol/water (1:1) For B. Tetrahydrofuran/chloroform (1:9) or chloroform
Flow-rate	1.0 ml min^{-1}; 0.5 ml min^{-1} (other fruits)
Derivatization	Pre-column, with *p*-nitrobenzyl bromide
Injection size	5–10 µl, 5–10 ng ml^{-1}
Detection	UV, λ = 305 nm; 0.05–0.5 AUFS
Retention times, min	(estimated from drawing)

Injection	direct		after derivatization
Column	A.	B.	B.
Thiabendazole	2.0 4.0*	11.0	
p-Nitrobenzyl derivative			5.3 7.5*

* other fruits

Ref.	Tafuri F, Marucchini C et al (1980) J. Agric. Food Chem. 28:1150

HPLC 5

Sample	Waste water
Apparatus	Altex M 110A LC pump, Perkin-Elmer M 650-105 spectrofluorimeter or LC-2A electrochemical detector (Bioanalytical Systems)
Column	Reversed-phase; Ultrasphere ODS, 10 μm, 250 × 4.6 mm ID
Pre-column	Normal phase; 50 × 2.1 mm ID
Mobile phase	Methanol/triethanolamine acetic acid buffer, pH 8.2 (70:30)
Injection size	100 or 250 μl containing 1 ng ml^{-1}
Detection	EC, GC electrode, + 1.05 V; fluorescence
Retention times, min	Maleic hydrazide 4.5
	Thiabendazole 5.5
	Ethoxyquin 6.4
Ref.	Victor DM, Hall RE et al (1984) J. Chromatogr. 283:383

211 Thiamine (Vitamin B_1)

CAS-No	59-43-8; 67-03-8 (hydrochloride)			
Chemical name	3-(4-Amino-2-methylpyrimidin-5-ylmethyl)-5-(2-hydroxyethyl)-4-methylthiazolium chloride			
Empirical formula	Thiamine	Hydrochloride		
	$C_{12}H_{17}ClN_4OS$	$C_{12}H_{17}ClN_4OS$, HCl		
Molecular mass	300.8	337.3		
Proprietary names	Aneurone; Benerva; Bisolvit			
Clinical use	Vitamin			
Solubility,	Water	alcohol	methanol	acetone
Hydrochloride	1 in 1	1 in 100	soluble	insoluble
Light absorption	λ_{max} = 232 nm in aqueous alkali			

HPLC 1

Sample	Cereals and urine
Extraction	5 mM $NH_4H_2PO_4$, pH 2.85/acetonitrile/phosphoric acid (3.9:1:0.1, v/v) on C_{18} RP column
Apparatus	Varian M 5060 LC pump and Varichrom-5 UV detector
Column	Anion-exchange; Micropak AX-5, 5 μ, 300 × 4.6 mm ID
Mobile phase	5 mM Ammonium dihydrogen phosphate, pH 2.85

Flow-rate	0.5 ml min^{-1}
Temperature	30 °C
Injection size	10 µl containing 0.5–10.0 ng
Detection	UV, λ = 245 nm; 0.01 AUFS
Retention time, min	Thiamine 4.0
Ref.	Hilker DM and Clifford AJ (1982) J. Chromatogr. Biomed. Appl. 231:433

HPLC 2

Sample	Cereals and urine
Extraction	As in Thiamine HPLC 1
Apparatus	As in Thiamine HPLC 1
Column	As in Thiamine HPLC 1
Mobile phase	A. Acetonitrile/5 mM $NH_4H_2PO_4$, pH 2.85 (87:13) B. 5 mM $NH_4H_2PO_4$, pH 2.85 C. 750 mM $NH_4H_2PO_4$, pH 4.0 Gradient programme: 100% A for 15 min, follow 0% B to 100% B at 4% min^{-1}; follow 0% C to 100% C at 2.5% min^{-1}
Flow-rate	0.4 ml min^{-1} for 15 min; increase to 1.0 ml min^{-1} in next 2 min
Temperature	30 °C
Injection size	10 µl, 0.15–0.3 mM
Detection	UV, λ = 245 nm; 0.05 AUFS
Retention times, min	Thiamine disulfide 27.0 Thiamine 28.5 Thiamine monophosphate 33.5 Thiamine diphosphate 62.0 Thiamine triphosphate 94.0 (estimated from drawing)
Ref.	Hilker DM and Clifford AJ (1982) J. Chromatogr. Biomed. Appl. 231:433

HPLC 3

Sample	Human whole blood and erythrocytes containing thiamine pyrophosphate
Extraction	6% Trichloroacetic acid
Apparatus	Gynkotek M 600 LC pump, Gynkotek RF-530 spectrofluorimeter and N & Z 190 injector
Column	Amino phase; µ-Bondapak NH_2, 10 µm, 300 × 4 mm ID
Pre-column	Amino phase; Lichrosorb NH_2, 10 µm, 30 × 4.6 mm ID
Mobile phase	Acetonitrile/0.07 M potassium dihydrogen phosphate, pH 7.8 (50:50, v/v), pH 8.5 with 3 N phosphoric acid
Flow-rate	2.0 ml min^{-1}
Derivatization	Pre-column; with cyanogen bromide

Injection size	100 µl containing 0.25–2 pmol standard vitamer
Detection	Fluorescence, λ = 375/435 nm
Retention time, min	Derivatised vitamer 7.8 (estimated from drawing)
Ref.	Floridi A, Pupita M et al (1984) Internat. J. Vit. Nutr. Res. 54:165

HPLC 4

Sample	Human blood serum
Extraction	Water-saturated diethyl ether after deproteinization with 30% trichloroacetic acid
Apparatus	Altex M 334-50 LC pump and Perkin-Elmer LS-4 fluorescence spectrophotometer
Column	Reversed-phase; PRP-1, 10 µm, 250 × 4.1 mm ID
Pre-column	Reversed-phase; Vydac-201, 30-44 µm, 32 × 4.2 mm ID
Mobile phase	I. 15 mM Sodium phosphate buffer, pH 8.5/ methanol (90:10) for thiamine phosphate esters II. 15 mM Sodium phosphate buffer, pH 8.5/ tetrahydrofuran (90:10) for thiamine
Flow-rate	0.5 ml min^{-1}
Derivatization	Pre-column; with potassium ferricyanide
Injection size	20 or 100 µl containing 180 fmol (ester) und 120 fmol thiamine
Detection	Fluorescence, λ = 365/433 nm
Recording	Shimadzu C-R1A Chromatopac
Retention times, min	(estimated from drawing)

Mobile phase	I.	II.
Thiamine		7.0
Thiamine triphosphate	7.6	
Thiamine pyrophosphate	8.8	
Thiamine monophosphate	12.8	

Ref.	Bettendorff L, Grandfils C et al (1986) J. Chromatogr. Biomed. Appl. 382:297

HPLC 5

Sample	Multivitamin tablets
Extraction	0.1 N HCl, pH 5.0
Column	Reversed-phase; µ-Bondapak C_{18}
Mobile phase	1.2 g Camphorsulfonic acid in 300 ml water, pH 7.0 with 0.5 N NaOH plus 275 ml methanol
Flow-rate	1.5 ml min^{-1}
Injection size	20 µl, 0-20 µg ml^{-1}
Detection	UV, λ = 280 nm
Retention times, min	(estimated from drawing)

Nicotinamide	2.12	Thiamine	7.00
Pyridoxine	3.20	Riboflavine	9.60

Ref.	Wittmer DP, Haney Jr. WG (1979) in: Tsuji K (ed) GLC and HPLC determination of Therapeutic agents, Part 3. Marcel Dekker, Inc., New York and Basel, p 1235

212 Thiamphenicol

CAS-No	15318-45-3; 2393-92-2 (glycinate hydrochloride)		
Chemical name	2,2-Dichloro-*N*-[(α*R*,β*R*)-β-hydroxy-α-hydroxymethyl-4-mesylphenethyl]acetamide		
Empirical formula	Thiamphenicol	Glycinate Hydrochloride	
	$C_{12}H_{15}Cl_2NO_5S$	$C_{14}H_{18}Cl_2N_2O_6S$, HCl	
Molecular mass	356.2	449.7	
Proprietary names	Flumucil; Urfamycin; Zambon		
Clinical use	Antibiotic		
Solubility,	Water	methanol	ether, ethyl acetate
Thiamphenicol	slightly soluble	soluble	slightly soluble
Glycinate-HCl	freely soluble		insoluble
Light absorption	λ_{max} = 266 nm in 95% ethanol		

HPLC 1

Sample	Serum and cerebrospinal fluid
Extraction	Methanol
Apparatus	Waters ALC/GPC M 204 LC with M 6000 pump, U6K injector and 440 UV detector
Column	Reversed-phase; μ-Bondapak C_{18}, 300 × 4 mm ID
Mobile phase	Water/methanol (80:20, v/v)
Flow-rate	1 ml min^{-1}
Injection size	100 μl containing 0.5 μg ml^{-1}
Detection	UV, λ = 254 nm; 0.005 AUFS
Retention times, min	Thiamphenicol 5.0
	Chloramphenicol 17.5
Ref.	Meulemans A, Manuel C et al (1981) J. Liq. Chromatogr. 4:145

HPLC 2

Sample	Chicken tissue
Extraction	Acetonitrile using n-hexane and then re-extraction with acetonitrile/diethyl ether (1:1) on Florisil cartridge
Apparatus	Shimadzu LC-3A pump and SPD-2A UV detector
Column	Reversed-phase; Zorbax ODS, 250 × 4.6 mm ID
Pre-column	Zorbax ODS, 50 × 4 mm ID

Mobile phase	Methanol/water (1:9)
Flow-rate	0.7 ml min^{-1}
Temperature	55 °C
Internal standard	*p*-Aminoacetophenone 1 µg ml^{-1}
Injection size	10 µl, 3 µg ml^{-1}
Detection	UV, λ = 230 nm; 0.01 AUFS
Retention times, min	(estimated from drawing)
	Thiamphenicol 25.2
	Internal standard 32.7
Ref.	Nagata T, Saeki M et al (1985) J. Fd. Hyg. Soc. Jap. 26:46

HPLC 3

Sample	Yellow tail fish
Extraction	Acetonitrile using n-hexane, then ethyl acetate and lastly re-extraction with acetone on Florisil column
Column	TSK-Gel, ODS-120 T, 150 × 4 mm ID
Mobile phase	Methanol/water (15:85, v/v)
Flow-rate	0.6 ml min^{-1}
Temperature	Ambient
Injection size	10 µl containing 20 ng
Detection	UV, λ = 225 nm; 0.02 AUFS
Retention time, min	Thiamphenicol 16.0 (estimated from drawing)
Ref.	Otsuka K, Horibe K et al (1986) J. Fd. Hyg. Soc. Jap. 27:263

HPLC 4

Sample	Raw materials
Column	Reversed-phase; Verspack C_{18}, 10 µm, 250 × 4.1 mm ID
Mobile phase	Methanol/water (40:60)
Flow-rate	1.7 ml min^{-1}
Detection	UV
Retention times, min	(estimated from drawing)
	Thiamphenicol 1.7
	Chloramphenicol 3.3
Ref.	Latek Labortechnik (1984) Katalog No. 2, Heidelberg, p 257

213 Thiopental

CAS-No	76-75-5; 71-73-8 (sodium salt)	
Chemical name	5-Ethyl-5-(1-methylbutyl)-2--thiobarbituric acid	
Empirical formula	Thiopental	Sodium Salt
	$C_{11}H_{18}N_2O_2S$	$C_{11}H_{17}N_2NaO_2S$

Molecular mass	242.3	264.3		
Proprietary names	Farmotal; Pentothal; Trapanal			
Clinical use	General Anaesthetic			
Solubility,	Water	alcohol	ether	light petroleum
Sodium salt	1 in 1.5	partly sol	insoluble	insoluble
Light absorption	λ_{max} = 303 nm in 1 M sodium hydroxide			

HPLC 1

Sample	Biological fluids
Extraction	Deproteinization with acetonitrile
Apparatus	Kontron T-414 LC pump, Uvikon 735-LC spectrophotometer and loop injector
Column	Reversed-phase; μ-Bondapak C_{18}, 10 μm, 300 × 4.0 mm ID
Mobile phase	0.05 M Perchlorate buffer, pH 2.9/acetonitrile (60:40 v/v)
Flow-rate	1.5 ml min^{-1}
Temperature	Ambient
Internal standard	Oxazepam
Injection size	20 μl, 0.1 μg ml^{-1}
Detection	UV, λ = 300 nm
Recording	Hewlett-Packard 3390A integrator
Retention times, min	Internal standard 5.0 Thiopental 8.3
Ref.	Marty P, Arnoux P and Cano JP (1988) Anal. Lett. 21:2241

HPLC 2

Sample	Rat blood plasma and brain tissue
Extraction	Chloroform under acidic condition
Apparatus	Waters M 6000A HPLC pump, 440 UV detector and U6K injector
Column	Reversed-phase; μ-Bondapak C_{18}, 10 μm, 300 × 3.9 mm ID
Mobile phase	0.01 M Phosphate buffer/acetonitrile/tetrahydrofuran, 78:22:0 (a), 78:22:2 (b); 78:22:4 (c)
Flow-rate	2.0 ml min^{-1}; 14 MPa
Internal standard	Pentobarbital sodium
Injection size	30 μl, 0–20 μg ml^{-1}
Detection	UV, λ = 254 nm; 0.01 AUFS
Recording	Shimadzu C-R1A Chromatopac
Retention times, min	(estimated from drawing)

Mobile phase	(a)	(b)	(c)
Pentobarbital	11.0	11.0	9.5
Thiopental	11.0	13.5	12.0

Ref.	Shiu GK and Nemoto EM (1982) J. Chromatogr. Biomed. Appl. 227:207

HPLC 3

Sample	Serum
Extraction	Phosphate buffer, 0.16 M, pH 6.6 using n-butyl chloride and re-extraction with dilute sodium hydroxide
Apparatus	Perkin-Elmer Series 2 LC pump with LC-85 ultraviolet detector and loop injector
Column	Reversed-phase; octadecyl sulfate C_{18}, 5 μm, 125 × 4.6 mm ID
Mobile phase	50% Methanol and 50% aqueous phosphate buffer pH 6.6 (containing 11 g KH_2PO_4 and 9 g Na_2HPO_4 l^{-1} water)
Flow-rate	2.0 ml min^{-1}
Internal standard	Barbital
Injection size	6 μl, 0–100 μg ml^{-1}
Detection	UV, λ = 240 nm
Retention times, min	(estimated from drawing) Thiopental 4.0; Barbital 1.0; Thiamylal 5.4; Pentobarbital 2.0
Ref.	Kelner M and Bailey DN (1983) Clin. Chem. 29:1097

HPLC 4

Sample	Serum
Extraction	Deproteinization with acetonitrile
Apparatus	Hewlett-Packard 1084A dual-pump and 1030A UV detector
Column	Reversed-phase; Spherisorb-5, C_{18}, 100 × 4.6 mm ID
Pre-column	Reversed-phase; Spherisorb-5, C_{18}, 30 × 4.6 mm ID
Mobile phase	A. Methanol/5 mM phosphate buffer, pH 6.5 (35:65) B. Methanol Isocratic: A:B (4:1, v/v)
Flow-rate	2 ml min^{-1}
Temperature	55 °C
Internal standard	Phenolphthalein
Detection	UV, λ = 280 nm
Retention times, min	Phenolphthalein 1.75 Thiopental 3.06
Ref.	Freeman DJ (1981) Clin. Chem. 27:1942

214 Ticarcillin

CAS-No	34787-01-4 (ticarcillin); 4697-14-7 and 29457-07-6 (both disodium salt)	
Chemically	Semisynthetic penicillin	
Empirical formula	Ticarcillin	Sodium Salt
	$C_{15}H_{16}N_2O_6S_2$	$C_{15}H_{14}N_2Na_2O_6S_2$
Molecular mass	384.4	428.4
Proprietary names	Aerugipen; Ticar; Ticarpen	
Clinical use	Antibiotic	
Solubility,	Water	
Sodium salt	freely soluble	

HPLC 1

Sample	Injectable dosage forms
Apparatus	Waters M 510 LC pump, 481 variable-wavelength detector and U6K injector valve
Column	Reversed-phase; μ-Bondapak C_{18}, 10 μm, 300 × 3.9 mm ID
Mobile phase	I. Acetonitrile/0.01 M KH_2PO_4 (12:88, v/v) II. Acetonitrile/methanol/0.01 M KH_2PO_4/ 0.02 M octanesulfonic acid sodium salt in 0.2% glacial acetic acid (12:7:41:40, v/v)
Flow-rate	1 ml min^{-1}
Internal standard	Propylthiouracil, 0.1 mg ml^{-1}
Injection size	20 μl, 20–90 μg ml^{-1}
Detection	UV, λ = 225 nm; 0.1 AUFS
Recording	Waters 730 integrating recorder

Retention times, min

Mobile phase	I.	II.
Ticarcillin	4.58	4.62
R-Ticarcillin		4.80
S-Ticarcillin		5.45
Propylthiouracil (IS)	9.66	9.75

Ref. Mopper B (1988) J. Assoc. Off. Anal. Chem. 71:390

HPLC 2

Sample	Serum
Extraction	Chloroform/butanol (3:1)
Apparatus	Varian M 5000 HPLC with UV-50 variable-wavelength detector
Column	Phenyl phase; Bondapak, 300 × 4.0 mm ID
Mobile phase	A. 10 mM Ammonium acetate, pH 4.2 B. 10 mM Ammonium acetate, pH 4.2 in methanol/ water (50:50)

	Linear gradient: 85% A and 15% B for 3.5 min; B 15% to 100% in 5.5 min; hold 7.0 min
Flow-rate	2.0 ml min^{-1}
Temperature	Ambient
Internal standard	Cefazolin
Injection size	100 µl containing 2.5 µg
Detection	UV, λ = 250 nm
Retention times, min	Ticarcillin 3.5 Cefazolin 9.2 Carbenicillin 5.5 Nafcillin 15.1
Ref.	Annesley T, Wilkerson K et al (1984) Clin. Chem. 30:908

HPLC 3

Sample	Serum and urine
Extraction	Ethyl acetate under acidic condition and then with methylene chloride at pH 6.8
Apparatus	Waters M 6000A LC pump, U6K injector and 450 UV detector
Column	Reversed-phase; RCM-10-Radial C_{18}, 100 × 8 mm ID
Pre-column	Wathman, Co-Pell ODS
Mobile phase	Acetonitrile/0.06 M sodium biphosphate (50.5:100), pH 2.05 with appr. 35 ml 85% H_3PO_4
Flow-rate	1.5 ml min^{-1}; 350 psi
Temperature	Ambient
Internal standard	Carbenicillin sodium
Injection size	10–50 µl
Detection	UV, λ = 210 nm; 0.04 AUFS
Recording	Water 730 data module
Retention times, min	(estimated from drawing) Ticarcillin appr. 5.3 Internal standard appr. 5.9
Ref.	Kwan RH, Macleod SM et al (1982) J. Pharm. Sci. 71:1118

HPLC 4

Sample	Diastereoisomers of penicillins
Apparatus	Waters M 6000A HPLC pump with 440 UV detector
Column	Reversed-phase; Zorbax C_8, 250 × 4.6 mm ID
Mobile phase	Water/methanol/5% of 0.2 M phosphate buffer, pH 7.0
Flow-rate	1 ml min^{-1}
Injection size	20 µl
Detection	UV, λ = 254 nm; 0.05 AUFS
Recording	Pye-Unicam DP88 integrator and Kipp & Zonen BD40 recorder

Retention times, min	Mobile phase	methanol %	RT
	Carbenicillin	5.0	10.0
	Ticarcillin	5.0	6.5
	Phenethicillin	37.5	13.0
	Propicillin	45.0	20.5
	Clometocillin	50.0	19.0
Ref.	Hoogmartens J, Roets E et al (1982) J. Chromatogr. 244:299		

215 Tobramycin

CAS No	32986-56-4
Chemically	Aminoglycoside antibiotic produced by Streptomyces tenebrarius or by other means
Empirical formula	$C_{18}H_{37}N_5O_9$
Molecular mass	467.5
Proprietary names	Nebcin; Obracin; Tobrex
Clinical use	Antibiotic
Solubility	Water 1 in 1.5; alcohol 1 in 2000; chloroform insoluble; ether insoluble

HPLC 1

Sample	Serum
Extraction	Acetonitrile on Bond-Elut RP C_{18} column after derivatization
Apparatus	Perkin-Elmer Series 3 LC with LC-75 ultraviolet detector and Rheodyne M 7105 injector
Column	Reversed-phase; Ultrasphere-octyl, 250×4.6 mm ID
Mobile phase	700 ml Acetonitrile and 300 ml 50 mM potassium dihydrogen phosphate, pH 3.5 with H_3PO_4
Flow-rate	3.0 ml min^{-1}
Temperature	50 °C
Derivatization	Pre-column; with 2,4,6-trinitrobenzenesulfonic acid
Internal standard	Sisomicin
Injection size	50 µl, 0–25 µg ml^{-1}
Detection	UV, $\lambda = 340$ nm
Retention times, min	Tobramycin 3.0 Sisomicin (IS) 3.8 (estimated from drawing)
Ref.	Kahra PM, Bhatnagar PK et al (1983) Clin. Chem. 29:672

HPLC 2

Sample	Serum
Extraction	Deproteinization with acetonitrile
Apparatus	HPLC with Shimadzu UV-200 spectrophotometer
Column	Reversed-phase; μ-Bondapak C_{18}, 10 μm, 300 × 3.9 mm ID
Mobile phase	Water/acetonitrile/acetic acid (340:660:1, v/v)
Flow-rate	3.0 ml min^{-1}
Temperature	Ambient
Derivatization	Pre-column; with 1-fluoro-2,4-dinitrobenzene
Internal standard	Gentamicin, 20 μg ml^{-1}
Injection size	150 μl, 0.5–16 μg ml^{-1}
Detection	UV, λ = 365 nm; 0.02 AUFS
Retention times, min	(estimated from drawing) Tobramycin derivate 4.4 Gentamicin C_{1a} derivate 10.4 Gentamicin C_1 and C_2 derivate 12.3
Ref.	Barends DM, Zwaan CL and Hulshoff A (1981) J. Chromatogr. Biomed. Appl. 225:417

HPLC 3

Sample	Serum
Extraction	Deproteinization with 0.078 N sulfosalicylic acid
Apparatus	Perkin-Elmer HPLC with multi-dimensional column switching system having ten-way-valve
Column	Multidimension and combination of A. Cation-exchange; 10 μm, 30 × 4.6 mm ID and B. Reversed-phase; C_{18}, 5 μm, 125 × 4.6 mm ID
Pre-column	Normal phase; 37–54 μm, 25 × 2.2 mm ID
Mobile phase	I. 10 mM Sodium phosphate buffer, pH 5.2 and II. 50 mM EDTA, pH 8.8
Flow-rate	1.5 ml min^{-1}
Internal standard	Post-column; with *o*-phthalaldehyde
Injection size	50 μl
Detection	Fluorescence, λ = 340/440 nm
Retention time, min	Tobramycin 5.0 (estimated from drawing)
Ref.	Meyer V (1984): Praxis der Hochleistungsfluessig-chromatographie, 3rd edn. Diesterweg Salle and Sauerlaender, Frankfurt am Main and Salzburg, p 183

216 Tocainide

CAS-No	41708-72-9; 35891-93-1 (hydrochloride)
Chemical name	2-Aminopropiono-2′,6′-xylidide

Empirical formula	Tocainide	Hydrochloride	
	$C_{11}H_{16}N_2O$	$C_{11}H_{16}N_2O$, HCl	
Molecular mass	193.2	228.7	
Proprietary name	Tonocard		
Clinical use	Antiarrhythmic		
Solubility,	Water	ethanol	chloroform
Tocainide	sparingly soluble	freely soluble	freely soluble
Hydrochloride	freely soluble	freely soluble	insoluble
Light absorption	λ_{max} = 262 nm in aqueous acid		

HPLC 1

Sample	Serum
Extraction	Deproteinization with acetonitrile
Apparatus	Beckman M 110A LC pump and variable-wavelength detector (LDC Spectromonitor II)
Column	Synchropak SCD-100, C_{18}, 5 μm, 150 × 4.6 mm ID
Mobile phase	Acetonitrile/25 mM amonium phosphate buffer, pH 7.3 (24:76)
Flow-rate	1.3 ml min^{-1}
Injection size	50 μl, 10 μg ml^{-1}
Detection	UV, λ = 200–230 nm
Retention time, min	Tocainide 6.0
Ref.	Shihabi ZK, Scaro J and Dyer RD (1988) J. Liq. Chromatogr. 11:2391

HPLC 2

Sample	Blood plasma	
Extraction	Dichloromethane	
Column	Reversed-phase; μ-Bondapak C_{18}, 300 mm	
Mobile phase	Acetonitrile/10 mM KH_2PO_4 (20:80, v/v)	
Flow-rate	1.5 ml min^{-1}	
Temperature	Ambient	
Injection size	50 μl, 43 μmol l^{-1}	
Detection	UV, λ = 214 nm	
Retention times, min	Tocainide	3.92
	N-Acetyl tocainide	6.20
	N-Chloroacetyl tocainide	11.82
Ref.	Boutagy J and O'Shea K (1987) Clin. Chem. 33:1069	

HPLC 3

Sample	Human blood plasma
Extraction	Dichloromethane under alkaline condition
Apparatus	Spectra-Physics 8000 LC, Schoeffel M 770 UV detector and Spectra-Physics 8010 autosampler

Column	Phenyl phase; μ-Bondapak, 10 μm, 300 × 4 mm ID
Mobile phase	1.5 mM Aqueous phosphoric acid/acetonitrile (95:5)
Flow-rate	1.0 ml min^{-1}
Temperature	50 °C
Internal standards	I and II (Astra Pharmaceutical)
Injection size	50 μl
Detection	UV, λ = 230 nm; 0.08 AUFS

Retention times, min		
	N-Acetylprocainamide	6.0
	Procainamide	6.5
	Internal standard I	7.2
	Theobromine	7.8
	Ethosuximide	8.4
	Tocainide	8.7
	Internal standard II	11.2
	Lidocaine	12.2
	Quinidine	15.0
	Primidone	16.1
	Caffeine	> 16.7
	Carbamazepine	> 16.7
	N-Desisopropylpropanol	> 16.7
	Dihydroquinidine	> 16.7
	Disopyramide	> 16.7
	4-Hydroxypropranolol	> 16.7
	Phenobarbital	> 16.7
	Phenytoin	> 16.7
	Propranolol	> 16.7

Ref. Reece PA and Stanley PE (1980) J. Chromatogr. Biomed. Appl. 183:109

HPLC 4

Sample	Serum
Extraction	1,2-Dichloromethane under alkaline condition
Apparatus	Waters M 510 LC pump, WISP 710B autosampler and M 481, variable-wavelength detector
Column	Cyano phase; CN 5 μm, 250 × 4.6 mm ID
Mobile phase	60 ml Acetonitrile, 7 ml methanol and 33 ml phosphate-TEA buffer (containing 1.361 g KH_2PO_4 in 1 liter water and 75 μl triethylamine), pH 5.7
Flow-rate	2 ml min^{-1}
Temperature	40 °C
Internal standard	*p*-Chlorodisopyramide
Injection size	10 μl containing 200 ng
Detection	UV, λ = 205 nm; 0.02 AUFS

Recording	Waters M 840 data-reduction and instrument control system	
Retention times, min	(estimated from drawing)	
	Lidocaine	2.8
	Tocainide	3.4
	Monoethylglycylxylidide	4.5
	N-Acetylprocainamide	5.8
	Procainamide	6.4
	Despropyldisopyramide	7.0
	Quinidine	7.5
	Disopyramide	8.0
	p-Chlorodisopyramide	8.5
Ref.	Proelss HF and Townsend TB (1986) Clin. Chem. 32:1311	

HPLC 5

Sample	Tablets
Column	Reversed phase; ODS 5–10 µm, 300 × 3.9 mm ID
Mobile phase	2.16 g Sodium 1-octanesulfonate in 500 ml 0.67 N acetic acid and 500 ml methanol
Flow-rate	2.0 ml min^{-1}
Injection size	Appr. 40 µl
Detection	UV, λ = 254 nm
Capacity factor, *k'*	Tocainide 1.6
Ref.	Anonym (1985): The United States Pharmacopeia XXI and The National Formulary XVI, Supplement 5, 1987. United States Pharmacopeial Convention Inc., Rockville, p 2448

217 Tolbutamide

CAS-No	64-77-7; 473-41-6 (sodium salt)			
Chemical name	1-Butyl-3-*p*-tolylsulfonylurea			
Empirical formula	Tolbutamide	Sodium Salt		
	$C_{12}H_{18}N_2O_3S$	$C_{12}H_{17}N_2NaO_3S$		
Molecular mass	270.3	292.3		
Proprietary names	Artosin; Pramidex; Rastinon			
Clinical use	Antidiabetic			
Solubility,	Water	alcohol	acetone	chloroform
Tolbutamide	insoluble	1 in 10	1 in 3	soluble
Sodium salt	freely soluble	soluble		soluble
Light absorption	λ_{max} = 257 nm in methanol			

HPLC 1

Sample	Sterile tolbutamide sodium
Column	Normal phase; 5–10 μm, 300 × 4 mm ID
Mobile phase	Hexane/water-saturated hexane/ tetrahydrofuran/alcohol/glacial acetic acid (475:475:20:15:9)
Internal standard	Tolazamide
Injection size	Appr. 20 μl
Detection	UV, $\lambda = 254$ nm
Retention times	(relative) Tolbutamide 0.6 Tolazamide 1.0
Ref.	Anonym (1985): The United States Pharmacopeia XXI and The National Formulary XVI. United States Pharmacopeial Convention Inc., Rockville, p 1070

HPLC 2

Sample	Commercial preparations and blood plasma
Apparatus	Waters M 90 LC pump with 450 UV detector
Column	Reversed-phase; Lichrosorb C_{18}, 5 μm, 150 mm
Mobile phase	A. Acetonitrile/water (50:50, v/v), for tablets B. Acetonitrile/0.05% H_3PO_4 in water (45:55, v/v), pH 2.6, for blood plasma
Flow-rate	1.2 ml min^{-1} (A) and 1.0 ml min^{-1} (B)
Internal standards	Paracetamol (A) and diazepam (B)
Injection size	10 μl, 25–600 μg ml^{-1}
Detection	UV, $\lambda = 226$ nm
Capacity factor k'	see below
Ref.	Vreugdenhil HJ (1983) Pharm. Weekbl. Sci. 5:1025

System	A.	B.
Paracetamol	0.9	
Tolbutamide	2.6	1.5
Diazepam		3.0

HPLC 3

Sample	Human blood plasma and urine
Extraction	Diethyl ether under acidic condition
Apparatus	Hewlett-Packard 1090A LC with photodiode array detector and HP 85B system master
Column	Reversed-phase; BST C_8 or BST C_{18}, 7 μm, 250 × 4 mm ID. (BST = Budapest, Hungary)
Mobile phase	Acetonitrile/2-propanol/0.1% ortho-phosphoric acid (17:17:66, v/v)

Flow-rate	1.2 ml min^{-1}
Temperature	Ambient
Internal standard	Chlorpropamide
Injection size	5 µl, 34.8 µg ml^{-1}
Detection	UV, λ = 235 nm
Recording	Hewlett-Packard 3390A recording integrator
Retention times, min	(estimated from drawing) Hydroxytolbutamide 4.4 Carboxytolbutamide 5.9 Chlorpropamide 8.8 Tolbutamide 11.0
Ref.	Csillag K, Vereczkey L and Gachalyi B (1989) J. Chromatogr. Biomed. Appl. 490:355

HPLC 4

Sample	Human blood plasma
Extraction	Ether under acidic condition
Apparatus	Waters M 6000 HPLC pump with 440 UV detector and U6K injector
Column	Reversed-phase; µ-Bondapak C_{18}
Mobile phase	35% Acetonitrile and 65% 0.05 M phosphoric acid buffer, pH 3.9
Flow-rate	2.5 ml min^{-1}; 2000 psi
Temperature	28 °C
Internal standard	Chlorpropamide
Injection size	5 µl, 2–80 µg ml^{-1}
Detection	UV, λ = 254 nm; 0.01 AUFS
Retention times, min	Carboxytolbutamide 1.8 Internal standard 5.6 Tolbutamide 7.8
Ref.	Raghow G and Meyer MC (1981) J. Pharm. Sci. 70:1166

HPLC 5

Sample	Test substance
Column	Normal phase; 10 µm, 500 × 1 mm ID
Mobile phase	Hexane/tetrahydrofuran/ethanol/glacial acetic acid (95:2:1.5:0.9)
Flow-rate	0.3 ml min^{-1}
Detection	UV
Retention times, min	(estimated from drawing) Tolbutamide 5.8 Tolazamide 10.0
Ref.	Latek Labortechnik (1984) Katalog No. 2, Heidelberg, p 273

218 Triamcinolone

CAS-No	124-94-7; 76-25-5 (acetonide)			
Chemical name	9α-Fluoro-16α-hydroxyprednisolone			
Empirical formula	Triamcinolone	Acetonide		
	$C_{21}H_{27}FO_6$	$C_{24}H_{31}FO_6$		
Molecular mass	394.4	434.5		
Proprietary names	Adcortyl; Aristocort; Kenacort			
Clinical use	Corticosteroid			
Solubility,	Water	alcohol	methanol	chloroform
Triamcinolone	1 in 500	1 in 240	slightly soluble	very slightly soluble
Acetonide	very slightly soluble	1 in 150	very sol	1 in 40
Light absorption	λ_{max} = 238 nm in methanol			

HPLC 1

Sample	Commercial ointment preparations
Apparatus	Varian M 4000 HPLC with UV detector
Column	Normal phase; Si 10, 250 × 2.1 mm ID
Mobile phase	5% Ethanol in methylene chloride
Flow-rate	0.8 ml min^{-1}; 500 psi
Injection size	1–2 µl
Detection	UV, λ = 254 nm; 0.08 AUFS
Retention times, min	(estimated from drawing)
	Propylpraben 1.02; Methylparaben 1.40; Triamcinolone acetonide 2.29
Ref.	Gaetani E and Laureri CF (1976) Liquid Chromatography at Work 29. Varian, California

HPLC 2

Sample	Steroid mixture	
Column	Amino phase; Nucleosil NH_2, 10 µm, 300 × 4 mm ID	
Mobile phase	Tetrahydrofuran	
Flow-rate	0.42 cm sec^{-1}; 1450 psi	
Temperature	Ambient	
Injection size	5 µl	
Detection	UV, λ = 254 nm	
Retention times, min	(estimated from drawing)	
	Triamcinolone hexacetonide	1.3
	Triamcinolone diacetate	1.7
	Triamcinolone acetonide	2.2
	Triamcinolone alcohol	3.5

Ref.	Sebestian I (1987) in: High Performance Liquid Chromatography. Application No. 459, Macherey-Nagel, Dueren, p 280

HPLC 3

Sample	Milk
Extraction	Ethyl acetate and then re-extraction with acetonitrile using hexane
Apparatus	Tracor 5000 LC with UV detector
Column	Reversed-phase; ODS 10 μm, 250 × 4.6 mm ID
Mobile phase	Water/acetonitrile (77:23)
Flow-rate	0.7 ml min^{-1}; 500 psi
Internal standard	Dexamethasone
Injection size	25 μl
Detection	UV, λ = 254 nm; 0.04 AUFS
Retention times, min	(estimated from drawing) Dexamethasone 9.0 Triamcinolone acetonide 11.0
Ref.	Depaolis AM, Schnabel G et al (1977) J. Assoc. Off. Anal. Chem. 60:210

HPLC 4

Sample	Test mixture
Column	Amino phase; Nucleosil 10 NH_2, 300 × 4.1 mm ID
Mobile phase	Tetrahydrofuran
Flow-rate	3.2 ml min^{-1}
Detection	UV
Retention times, min	(estimated from drawing) Triamcinolone hexacetonide 1.2 Triamcinolone diacetate 1.5 Triamcinolone acetonide 2.0 Triamcinolone alcohol 3.3
Ref.	Latek Labortechnik (1984) Katalog No. 2, Heidelberg, p 260

HPLC 5

Sample	Raw materials
Column	Reversed-phase; Hypersil ODS 5 μm, 200 × 4 mm ID
Mobile phase	Methanol/water (50:50) or adjusted for a retention time appr. five minutes
Flow-rate	2 ml min^{-1}
Detection	UV, λ = 238 nm
Retention time, min	Triamcinolone 5.00 (adjusted by mobile phase)

Ref.	Anonym (1980): British Pharmacopoeia, Addendum 1983. London Her Majesty's Stationary Office, p 259

219 Triclabendazole

CAS-No	68786-66-3
Chemical name	5-Chloro-6-(2,3-dichlorophenoxy)-2-methylthio-1*H*-benzimidazole
Empirical formula	$C_{14}H_9Cl_3N_2OS$
Molecular mass	359.7
Proprietary names	Fasinex
Clinical use	Anthelmintic

HPLC 1

Sample	Sheep blood plasma
Extraction	Deproteinization with acetone, and then with methanol on C_{18} Sep-Pak column
Apparatus	Waters M 6000A LC pump, 481 spectrophotometer detector and 710B WISP sample processor
Column	Reversed-phase; μ-Bondapak C_{18}, 10 μm, 300 × 3.9 mm ID
Mobile phase	Methanol/water/reagent 1 (120 ml CH_3COOH + 200 g CH_3COONH_4 in 1 liter water)/chloroform (750:200:25:10, v/v)
Flow-rate	1 ml min^{-1}
Injection size	40 μl, 0.1–10 μg ml^{-1}
Detection	UV, $\lambda = 300$ nm
Recording	Hewlett-Packard 3390A integrator
Retention times, min	Triclabendazole sulfone 6.0 Triclabendazole sulfoxide 7.0 Triclabendazole 9.0
Ref.	Bull MS and Shume GRE (1987) J. Pharm. Biomed. Anal. 5:527

HPLC 2

Sample	Cattle and sheep blood plasma
Apparatus	Waters M 6000A LC pump and ultraviolet detector
Column	Reversed-phase; μ-Bondapak C_{18}, 10 μm, 300 × 3.9 mm ID
Mobile phase	Acetonitrile/0.04 M $(NH_4)_2HPO_4$ (41:59, v/v), pH 7.5–7.6 with 20% H_3PO_4

Flow-rate	1.5 ml min^{-1}
Injection size	100 µl, 0.05–5 µg ml^{-1}
Detection	UV, λ = 300 nm
Recording	Hewlett-Packard 3390A integrator
Retention times, min	Triclabendazole sulfone 5.2 Triclabendazole sulfoxide 9.0 Triclabendazole 40.0
Ref.	Bull MS and Shume GRE (1987) J. Pharm. Biomed. Anal. 5:501

HPLC 3

Sample	Blood plasma
Extraction	Ethyl acetate
Apparatus	Waters M 6000A LC pump with WISP 710B autosampler and 481 spectrophotometer
Column	Normal phase; Partisil silica 5 µm, 250 × 4.6 mm ID
Mobile phase	Hexane/glacial acetic acid/ethanol (500:0.6:50)
Flow-rate	1.5 ml min^{-1}; 1088 psi
Temperature	Ambient
Internal standard	Oxfendazole
Injection size	80 µl
Detection	UV, λ = 215 nm
Recording	Perkin-Elmer LCI 100 laboratory computing integrator
Retention times, min	Triclabendazole 5.69 Triclabendazole sulfone 6.80 Triclabendazole sulfoxide 9.92 Oxfendazole 13.04
Ref.	Alvinerie M and Galtier P (1986) J. Chromatogr. Biomed. Appl. 374:409

220 Trimethoprim

CAS-No	738-70-5
Chemical name	5-(3,4,5-Trimethoxybenzyl)pyrimidine-2,4-diyldiamine
Empirical formula	$C_{14}H_{18}N_4O_3$
Molecular mass	290.3
Proprietary names	Borgal (vet); Cotrimoxazol; Monotrim
Clinical use	Chemotherapeutic

Solubility	Water	alcohol	chloroform	methanol	ether
	1 in 2500	1 in 300	1 in 55	1 in 80	insoluble
Light absorption	λ_{max} = 287 nm in aqueous alkali				

HPLC 1

Sample	Biological fluids
Extraction	Methanol using rapid sorbent extraction column
Apparatus	Waters M 501 LC pump with 440 UV detector and WISP 710B autosampler
Column	Reversed-phase; μ-Bondapak C_{18}, 150 mm
Mobile phase	0.01 M Sodium acetate in water/acetonitrile (70:30)
Flow-rate	2 ml min^{-1}
Temperature	25 °C
Injection size	5–40 μl, 0.2–100 μg ml^{-1}
Detection	UV, λ = 254 nm; 0.01–0.05 AUFS
Retention time, min	Trimethoprim 2.00
Ref.	Svirbely JE and Pesce AJ (1988) J. Liq. Chromatogr. 11:1075

HPLC 2

Sample	Blood plasma and urine
Extraction	Chloroform/2-propanol (95:5) under alkaline condition
Apparatus	LDC/Milton Roy reciprocating piston pump, Rheodyne M 7120 loop injector and Schoeffel FS-970 fluorimetric detector
Column	Reversed-phase; 10 μm, C_{18}, 250 × 4 mm ID
Mobile phase	Methanol/0.015 M sodium borate, pH 9.0 (35:65)
Flow-rate	1.0 ml min^{-1}
Temperature	Ambient
Injection size	20 μl, 0.25–2.0 μg ml^{-1}
Detection	Fluorescence, λ = 279/370 nm
Retention time, min	Trimethoprim 12.0
Ref.	Gautam SR, Chungi VS et al (1978) Anal. Lett. B 11:967

HPLC 3

Sample	Serum and urine
Extraction	Serum: Ethyl acetate after deproteinization with acetonitrile Urine: Acetonitrile and Sorensen's buffer, pH 7.2
Apparatus	Perkin-Elmer 601 LC, LC 55 variable-wavelength detector and Rheodyne M 70-10 sample injection valve

Column	Reversed-phase; μ-Bondapak C_{18}, 10 μm, 300 × 3.9 mm ID
Mobile phase	Methanol/1% acetic acid (20:80), pH 2.9
Flow-rate	1.5 ml min^{-1}; 1450 psi
Temperature	30 °C (serum); 20–24 °C (urine)
Internal standard	Sulfafurazole
Injection size	1–10 μl
Detection	UV, λ = 230 nm
Retention times, min	(estimated from drawing) (serum)

Trimethoprim	5.9	Sulfafurazole	9.5
Sulfamethoxazole	7.5	N^4-Acetyl-sulfa-methoxazole	17.0

Ref. Gochin R, Kanfer I and Haigh JM (1981) J. Chromatogr. Biomed. Appl. 223:139

HPLC 4

Sample	Phamaceuticals
Apparatus	Knauer FR-30 dual-piston HPLC pump, 87-variable-wavelength monitor and Rheodyne M 7010 loop injector
Column	Du Pont Zorbax TMS 7 μm, 250 × 4.6 mm ID
Mobile phase	Acetonitrile/1-propanol/methanol/tetrahydro-furan/acetic acid/water (5:20:15:25:1:34)
Flow-rate	2 ml min^{-1}
Injection size	20 μl, 0.16–0.24 mg ml^{-1}
Detection	UV, λ = 271 nm
Recording	Columbia Scientific Supergrator 3A integrator

Retention times, min

Sulfamethoxazole	1.21	Degradation products (4)	2.22, 2.25, 2.49, 2.71
Methylparaben	1.40		
Propylparaben	1.52	Trimethoprim	4.37

Ref. Bergh JJ and Breytenbach JC (1987) J. Chromatogr. 387:528

HPLC 5

Sample	Blood plasma
Extraction	Ethyl acetate
Apparatus	Waters M 510 LC pump, U6K injector and Schoeffel SF 770 ultraviolet detector
Column	Reversed-phase, Lichrosorb C_{18}, 5 μm, 250 × 4.0 mm ID
Mobile phase	0.15 M Ammonium phosphate buffer, pH 4.85/acetonitrile (87.4:12.6)
Flow-rate	1.5 ml min^{-1}
Internal standard	Sulfafurazole
Injection size	100 μl

Detection	UV, λ = 254 nm
Recording	Shimadzu C-R3A integrator
Retention times, min	Trimethoprim 10.0 Sulfamethoxazole 20.0 N-Acetyl-sulfamethoxazole 21.0 Sulfafurazole (IS) 30.5 (estimated from drawing)
Ref.	Van der Steuijt K and Sonneveld P (1987) J. Chromatogr. Biomed. Appl. 422:328

HPLC 6

Sample	Biological fluids
Extraction	Diethyl ether at pH 6.8
Apparatus	Altex M 110A LC pump, Kontron Uvikon 725 spectrophotometric detector and Rheodyne M 7120 injector
Column	Normal phase; Lichrosorb Si 60, 10 μm, 250 × 4 mm ID
Pre-column	Normal phase; Lichroprep Si 60, 25–40 μm
Mobile phase	Dichloromethane/methanol/25% aqueous ammonia (80:19:1)
Flow-rate	1.5 ml min^{-1}; 725 psi
Internal standard	Sulfadimethoxine
Injection size	20 μl, 0.03–10 μg ml^{-1}
Detection	UV, λ = 289 nm; 0.1 AUFS
Recording	Perkin-Elmer Sigma 10 chromatographic computer
Retention times, min	Trimethoprim 2.6 Sulfadiazin 6.7 Internal standard 3.7 N^4-Acetyl-sulfadiazin appr. 10.4
Ref.	Ascalone V (1981) J. Chromatogr. Biomed. Appl. 224:59

221 Tripelennamine

CAS-No	91-81-6; 154-69-8 (hydrochloride)	
Chemical name	*N*-Benzyl-*N'N'*-dimethyl-*N*-(2-pyridyl)ethelene-diamine	
Empirical formula	Tripelennamine $C_{16}H_{21}N_3$	Hydrochloride $C_{16}H_{21}N_3$, HCl
Molecular mass	255.4	291.8
Proprietary names	Azaron; PBZ; Pyribenzamine	
Clinical use	Antihistamine	

Solubility,	Water	alcohol	chloroform	ether
Tripelennamine	miscible			
Hydrochloride	1 in 1	1 in 6	1 in 6	insoluble
Light absorption	λ_{max} = 249 nm in aqueous alkali			

HPLC 1

Sample	Bovine blood plasma and milk
Apparatus	Waters M 45 LC pump with gradient controller, U6K manual injector and Shimadzu SPD-6A UV detector
Column	Cyano phase; Techsphere 3CN, 100 × 4.6 mm ID
Pre-column	Reversed-phase; Corasil C_{18}, 37–50 µm, 10 × 1.5 mm ID
Mobile phase	A. Water B. Acetonitrile/0.05 M acetate buffer, pH 7.2 with acetic acid (70:30, v/v)
Flow-rate	Start: 0.8 ml min^{-1} (A) for 3 min; next 0.9 ml min^{-1} (B)
Internal standard	Protriptyline hydrochloride, 4 µg ml^{-1}
Injection size	250 µl, 0.04–6 µg ml^{-1}
Detection	UV, λ = 246 nm
Retention times, min	Tripelennamine 5.2 Internal standard 6.8
Ref.	Dadgar D and Power A (1987) J. Chromatogr. Biomed. Appl. 421:216

HPLC 2

Sample	Animal feed, human urine and waste water
Extraction	Dichloromethane
Column	Normal phase; Ultrasphere Si 5 µm, 250 × 4.6 mm ID
Mobile phase	0.005 M Triethylamine in dichloromethane/ 2-propanol (99:1, v/v)
Flow-rate	2.0 ml min^{-1}
Injection size	50 µl, 0.5 µg ml^{-1}
Detection	Fluorescence, λ = 310/360 nm
Retention times, min	Tripelennamine 6.5 Thenyldiamine 7.4 Chlorothen 8.1
Ref.	Thompson HC Jr, Holder CL and Althaus JR (1983) Talanta 30:251

HPLC 3

Sample	Raw materials
Apparatus	Spectra-Physics 3500B LC pump with M 770 ultraviolet detector

Column	Reversed-phase; 10 μm Bondapak C_{18}, 300 × 3.9 mm ID
Mobile phase	Methanol/acetic acid/triethylamine/water (50:1.5:0.5:48)
Flow-rate	1.5 ml min^{-1}
Injection size	10 μl, 0.25 mg ml^{-1}
Detection	UV, $\lambda = 254$ nm; 0.16 AUFS
Recording	Hewlett-Packard 3380A integrator
Retention times, min	(estimated from drawing)

Pheniramine	4.4	Brompheniramine	8.8
Thozylamine	5.0	Phenindamine	10.0
Tripelennamine	6.0	Phenyltoxamine	11.8
Chlorpheniramine	7.6	Clemizole	16.2

Ref. Roos RW and Lau-Cam CA (1986) J. Chromatogr. 370:403

222 Tubocurarine Chloride

CAS-No	57-95-4 (tubocurarine); 6989-98-6 (chloride, pentahydrate)
Chemical name	(+)-7′,12′-Dihydroxy-6,6′-dimethoxy-2,2′,2′-trimethyltubocuraranium dichloride pentahydrate
Empirical formula	$C_{37}H_{42}Cl_2N_2O_6$, $5H_2O$
Molecular mass	771.7
Proprietary names	Curarin; Jexin; Tubarine
Clinical use	Muscle Relaxant

Solubility

Water	ethanol	chloroform, ether	aq NaOH
1 in 20	1 in 30	insoluble	soluble

Light absorption: $\lambda_{max} = 280$ nm in water

HPLC 1

Sample	Plant extract
Apparatus	Waters M 6000 HPLC pumps with 660 solvent programmer and 440 UV detector
Column	Reversed-phase; μ-Bondapak C_{18}, 10 μm, 300 mm
Pre-column	Reversed-phase; Corasil C_{18}, 37–50 μm, 50 × 4.6 mm ID
Mobile phase	A. 0.025 M Tetramethylammonium phosphate in water/methanol (75:25), pH 4 with concentrated ortho-H_3PO_4

	B. Same buffer and pH as in A/methanol (55:45)
	Linear gradient: B 10% to 85% in 30 min
Flow-rate	1 ml min^{-1}; 1500–2000 psi
Temperature	20 °C
Injection size	20 µl
Detection	UV, λ = 280 nm
Recording	Spectra-Physics 4000 computing integrator
Retention times, min	(estimated from drawing)

D-Tubocurarine chloride	11.0	Curarine chloride	17.0
		Isochondrodendrine	22.5
Chondrocurine	13.0	Curine	23.0

Ref.	Van der Maeden FPB, Van Rens PT and Buytenhuys FA (1977) J. Chromatogr. 142:715

HPLC 2

Sample	Human blood plasma
Extraction	Ethylene dichloride
Apparatus	Waters M 6000A LC pump with U6K injector and 440 UV detector
Column	Reversed-phase; Rad-Pack A C_{18}, 10 µm, 150 × 8 mm ID
Mobile phase	40% Methanol in aqueous mixture of 10 g triethylamine, 1 ml Pic B5 (Waters), 2 ml phosphoric acid and distilled water to make 1 liter, pH 3.4
Flow-rate	2 ml min^{-1}
Injection size	120 µl, 25–500 ng ml^{-1}
Detection	UV, λ = 280 nm; 0.005 AU
Recording	Kipp & Zonen potentiometric recorder
Retention time, min	D-Tubocurarine appr. 4.4
Ref.	Meulemans A, Mohler J et al (1981) J. Chromatogr. Biomed. Appl. 226:255

---

223 Tylosin

CAS-No	1401-69-0; 1405-54-5 (tartrate)	
Chemically	Mixture of macrolide antibiotics produced by Streptomyces fradiae or by other means	
Empirical formula	Tylosin	Tartrate
	$C_{46}H_{77}NO_{17}$	$(C_{46}H_{77}NO_{17})_2$, $C_4H_6O_6$
Molecular mass	916.1	1982

Proprietary names	Tylamix (vet); Tylan (vet); Pharmachim (vet)			
Clinical use	Antibiotic/Growth Promoter			
Solubility,	Water	ethanol	methanol	0.1N HCl
Tylosin	1 in 400	1 in 15	1 in 6	soluble
Tartrate	1 in 10	slightly sol		
Light absorption	λ_{max} = 288 nm in water			

HPLC 1

Sample — Fermentation broth
Extraction — Chloroform after filtration, at pH 5.0
Apparatus — Du Pont 841 HPLC with ultraviolet detector
Column — A. Du Pont ODS Permaphase
B. Du Pont Sorbax SIL
Mobile phase — For A. Methanol/water (75:25)
For B. Chloroform/isopropanol (65:35)
Flow-rate — 1.0 ml min^{-1}; 1200 psi
Temperature — Ambient
Injection size — 5 µl
Detection — UV, λ = 254 nm; 0.04 AUFS
Retention times, min

Column	A.		B.
Tylosin A, C, D	3.2	Tylosin A	10.4
Tylosin B	3.7	Tylosin C	12.4
(estimated from		Tylosin D	13.2
drawing)		Tylosin B	46.0

Ref. — Bhuwapathanapun S and Gray P (1977) J. Antibiotics 30:673

HPLC 2

Sample — Fermentation products
Apparatus — Spectra-Physics 8000 LC, 8310 UV detector and loop injector
Column — A. Reversed-phase; Spherisorb ODS, 150 × 4.6 mm ID
B. Reversed-phase; Zorbax C_8, 150 × 4.6 mm ID
Mobile phase — Ion-pair solution (250 mg sodium 1-pentanesulfonate in 1 liter water with 1% acetic acid)/tetrahydrofuran/acetonitrile (47:24:29), pH 3.5
Flow-rate — 1.5 ml min^{-1}
Temperature — 25 and 55 °C (A); 55 °C (B)
Injection size — 10 µl, 0.2 mg ml^{-1} of tylosin and 0.02 mg ml^{-1} of other macrolides
Detection — UV, λ = 280 nm
Recording — Spectra-Physics 4000 computing integrator

Retention times, min (estimated from drawing)

Colums	A. 25 °C	55 °C	B. 55 °C
Lactenocin	6.3	7.2	5.0
o-Demethylmacrocin	7.0	7.5	5.3
5-O-Mycaminosyltylonolide	8.2	8.5	5.7
20-Dihydrodesmycosin	8.2	9.0	6.4
Macrocin	8.7	9.3	6.7
Desmycosin	9.2	10.8	7.2
Relomycin	10.3	11.5	8.0
Demycinosyltylosin	10.8	11.2	7.5
Tylosin	12.5	14.5	9.4

Ref. Kennedy JH (1983) J. Chromatogr. 281:288

HPLC 3

Sample	Blood and tissue from pigs
Extraction	Deproteinization with acetonitrile and then with methylene chloride
Apparatus	Varian 5000 LC with automatic injection loop and UV-50 detector
Column	Varian Micropak MCH-10-N-Cap, 300 × 4.6 mm ID
Mobile phase	0.005 M $NH_4H_2PO_4$/acetonitrile/methanol (10:60:30). The concentrations and proportions are different for serum and tissue
Flow-rate	1.5–2.0 ml min^{-1}
Injection size	200 µl containing 10 µg
Detection	UV, $\lambda = 280$ nm
Ref.	Moats, WA, Harris EW and Steele NC (1985) J. Assoc. Off. Anal. Chem. 68:413

HPLC 4

Sample	Raw materials
Column	Reversed-phase; Nucleosil ODS 5 µm, 200 × 5 mm ID
Mobile phase	0.85 M Sodium perchlorate in 40% (v/v) acetonitrile, pH 2.5 with 1 M HCl
Flow-rate	1 ml min^{-1}
Temperature	Ambient or up to 50 °C
Detection	UV, $\lambda = 290$ nm
Retention times	Desmycinosyltylosin, tylosin C, tylosin B, tylosin D, aldol impurity and tylosin A, in ascending order
Ref.	Anonym (1985) British Pharmacopocia (Veterinary), Her Majesty's Stationary Office London, p 42

HPLC 5

Sample	Raw materials
Apparatus	Varian M 5000 LC with UV-50 detector and Valco automatic loop injector

Column	Reversed-phase; Micropak SPC-18, 4 µm, 150 × 4.6 mm ID
Pre-column	Special
Mobile phase	A. 0.004 M Ammonium dihydrogen phosphate/acetonitrile/methanol (10:60:30) B. As in A but (5:70:25)
Flow-rate	1 ml min^{-1}
Injection size	200 µl containing 2 µg
Detection	UV, λ = 280 nm; 0.1 AUFS
Retention times, min	(estimated from drawing) Tylosin 9.5 (A and B)
Ref.	Moats WA (1986) J. Chromatogr. 366:69

224 Vancomycin

CAS-No	1404-90-6; 1404-93-9 (hydrochloride)			
Chemicaly	Amphoteric glycopeptide produced by Streptomyces orientalis			
Empirical formula	Vancomycin $C_{66}H_{75}Cl_2N_9O_{24}$			
Molecular mass	1449.2			
Proprietary names	Diatracin; Vancocin			
Clinical use	Antibiotic			
Solubility,	Water	ethanol	ether	chloroform
Hydrochloride	1 in 10	1 in 700	slightly soluble	insoluble
Light absorption	λ_{max} = 281 nm aqueous acid			

HPLC 1

Sample	Blood plasma, bone, atrial appendage tissue and pericardial fluid
Extraction	Methanol/water (1:1) using C_8 Bond Elut sample clean-up column
Apparatus	Waters M 6000 LC pump, 710B WISP autoinjector and 440 UV detector
Column	Reversed-phase; Supelcosil C_8, 5 µm, 250 × 4.6 mm ID
Pre-column	Reversed-phase; Supelcosil C_8, 40 µm, 50 × 4.6 mm ID
Mobile phase	0.05 M Potassium dihydrogen phosphate/acetonitrile (90:10, v/v), pH 5.0 with 4 M sodium hydroxide
Flow-rate	1.5 ml min^{-1}; 1998 psi
Temperature	Ambient (22–24 °C)

Internal standard	Cefazolin sodium
Injection size	100–200 µl, 3.13–100 µg ml^{-1}
Detection	UV, λ = 254 nm; 0.05 AUFS
Recording	Waters M 840 data and chromatography control system
Retention times, min	Vancomycin 6.0– 6.5 Internal standard 9.0–11.5
Ref.	Greene SV, Abdalla T and Morgan SL (1987) J. Chromatogr. Biomed. Appl. 417:121

HPLC 2

Sample	Raw materials stored at different temperatures
Apparatus	LDC Constametric I and Constametric II G LC pumps with M 1601 gradient master, Cecil M 272 spectrophotometer and Rheodyne M 7125 injector
Column	PLRP-S 100A, 8 µm, 150 × 4.6 mm ID
Mobile phase	0.02 M Borate buffer, pH 8.6/acetonitrile (92:8, v/v) Linear-gradient: Acetonitrile 8% to 16% in 17.5 min; hold up to 35 min
Flow-rate	0.5 ml min^{-1}
Injection size	20 µl, 0.25 mg ml^{-1}
Detection	UV, λ = 235 nm; 0.05 AUFS
Recording	Spectra-Physics 4270 computing integrator
Retention times, min	(estimated from drawing) (stored at 56°C for 15 years)

Degradation product-I minor	11.0	Dimethylvancomycin	18.0
		Vancomycin	20.5
Degradation product-I major	13.8	Aglucovancomycin	45.0

Ref.	Thomas AH and Newland P (1987) J. Chromatogr. 410:373

HPLC 3

Sample	Blood plasma
Extraction	Deproteinization with acetonitrile
Apparatus	Shimadzu LC-6A pump, SPD-6A ultraviolet detector and Rheodyne M 7125 injector
Column	Normal phase; Shim-pak CLC-NH_2 5 µm, 150 × 6.0 mm ID
Mobile phase	0.07 M Phosphate buffer, pH 2.5/acetonitrile (40:70, v/v)
Flow rate	1.5 ml min^{-1}
Temperature	Ambient
Injection size	20 µl, 28.4 µg ml^{-1}
Detection	UV, λ = 240 nm; 0.01–0.002 AUFS

Recording	Shimadzu C-R4A Chromatopac chromatographic data processor
Retention time, min	Vancomycin 8.3
Ref.	Hosotsubo H (1989) J. Chromatogr. Biomed. Appl. 487:421

HPLC 4

Sample	Serum
Extraction	Methanol
Apparatus	Perkin-Elmer Series 10 HPLC, LC-85B ultraviolet detector and LC-600 autosampler
Column	Supelcosil LC-18, 5 μm, 150 × 4.6 mm ID
Mobile phase	Buffer (containing 1.528 g sodium octanesulfonate and 131 mg disodium EDTA in 600 ml water, pH 2.0 with 1 M H_3PO_4 and readjust at 2.6 with concentrated KOH), 250 ml acetonitrile and 50 ml methanol, and make 1 liter with water
Flow-rate	1 ml min^{-1}
Internal standard	3,4,5-Trimethoxyphenylacetonitrile
Injection size	20 μl, 30 μg ml^{-1}
Detection	UV, λ = 230 nm
Recording	Perkin-Elmer 3600 computing integrator
Retention times, min	Vancomycin 3.72–3.80 Internal standard 7.05–7.20
Ref.	Rosenthal AF, Sarfati I and A'Zary E (1986) Clin. Chem. 32:1016

HPLC 5

Sample	Serum and tissue
Extraction	Acetonitrile/50 mM KH_2PO_4, pH 4.0 (30:70, v/v) on Sep-Pak C_{18} column
Apparatus	Waters M 6000A HPLC pump, 440 UV detector and WISP 710B autosampler
Column	Reversed-phase; Nucleosil C_{18}, 5 μm, 150 × 4.6 mm ID
Mobile phase	5 mM Potassium dihydrogen phosphate, pH 2.8 with ortho-phosphoric acid containing 22% methanol
Flow-rate	1.5 ml min^{-1}
Temperature	Ambient
Internal standard	Trimethoprim, 5 μg ml^{-1}
Injection size	40–200 μl, 20 μg ml^{-1}
Detection	UV, λ = 229 nm; 0.002 AU
Retention times, min	Trimethoprim 2.5 Vancomycin 4.5
Ref.	Bauchet J, Pussard E and Garaud JJ (1987) J. Chromatogr. Biomed. Appl. 414:472

HPLC 6

Sample	Blood
Apparatus	Hewlett-Packard 1084B LC
Column	Cation exchange; CM Sphadex, C-25
Mobile phase	Methanol-borax buffer (10:990, v/v)
Internal standard	Ristocetin
Injection size	500 µl, 100 µg ml^{-1}
Detection	UV, λ = 200 nm; 0.05 AUFS
Retention times, min	Ristocetin 5.44 Vancomycin 7.86
Ref.	Bever FN, Finley PR et al (1984) Clin. Chem. 30:1586

HPLC 7

Sample	Serum
Extraction	Borax-sodium sulfate buffer, pH 9.45 on CM-Sephadex extraction column
Column	Reversed-phase; µ-Bondapak C_{18}, 300 × 3.9 mm ID
Mobile phase	0.05 M Phosphate buffer/acetonitrile (91:9, v/v)
Flow-rate	2.0 ml min^{-1}
Temperature	Ambient
Internal standard	Ristocetin
Injection size	25 µl
Detection	UV, λ = 210 nm
Ref.	Ristuccia PA (1987) J. Liq. Chromatogr. 10:241

225 Verapamil

CAS-No	52-53-9; 152-11-4 (hydrochloride)				
Chemical name	5-[*N*-(3,4-Dimethoxyphenethyl)-*N*-methylamino]-2-(3,4-dimethoxyphenyl)-2-isopropylvaleronitrile				
Empirical formula	Verapamil	Hydrochloride			
	$C_{27}H_{38}N_2O_4$	$C_{27}H_{38}N_2O_4$,HCl			
Molecular mass	454.6	491.1			
Proprietary names	Cordilox; Isoptin; Vasolan				
Clinical use	Antiarrhythmic				
Solubility,	Water	ethanol	chloroform	ether	aq HCl
Verapamil	insoluble	soluble	soluble	soluble	soluble
Hydrochloride	1 in 20	1 in 25	1 in 1.5	insoluble	
Light absorption	λ_{max} = 278 nm in aqueous acid				

HPLC 1

Sample	Blood plasma
Extraction	Methyl tert.-butyl ether, pH 14.0
Apparatus	HPLC with fluorescence detector
Column	Normal phase; Spherisorb S5W, silica, 250 × 5 mm ID
Mobile phase	7 mM Methanolic ammonium perchlorate, pH 1.3 with 0.1 M methanolic perchloric acid
Flow-rate	2 ml min^{-1}
Internal standard	Benzoquinoline
Injection size	100 µl, 72 µg ml^{-1}
Detection	Fluorescence, λ = ex 203 mm
Retention times, min	(estimated from drawing) Verapamil metabolite I 3.4 Verapamil 5.8 Norverapamil 3.6 IS 8.1 Verapamil metabolite II 4.7
Ref.	Flanagan RJ, Bhamra RK et al (1988) J. Liq. Chromatogr. 11:1015

HPLC 2

Sample	Serum
Extraction	Diethyl ether under alkaline condition
Apparatus	LDC/Milton Roy Constametric III pump, Rheodyne M 7126 injection valve and Schoeffel FS-970 fluorescence detector
Column	Normal phase; µ-Porasil 10 µm silica, 300 × 3.9 mm ID
Mobile phase	0.6 mM NaBr in methanol
Flow-rate	2.0 ml min^{-1}; 2100 psi
Temperature	Ambient
Internal standard	2-Hydroxydesipramine
Injection size	50 µl containing 1 ng
Detection	Fluorescence, λ = ex 203 nm
Retention times, min	Verapamil 4.6 N-Dealkylnorverapamil 13.4 Norverapamil 9.4 Internal standard 15.7
Ref.	Pieper JA and Rutledge DR (1988) J. Chromatogr. Sci 26:473

HPLC 3

Sample	Human blood plasma
Extraction	Diethyl ether
Apparatus	Altex M 100A pump, Rheodyne M 7120 injector and Schoeffel FS-970 fluorescence detector
Column	Reversed-phase; Nucleosil 5 C_{18}, 200 × 4.6 mm ID

Mobile phase	Aqueous buffer (containing 0.71 g sodium hydrogen phosphate, 300 µl H_3PO_4 and 2 g 1-heptanesulfonic acid sodium in 1 liter water)/acetonitrile (1:1, v/v), pH 4.6 with 0.1 M H_3PO_4
Flow-rate	1.5 ml min^{-1}
Temperature	30 °C
Internal standard	D 600 (Knoll, Ludwigshafen, West Germany)
Injection size	200 µl
Detection	Fluorescence, λ = 203/320 nm
Retention times, min	(seven metabolites of verapamil supplied under numbers by Knoll, Ludwigshafen, Germany)

PR 25	3.5	PR 23	7.4
D 620	4.0	PR 24	7.7
D 617	4.9	Verapamil	10.7
PR 21	5.5	Internal standard	11.5
PR 22	6.7		

Ref. Kuwada M, Tateyama T and Tsutsumi J (1981) J. Chromatogr. Biomed. Appl. 222:507

HPLC 4

Sample	Human blood plasma
Extraction	Ethyl acetate under alkaline condition
Apparatus	Hewlett-Packard 1090 LC pump with automatic injector and Schoeffel fluorescene detector
Column	Cyanopolysilane; Du Pont, 150 × 4.6 mm ID
Mobile phase	Acetonitrile/0.02 M sodium acetate, pH 7.0 (65:35, v/v)
Flow-rate	3.0 ml min^{-1}
Temperature	40 °C
Internal standard	Trimipramine maleate
Injection size	30 µl
Detection	Fluorescence, λ = 201/320 nm
Recording	Hewlett-Packard 3392A integrator
Retention times, min	Verapamil 3.6
	Norverapamil 6.7
	Trimipramine 8.3
	(estimated from drawing)
Ref.	Johnson SM, Khalil SKW (1987) J. Liq. Chromatogr. 10:1187

HPLC 5

Sample	Biological fluids
Extraction	Pentane/amyl alcohol (20:1) under alkaline condition
Apparatus	Altex M 110 pump, Rheodyne M 7120 injector and Schoeffel FS-970 fluorescence detector

Column	Ion-exchange; Partisil 10 SCX, 250 × 4.6 mm ID
Pre-column	Ion-exchange; Partisil 10 SCX, 40 × 3.2 mm ID
Mobile phase	Acetonitrile/diethylamine/glacial acetic acid/ water (200:0.86:0.57:798.57, v/v)
Flow-rate	2 ml min^{-1}
Temperature	Ambient
Internal standard	Glaucin
Injection size	50–90 μl, 10–1000 ng ml^{-1}
Detection	Fluorescene, λ = 206 nm (ex)
Retention times, min	(estimated from drawing) Metabolite III (D 620) 5.5; Metabolite II (D 617) 6.3; Norverapamil 8.0; Verapamil 10.2; Glaucin 13.3
Ref.	Piotrovskii VK, Rumiantsev DO and Metelitsa VI (1983) J. Chromatogr. Biomed. Appl. 275:195

226 Vitamin A

CAS-No	68-26-8; 79-81-2 (palmitate)		
Chemical name	15-Apo-β-caroten-15-ol		
Empirical formula	Vitamin A	Palmitate	
	$C_{20}H_{30}O$	$C_{36}H_{60}O_2$	
Molecular mass	286.5	524.9	
Proprietary names	Arovit; Euvitol; Ro-A-vit		
Clinical use	Vitamin		
Solubility,	Water	alcohol, ether	fixed oils
Vitamin A	insoluble	freely soluble	freely soluble
Palmitate	insoluble		soluble
Light absorption	λ_{max} = 224-225 nm in isopropanol		

HPLC 1

Sample	Serum
Extraction	Hexane and ethanol (2:1)
Apparatus	Kontron HPLC system with M 414 pumps, 881 pulse dampener, SFM 23 fluorescence detector and Rheodyne M 7125 injector
Column	Reversed phase; Spherisorb ODS C_{18}, 5 μm, 250 × 5 mm ID
Pre-column	Reversed-phase; C_{18}, Spherisorb 10
Mobile phase	Methanol
Flow-rate	2 ml min^{-1}
Temperature	Ambient

Injection size	50 μl
Detection	Fluorescence, λ = 343/480 nm
Recording	Shimadzu C-R1B Chromatopac
Retention times, min	(estimated from drawing)

Retinol	2.0	Retinyl linoleate	7.6
Retinyl propionate	3.0	Retinyl palmitate	9.8
Retinyl laurate	6.0	Retinyl stearate	12.4
Retinyl myristate	7.2		

Ref.	Schindler R, Klopp A et al (1985) Internat. J. Vit. Nutr. Res. 55:25

HPLC 2

Sample	Human serum
Extraction	n-Hexane
Apparatus	Waters M 6000A LC pump, U6K injector and Shimadzu RF-530 fluorescence spectrophotometer
Column	Polygosil 60-5, 5 μm, 80 × 4.6 mm ID
Mobile phase	n-Hexane/methylene chloride/2-propanol, (90:10:1, v/v)
Flow-rate	1.2 ml min^{-1}
Injection size	100 μl containing 50 nmol l^{-1}
Detection	Fluorescence, λ = 333/470 nm
Retention time, min	Vitamin A 6.2 (estimated from drawing)
Ref.	Speek AJ. Wongkham C et al (1986) J. Chromatogr. Biomed. Appl. 382:284

HPLC 3

Sample	Retinyl esters
Apparatus	LDC gradient HPLC system, Spectromonitor II UV detector and Rheodyne M 7125 injector
Column	A. Reversed-phase; Resolve 5 μm, C_{18}, 150 × 3.9 mm ID B. Reversed-phase; Partisil 10/25 ODS-2, C_{18}, 10 μm, 250 × 4.6 mm ID
Mobile phase	Acetonitrile/dichloromethane (80:20)
Flow-rate	1.5 ml min^{-1}
Temperature	Ambient
Injection size	10 μl containing 3–300 ng (retinyl esters)
Detection	UV, λ = 325 nm; 0.1–0.02 AUFS
Recording	Shimadzu C-R2AX integrator
Retention times	(relative, using column A)

Retinol	0.26	Retinyl pentadecanoate	0.87
Non-retinoid	0.36		
Retinyl linolenate	0.56	Retinyl oleate	0.93
Retinyl laurate	0.59	Retinyl palmitate	1.00
Non-retinoid	0.69		

	Retinyl linoleate	0.70	Retinyl heptadecanoate	1.16
	Retinyl myristate and retinyl palmitoleate	0.75	Non-retinoid	1.20
			Retinyl stearate	1.35
Ref.	Furr HC, Cooper DA and Olson JA (1986) J. Chromatogr. Biomed. Appl. 378:45			

HPLC 4

Sample	Fat soluble vitamins	
Apparatus	Azuma Electric Microfeeder MF-2, ML 422 injector and Jasco Uvidec 100 detector	
Column	Silica capillaries of 13 various bonded phases, 500 × 0.53 mm ID	
Mobile phase	Acetonitrile/ethanol/methanol (35:45:20)	
Flow-rate	4 µl min^{-1}	
Temperature	20 °C	
Detection	UV, λ = 280 nm	
Retention times, min	Vitamin K_3	20–37
	Vitamin D_2	22–57
	Vitamin D_3	22–62
	Vitamin E	22–62
	Vitamin E_a	23–65
	Vitamin K_1	23–82
Ref.	Wahyuni WT and Jinno K (1987) Chromatographia 23:320	

HPLC 5

Sample	Human buccal mucosal cells	
Extraction	Chloroform	
Apparatus	Beckman M 320 LC pump and Hitachi F-1000 fluorescence detector	
Column	Normal phase; Ultrasphere Si 5 µm, 250 × 4.6 mm ID	
Pre-column	Spherisorb Si 10 µm, 50 × 4.6 mm ID	
Mobile phase	Heptane/dioxane (92:8)	
Flow-rate	2.0 ml min^{-1}	
Internal standard	1-Naphthol	
Injection size	250 µl	
Detection	Fluorescence, λ = 326/470 nm	
Retention times, min	1-Naphthol	5.8
	Retinol	7.9
Ref.	Badcock NR and O'Reilly DA (1986) J. Chromatogr. Biomed. Appl. 382:290	

HPLC 6

Sample	Pharmaceutical preparations (capsules) and blood
Apparatus	Knauer HPLC with ultraviolet detector

Column	Reversed-phase; Lichrosorb C_{18}, 7 µm, 100 × 4.6 mm ID
Mobile phase	Methanol
Flow-rate	2 ml min^{-1}; 1160 psi
Temperature	22–25 °C
Injection size	5–50 µl, 7.5–50 ng
Detection	UV, λ = 280–350 nm; 0.04 AUFS
Retention time, min	Vitamin A acetate 0.90 (estimated from drawing)
Ref.	Amin M (1987) J. Liq. Chromatogr. 10:3127

HPLC 7

Sample	Retinol mixtures
Apparatus	Perkin-Elmer HPLC Series 4 pump, ISI 100 auto-sampler with PE-7000 laboratory computer and Kratos M 773 UV detector
Column	Normal phase; spherical silica 5 µm, 250 × 4.5 mm ID
Mobile phase	1,1,2-Trichlorofluoroethane (freon 113)/methyl-tert.-butyl ether (97:3, v/v)
Flow-rate	0.7 ml min^{-1}
Detection	UV, λ = > 380 nm
Recording	Spectra-Physics 4100 computing integrator
Retention times, min	Retinols 1–17.00 Retinol isomers 1–25.00
Ref.	Bruening RC, Derguini F and Nakanishi K (1986) J. Chromatogr. 361:437

HPLC 8

Sample	Blood plasma
Extraction	Hexane after diluting with ethanol
Apparatus	Waters M 45 LC pump, U6K injector and 440 UV or Bioanalytical Systems amperometric detector
Column	Reversed-phase; Supelcosil LC-18, 5 µm, 150 × 4.6 mm ID
Pre-column	Supelcosil 40 µm, 50 mm
Mobile phase	94% Methanol in 0.05 M sodium perchlorate
Flow-rate	1.5 ml min^{-1}
Temperature	31 °C
Injection size	30 µl
Detection	UV, λ = 313 nm; EC, GC electrode, + 0.6 V vs Ag/AgCl RE
Recording	Hewlett-Packard 3390A integrator
Retention times, min	all-trans-Retinol 3.0; β-Tocopherol 11.1 Tocol 7.1; α-Tocopherol 13.4 δ-Tocopherol 9.0
Ref.	Huang M-L, Burckart GJ and Venkataramanan R (1986) J. Chromatogr. Biomed. Appl. 380:331

227 Vitamin D_3

CAS-No	67-97-0
Chemical name	(5*Z*, 7*E*)-9,10-Secocholesta-5,7,10(19)-trien-3β-ol
Empirical formula	$C_{27}H_{44}O$
Molecular mass	384.0
Proprietary names	Actifral D_3; D-Mulsin; Vi-De-3
Clinical use	Vitamin
Solubility	Water: insoluble; alcohol, acetone and vegetative oils: freely soluble
Light absorption	λ_{max} = 265 nm in ethanol

HPLC 1

Sample	Pharmaceutical formulations and blood
Extraction	Chloroform/methanol (1:1) after treating with acetone/methanol (1:1)
Apparatus	Knauers HPLC pump and UV detector
Column	Reversed-phase; Vertex Lichrosorb 7 µm, C_{18}, 100 × 4.6 mm ID
Mobile phase	Methanol
Flow-rate	2 ml min^{-1}; 1015 psi
Temperature	22–25 °C
Internal standard	Vitamin A-acetate, 2.5 µg ml^{-1}
Injection size	10–20 µl containing 25 ng
Detection	UV, λ = 270 nm; 0.08 AUFS and λ = 320 nm; 0.04–0.16 AUFS
Retention times, min	(estimated from drawing) Vitamin D_3 0.5 Vitamin A-acetate 1.2
Ref.	Amin M (1988) J. Liq. Chromatogr. 11:1347

HPLC 2

Sample	Fortified milk and infant formulas
Extraction	Benzene and by clean-up method with LC
Apparatus	LDC Constametric III pump, Rheodyne injector and LDC spectromonitor D
Column	Reversed-phase; Shandon ODS 5 µm, 250 × 4.6 mm ID
Mobile phase	Methanol
Injection size	120 µl containing 250 pg
Detection	UV, λ = 265 nm; 0.004 AUFS
Recording	Shimadzu C-R2AX integrator
Retention time, min	Vitamin D_3 21.3

Ref.	Johnsson H and Hessel H (1987) Internat. J. Vit. Nutr. Res. 57:357

HPLC 3

Sample	Pharmaceutical formulations with hydroxylated products (metabolites) of Vitamin D_3
Apparatus	Waters M 6000A HPLC pump with 440 UV detector and U6K injector
Column	A. Normal phase; Lichrosorb Si 60, 5 μm, 250 × 4 mm ID B. Reversed-phase; μ-Bondapak C_{18}, 10 μm, 300 × 3.9 mm ID
Mobile phase	I. For A: Light petroleum (b.p. 60–80°)/dichloroethane/tetrahydrofuran/n-propanol (50:25:24:5, v/v) II. For A: Light petroleum (b.p. 60–80°)/dichloroethane/tetrahydorfuran/n-propanol (40:5:15:0.6, v/v) III. For B: Ethanol 94%/water (8:2, v/v) IV. For B: Ethanol 94%/water (7:3, v/v)
Flow-rate	1–1.5 ml min^{-1}; 900–2500 psi
Temperature	Ambient
Injection size	25–50 μl, 1–100 μg ml^{-1}
Detection	UV, λ = 254 nm
Recording	Kipp & Zonen recorder with Varian integrator
Retention times, min	

Mobile phase	I.	I.	II.	III.	IV.
Flow rate, ml min^{-1}	1.0	1.5	1.0	1.5	1.5
1α-OH-D_3	4.5	4.7			
pre-1α-OH-D_3	5.6	5.7			
Pre-25-OH-D_3			7.1		
25-OH-D_3			7.6		
1α,25-$(OH)_2$-D_3		6.8			
Pre-1α,25-$(OH)_2$-D_3		9.1			
1α-25-$(OH)_2$-D_3 + pre-1α,25-$(OH)_2$-D_3					5.1
25-OH-D_3 + pre-25-OH-D_3				4.5	

Ref.	Vanhaelen-Fastre R and Vanhaelen M (1979) J. Chromatogr. 179:131

HPLC 4

Sample	Mixture of Vitamin A, D_3 and E
Column	HC-ODS-Sil-X, 250 × 2.6 mm ID
Mobile phase	Acetonitrile/water (90:20)
Flow-rate	1 ml min^{-1}; 508 psi

Temperature	65 °C
Detection	UV, λ = 220 nm and 293 nm
Retention times, min	(estimated from drawing) Vitamin D_3 5.0 Vitamin A alcohol 1.4 Vitamin E 6.8
Ref.	Meyer V (1984): Praxis der Hochleistungsfluessigchromatographie, 3rd edn. Diesterweg Salle and Sauerlaender, Frankfurt am Main and Salzburg, p 204

HPLC 5

Sample	Vitamin AD_3E
Apparatus	Merck HPLC low-pressure ternary gradient system
Column	Reversed-phase; Lichrosorb C_{18}, 250 × 4 mm ID
Mobile phase	Methanol/acetonitrile/phosphoric acid, pH 2.8 (48:48:4, v/v) with 5 mM sodium dodecylsulfate
Flow-rate	1.2 ml min^{-1}; 1100 psi
Injection size	20 μl
Detection	UV, λ = 265 nm
Retention times, min	(estimated from drawing) Vitamin A 6.03 Vitamin E 20.53 Vitamin D_3 17.93 Vitamin E-acetate 26.96
Ref.	Merck Application Manual on HPLC, TLC and HPTLC (1986). E. Merck, Darmstadt, p 13

HPLC 6

Sample	Fortified milk
Extraction	Ethanolic pyrogallol and aqueous KOH, and then with hexane
Apparatus	LDC Constametric III pump, Spectromonitor III UV detector and Rheodyne M 7105 injector
Column	Normal phase; Spherisorb 3 μm, 250 × 2 mm ID
Mobile phase	Hexane/ethyl acetate/methanol (97:2.5:0.05)
Flow-rate	0.7 ml min^{-1}
Injection size	20 μl containing 0.125–1.25 μg
Detection	UV, λ = 301 nm; 0.01 AUFS
Recording	Hewlett-Packard 3390A integrator
Retention times, min	Isotachysterol D_3 13–14.0 Vitamin D_3 13.2 (estimated from drawing)
Ref.	Agarwal VK (1988) J. Assoc. Off. Anal. Chem. 71:19

228 Vitamin E (α-Tocopherol)

CAS-No	59-02-9 (d-alpha tocopherol); 7695-91-2 (acetate)	
Chemical name	5,7,8-Trimethyltocol	
Empirical formula	α-Tocopherol	Acetate
	$C_{29}H_{50}O_2$	$C_{31}H_{52}O_3$

Molecular mass	430.7		472.8		
Proprietary names	Ephynal; Evion; Tokols				
Clinical use	Vitamin				
Solubility,	Water	alcohol	acetone	chloroform	fixed oils
α-Tocopherol	insol	soluble	soluble	miscible	miscible
Acetate	insol	freely soluble		freely soluble	
Light absorption	λ_{max} = 284 nm in methanol				

HPLC 1

Sample	Pharmaceutical preparations and blood
Extraction	Chloroform/methanol (1:1) after treating with 0.9% NaCl solution and acetone/methanol (1:1)
Apparatus	Knauer compact HPLC system and spectrophotometric detector
Column	Reversed-phase; Vertex lichrosorb 7 μm, C_{18}, 100 × 4.6 mm ID
Mobile phase	Methanol
Flow-rate	2 ml min^{-1}; 1015 psi
Injection size	50 μl containing 37.5 ng
Detection	UV, λ = 288; 0.04 AUFS
Retention time, min	Vitamin E acetate 2.0 (estimated from drawing)
Ref.	Amin M (1988) J. Liq. Chromatogr. 11:1335

HPLC 2

Sample	Dairy products, foods and tissues		
Extraction	Light petroleum/diisopropyl ether (3 + 1)		
Apparatus	Isocratic HPLC system and fluorescence detector		
Column	A. Reversed-phase; Rad-Pak cartridge 5 μm, C_{18} B. Normal phase; Rad-Pak cartridge 5 μm silica		
Mobile phase	For A. Methanol 100% For B. Hexane/isopropanol (99:1)		
Flow-rate	1 ml min^{-1}		
Injection size	10 μl		
Detection	Fluorescence, λ = 295/330 nm		
Retention times, min	Columns	reversed-phase	normal phase
	δ-Tocopherol	10.0	17.5
	β-Tocopherol	12.3	10.4
	γ-Tocopherol	12.3	12.0
	α-Tocopherol	14.0	7.4
		(estimated from drawing)	
Ref.	Indyk HE (1988) Analyst. 113:1217		

HPLC 3

Sample	Diets and blood plasma
Extraction	n-Hexane

Apparatus	Waters M 6000A LC pump, Rheodyne M 7125 injector and Perkin-Elmer M 3000 fluorimeter
Column	Normal phase; Lichrosorb Si 60, 5 μm, 250 × 4 mm ID
Pre-column	Corasil II (Waters)
Mobile phase	n-Hexane/diisopropylether (93:7)
Flow-rate	2.1 ml min^{-1}
Temperature	37 and 34 °C
Injection size	20–100 μl
Detection	Fluorescence, λ = 292/324 nm
Recording	Hewlett-Packard 3390 integrator
Retention times, min	(estimated from drawing)

Temperature	37 °C	34 °C
α-Tocopherol	8.5	9.0
α-Tocotrienol	11.0	
β-Tocopherol	13.8	14.0
γ-Tocopherol	17.5	18.5
β-Tocotrienol	18.0	
Butylhydroxyanisol	20.0	
γ-Tocotrienol	23.0	
δ-Tocopherol	27.5	28.2

Ref. Piironen V, Varo P et al (1984) Intern. J. Vit. Nutr. Res. 53:35

HPLC 4

Sample	Human buccal mucosal cells
Extraction	Chloroform
Apparatus	Beckman M 320 LC pump and Hitachi F 1000 fluorescence detector
Column	Reversed-phase; Ultrasphere ODS 5 μm, 250 × 4.6 mm ID
Pre-column	Ultrasphere 20 μm ODS, 50 × 4.6 mm ID
Mobile phase	Methanol
Flow-rate	2.0 ml min^{-1}
Internal standard	d-Tocol
Injection size	250 μl
Detection	Fluorescence, λ = 295/330 nm
Retention times, min	d-Tocol 5.0
	α-Tocopherol 7.5
Ref.	Badcock NR and O'Reilly DA (1986) J. Chromatogr. Biomed. Appl. 382:290

HPLC 5

Sample	Blood plasma
Extraction	n-Hexane or n-hexane/methanol (4:1, v/v)

Apparatus	Trirotar-V-pumping system with Rheodyne M 7125 injector and Jasco Uvidec 100V UV detector
Column	A. Normal phase; Lichrosorb Si 60, 5 μm, 250 × 4.0 mm ID B. Reversed-phase; Shodex F-511B, ODS 5 μm, 250 × 4 mm ID
Mobile phase	For A: 0.5% (v/v) ethanol in n-hexane For B: Methanol
Flow-rate	1.0 ml min^{-1}
Injection size	10 μl
Detection	UV, $\lambda = 264$ nm; 0.32 AUFS

Retention times, min	Column	A.	B.
	α-Tocopherol nicotinate	12.4	5.2
	γ-Toxopherol nicotinate	14.5	4.0

Ref.	Ijitsu T and Ueno M (1988) J. Chromatogr. Biomed. Appl. 427:29

HPLC 6

Sample	Human serum
Extraction	n-Hexane
Apparatus	Waters M 510 LC pump with 440 UV detector and U6K injector
Column	Reversed-phase; Nucleosil 5 C_{18}, 120 × 3 mm ID
Pre-column	PIN 080040 RCSS C_{18} cartridge (Machery-Nagel)
Mobile phase	Methanol/water (95:5)
Flow-rate	2 ml min^{-1}; 2697 psi
Internal standard	Retinyl acetate
Injection size	50 μl, 1.14–72.94 μg ml^{-1}
Detection	UV, $\lambda = 340$ nm (first 3 min), thereafter 280 nm (3 min); 0.01–0.02 AUFS
Recording	Waters M 730 integrator
Retention times	Retinol, retinyl acetate and α-tocopherol, in ascending order in ca 7.0 min
Ref.	Sanz DC and Santa-Cruz MC (1986) J. Chromatogr. Biomed. Appl. 380:140

229 Vitamin K_1

CAS-No	84-80-0
Chemical name	2-Methyl-3-phytyl-1,4-naphthoquinone
Empirical formula	$C_{31}H_{46}O_2$
Molecular mass	450.7
Proprietary names	Kaywan; Konakion; Mephyton

Clinical use	Vitamin
Solubility	Water insol; alcohol 1 in 70; ether, chloroform, fixed oils freely soluble
Light absorption	λ_{max} = 244 nm and 249 nm in methanol

HPLC 1

Sample	Powdered infant formula
Apparatus	Varian M 8500 LC pump with electrochemical detector
Column	Reversed-phase; μ-Bondapak C-18, 10 μm, 150 × 3.9 mm ID
Mobile phase	Acetonitrile/methylene chloride/0.025 M $NaClO_4$ (90:5:5, v/v)
Flow-rate	2 ml min^{-1}
Temperature	Ambient
Injection size	20 μl containing 8–80 ng
Detection	EC, silver electrode at -1.1 V vs AgCl RE
Retention time, min	Vitamin K_1 8.0
Ref.	Schneiderman MA, Sharma AK et al (1988) J. Assoc. Off. Anal. Chem. 71:815

HPLC 2

Sample	Powdered infant formula
Apparatus	Perkin-Elmers 4000 LC system and UV detector
Column	Reversed-phase; Whatman C-18, 150 × 3.9 mm ID
Mobile phase	Acetonitrile/CH_2Cl_2/water (90:5:5, v/v)
Flow-rate	2 ml min^{-1}
Temperature	Ambient
Detection	UV, λ = 248 nm; 0.10 AUFS
Retention times, min	Vitamin K_1 7.5
Ref.	Schneiderman MA, Sharma AK et al (1988) J. Assoc. Off. Anal. Chem. 71:815

HPLC 3

Sample	Neonatal plasma
Extraction	Hexane
Apparatus	Waters M 510 LC pump, U6K injector and Lambdamax 481 ultraviolet detector
Column	Reversed-phase; Novapak C_{18}, Radial-Pak RCS, 100 × 8 mm ID
Pre-column	Guard-Pak, μ-Bondapak C_{18}
Mobile phase	Acetonitrile/propan-2-ol/dichloromethane (68.5:22.2:9.3, v/v)
Flow-rate	1.8 ml min^{-1}; 6.8 MPa
Internal standard	Vitamin K_1 (25)

Injection size	25 μl, 0.025–1.0 μg ml^{-1}
Detection	UV, λ = 270 nm; 0.002 AUFS
Retention times, min	

Vitamin $K_2(20)$	3.49	Vitamin $K_2(30)$	6.06
Vitamin K_1 2,3-epoxide	4.28	Vitamin $K_2(35)$	8.54
Vitamin K_1	5.96	Vitamin $K_1(25)$	10.55

Ref.	Lucock M, Hartley R and Wild NJ (1987) J. Liq. Chromatogr. 10:191

HPLC 4

Sample	Rat liver
Extraction	Hexane
Apparatus	Altex M 110A LC pump with 160-UV detector
Column	Normal phase; Partisil 10 ODS, 10 μm, 250 × 4.5 mm ID
Pre-column	Partisil 10 Si, 25 × 4.5 mm ID
Mobile phase	0.23 % Acetonitrile in hexane
Flow-rate	2.0 ml min^{-1}; 479–1450 psi
Injection size	50 μl
Detection	UV, λ = 254 nm; 0.002 AUFS
Recording	Philips PM 8251 single-pen-recorder
Retention times, min	cis-Vitamin K_1 6.9
	trans-Vitamin K_1 7.8
	Vitamin K_1 2,3-epoxide 8.7
Ref.	Cholerton S and Park BK (1986) J. Chromatogr. Biomed. Appl. 375:147

HPLC 5

Sample	Rat liver
Extraction	Hexane
Apparatus	Spectra-Physics 8700 solvent delivery system and Altex M 110A LC pump and Kratos M 773 UV detector
Column	Reversed-phase; Ultrasphere ODS C_{18}, 5 μm, 250 × 4.5 mm ID
Pre-column	Reversed-phase; Ultrasphere ODS C_{18}, 5 μm, 25 × 4.5 mm ID
Mobile phase	12.5% Dichloromethane in acetonitrile
Flow-rate	1 ml min^{-1}; 20 min after injection, 3 ml min^{-1} for 6 min
Injection size	50 μl
Detection	UV, λ = 254 nm; 0.002 AUFS
Recording	Philips PM 8251 single-pen-recorder
Retention times, min	Vitamin K_1 2,3-epoxide 10.4
	Vitamin K_1 14.8

Ref.	Cholerton S and Park BK (1986) J. Chromatogr. Biomed. Appl. 375:147

HPLC 6

Sample	Blood plasma
Extraction	Hexane/acetonitrile (1:4, v/v) containing 70 mmol zinc chloride l^{-1}
Apparatus	Waters M 510 reciprocating pump, WISP 710B autoinjector and Kratos spectroflow 980 fluorescence detector
Column	Reversed-phase; Hypersil ODS 5 μm, 250 × 4.6 mm ID
Mobile phase	200 ml Dichloromethane, 800 ml methanol per liter, to each liter 5 ml of a solution containing 2.0 mol $ZnCl_2$, 1.0 mol sodium acetate and 1 mol acetic acid per liter
Flow-rate	1.0 ml min^{-1}
Internal standard	Dihydro-vitamin K_1
Reactor	Solid-phase; post-column, 20 × 3.9 mm ID with zinc particles
Injection size	100 μl
Detection	Fluorescence, λ = 248/418 nm
Recording	Waters 730 data module
Retention times, min	Vitamin K_1 epoxide 7.2
	Vitamin K_1 9.4
	Vitamin K_1 (1-H_2) 10.5
Ref.	Haroon Y, Bacon DS and Sadowski JA (1986) Clin. Chem. 32:1925

230 Warfarin

CAS-No	81-81-2; 129-06-6 (sodium salt)			
Chemical name	4-Hydroxy-3-(3-oxo-1-phenylbutyl)coumarin			
Empirical formula	Warfarin	Sodium Salt		
	$C_{19}H_{16}O_4$	$C_{19}H_{15}NaO_4$		
Molecular mass	308.3	330.3		
Proprietary names	Athrombin-K; Coumadin; Marevan			
Clinical use	Anticoagulant			
Solubility,	Water	alcohol	ether	aq NaOH
Warfarin	insoluble	moderately sol		soluble
Sodium salt	1 in < 1	1 in < 1	soluble	
Light absorption	λ_{max} = 293 nm and 308 nm in aqueous alkali			

HPLC 1

Sample	Tablets
Apparatus	Du Pont M 8000 LC pump and ultraviolet detector
Column	Reversed-phase; μ-Bondapak C_{18}, 10 μm, 300 × 3.9 mm ID
Mobile phase	Tetrahydrofuran/methanol/water/glacial acetic acid (35:10:65:0.1, v/v)
Flow-rate	1.5 ml min^{-1}
Temperature	Ambient
Injection size	20 μl, 0.2 mg ml^{-1}
Detection	UV, λ = 311 nm; 16 AUFS
Ref.	Moore ES (1987) J. Assoc. Off. Anal. Chem. 70:834

HPLC 2

Sample	Serum with warfarin enantiomers	
Extraction	Deproteinization with acetonitrile under alkaline condition	
Apparatus	Beckman M 110B LC pump, Spectra-Physics 8480 XR UV detector and Rheodyne M 7125 injection valve	
Column	Pinkerton internal surface reversed-phase (ISRP); 50 × 4.6 mm ID	
Mobile phase	0.2 M Phosphate buffer, pH 6.5/1-propanol (99:1)	
Flow-rate	1 ml min^{-1}	
Temperature	Ambient	
Internal standard	Benoxaprofen	
Injection size	7 μl, 1.25–20 μg ml^{-1}	
Detection	UV, λ = 225 nm	
Retention times, min	RS-Warfarin	16.00
	Internal standard	27.00
Ref.	Chu Y-Q and Wainer IW (1988) Pharm. Res. 5:680	

HPLC 3

Sample	Serum with warfarin enantiomers
Apparatus	Beckman M 110B LC pump, Spectra-Physics 980 fluorescence detector and Rheodyne M 7010 switching valve
Column	Chiral-Phase with bovine serum albumin, 10 μm, silica connected with internal surface reversed-phase
Mobile phase	0.2 M Phosphate buffer, pH 7.5/1-propanol (97:3) modified with 3 mM trichloroacetic acid
Flow-rate	1 ml min^{-1}
Temperature	30 °C
Injection size	7 μl, 1.25–20 μg ml^{-1}
Detection	Fluorescence, λ = 310/370 nm
Recording	Shimadzu C R6A integrator

Retention times, min	S-Warfarin 37.5 R-Warfarin 40.0
Ref.	Chu Y-Q and Wainer IW (1988) Pharm. Res. 5:680

HPLC 4

Sample	Microbial culture containing warfarin metabolites
Extraction	First 10% dichloromethane in cyclohexane, then re-extraction with 0.1 M KOH and lastly with diethyl ether
Apparatus	Beckman M 110A LC pump and Schoeffel FS 970 fluorimetric detector and Hitachi M 100 – 10 UV detector
Column	Reversed-phase; Beckman ODS 5 μm, C_{18}, 250 × 4.5 mm ID
Mobile phase	Methanol/tetrahydrofuran/1.0 M aqueous tetrabutylammonium ion/5 mM aqueous ammonium phosphate, pH 7.5 (30:7:1:62)
Flow-rate	1 ml min^{-1}
Internal standard	4′-Hydroxywarfarin alcohols
Injection size	50–200 μl, 100 ng ml^{-1}
Detection	Fluorescence, $\lambda = 290$ nm (ex) and UV, $\lambda = 310$ nm
Recording	Hewlett-Packard 3390A reporting integrator
Retention times, min	(estimated from drawing)

Internal standard	10.0, 13.0
4′-Hydroxywarfarin	12.0
3′-Hydroxywarfarin	15.5
Dehydrowarfarin	16.5
6-Hydroxywarfarin	18.0
7-Hydroxywarfarin	19.8
8-Hydroxywarfarin	21.8
RR/SS-Warfarin alcohol	23.2
Warfarin	27.6
RS/SR-Warfarin alcohol	32.0

Ref.	Wong YWJ and Davis PJ (1989) J. Chromatogr. 469:281

HPLC 5

Sample	Blood plasma
Extraction	Ethylene dichloride under acidic condition
Column	ODS-Permaphase, 1000 × 2.1 mm ID
Mobile phase	10% Dioxane in water, pH 4.0
Flow-rate	0.75 ml min^{-1}; 1000 psi
Temperature	25 °C
Injection size	6 μl, 1–10 μg ml^{-1}
Detection	UV, $\lambda = 254$ nm; 0.02 AUFS
Retention times, min	(estimated from drawing)

7-Hydroxywarfarin	4.3
Warfarin	7.5

Ref.	Midha KK and Butterfield AG (1979) in: Tsuji K (ed) GLC and HPLC Determination of Therapeutic Agents, Part 3. Marcel Dekker Inc., New York and Basel, p 1049

231 Xylazine

CAS-No	7361-61-7; 23076-35-9 (hydrochloride)		
Chemical name	*N*-(5,6-Dihydro-4*H*-1,3-thiazin-2-yl)-2,6-xylidine		
Empirical formula	Xalazine	Hydrochloride	
	$C_{12}H_{16}N_2S$	$C_{12}H_{16}N_2S,HCl$	
Molecular mass	220.3	256.8	
Proprietary name	Rompun		
Clinical use	Sedative Analgesic/Muscle Relaxant		
Solubility,	Water,	alcohol, methanol	acetone, chloroform
Xylazine	insoluble	soluble	soluble
Hydrochloride	readily sol	readily sol	very slightly sol

HPLC 1

Sample	Horse blood
Extraction	Diethyl ether
Apparatus	Waters M 6000A LC pump and ultraviolet detector
Column	Reversed-phase; μ-Bondapak C_{18}, 300 × 2.9 mm ID
Mobile phase	Acetonitrile/0.001 M aqueous HCl/methanol (65:30:5)
Flow-rate	1.0 ml min^{-1}
Temperature	Ambient
Injection size	50 μl, 20–1000 ng ml^{-1}
Detection	UV, $\lambda = 254$ nm
Retention time, min	Xylazine 7.2
Ref.	Akbari A, Gordon BJ et al (1988) J. Chromatogr. Biomed. Appl. 426:207

HPLC 2

Sample	Swine kidney
Extraction	Acidic acetonitrile on a Sep-Pak C_{18} cartridge
Apparatus	Millipore M 6000A HPLC with WISP 710B autosampler and UV-VIS and fluorescence detectors
Column	Reversed-phase; Bondapak C_{18}, 300 × 3.9 mm ID
Pre-column	Reversed phase; Bondapak C_{18}, 37–50 μm, 10 × 2.1 mm ID

Mobile phase	2.46 g Anhydrous sodium acetate, 450 ml water and 550 ml acetonitrile, pH 6.5 with acetic acid
Flow-rate	1.2 ml min^{-1}
Injection size	50 μl, 0.4 μg ml^{-1}
Detection	UV, λ = 240 nm; 0.002 AUFS Fluorescence; λ = 246/351 nm
Retention times, min	(estimated from drawing) Carazolol 5.8; Acepromazine 13.0 Xylazine 6.5; Propionylpromazine 18.0 Azaperone 7.8; Chlorpromazine 25.5 Haloperidol 8.2
Ref.	Keukens HJ and Aerts MML (1989) J. Chromatogr. 464:149

HPLC 3

Sample	Blood plasma
Extraction	Chloroform
Apparatus	Waters M 6000A HPLC pump, U6K injector and Schoeffel GM 770 UV detector
Column	Reversed-phase; μ-Bondapak C_{18}, 10 μm, 300 × 4 mm ID
Mobile phase	2% Glacial acetic acid in water/methanol/heptane-sulfonic acid, Pic B7 (Waters) (55:45:0.2, v/v)
Flow-rate	2 ml min^{-1}
Temperature	Ambient
Internal standard	Doxapram
Injection size	10 μl containing 0.01–1.0 μg
Detection	UV, λ = 255 nm
Retention times, min	Xylazine 4.0 Doxapram 5.5
Ref.	Alvinerie M and Toutain PL (1981) J. Chromatogr. Biomed. Appl. 222:308

HPLC 4

Sample	Greyhound urine
Extraction	Methanolic hydrochloric acid/acetonitrile (50:50, v/v) on Bond-Elut cyanopropyl column
Apparatus	Kratos M 400 LC pump with Pye-Unicam 4025 UV detector
Column	Reversed-phase; ODS 5 μm, C_{18}, 250 × 4.6 mm ID
Mobile phase	1 g Tetramethylammonium hydroxide in (250 ml) water/acetonitrile/methanol (250:150:10, v/v)
Flow-rate	2 ml min^{-1}
Temperature	Ambient (20 °C)
Internal standard	Diazepam
Injection size	20 μl, 10–100 ng ml^{-1}

Detection	UV, $\lambda = 225$ nm; 0.02 AUFS
Retention times, min	Xylazine 6.5 Diazepam 9.4
Ref.	Moore CM and Oliver JS (1989) J. Chromatogr. Biomed. Appl. 491:519

232 Yohimbine

CAS-No	146-48-5; 65-19-0 (hydrochloride)			
Chemical name	Methyl 17α-hydroxy-yohimban-16α-carboxylate			
Empirical formula	Yohimbine $C_{21}H_{26}N_2O_3$	Hydrochloride $C_{21}H_{26}N_2O_3$,HCl		
Molecular mass	354.4	390.9		
Proprietary names	Aphrodyne; Vikonon; Yocon			
Clinical use	Aphrodisiac			
Solubility,	Water	alcohol	chloroform	ether
Yohimbine	very slightly soluble	soluble	soluble	sparingly sol
Hydrochloride	1 in 100	soluble		
Light absorption	$\lambda_{max} = 271$ nm in aqueous acid			

HPLC 1

Sample	Horse serum
Extraction	Ethyl acetate on Sep-Pak C_{18} extraction column
Apparatus	Waters M 6000A LC pump and 440 ultraviolet detector
Column	Reversed-phase; μ-Bondapak C_{18}, 300 × 3.9 mm ID
Mobile phase	Acetonitrile/$1.2 \cdot 10^{-3}$ M aqueous HCl/methanol (60:10:30)
Flow-rate	1.0 ml min^{-1}
Temperature	Ambient
Injection size	40 μl, 10–1000 ng ml^{-1}
Detection	UV, $\lambda = 280$ nm
Retention time, min	Yohimbine 6.6
Ref.	Akbari A, Jernigan AD et al (1986) J. Chromatogr. 361:400

HPLC 2

Sample	Blood
Extraction	Ethyl acetate under alkaline condition
Apparatus	Beckmann M 110A LC pump, Valco C6U injector and Shimadzu RF-530 spectrofluorimeter

Column	Normal phase; Partisil-5, 250 × 2.1 mm ID
Mobile phase	Methanol/water (95:5)
Internal standard	Eserine sulfate
Injection size	20 µl, 50–1000 pg ml^{-1}
Detection	Fluorescence, λ = 280/360 nm
Recording	Hewlett-Packard 3390 electronic integrator
Retention times, min	(estimated from drawing)
	Yohimbine 4.1
	Eserine 7.8
Ref.	Owen JA, Nakatsu SL et al (1985) J. Chromatogr. Biomed. Appl. 342:333

HPLC 3

Sample	Alkaloids
Apparatus	Laboratory Instruments HPP 4001 HPLC, Pye-Unicam LC-871 UV/VIS detector and LCI 20 stop-flow injector
Column	Reversed-phase; Separon SIX C_{18}, 7 µm, 250 × 4 mm ID
Mobile phase	80% (v/v) Methanol and 20% (v/v) 0.025 M NaH_2PO_4, pH 4.0, 5.0, 6.0, 7.0 and 8.0 with 10% H_3PO_4 or 10% NaOH solution
Injection size	1–3 µl
Detection	UV, λ = 190–380 nm

Capacity factor, *k'*

pH, mobile phase	4.0	5.0	6.0	7.0	8.0
Caffeine	0.31	0.31	0.35	0.35	0.32
Emetine	0.49	0.68	1.4	2.2	2.1
Codeine	0.51	0.68	1.2	1.3	1.3
Pilocarpine	0.55	0.58	0.50	0.50	0.59
Yohimbine	0.71	0.85	0.96	0.90	0.95
Papaverine	0.81	0.81	0.86	0.76	0.80
Quinine	0.81	1.0	2.6	3.3	3.4
Quinidine	0.91	1.2	3.1	3.4	3.5
Dihydroergocornine	1.0	1.4	1.5	1.6	1.5
Dihydroergotamine	1.1	1.5	1.6	1.5	1.5
Reserpine	1.1	2.2	2.5	2.5	2.7
Dihydro(α,β)-ergocryptine	1.2	1.8	1.9	1.9	1.9
Dihydroergocristine	1.4	1.7	2.2	2.4	2.4

Ref. Popl M, Ky LD and Strnadova J (1985) J. Chromatogr. Sci. 23:95

HPLC 4

Sample	Standard solution
Column	Ion-exchange; Ion-X-SC, 1000 mm

Mobile phase	0.02 M Ammonium monohydrogen phosphate in methanol/water (3:7), pH 7.0			
Flow-rate	1.0 ml min^{-1}			
Temperature	60 °C			
Detection	UV, λ = 254 nm			
Retention times, min	Raupin	9.0	Reserpine	28.0
	Ajmaline	11.0	Rescinnamine	37.0
	Yohimbine	15.0		
Ref.	de Silva JAF (1978) in: Tsuji K (ed) GLC and HPLC Determination of Therapeutic Agents, Part 2. Marcel Dekker Inc., New York and Basel, p 581			

Table 1. Conversion of Imperial and SI Unit Equivalents of other Units

LENGTH	
1 ångström (Å)	$= 10^{-10}$ meter $= 10^{-1}$ nanometer
1 micron	$= 10^{-6}$ meter
1 inch (in)	$= 2.54 \times 10^{-2}$ meter
1 foot (ft)	$= 3.048 \times 10^{-1}$ meter
1 yard (yd)	$= 9.144 \times 10^{-1}$ meter
1 mile (mi)	$= 1.6094 \times 10^{3}$ meters
1 meter (m)	$= 10^{10}$ ångströms 10^{6} microns 39.3701 inches 3.28084 feet 1.09361 yards
1 kilometer (km)	= 0.621372 mile (statue)
MASS	
1 grain (gr)	$= 6.47989 \times 10^{-2}$ gram
1 ounce (avoirdupois) (oz) (= 437.5 grains)	$= 2.83495 \times 10$ grams
1 ounce (apothecaries') (= 480 grains)	$= 3.11035 \times 10$ grams
1 pound (avoirdupois)	$= 4.53592 \times 10^{2}$ grams
1 ton	$= 1.01605 \times 10^{3}$ kilograms
1 gram (g)	= 15.4324 grains 0.032151 ounce (apothecaries') 0.035274 ounce (avoirdupois)
1 kilogram (kg)	= 35.274 ounces (avoirdupois) 2.20462 pounds
AREA	
1 are (a)	= 100 square meters
1 square inch (in^2)	$= 6.4516 \times 10^{-4}$ square meter
1 square foot (ft^2)	$= 9.29030 \times 10^{-2}$ square meter
1 square yard (yd^2)	$= 8.36127 \times 10^{-1}$ square meter
1 square mile (mi^2)	$= 2.58999 \times 10^{6}$ square meters
1 square meter (m^2)	= 1550 square inches 10.7639 square feet 1.196 square yards 3.86102×10^{-7} square mile
VOLUME	
1 milliliter (ml)	= 1 cubic centimeter
1 liter (l)	= 1 cubic decimeter
1 cubic inch (in^3)	$= 1.63871 \times 10^{-5}$ cubic meter
1 cubic foot (ft^3)	$= 2.83168 \times 10^{-2}$ cubic meter
1 cubic yard (yd^3)	$= 7.64555 \times 10^{-1}$ cubic meter
1 minim (UK)	$= 5.91939 \times 10^{-8}$ cubic meter
1 fluid ounce (UK) (fl oz)	$= 2.8413 \times 10^{-5}$ cubic meter
1 fluid ounce (US)	$= 2.95735 \times 10^{-5}$ cubic meter

Table 1. (Continuation)

1 pint	$= 5.68261 \times 10^{-4}$ cubic meter
1 gallon (UK)	$= 4.54609 \times 10^{-3}$ cubic meter
1 liquid pint (US)	$= 4.73176 \times 10^{-4}$ cubic meter
1 gallon (US)	$= 3.78541 \times 10^{-3}$ cubic meter
1 cubic meter (m^3)	$= 6.10236 \times 10^{4}$ cubic inches 35.3147 cubic feet 1.30795 cubic yards
1 cubic centimeter (cm^3)	= 1 milliliter 16.8934 minims (UK)
1 cubic decimeter (dm^3)	= 1 liter 35.1952 fluid ounces 1.75976 pints 0.21997 gallon (UK)
ENERGY	
1 kilocalorie, thermo-chemical (kcal)	$= 4.1840 \times 10^{3}$ joules
1 erg (erg)	$= 10^{-7}$ joule
1 electronvolt (eV)	$= 1.60219 \times 10^{-19}$ joule
1 thousand (kilo) electronvolt (keV)	$= 1.60219 \times 10^{-16}$ joule
1 million (mega) electronvolt (MeV)	$= 1.60219 \times 10^{-13}$ joule
1 British thermal unit (Btu)	$= 1.05506 \times 10^{3}$ joules
1 joule (J)	$= 2.39006 \times 10^{-4}$ kilocalorie 10^{7} ergs 6.2415×10^{18} electronvolts 9.47813×10^{-4} British thermal unit
FORCE	
1 newton (N)	$= 10^{5}$ dynes 1.01972×10^{2} gram-force 1.01972×10^{-1} kilogram-force 7.23301 poundals
PRESSURE	
1 millimeter of mercury (mmHg)	$= 1.33322 \times 10^{2}$ pascals
1 bar (bar)	$= 10^{5}$ pascals
1 pound-force per square inch	$= 6.89476 \times 10^{3}$ pascals
1 atmosphere (atm)	$= 1.01325 \times 10^{5}$ pascals
1 kilogram per square centimeter	$= 9.80392 \times 10^{4}$ pascals
1 pascal (Pa)	$= 7.50064 \times 10^{-3}$ millimeter of mercury 1.45038×10^{-4} pound-force per square inch 9.86923×10^{-6} atmosphere

Table 1. (Continuation)

1 kilopascal (kPa)	= 7.50064 millimeters of mercury 0.01000 bars 0.01020 kilogram per square centimeter
VISCOSITY, DYNAMIC	
1 poise (P)	= 10^{-1} pascal second 10^{-1} newton second per square meter
1 pascal second (Pa s)	= 10.0 poises
VISCOSITY, KINEMATIC	
1 centistokes (cSt)	= 10^{-6} square meter per second
1 stokes (St)	= 10^{-4} square meter per second
1 square meter per second (m^2s^{-1})	= 10^4 stokes
TEMPERATURE	
1 degree Fahrenheit (°F)	= 5/9 kelvin (K) or ° Celsius (C)
1 kelvin (K)	= 9/5 degrees Fahrenheit
ACTIVITY OF A RADIOACTIVE SOURCE	
1 nanocurie (nCi)	= 3.7×10 becquerels
1 microcurie (μCi)	= 3.7×10^4 becquerels
1 millicurie (mCi)	= 3.7×10^7 becquerels
1 curie (Ci)	= 3.7×10^{10} becquerels
1 becquerel (Bq)	= 2.7027×10^{-2} nanocurie 2.7027×10^{-5} microcurie 2.7027×10^{-8} millicurie 2.7027×10^{-11} curie

Table 2. Atomic Weights of the Elements

Name	Symbol	Atomic Weight	Atomic Number
Actinium	Ac	227.0278	89
Aluminium	Al	26.98154	13
Americium	Am	243	95
Antimony	Sb	121.75	51
Argon	Ar	39.948	18
Arsenic	As	74.9216	33
Astatine	At	210	85
Barium	Ba	137.34	56
Berkelium	Bk	247	97
Beryllium	Be	9.01218	4
Bismuth	Bi	208.9804	83
Boron	B	10.811	5
Bromine	Br	79.904	35
Cadmium	Cd	112.41	48
Caesium	Cs	132.9054	55
Calcium	Ca	40.078	20
Californium	Cf	251	98
Carbon	C	12.011	6
Cerium	Ce	140.12	58
Chlorine	Cl	35.453	17
Chromium	Cr	51.9961	24
Cobalt	Co	58.9332	27
Copper	Cu	63.546	29
Curium	Cm	247	96
Dysprosium	Dy	162.50	66
Einsteinium	Es	252	99
Erbium	Er	167.26	68
Europium	Eu	151.96	63
Fermium	Fm	257	100
Fluorine	F	18.998403	9
Francium	Fr	223	87
Gadolinium	Gd	157.25	64
Gallium	Ga	69.72	31
Germanium	Ge	72.59	32
Gold	Au	196.9665	79
Hafnium	Hf	178.49	72
Helium	He	4.002602	2
Holmium	Ho	164.9304	67
Hydrogen	H	1.00794	1
Indium	In	114.82	49
Iodine	I	126.9045	53
Iridium	Ir	192.22	77
Iron	Fe	55.847	26
Krypton	Kr	83.80	36
Lanthanum	La	138.9055	57

Table 2. (Continuation)

Name	Symbol	Atomic Weight	Atomic Number
Lawrencium	Lr	260	103
Lead	Pb	207.2	82
Lithium	Li	6.941	3
Lutetium	Lu	174.967	71
Magnesium	Mg	24.305	12
Manganese	Mn	54.9380	25
Mendelevium	Md	258	101
Mercury	Hg	200.59	80
Molybdenum	Mo	95.94	42
Neodymium	Nd	144.24	60
Neon	Ne	20.179	10
*Neptunium	Np	237.0482	93
Nickel	Ni	58.70	28
Niobium	Nb	92.9064	41
Nitrogen	N	14.0067	7
Nobelium	No	259	102
Osmium	Os	190.2	76
Oxygen	O	15.9994	8
Palladium	Pd	106.4	46
Phosphorus	P	30.97376	15
Platinum	Pt	195.08	78
Plutonium	Pu	244	94
Polonium	Po	209	84
Potassium	K	39.0983	19
Praesodymium	Pr	140.9077	59
Promethium	Pm	145	61
*Protactinium	Pa	231.0359	91
*Radium	Ra	226.0254	88
Radon	Rn	222	86
Rhenium	Re	186.207	75
Rhodium	Rh	102.9055	45
Rubidium	Rb	85.4678	37
Ruthenium	Ru	101.07	44
Samarium	Sm	150.4	62
Scandium	Sc	44.95591	21
Selenium	Se	78.96	34
Silicon	Si	28.0855	14
Silver	Ag	107.8682	47
Sodium	Na	22.98977	11
Strontium	Sr	87.62	38
Sulphur	S	32.066	16
Tantalum	Ta	180.9479	73
Technetium	Tc	98	43
Tellurium	Te	127.60	52
Terbium	Tb	158.9254	65

Table 2. (Continuation)

Name	Symbol	Atomic Weight	Atomic Number
Thallium	Tl	204.383	81
*Thorium	Th	232.0381	90
Thulium	Tm	168.9342	69
Tin	Sn	118.710	50
Titanium	Ti	47.88	22
Tungsten	W	183.85	74
Unnilhexium	Unh	263	106
Unnilpentium	Unp	262	105
Unnilquadium	Unq	261	104
Uranium	U	238.029	92
Vanadium	V	50.9415	23
Xenon	Xe	131.29	54
Ytterbium	Yb	173.04	70
Yttrium	Y	88.9059	39
Zinc	Zn	65.39	30
Zirconium	Zr	91.224	40

(*) is that of the radioisotope of longest half-life.

Table 3. Miscibility of Solvents

□ miscible

■ immiscible

	Acetic acid	Acetone	Acetonitrile	Benzene	Butyl alcohol	Carbon tetrachloride	Chloroform	Cyclohexane	Cyclopentane	Dichloroethane	Dichloromethane	Dimethylformamide (DMF)	Dimethyl sulfoxide (DMSO)	Dioxane	Ethyl acetate	Ethyl alcohol
Acetone	□															
Acetonitrile	□	□														
Benzene	□	□	□													
Butyl alcohol	□	□	□	□												
Carbon tetrachloride	□	□	□	□	□											
Chloroform	□	□	□	□	□	□										
Cyclohexane	□	□	□	□	□	□	□									
Cyclopentane	□	□	■	□	□	□	□	□								
Dichloroethane	□	□	■	□	□	□	□	□	□							
Dichloromethane	□	□	□	□	□	□	□	□	□	□						
Dimethylformamide (DMF)	□	□	□	□	□	□	□	□	□	□	□					
Dimethyl sulfoxide (DMSO)	□	□	□	□	□	□	□	■	■	□	□	□				
Dioxane	□	□	□	□	□	□	□	■	■	□	□	□	□			
Ethyl acetate	□	□	□	□	□	□	□	□	□	□	□	□	□	□		
Ethyl alcohol	□	□	□	□	□	□	□	□	□	□	□	□	□	□	□	
Diethyl ether	□	□	□	□	□	□	□	□	□	□	□	□	□	□	□	□

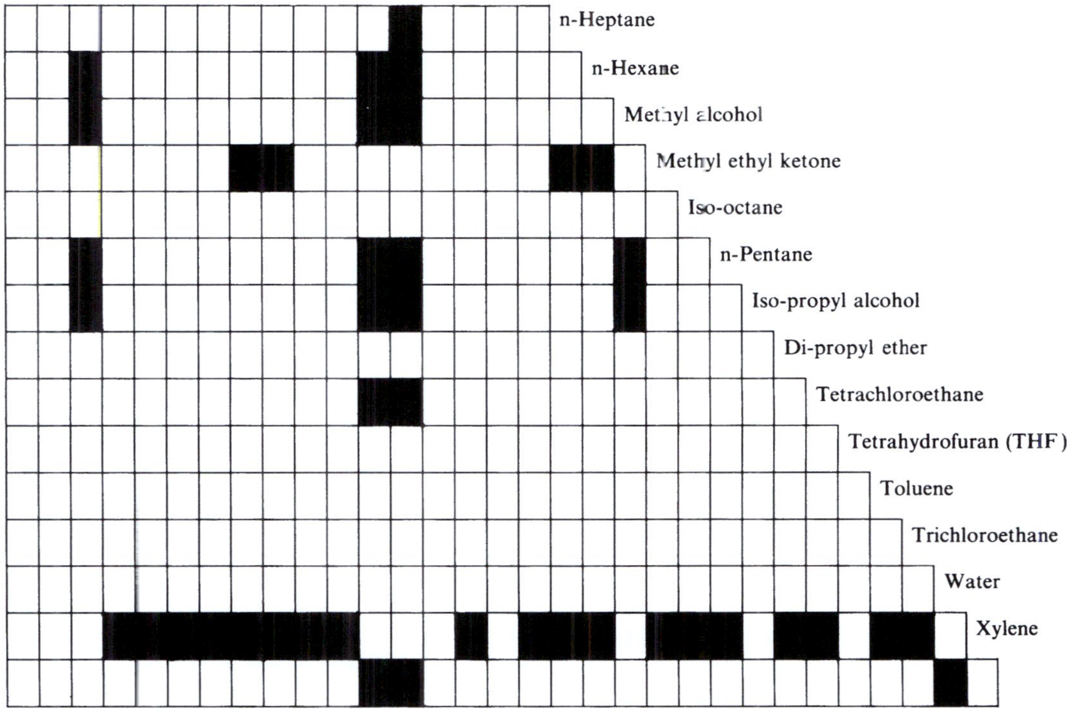

From: Godfrey NB (1972): Chem. Technol. 2:359
Hildebrand JH and Scott RL (1950): The Solubility of Nonelectrolytes. Reinhold Publ. Corp. new York, N.Y.

Table 4. Characteristics of the Solvents

Solvents	Polarity	Viscosity mPa (20 °C)	Refractive Index	UV^{-2} Cutt off	Boiling Point, °C
Fluoroalkane	– 0.25		1.25	210	
n-Pentane	0.00	0.23	1.3575	195	36
n-Hexane	0.00	0.33	1.3749	190	69
Isooctane	0.01	0.50	1.3914	200	99
Petroleum ether	0.01	0.3		210	
n-Decane	0.04	0.92	1.4119	200	174
Cyclohexane[3]	0.04	1.00	1.4262	200	81
Cyclopentane	0.05	0.47	1.4064	200	49
Di-isobutylene	0.06		1.411	210	
1-Penten	0.08	0.24 (°C)	1.3715	210	30
1,1,2-Trichlorotrifluoro-ethane	0.14	0.71	1.3588	230	48
Carbon disulfide	0.15	0.37	1.6280	380	46
Carbon tetrachloride	0.18	0.97	1.4652	265	77
1,1,1-Trichloroethane	0.19	0.85	1.4379	250	74
tert.-Butyl methyl ether	0.20	0.35	1.3689	220	53
Amyl chloride	0.26	0.43	1.4120	225	108
n-Butyl chloride	0.26	0.47	1.4021	220	78
Xylene	0.26	0.62–0.81	≈ 1.50	290	138–144
Isopropyl ether	0.28	0.37	1.3681	220	68
Isopropyl chloride	0.29	0.33	1.3777	225	36
Toluene	0.29	0.59	1.4969	285	111
n-Propyl chloride	0.30	0.35	1.3879	225	47
Chlorobenzene	0.30	0.80	1.5248	290	132
Benzene	0.32	0.65	1.5011	280	80
Ethyl bromide	0.37	0.39	1.4239		38
Diethyl ether	0.38	0.24	1.3524	205	34.5
Ethyl sulfide	0.38	0.45	1.4429	290	92
Chloroform	0.40	0.57	1.4457	245	61
Methylene chloride	0.42	0.44	1.4242	230	40
Methyl-isobutyl ketone	0.43	0.54	1.3957	330	116.5
Tetrahydrofuran	0.45	0.46	1.4072	220	66
1,2-Dichloroethane	0.49	0.79	1.4448	230	83
Methyl-ethyl ketone	0.51	0.4	1.3788	330	80
1-Nitropropan	0.53	0.77	1.4016	380	131
Acetone	0.56	0.32	1.3587	330	56
Dioxane	0.56	1.54	1.4224	220	101
Ethyl acetate	0.58	0.45	1.3724	260	77
Methyl acetate	0.60	0.37	1.3614	260	56
Amyl alcohol	0.61	4.1	1.4100	210	138
Dimethyl sulfoxide	0.62	2.24	1.4783	270	189
Aniline	0.62	4.4	1.5863		184
Diethyl amine	0.63	0.38	1.3854	275	55
Nitromethane	0.64	0.67	1.3819	380	101
Acetonitrile	0.65	0.37	1.3441	190	82
Pyridine	0.71	0.94	1.5102	305	115
Butyl cellosolve	0.74	3.1	1.4198	220	170

Table 4. (Continuation)

Solvents	Polarity	Viscosity mPa (20 °C)	Refractive Index	UV[2] Cutt off	Boiling Point, °C
Iso-+n-propanol	0.82	2.3	1.3772, 1.3856	210	82.97
Ethanol	0.88	1.20	1.3614	210	78
Methanol	0.95	0.60	1.3284	205	65
Ethylene glycol	1.11	19.9	1.4318	210	197
Acetic acid	high	1.26	1.3719		118
Water	higher	1.00	1.3330	< 190	100
Salt solutions and buffer	very high				

[2] Wavelength in nm for the measurement of extinction with 1 cm against air
[3] Solid at 350 bar

Meyer V (1984): Praxis der Hochleistungsfluessigchromatographie, 3 edn. Diesterweg Salle Sauerlaender, Frankfurt am Main and Salzburg, p 102

Table 5. Characteristics of the Common HPLC Separation Systems (columns)

Name	Chemical nature/ functional groups	Pore-size	Particle shape	Particle size, μm	pH-range/ reaction	Solvents as eluent
1. Adsorption Chromatography (Normal phase)	Silica gel Si 40-Si 4000	4–400 nm	round or irregular	3–12	non polar	Acetonitrile, benzene, chloroform, ethanol, ethyl acetate, dichloromethane, hexane, methanol, pentane and tetrahydrofuran
	Aluminium oxide	6–15 60 Å	round or irregular	3-12	acid/ base	Aqueous and non-aqueous solvents
2. Reversed-phase	Silica gel with butyl(C_4), ethyl(C_2), hexyl(C_6), octyl(C_8), octadecyl(C_{18}), amino (NH_2), cyano or nitrile (CN), diol (OH), methyl(CH_3), dimethyl amino ($N(CH_3)_2$),trimethyl$(CH_3)_3$, nitro(NO_2), phenyl (C_6H_5), diphenyl $(C_6H_5)_2$, sulfonic acid (SA), quarternary ammonium (SB), dodecyl, docosyl and cellulose(acetyl)		round or irregular	3,4,5,7 10 and 30–50 30-50	2–9 2–14	Acetic acid, acetone, acetonitrile, buffer solutions ethanol, formic acid, methanol, tetrahydrofuran and water

3. Ion-exchange Chromatography	Silica gel Anion (base),	60 Å	round or irregular	2.5, 3.5	1–13	HCl, CH_3COOH.
	cation (acid) (aminopropyl, tetraalkylammonium, carboxylic acid or sulfonic acid groups	1000 Å	irregular	10, 5–12,20	1.5–7.5 1–9 1.5–7 7–10	boric acid buffer, tris HCl-buffer, 0.5 M NaCl acetate buffer and NaOH solution
4. Ion-pair Chromatography	Reversed-phase systems modified by ion-pair reagents to anionic		round or irregular	3–10	1–9	Acetonitrile, buffer solutions, methanol and water with modifiers, or modifiers
5. Partition Chromatography	Liquids; liquid propane, dimethyl sulfoxide, nitromethane and CCl_4, coated on inert solid as silica and bonded with various functional groups (amino, cyano, diol and nitro)					Acetonitrile, acetone, chloroform, dioxane, isooctane, hexane, methanol/water and pentane
6. Gelchromatography	Silica with glyceropropyl, ether and styrene-divinyl, polystyrene, polyacrylamide, methacrylate, glycerin, co-polymer with functional groups (amino propyl, tetraalkyl ammonium or sulfonic acid	$50–4.10^8$		3-17		Chloroform, tetrahydrofuran, buffer solutions and water

Table 5. (Continuation)

Name	Chemical nature/ functional groups	Pore-size	Particle shape	Particle size, µm	pH-range/ reaction	Solvents as eluent
7. Size-Exclusion Chromatography	Gel-Permeation Chromatography (GPC) (non aqueous)			9-20		Acetone, dimethylformamide, dioxane, methyl ethyl ketone, and tetrahydrofuran
	Gel-Filtration Chromatography (GFC) (aqueous)	80– > 2000 Å		6–30	acid/ base/ neutral	Buffer solutions, salt solutions and water
8. Affinity Chromatography	Polymer based	1000 Å		10–15	2–7	Acetonitrile, buffer solutions, salt solutions, tris-buffer and water
9. Miscellaneous						
a) For Biopolymer	Silica gel with anion-exchange, silica gel with cation-exchange, silica gel with different reversed-phases	10,30 33		5,8,10		Acetonitrile, buffer solutions, salt solutions, tris-buffers and water

b) For Sugars	Amino or octadecyl phase on silica gel, cation-exchange on styrene-divinyl benzene, cation-exchange on styrene-divinyl benzene and or Ca-form or Pb-form	5,8,10	80% Acetonitrile and water, sodium phosphate and sodium acetate solutions
c) For amino acids and nucleotides	Silica gel with amino, octyl and nitrile phases,	5,9,10	Acetonitrile, buffer solutions, 0.1% trifluoro acetic acid, 0.1 M triethyl amine and water
	cation-exchange on styrene-divinyl benzene,	5,10	
	cation-exchange, H-form, proline or valine-copper-complex, amino-phase with ionic amino acids or albumin	5,10	
d) For enantiomers	Chiral phase-polyacrylamide on silica gel	5	n-Hexane, dioxane, n-propanol, tetra-hydrofuran and various other organic solvents
	Cellulose acetate	10,15-24 25-40	Ethanol, methanol, 80% n-hexane, n-propanol, water and various other organic solvents

Table 6. Sorbent Selection Guides

6a. Molecular Weight of the Sample is less than 2000 (HPLC)

Sample	Solubility	Sample type	Method 1	Method 2	Method 3
sample MG < 2000	soluble in organic solvents	non-polar	**Reversed Phase** RP-2, RP-8, RP-18 CH_3CN/H_2O $MeOH/H_2O$	**Adsorption** Si 40, Si 60, Si 100, Alox T Heptane, $CHCl_3$, CH_2Cl_2 CH_3CN etc.	**Partition** **(polar/hydrophilic)** NH_2, DIOL, CN CH_3CH, CH_2Cl_2, MeOH
	soluble in organic solvents		**Gel-permeation** PS 1, PS 4, PS 20 THF, CH_2Cl_2, Toluene	**Gel-permeation** Si 40, Si 60 THF, Toluene, CH_2Cl_2	
	soluble in organic solvents	polar	**Reversed Phase** **pH-control** RP-2, RP-8, RP-18 CH_3CN/aq buffer MeOH/aq buffer	**Partition** **(polar/hydrophilic)** NH_2, DIOL, CN THF, MeOH, H_2O buffer solution	**Reversed Phase** **Ion-pair-Partition** RP-2, RP-8, RP-18 CH_3CN/aq buffer + Ion-pair reagent
	soluble in water	non-polar	**Reversed Phase** RP-2, RP-8, RP-18 CH_3CN/H_2O $MeOH/H_2O$	**Partition** **(polar/hydrophilic)** NH_2, DIOL, CN MeOH, CH_3CN, H_2O	
	soluble in water	acid base	**Reversed Phase** **pH-control** RP-2, RP-8, RP-18 CH_3CN/aq buffer MeOH/aq buffer	**Reversed Phase** **Ion-pair Partition** RP-2, RP-8, RP-18 CH_3CN/aq buffer + Ion-pair reagent	**Ion-exchange** AX, CX aq buffer
	soluble in water	polar amphoteric	**Reversed Phase** **Ion-pair Partition** RP-2, RP-8, RP-18 CH_3CN/aq buffer + Ion-pair reagent	**Ion-exchange** AX, CX aq buffer	

Table 6b. Molecular Weight of the Sample is greater than 2000 (HPLC/GPC)

Sample MG > 2000	Soluble in organic solvents	**Non-aqueous Gel-permeation Chromatography** LiChrogel PS 1, PS 4, PS 20, PS 40, PS 400, PS 4000, PS 40000 100 CH-8, 500 CH-8, 1000 CH-8, 4000 CH-8 Si 40, SI 60, Si 100, Si 500, Si 1000, Si 4000 100 DIOL, 500 DIOL, 1000 DIOL, 4000 DIOL THF, Toluene, $CHCl_3$, CH_2Cl_2
	Soluble in water	**Aqueous Gel-filtration Chromatography** 100 DIOL, 500 DIOL, 1000 DIOL, 4000 DIOL Fractogel TSK HW-40, TSK SW, TSK phenyl 5-PW, HW-50, HW-55, HW-65, HW-75 water, buffer solutions
		Ion-exchange pH 2–12: TSK SP-5PW, TSK DEAE-SPW pH 2–7.5: Carboxymethyl (CM), WAX 300, Sulfopropyl (SP), quarternary ammonium (QA)

From: Mercks Chromatographie in der Pharmazeutischen Chemie, Manual für den Praktiker. E. Merck Darmstadt, p 9
Hewlett-Packard HPLC Column Selection Chart, Publication Number 12-5954-9883. Hewlett-Packard, Waldbronn

D. A. W. Wendisch,
Leverkusen-Bayerwerk, FRG

Acronyms and Abbreviations in Molecular Spectroscopy

An Encyclopedic Dictionary

1990. V, 315 pp. 10 figs. Hardcover
ISBN 3-540-51348-5

Acronyms are extensively used to name new spectroscopic methods. The uninitiated reader of research papers is often confused with terms like ISIS, NERO, TANGO. This dictionary gives correct definitions of acronyms frequently used in molecular spectroscopy and imaging methods, descriptions of physical effects, practical applications and references to the scientific literature for further reading.
The bulk of the more than 450 acronyms explained are from NMR and MRI, IR, RAMAN and ESR methods. The dictionary is arranged alphabetically and indexes, e.g. a subject index, allow easy access to the information. This book is an invaluable source for the spectroscopist.

Springer-Verlag
Berlin
Heidelberg
New York
London
Paris
Tokyo
Hong Kong
Barcelona
Budapest

E. Pretsch, T. Clerc, J. Seibl, W. Simon

Tables of Spectral Data for Structure Determination of Organic Compounds

Translated from the German by K. Biemann

2nd ed. 1989. XIII, 415 pp. (Chemical Laboratory Practice)
Softcover ISBN 3-540-51202-0

This book represents a compilation of spectroscopic reference data in the format of tables and charts and their correlation to molecular structure. It is intended to aid the interpretation of UV-, IR-, NMR- and mass spectra and complements text- and reference books on these techniques. It is designed to be of use to students as well as to the every day practitioner and expert in the field.
This second edition has been improved by among other things adding data on new compound classes and on C13-NMR-spectra.

From the contents: Introduction. - Abbreviations and Symbols. - Summary Tables. - Combination Tables. - ^{13}C-Nuclear Magnetic Resonance Spectroscopy. - Proton Resonance Spectroscopy. - Infrared Spectroscopy. - Mass Spectrometry. - UV/VIS (Spectroscopy in the Ultraviolet or Visible Region of the Spectrum). - Subject Index.

Springer-Verlag
Berlin
Heidelberg
New York
London
Paris
Tokyo
Hong Kong
Barcelona
Budapest

MIX
Papier aus verantwortungsvollen Quellen
Paper from responsible sources
FSC® C105338

If you have any concerns about our products,
you can contact us on
ProductSafety@springernature.com

In case Publisher is established outside the EU,
the EU authorized representative is:
Springer Nature Customer Service Center GmbH
Europaplatz 3, 69115 Heidelberg, Germany

Printed by Libri Plureos GmbH
in Hamburg, Germany